S0-CDN-092

SELECTED WRITINGS OF HERMANN VON HELMHOLTZ

SELECTED WRITINGS OF

Hermann von Helmholtz

Edited, with Introduction, by

RUSSELL KAHL

WESLEYAN UNIVERSITY PRESS

Middletown, Connecticut

ISBN: 0-8195-4039-0
Library of Congress catalog card number: 70-105503
Manufactured in the United States of America
First edition

CONTENTS

PREFACE

Prefaces can be valuable in a number of different ways. Among others, they provide an author or editor with the opportunity to say something about the work he has done, about the book he has produced. They also give him a welcome opportunity to acknowledge debts, to express appreciation to friends and colleagues for their assistance, and to express hope that the work done is at least of some little merit. I should like to do each of these things here.

The present work is easily characterized: it is a collection of those of Helmholtz' shorter papers which, in my judgment, best express his views on the nature of science and philosophy, along with his scientific papers of the most enduring value—in short, all of his papers which seem most clearly to deserve a permanent place in our libraries, alongside his books *Handbook of Physiological Optics* and *Sensations of Tone.*

With respect to the present form of the papers, they fall rather neatly into four categories (and here is certainly one place where I must acknowledge my debts and express my hopes):

I. *Papers which appeared originally in English and are here moderately revised in spelling, syntax, phrasing, and terminology:*

"The Application of the Law of the Conservation of Force to Organic Nature"

"The Origin and Meaning of Geometric Axioms (II)"

"The Modern Development of Faraday's Conception of Electricity"

II. *Revisions of translations appearing in earlier collections of Helmholtz' essays and addresses:*

"The Scientific Researches of Goethe" (original translation by H. W. Eve)

"The Physiological Causes of Harmony in Music" (original translation by A. J. Ellis)

"Recent Progress in the Theory of Vision" (original translation by Philip H. Pye-Smith)

"The Origin and Meaning of Geometric Axioms (I)" (original translation by E. Atkinson)

III. *Papers appearing here very substantially changed from the way in which they appeared in earlier translations (the changes are so extensive that the present translations might well be considered new):*

"The Conservation of Force: A Physical Memoir" (original translation by John Tyndall)

"The Relation of the Natural Sciences to Science in General" (original translation by H. W. Eve)

"The Aim and Progress of Physical Science" (original translation by W. Flight)

"The Origin of the Planetary System" (original translation by E. Atkinson)

"The Relation of Optics to Painting" (original translation by E. Atkinson)

"Thought in Medicine" (original translation by E. Atkinson)

"An Autobiographical Sketch" (original translation by E. Atkinson)

IV. *Papers appearing here newly translated or translated for the first time:*

"The Endeavor to Popularize Science"

"The Facts of Perception"

"An Epistemological Analysis of Counting and Measurement"

"Goethe's Anticipation of Subsequent Scientific Ideas"

"The Origin and Correct Interpretation of Our Sense Impressions"

Introduction to the Lectures on Theoretical Physics [Introduction and Part I]

I must thus assume responsibility for the translations of thirteen papers, and for the modification or revision of seven others.

In preparing this collection I have, over a period of years, made use of four libraries, and I am most grateful to the staff of each for making my visits so pleasant and profitable. These are the Columbia University Library, where I spent some of my student days and where I first read Helmholtz; the Library of the University of California at Berkeley, where my ramblings in the stacks have for many years been a great pleasure; the Science Museum Library in

London, which remains fixed in my memory as a lovely part of a lovely summer; and the San Francisco State College Library, where I have always felt at home, having as it were married into it.

A number of individuals have assisted in this work by making suggestions, by reading one or more of the papers, or just by their interest. Among these should surely be mentioned Ernest Nagel (who first brought Helmholtz to my attention), Lloyd Elliott, Ernest H. Halliwell, III, Christy Taylor, and Rudolph Weingartner. None of them, of course, is responsible for any of the flaws in the collection that appears here.

I owe, and this collection owes, a very special debt to Mr. Alan Levensohn, who did the editorial work for the Wesleyan University Press. It was my original plan to redo or revise only those existing translations which were hopelessly inadequate in form and/or in fidelity to the original German text. Thus I planned to leave the translations of the papers in categories I and II substantially in their original form, although they were frequently most inelegant and contained occasional mistranslations. I justified my decision by the fact that three of the papers appeared originally in English, while the other four were well-known and indeed standard translations. Mr. Levensohn was bolder than I was, his standards of acceptable prose were higher, and he was insistent that all seven translations be revised in order to come at least somewhat closer to meeting those standards. The result, and the result of his general editorial work, is a collection far superior to what it might otherwise have been. While he thus should receive much credit for improving many translations, he should naturally not be held responsible for any flaws that remain.

I am also grateful to the California State Colleges for granting me one of the first Faculty Development Leaves. The present collection, as well as other work, has been substantially furthered by a semester of freedom.

The task of typing the manuscript has been shared gracefully by a sequence of four young ladies, Toby Young, Anne Ramacher, Marie Fedaleo, and Gayle Àlstrom. I am most grateful to each of them for her interest and helpfulness. I wish each happiness and, should it be necessary, more lucrative employment in the future.

I should like to dedicate this collection to my parents, and to others of their generation, who have managed over a period of years to sustain a genuine interest in what I am doing.

INTRODUCTION

In the autumn of 1894 a London weekly, noted more for its caricatures of British statesmen and its satires on English manners and morals than for its reflections upon the passing of great men, joined with many others in paying tribute to the memory of the German scientist, mathematician, and philosopher Hermann von Helmholtz. In an interval of sobriety between "A Very Un-Virgilian Pastoral Eclogue" and an "Ode on a Distant Partridge," *Punch, or The London Charivari*, presented a memorial whose lack of grace was more than balanced by the appropriateness of its judgment:

What matter titles? HELMHOLTZ is a name
That challenges, alone, the award of Fame!
When Emperors, Kings, Pretenders, shadows all,
Leave not a dust-trace on our whirling ball,
Thy work, oh grave-eyed searcher, shall endure,
Unmarred by faction, from low passion pure.
To bridge the gulf 'twixt matter-veil and mind
Perchance to mortals, dull-sensed, slow, purblind,
Is not permitted–yet; but patient, keen,
Thou on the shadowy track beyond the Seen,
Didst dog the elusive truth, and seek in sound
The secret of soul-mysteries profound,
Essential Order, Beauty's hidden law!
Marvels to strike more sluggish souls with awe,
Great seekers, lonely-souled, explore that track,
We welcome the wild wonders they bring back
From ventures stranger than an earthly Pole
Can furnish. Distant still that mental goal
To which great spirits strain; but when calm Fame
Sums its bold seekers, Helmholtz, thy great name

Among the foremost shall eternal stand,
Science's pride, and glory of thy land.[1]

More serious periodicals were equally high in their estimation and lavish in their praise. In the *Fortnightly Review* Arthur W. Rücker, who at the time was working on a longer study of Helmholtz, after a review of the scientist's many important accomplishments in a career of more than fifty years, concluded:

> In a brief and imperfect sketch such as this it is barely possible to give an idea of the extent of the work of von Helmholtz; it is certainly impossible to do justice to its fulness and depth. I have mapped the direction of the main streams of his thought. Only those who follow them in detail can count the fields they have fertilized. In the course of his investigations all sorts of side issues were studied, a vast number of subsidiary problems solved. The alertness of his intellect, the readiness with which he turned from one science to another, the extraordinary ease with which he handled weapons the most diverse and the most difficult to master, these are not less wonderful than the catalogue of his main achievements.
>
> The technical merits of his work will, of course, be appreciated chiefly by experts. Special knowledge is not necessary to understand its importance. He was one of the first to grasp the principle of the conservation of energy. He struck, independently and at a critical moment, a powerful blow in its defense. He penetrated further than any before him into the mystery of the mechanism which connects us with external nature through the eye and the ear. He discovered the fundamental properties of vortex motion in a perfect liquid, which have since not only been applied in the explanation of all sorts of physical phenomena, of ripple marks in the sand and of cirrus clouds in the air, but have been the bases of some of the most advanced and pregnant speculations as to the constitution of matter and of the luminiferous ether itself.
>
> These scientific achievements are not, perhaps, of the type which most easily commands general attention. They have not been utilized in theological warfare; they have not revolutionized the daily business of the world. It will, however, be universally admitted that such tests do not supply a real measure of the greatness of a student of nature. That must finally be appraised by his power of detecting beneath the complication of things as they seem something of the order which rules things as they are. Judged by this standard, few names will take a higher place than that of Hermann von Helmholtz.[2]

---

1. *Punch, or The London Charivari*, September 22, 1894.

2. Arthur W. Rücker, "Hermann von Helmholtz," *Fortnightly Review*, LVI, New Series, 135 (November 1894), p. 660.

On the continent the response was the same. Germany had lost her greatest scientist, one of the most versatile and productive thinkers in her history. Helmholtz was compared frequently with Goethe and Leibnitz, but like both of these illustrious predecessors, he was also judged to be unique, defying comparison with anyone. Leo Koenigsberger, professor of mathematics and prorector of the University of Heidelberg, who was later to write the standard biography of Helmholtz, called him the "pride and ornament of our university and our fatherland." The president of the French Academy paid tribute to him as "one of the most illustrious men of the times" and noted that "even among the most favored nations men gifted with such powerful intellectual faculties appear but rarely in the course of several generations."[3]

All the tributes were in agreement on a number of points. There were frequent references to "the qualities of his fine nature," his astonishing versatility, his enormous capacity for work, and the profound importance of the unusually large number of contributions he made to an unusual range of sciences. Many paid tribute to his greatness as a teacher and noted his commanding position as a statesman of science, one who did much to lay out the paths that science followed during the second half of the nineteenth century. Many, too, confessed that it was impossible for any one follower to write with a high level of competence on all phases of Helmholtz' life and work.

EARLY YEARS[4]

Hermann von Helmholtz (the "von" was added in 1882 when he was raised by Wilhelm I to the rank of the hereditary nobility) was born at Potsdam on August 31, 1821, the eldest child of August and Caroline Helmholtz. His father, who had fought in the campaigns of 1813-14 against Napoleon, also showed inclusive interests: he taught philology and classical languages and literature at the Potsdam Gymnasium. His mother, whose maiden name was Penne, was the daughter of a Hanoverian artillery officer and a direct descendant of William Penn. As Helmholtz himself recalled in 1891 at the celebration

3. M. Loewy, "The Institute of France in 1894," *Smithsonian Report,* 1894, p. 704.

4. The two best sources of biographical information in English are Helmholtz' own "An Autobiographical Sketch"–the address he gave on the occasion of his seventieth birthday (1891)–and Leo Koenigsberger, *Hermann von Helmholtz*, translated by Francis A. Welby, with a Preface by Lord Kelvin (Oxford, 1906; New York: Dover [reprint], 1965). Both are used here.

in honor of his seventieth birthday, his health was rather delicate during his first years. Through the use of pictures and blocks, however, and under his father's careful guidance and instruction, he acquired at an early age some knowledge of geometry and physics, as well as of poetry and the classical languages. Later, after he entered the Potsdam Gymnasium, his interest in physics—the science that was to become his greatest love—increased greatly. It was also during his student days at the Gymnasium that his ideas on the relationship between science and philosophy, which were to be reflected so frequently in his work throughout his career, first began to take form.

Although the physical sciences and mathematics came to dominate his early interests, Helmholtz' entrance into science took quite a different path. His father was unable to finance a scientific education for the boy out of his very small salary. Moreover, "at that time," as Helmholtz later recalled, "physics was not considered a profession at which one could make a living." The Prussian government, however, offered a complete course of study in medicine to qualified students in exchange for several years of service as an army surgeon, and under this arrangement Helmholtz attended the Royal Friedrich-Wilhelm Institute of Medicine and Surgery, in Berlin. In 1842, at the age of twenty-one, he received his doctor's degree in medicine with a dissertation (*De Fabrica Systematis nervosi Evertebratorum*) establishing that the nerve cells of ganglia, discovered by von Ehrenberg in 1833, are individually connected with the nerve fibers leading from them.[5] Among his teachers were Johannes Müller, whose work on the physiology of the senses was to be one of the foundations of his own work in physiological psychology, and Gustav Magnus, whom he succeeded many years later as professor of physics at the University of Berlin.

After his graduation Helmholtz spent a year as a physician at the Charité in Berlin and then, late in 1843, took up his duties as an army surgeon at Potsdam, a post which he held until 1848. While he was officially occupied in this capacity, he began scientific investigations that were to lead to a number of important discoveries.

"THE CONSERVATION OF FORCE"

Helmholtz' famous paper "The Conservation of Force" (*Über die Erhaltung der Kraft*), which he read before the young Physical Society of Berlin

5. E. G. Boring, in *A History of Experimental Psychology* (2nd ed.; New York, 1950), writes that this discovery "foreshadows the neuron theory" (p. 299).

on July 23, 1847, is rich and complex in origin, content, and influence. According to his own account, it originated in connection with certain problems in physiology, yet it contains only the most cursory discussion of the application of the principle of the conservation of force to physiological phenomena. (His analysis of energy transformations in living organisms occupies only one moderately long paragraph near the end of the text.) The paper is explicitly addressed to physicists, and the short summary at the end emphasizes only the importance of the principle for physics. One is left to draw the conclusion, amply justified in his later writings, that energy transformations involving living organisms are simply special cases of more general and more fundamental physico-chemical processes. Helmholtz also announces at the beginning of the paper that the principle is presented, and should be understood, as a hypothesis within physics, totally divorced from philosophical considerations. He then turns, however, to a philosophical Introduction which is a marvelous combination of epistemological points, clearly reflecting the influence of Kant, with some theoretical physics which reads at times like a priori speculation.

According to his own account of the origin of the paper, he, along with a number of other students under the powerful influence of Johannes Müller, became profoundly interested in the vitalism-mechanism controversy. Their interest centered on the question whether there is, in addition to the obvious chemical and physical agents and processes in the body, a vital force which controls these other forces during life. While there was considerable disagreement, most physiologists, especially the older ones, including Müller, followed G. E. Stahl (who also formulated the phlogiston theory) in maintaining the existence of such a vital force. Later the mechanism–vitalism issue was partially reformulated by the chemist Liebig into problems involving the analysis of physiological heat. Among Helmholtz' earliest scientific works were a number of related experimental investigations of, and reports on, animal heat, metabolism during muscular activity, fermentation, and putrefaction.

As was typical of all of his work in physiology, Helmholtz approached these problems with the mentality and techniques of a physicist, trying constantly to analyze them in terms of the principles of physics. He soon came to believe that Stahl's theory presupposed the existence of some kind of physiological *perpetuum mobile*, and to investigate this possibility, he worked throughout his years as an army surgeon on a systematic study of perpetual motion and energy transformations, both in physics and in physiology. Becoming convinced that a *perpetuum mobile*, long known to be impossible according to the principles of mechanics, was impossible in general, he confronted the inevitable question: If perpetual motion is impossible, what energy relations actually do hold in nature? His answer was "The Conservation of Force."

Helmholtz was not, of course, the only scientist interested in energy transformations and conservation principles. In the eighteenth and early nineteenth centuries a number of mathematicians and physicists had devoted considerable time to the analysis of conservation principles and to various problems within physics involving maxima and minima. Later in the nineteenth century, at the same time that Helmholtz was working on his memoir, other scientists, such as Joule, Mayer, and Colding, were experimenting with or speculating on the relationship between heat and mechanical energy. Thus Helmholtz' investigations were not totally unique or original. They took on great significance, however, because of their scope—the many types of energy transformations considered—and because of the philosophical and mathematical unity of the presentation. Helmholtz not only had the good fortune to see the problems raised by the vitalistic theory; he also had the learning and the mathematical power to appreciate and expand on the work of earlier physicists, as well as the vision and technical knowledge to bring the entire subject into focus.

Two basic and powerful tendencies of his thought are evident in this first major work—tendencies which, combined with his astonishing genius, were to make him and his contributions so outstanding. The first was his drive to seek that which is absolutely basic or fundamental. One feels, for example, in coming to know his thought, that it would have been impossible for him not to have written the philosophical Introduction to the memoir on "The Conservation of Force." The other tendency was toward encyclopedic inclusiveness. One feels that there were few, if any, aspects of the problems before him that were not considered with care.

The immediate reception of this epoch-making memoir indicates to some extent the state of science in Germany near the middle of the nineteenth century. It was eagerly accepted and acclaimed by the younger members of the Physical Society, among whom were a number of men who had studied with Müller, including Helmholtz' friends Emil du Bois-Reymond, Virchow, Ludwig, and Brücke.[6] The older generation, however, were much more hesi-

---

6. It was only to be expected that these men should greet Helmholtz' memoir with enthusiasm. Their general attitude is best expressed by a letter written by du Bois-

tant about his conclusions and about the general philosophical tone of the paper. To them, and perhaps not without reason, the generality and scope of the Introduction and the conclusions were too reminiscent of the *Naturphilosophie* of the followers of Hegel and Schelling, against which they had been fighting for a number of years and against which Helmholtz himself carried on a long campaign. With the exception of the mathematician Jacobi, they were perplexed, too, by the mathematics in the paper. Because of these factors and because there were no new experimental findings in the paper, it was rejected when it was submitted to *Poggendorff's Annals* for publication, and it was only through the influence of du Bois-Reymond, who convinced a publisher of its importance, that it finally made its appearance in print.

As to the long-term reception of the memoir, Maxwell perhaps best characterized its importance:

> . . . [to] appreciate to the full the scientific value of Helmholtz's little paper on the Conservation of Force, we should have to ask those to whom we owe the greatest discoveries in thermodynamics and other branches of modern physics, how many times they have read it over, and how often during their researches they felt the weighty statement of Helmholtz acting on their minds like an irresistible driving force.[7]

"The Conservation of Force" was important, too, in Helmholtz' personal history. In 1848, upon the recommendation of Alexander von Humboldt, he was released from the last three years of his contract as an army surgeon and took the post of lecturer on anatomy at the Academy of Arts in Berlin. At the same time he gave up the practice of medicine. From that point on, he devoted his energies to teaching and to experimental and theoretical science.[8]

Reymond in 1842: "Brücke and I pledged a solemn oath to put in effect this truth: no other forces than the common physical-chemical ones are active within the organism. In those cases which cannot at the time be explained by these forces, one has either to find the specific way or form of their action by means of the physical-mathematical method, or to assume new forces equal in dignity to the chemical-physical forces inherent in matter, reducible to the force of attraction and repulsion." (See S. Bernfeld, "Freud's Earlier Theories and the School of Helmholtz," *Psychoanalytic Quarterly, XIII* [1944], p. 348.)

7. Quoted in Robert H. Murray, *Science and Scientists in the Nineteenth Century* (London, 1925), pp. 96-97.

8. The increase in salary that accompanied this new position also enabled him to marry his fiancée of several years, Olga van Velten. Their marriage ended with her death in 1859. In 1861 he married Anna von Mohl, who survived him by five years. According to Koenigsberger (see note 4, above), both marriages were very happy; the second was brilliant.

PHILOSOPHY OF SCIENCE

The Introduction to "The Conservation of Force" is Helmholtz' first attempt to state his philosophy of science. It shows strongly the influence of Kant and of the philosophy of physics developed in the eighteenth century by various students of Newtonian mechanics. Later, in other essays and popular addresses, Helmholtz amplified and modified a number of the points he first presented in this brief, terse statement of 1847.

According to Helmholtz, the task of experimental physics is to discover laws under which particular properties and processes may be subsumed and from which they may be deduced. The task of theoretical physics, on the other hand, is to search for the causes of natural processes; it seeks to comprehend these processes according to the law of causality.

> We are compelled to and justified in this undertaking by the fundamental principle that every change in nature must have a sufficient cause. . . . Thus the final goal of the theoretical natural sciences is to discover the ultimate invariable causes of natural phenomena. Whether all processes may actually be traced back to such causes, in which case nature is completely comprehensible, or whether on the contrary there are changes which lie outside the law of necessary causality and thus fall within the region of spontaneity or freedom, will not be considered here. In any case it is clear that science, the goal of which is the comprehension of nature, must begin with the presupposition of its comprehensibility and proceed in accordance with this assumption until, perhaps, it is forced by irrefutable facts to recognize limits beyond which it may not go.

Helmholtz never departed from this general position. It is to be found—amplified, of course, and related to the parts of his epistemology and philosophy of science which he developed later—in "The Aim and Progress of Physical Science" (1869), "The Facts of Perception" (1878), the *Introduction to the Lectures on Theoretical Physics* (1894), and other papers.

Within this general, programmatic position there are some points which, in light of the development of the sciences and of his own increased awareness of philosophical difficulties, Helmholtz did qualify or reject at a later time. According to the Introduction to "The Conservation of Force," science uses two basic abstractions, matter and force, in dealing conceptually with the external world. Matter is pictured as existing in space and time; material objects have the properties of spatial location (which may and does vary) and mass (which is constant). Matter itself is inert. It is obvious, however, that objects affect one another in various, sometimes complex ways and that they affect our sense organs. Hence we must ascribe forces, or various capacities to produce effects, to matter. The concepts of matter and force are inseparable: matter alone is insufficient to account for phenomena, while force without a material basis is a contradiction in terms.

Much of this general position Helmholtz retained throughout his career. He thought always in terms of forces which are in some fashion related to a material substratum and which bring about, or cause, that which occurs, including the excitation of our sense organs. As the goal of the theoretical sciences is to trace phenomena back to causes, and as the causes of phenomena are forces, it followed that the task of those sciences is to investigate forces, their effects, and the relationships (in which force or energy is always conserved) among them.

In "The Conservation of Force" these forces were given an unquestioned objective status. Helmholtz apparently did not even consider the possibility that someone might challenge this position; certainly he did not defend it. Soon, however—that is, by the time he had gotten a good start on his investigation of human vision—he realized clearly that he had to offer some justification for ascribing an objective status to forces and their material substratum. On the whole he continued to be a consistent realist, arguing, among other things, that a correct analysis of perception, while it does not provide conclusive evidence, does support the realistic position. That hypothesis is the most sensible one to follow.

He used, too, the Fichtean distinction between the Ego and the Non-Ego to support his realism, arguing that as we become aware of what we can and what we cannot do, it is impossible not to think of the Non-Ego—that is, the world about us—as an objective power acting independently of and often in opposition to us. The reality of the external world forces itself upon us. Helmholtz also argued frequently that the fact that the same things always happen whenever the same sets of specific necessary conditions are present indicates the existence of forces ready to act whenever the conditions required for their action are realized.

Helmholtz recognized, of course, that these arguments were not conclusive. Sometimes, especially in the *Introduction to the Lectures on Theoretical Physics*, his scientific realism faltered for a moment and he favored the position that to use abstract substantives such as *force* was only a *façon de parler*. It was convenient and consistent with the natural, common-sense realism of everyday life; but a formulation, for example, of a law of motion using the term *force* ("the force of gravity") was, he maintained at these times, empirically no more than a description of the motion.

It was in connection with the kind of force that Helmholtz took to be fundamental that the greatest change in his thinking came over the years. Perhaps the best index of the change is that whereas in 1847 and for a number of years after, Helmholtz was committed to one pair of forces as basic, in his later writings he ceased to give any force a favored status. In 1847 he thought of the universe as consisting of material elements with unchanging qualities. Natural objects were made up of, and were resolvable into, elementary parts; hence the forces of natural objects were made up of, and were resolvable into, the forces of these elementary parts. Ultimately the only changes possible were changes in relative spatial position, that is, movements. Consequently, the ultimate forces were those which bring about changes in spatial relations, that is, forces which produce movements. The only possible change in the relative spatial positions of two bodies, however (here the a priori physics is at its height!), was a change in their distance from one another. Thus the ultimate forces were those which produce such changes—in short, the central forces of attraction and repulsion, which, like all central forces, vary in intensity with distance. In the Introduction to "The Conservation of Force" he wrote confidently:

> Thus we see that the problem of the physical sciences is to trace natural phenomena back to inalterable forces of attraction and repulsion, the intensity of the forces depending upon distance. The solution of this problem would mean the complete comprehensibility of nature. . . .
>
> Theoretical natural science, therefore, if it is not to rest content with a partial view of the nature of things, must take a position in harmony with the present conception of simple forces and the consequences of this conception. Its task will be completed when the reduction of phenomena to simple forces has been completed and when, at the same time, it can be proved that the reduction is the only possible one which the phenomena will allow. This will then be established as the conceptual form necessary for understanding nature, and we shall be able to ascribe objective truth to it.

In "The Aim and Progress of Physical Science" (1869), the reduction of all forces to central ones and the explanation of all phenomena by reference to these forces was still his ideal. This was for him the aim of physical science. A short time later he began to work intensively on the theory of electricity and magnetism, and his mechanistic ideal came to seem increasingly remote. He continued to work on the fundamental principles of physics, but he had to be content with a plurality of forces, interrelated but not reducible to any one fundamental force or pair of forces.

Helmholtz' Kantianism, which is evident in "The Conservation of Force" primarily in his interpretation of the principle of causality as the supreme regulative principle to be followed in the ordering of scientific knowledge, was explicitly acknowledged in the first of the appendices which he added to the memoir in 1881. In philosophy he always considered himself a follower of Kant, even when he was arguing most strenuously against some aspect of Kant's thought, such as the philosopher's interpretation of the axioms of geometry, which Helmholtz took to be not really essential to Kant's basic position.[9]

In addition to emphasizing the principle of causality, Helmholtz identified himself with Kant on a number of other general points. One was in his basic conception of philosophy. "The task of philosophy," he wrote in his first paper dealing with philosophical issues and repeated many times later, "is to investigate the sources of our knowledge and the extent of its qualifications, a task which will always remain that of philosophy and which no age can avoid with impunity."[10]

Another way in which Helmholtz felt he was following Kant was in his unqualified opposition to metaphysical speculation, especially the theorizing of Hegel and Schelling. He insisted that after Kant (and Fichte, whom Helmholtz frequently linked with Kant when he was discussing these issues) metaphysicians drew philosophy away from its true task. Philosophy after Fichte was, for Helmholtz, completely aberrant, deviating so far from epistemology as no longer to be philosophy at all. Moreover, the speculations of metaphysicians were totally barren; all that they had succeeded in doing was to create an unfortunate and destructive antagonism between philosophy and the natural sciences.

Helmholtz never tired of returning to these themes. In a letter to his father in 1857 he wrote:

> I am delighted at what you write about your present life; I think you will be more and more interested in philosophy, the more you give your-

---

9. In the Preface to the third edition of *Vorträge und Reden* (1884), Helmholtz wrote: "I was a faithful Kantian in the beginning of my career as I am now; or rather, I believed then that that which I wished to see altered in Kant was an insignificant side issue which, next to that which I still esteem to be his chief result, need not come up for consideration" (I, vii).

10. "Über das Sehen des Menschen," *ibid.* (1st edition, 1884), I, 95. Ernest Cassirer claims that Helmholtz was one of the first to raise the cry of "Back to Kant!" which ushered in the neo-Kantian movement in Germany in the latter half of the nineteenth century (see *The Problem of Knowledge* [New Haven, 1950], pp. 3–5).

> self up to it. It seems to me a favourable moment for voices of the old school of Kant and the elder Fichte to obtain a hearing once more. The philosophical vapouring and consequent hysteria of the "nature-systems" of Hegel and Schelling seem to have exploded, and people are beginning to interest themselves in philosophy again. . . . Philosophy finds its great significance among the sciences as the theory of the source and functions of knowledge, in the sense in which Kant, and, so far as I have understood him, the elder Fichte, took it. Hegel, however, wanted it to replace all the other sciences, and to find out by its means what is perhaps denied to man, by which he diverted philosophy from its proper scope, and gave it tasks it never can accomplish. The majority of educated men at first believed in him, and then rejected philosophy altogether, seeing that nothing came of it. . . .[11]

In another letter, approximately twenty years later, he asserted:

> I believe that philosophy will only be reinstated when it turns with zeal and energy to the investigation of epistemological processes and of scientific methods. There it has a real and a legitimate task. The construction of metaphysical hypotheses is vanity. Most essential of all in this critical investigation is the exact knowledge of the processes of sense-perception. . . .Philosophy has been at a standstill because it was exclusively in the hands of philologists and theologians, and has so far imbibed no new life from the vigorous development of the natural sciences. Hence it has been almost entirely confined to the history of philosophy. I believe that any German University that had courage to appoint a scientific man with an inclination for philosophy to its Chair of Philosophy would confer a lasting benefit on German science.[12]

It soon becomes evident in reading his works in philosophy, however, that Helmholtz was not a slavish follower of Kant. Quite early in his career he saw that the Kantian position, while basically sound as a general framework, could not be accepted without alterations and corrections. The establishment of new sciences and the expansion of older ones compelled such changes. As the years passed, as he worked through physiological optics and on to the foundations of geometry and arithmetic, he argued increasingly for these alterations and corrections, while retaining what he took to be the basic outlines of Kant's epistemology. Thus the advances made in physiological optics, based on Helmholtz' extension of Johannes Müller's principle of spe-

---

11. Quoted in Leo Koenigsberger, *Hermann von Helmholtz* (see note 4, above), p. 159.

12. Quoted in *ibid.*, p. 139.

cific energies, necessitated a change in Kant's conception of human knowledge to allow a greater role in perception to experience and to unconscious inferences, or the "lower psychical processes." The concept of space had to be reconsidered, with more emphasis given to the role that experience plays in the perception of spatial relations. This in turn necessitated a reconsideration of the status of geometrical axioms, which led Helmholtz to an attempt to establish them empirically and to divorce them from any relationship to a Kantian pure form of intuition.

PHYSIOLOGICAL PSYCHOLOGY

Helmholtz had been at the Academy of Arts in Berlin for only one year when he was called in 1848 to fill the post of professor of physiology at the University of Königsberg. For the next twenty years he occupied the position of professor of physiology and anatomy at the universities of Königsberg (1848-55), Bonn (1855-58), and Heidelberg (1858-71). Although he contributed occasionally to physics and continued to emphasize in public lectures the importance of the principle of the conservation of energy, his main areas of research during this period were physiological optics and physiological acoustics. Helmholtz did more to establish these two sciences than any other man.

His career at Königsberg was marked from the beginning by a series of important investigations and discoveries. Relying on his knowledge of experimental physics and his ability in designing instruments, he devoted himself to the problem of the rate of transmission of impulses or excitations in nerves. After working out for the first time the techniques by which the rate could be determined, he made the first measurements. It had been almost universally believed that the velocity of nerve impulses was almost instantaneous or at least comparable to the speed of light. Helmholtz succeeded in showing that the velocity was clearly finite and far less than that of light.

It was also during his earliest years at Königsberg that he began to investigate the physiology of the senses, a subject to which he had been introduced by Müller. Among his first contributions was the invention of the ophthalmoscope, which was to do more than his work on the conservation of energy to bring his name into prominence. As he recalled in 1891 in "An Autobiographical Sketch":

> The construction of the ophthalmoscope had a most decisive effect on my position in the eyes of the world. From that time on I met with im-

> mediate recognition from the authorities and my colleagues, and with an eagerness to satisfy my wishes. Thus I was able to follow far more freely the impulses of my desire for knowledge. I must, however, say that I attribute my success in great measure to the fact that, possessing some geometric understanding and equipped with a knowledge of physics, I had the good fortune to be thrown into medicine, where I found in physiology a virgin territory of great fertility. Furthermore, I was led by my knowledge of vital processes to questions and points of view which are usually foreign to pure mathematicians and physicists.

While considering the color theories of his predecessors, he made another discovery that he thought should have been made long before. He discovered the simple yet profoundly important fact that the compound color which results from mixing two colored pigments is different from that which results from mixing the same two colors of the spectrum. Concerning this discovery, he remarked in his speech upon receiving the first Graefe Medal (1886):

> Not being inclined to describe in my lectures things I had not myself seen, I made experiments in which I blended the colors of the spectrum in pairs. To my astonishment, I found that yellow and blue gave not green, as was then supposed, but white. Yellow and blue pigments, when mixed, no doubt give green, and until then the mixing of pigments was supposed to produce the same effect as the mixing of the colors of the spectrum. This observation not only at once produced an important change in all the ordinarily accepted notions of color mixture, but it also had an even more important effect on my views. Two master minds of the first rank, Goethe and David Brewster, were of the opinion that yellow and blue could be directly seen in green. Their observations were made with pigments, and they thought they could divide their perception of the resulting color into two parts, yellow and blue, while in reality, as I was able to show, neither was present.[13]

Much of Helmholtz' theory of color vision resulted from his attempt to understand these and related phenomena of color mixture.

The cornerstone of Helmholtz' work on the physiology and psychology of vision and hearing was the principle of the specific energies of the nerves, which was first explicitly formulated by Johannes Müller in 1826. The principle was implicit in many earlier works in philosophy and physiology, most notably in the writings of Locke and other British empiricists and in the physiological studies of Sir Charles Bell. Müller's contribution consisted in

---

13. Quoted in John G. McKendrick, *Hermann von Helmholtz* (London, 1899), p. 139.

giving the principle a more detailed, systematic exposition than it had previously received and in emphasizing its importance. In addition he was able to popularize it because of the wide use of his *Handbuch der Physiologie des Menschens* (1833–38).

Fundamental to this principle is the assumption that all we know about the external world enters our consciousness as the result of external causes. These causes affect the sense organs, producing excitations which are transmitted by the nerves to the brain. It is in the brain that they first become conscious sensations, and there they are combined to produce perceptions of the objects around us. If the nerves that carry these excitations to the brain are cut, the sensations and the accompanying perceptions immediately cease. Thus, in the case of the eye, visual perceptions are produced, not directly in each retina, but in the brain itself as a result of excitations transmitted to it from the eye.

The sensations caused by external objects fall into five distinct groups, corresponding to the five senses, and the differences among them are such that it is impossible to compare a sensation of one sense with that of another—a sensation of light, for example, with one of sound or of smell. Helmholtz called this a difference of *mode* or *kind*, and the sensations belonging to a particular sense he thought of as comprising the *circle of quality* of that sense. Thus whether a muscular movement, a secretion, or a sensation is produced by the excitation of a nerve depends entirely upon whether a motor, a glandular, or a sensory nerve is excited, not upon the source of excitation. In the same way, the kind or mode of sensation which is produced when a sensory nerve is excited depends solely upon which sense the nerve subserves. Excitations of the optic nerve, for example, produce only the sensations of light, whether the optic nerve is excited by objective light, by a blow on the eyeball, or by strain in the nerve trunk during rapid movement of the eye.

The same outer stimulus, therefore, if it excites different kinds of nerves, produces different kinds of sensations. These sensations, however, are always grouped within the circles of quality of the excited nerves. In the lecture "The Scientific Researches of Goethe" (1853), Helmholtz explains:

> The same electric current whose existence is indicated by the optic nerve as a flash of light, or by the organ of taste as an acid flavor, excites in the nerves of the skin the sensation of burning. The same ray of sunshine, which is called light when it falls on the eye, we call heat when it falls on the skin.

It is only because objective light usually excites the optic nerve that we are led to believe that there is objective light similar to our subjective sensations of light.

Helmholtz once wrote that he considered Müller's principle of specific nervous energies to be second in importance only to Newton's theory of gravitation. Over a period of years he extended it in several directions and used it as the basis of and the central principle in an enormous body of experimental, theoretical, and historical work in physiological optics and physiological acoustics. The principle also led him more deeply into the philosophical aspects of perception and into investigations concerning the foundations of geometry. Indeed, it led him ultimately to consider more carefully the foundations of the sciences in general.

The first way in which the principle of specific energies was extended was in the development, which Helmholtz began in 1852, of Thomas Young's hypothesis concerning the subjective effects of colors. Young was by training and profession a physician; his interests, however, were broad, and he is now best known for his work in connection with the reintroduction of the wave theory of light. He also did notable work on ocular accommodation and on the theory of color, although his investigations in physiological optics were neglected until Helmholtz and Maxwell developed them just after the middle of the century.

The basic idea in Young's theory of color sensation can be traced back to Newton who, in a communication to the Royal Society in 1675, suggested that various rays of light excite different vibrations in the retina:

> . . .the biggest, strongest, and most potent rays, the largest vibrations; and others shorter, according to their bigness, strength, or power. . . . [These vibrations] will run . . . through the optic nerves, into the sensorium;– and there, I suppose, affect the sense with various colours, according to their bigness and mixture; the biggest with the strongest colours, reds and yellow, the least with the weakest, blues and violets; the middle with green, and a confusion of all with white;–much after the manner that, in the sense of hearing, nature makes use of aerial vibrations of several bignesses, to generate sounds of diverse tones; for the analogy of nature is to be observed.[14]

Newton was thinking of seven colors, analogous to the tones in an octave. Young, however, knew that there was a very large number of waves of differ-

---

14. H. S. Thayer (ed.), *Newton's Philosophy of Nature* (New York, 1953), pp. 97–98.

ent frequencies in the spectrum, and he felt that it was extremely unlikely that there were retinal points to correspond to each of these frequencies. He was led therefore to suppose that there must be a limited number of kinds of retinal fiber; following Newton's principle of color mixture, he suggested that there were three, corresponding to the colors red, green, and violet (or blue). Each of these retinal particles or fibers could be excited by any frequency in the visible spectrum, but they were selectively activated by different wave lengths of light, the intensity of their response varying with the frequency. The "undulations" of green light, for example, according to Young, "affect equally the particles in unison with yellow and blue, and produce the same effect as light composed of those two species."[15]

It was in connection with his investigation of complementary colors and color mixtures that Helmholtz first came across Young's work, "buried," as he said, in the *Transactions of the Royal Society*. He quickly recognized that Young's hypothesis could be treated as an extension of Müller's principle of specific energies. Just as the difference between the sensations of light and warmth depends upon whether the rays of the sun, for example, fall upon nerves of sight or nerves of feeling, so the differences in the sensation of color might depend simply upon which of the nerve fibers sensitive to light is most strongly affected. When all three are equally excited, the result is the sensation of white.

This hypothesis explained to Helmholtz' satisfaction the phenomena with which he had been working, those of color mixing and complementary colors. It explained, too, the differences between the effects when light rays of different wave lengths are mixed and the effects when a painter mixes different pigments. He was also able to account for color blindness and accepted this as a confirmation of the hypothesis. As a further substantiation he used, among others, the phenomenon of retinal fatigue, in which after the retina has been subjected to extended stimulation by a given color, an afterimage of the complementary color appears. If part of the retina is overstimulated by a wave length corresponding to red, for example, it will be red-blind for a time, and a bluish-green afterimage will be experienced.

In 1856, when the first part of his *Handbook of Physiological Optics* made its appearance, Helmholtz began his study of physiological acoustics; in 1863, after seven years of work, he published *The Theory of the Sensations of*

---

15. Young, "On the Theory of Light and Colours," quoted in *Color Vision*, edited by Richard C. Teevan and Robert C. Birney (Princeton, N.J.: Van Nostrand, 1961), p. 5.

Tone as a Physiological Basis of the Theory of Music. Here, as elsewhere, the range of his accomplishment and understanding is amazing. The book shows a thorough knowledge of physiology, physical acoustics, and the history of music. It is the work of a physiologist-physicist-musician and may be interpreted as an attempt to deal scientifically and experimentally with problems, not only in physics and physiology, but also in aesthetics.

Helmholtz' central hypothesis in physiological acoustics is a combination of elements established within mathematics and physics during the first half of the nineteenth century, on the one hand, and another extension of the principle of specific nervous energies, on the other. The mathematics and physics are the results of work done by Fourier and Ohm. The former showed that certain continuous periodic functions or curves (which can be used to represent sound waves), provided they satisfy certain conditions (which sound waves, represented in this way, clearly satisfy), can be thought of as the summation or synthesis of a series of sine curves or uniform waves. In general, the separate members of this series vary in height (amplitude), phase relation, and wave length, the wave lengths of the members of the series being even fractions (1/1, 1/2, 1/3, 1/4, etc.) of the length of the wave being represented. In the case of sound, this corresponds to the fundamental note and its several harmonics.

Fourier's principle can be used both synthetically and analytically. It is not only theoretically possible to construct a complex curve or to represent it as being constructed (if it satisfies certain conditions) from a series of sine curves; it is also possible to analyze such a complex curve into a series of simple sine curves. Moreover, with sound waves this analysis can be performed not only mathematically but also by using various kinds of resonators—objects, such as tuning forks or stretched strings, which vibrate with or respond sympathetically to a sounding body. Given an adequate set of resonators, it is possible to analyze very complex sound waves. Ohm gave this kind of analysis a psycho-physiological interpretation in 1843 by arguing that the ear performs this function—that is, separates out the simple harmonic components—when we listen to complex notes or sounds.

Helmholtz began his work at this point with studies on vowel sounds, timbre, and combination tones. It was clear to him that the ear does make the kind of analysis indicated by Ohm's acoustical law, that is, the same kind of analysis as a large number of resonators. The ear should therefore be considered a resonating organ or a complex of resonating mechanisms. It remained for him only to find within the ear the actual mechanism and to give a general explanation of the whole process of hearing.

The mechanism he chose in his earliest presentations of his acoustical theory, "The Physiological Causes of Harmony in Music" (1857) and in the first two editions of *The Sensations of Tone* (1863, 1865), were the microscopic plates or rods of Corti (discovered in 1851), which Helmholtz treated as a graded series of resonators–an extension of Müller's principle. In physiological optics Helmholtz restricted himself to three specific energies, each related to a range of wave lengths. In physiological acoustics he turned in the other direction, boldly suggesting a very large number of specific energies, each correlated specifically with a frequency in the range to which the ear is sensitive. Indeed, he postulated almost five thousand specific energies, with a separate nerve fiber for each energy. On this hypothesis he based an explanation of the difference between consonance and dissonance, the perception of the quality of sounds, the formation of musical scales, and other acoustical phenomena.

Since its publication in 1863, Helmholtz' resonance theory of hearing has been central to physiological acoustics. He modified it himself in a paper published in 1869 and in the third edition of *The Sensations of Tone* (1870), replacing the rods of Corti with a membrane in the cochlea consisting of a series of fibers of graded length, like the strings of a piano. This was a change only in the physiological aspect of the theory; it remained a resonance theory allied with the principle of specific nervous energies.

PERCEPTION

These developments of the principle of specific energies by no means exhaust its implications for Helmholtz. It is obvious that except in unusual cases, or when we are experimenting with them, or when we are experiencing pleasure or pain, we are generally not conscious of our sensations. Instead, we are conscious of the various objects around us. We become aware of these objects by means of the sensations they cause, but the sensations alone are not sufficient to account for our awareness of and ideas about the objects themselves.

For Helmholtz this conclusion follows directly from the principle of specific energies, according to which there is no one-to-one or perfect correspondence between our sensations and any objective cause of them. The same objective cause may excite different sense nerves, and different causes may excite the same one; the sense nerves, however, always respond in their characteristic manner, no matter what the cause of the excitation may be.

Our ideas of the existence, the form, and the spatial location of external objects Helmholtz called perceptions. The insufficiency of sensations alone to account for our perceptions of objects raised in his mind three questions of fundamental importance. The first was how we actually come to have perceptions or ideas of external objects. What must be added to, or what must happen to, sensations in order for perceptions to exist? His answer to this question—which he presented for the first times in 1852 and 1855 at the University of Königsberg, where Kant had once taught, and which he developed and defended in a series of publications down to "The Origin and Correct Interpretation of Our Sense Impressions" (1894)—shows his affinity to British empirical philosophy and psychology. This is one of the places where he broke most sharply from the philosophy of Kant.

This problem, coupled with the principle of specific energies, led to a second one, a problem in the theory of knowledge. If, as the principle of specific energies seems to indicate, there is a radical difference between sensations and the objects which excite them, in what way can we say that our sense perceptions or ideas of external objects are true? In what way do our ideas correspond to reality?

The third problem has to do with the perception of space, that is, with the perception of spatial relations. Helmholtz' investigations in this area led him into non-Euclidean geometry and from there to still other fundamental questions in the theory of knowledge.

For Helmholtz the fact that sensations are caused by external objects but do not correspond to them could mean only one thing: our sensations are only subjective signs of the objects. Like all signs, what they mean requires interpretation or is the result of interpretation. Perceptions, that is, our ideas of objects, result from an interpretation of sensations, this interpretation being carried out by unconscious mental processes or, more specifically, unconscious inferences. These inferences are aided by various, incessant movements and feelings of movement on the part of the perceiver. In general and in essence, perceptions arise because of the unconscious activity of the mind in comparing and associating sensations; they are the product of a process of interpretation which begins at birth and continues throughout life.

The role of mental activities or processes in perception had been the subject of debate long before Helmholtz' time and is still a subject of controversy. In the nineteenth century those thinkers, like Helmholtz, who attributed as much as possible—especially all perceptions of spatial relations—to experience and to the "lower psychical activities" were said to follow the empirical or empiricist theory of sense perception. According to this general point of view, no other mechanisms or functions were necessary to account for the origin of perceptions beyond certain elementary activities of the mind, even though these activities themselves were not fully understood.

There were other thinkers, however, who, while willing to admit the influence of experience in certain classes of perceptions, believed that it was also necessary to assume a system of native or innate perceptions or ideas which were not the result of experience. In particular, this was held to be necessary in the case of certain elementary perceptions common to all observers, especially with reference to spatial relations. All those who maintained such a position were said to hold an intuitive or nativistic theory of sense perception. The cardinal fact about all of these theories, Helmholtz wrote in the *Handbook of Physiological Optics,*

> is that the localization of the impressions in the field of view is derived through some innate contrivance, and either the mind is supposed to have some direct knowledge of the dimensions of the retina, or it is assumed that, as the result of the stimulation of definite nerve fibres, certain apperceptions of space arise by virtue of an innate mechanism which cannot be further defined.[16]

Helmholtz acknowledged that the question of the ultimate validity of the two competing theories (or general points of view) could not be settled. Neither theory could be refuted, and both could be used in many instances to explain the same phenomenon. In spite of this, however, he considered all intuitive or nativistic theories to be based upon unnecessary and gratuitous hypotheses, which in many cases were contrary to fact. In opposition to them, he defended consistently the empirical theory of sense perception, insisting always on the tremendous influence of experience and what he considered the lower mental functions in perception:

> In my judgment, many natural philosophers have been far too ready to presuppose all kinds of anatomical structures in the theory of the perceptions of vision and also to postulate new qualities of the nervous substance that are contrary to what we actually know about the physical and chemical properties of bodies in general and about the nerves in particular. Hypotheses of this kind never attempt to do more than account for some one, or perhaps for a few, of the phenomena of vision.[17]

---

16. *Helmholtz's Treatise on Physiological Optics* (New York, 1924–25; New York: Dover [reprint], 1962), III, 541–42.

17. *Ibid.*, pp. 531–32.

His preference for the empirical theory was due partially to the simplicity of the explanations he felt it offered. In addition, he held that the nativistic theories either ignored entirely the unquestionable influence of the so-called lower psychical processes or else dismissed them as of relatively slight importance. He was convinced that this view was incorrect. No matter how difficult it might be to explain these activities or processes, there could be no denying their existence. Further, from daily experience we are familiar, to a certain extent, with their laws.

> [It] is safer, in my opinion, to connect the phenomena of vision with other processes that are certainly present and actually effective, although they may require further explanation themselves, instead of trying to base these phenomena on perfectly unknown hypotheses as to the mechanism of the nervous system and the properties of the nervous tissue, which have been invented for the purpose and have no analogy of any sort. The only justification I can see for proceeding in this way would be after all attempts had failed to explain the phenomena by known facts.[18]

The learning of a native language is, for Helmholtz, a good example of the functioning of the mental processes that are important in perception. A native language is learned largely through the *use* of words. A child hears the common names of objects pronounced when these objects are shown or given to him, and in this way the word and the object are linked in his memory in some fashion. This union of word and object is fixed by continued repetition. The child, however, is either unconscious of this process or at least soon forgets how he learned any particular word. Thus most native language is learned unconsciously by experience in the use of language and only exceptionally by the use of explicit definitions.

The processes involved in the interpretation of sensations is similar; the only difference is that words are artificial or conventional symbols, while sensations are natural signs. The mental activity is the same—an unconscious process by which perceptions of external objects are developed by association and experience out of our countless sensations.

Concerning our visual perceptions of spatial relations and spatial location, Helmholtz, following Lotze, felt it necessary to make one small concession to nativism by assuming certain differences among the sensations on various parts of the retina. The sensation of red, for example, on the right-hand side of the retina must be different in some fashion from the sensation of the same

18. *Ibid.*, p. 532.

red on the left-hand side. Otherwise, he argued, it would be impossible to account for any local differences in the field of vision, hence impossible to account for our perceptions of spatial relations. Thus visual sensations are not just signs; they are local signs (the term is also Lotze's). Like all signs, however, their constantly varying significance can only be learned by experience gained through movement and through the use of our eyes.

Helmholtz' interpretation of sensations as signs for relations in the external world led him to the conclusion that it is impossible to speak of our perceptions or ideas being *similar to* objects in the external world. Moreover, all the properties we ascribe to external objects must be thought of either as these objects' effects on our sense organs or as their effects upon one another. Thus properties are, in reality, relations. The only respect in which he could see any agreement between our perceptions and the external world was in temporal sequence. External events and our perceptions of them occur in time, and the temporal sequence of the perceptions can be a faithful reproduction of the sequence of the events. All other notions of similarity, according to the findings of physiological optics, are impossible or "unthinkable."

Helmholtz cautioned, however, against concluding that all our ideas of things are false, simply because they are not copies of things themselves. We must be on guard against the false belief that we are unable to have any knowledge about the true nature of things.

> Every image is the image of a thing merely for him who knows how to read it, and who is enabled by the aid of the image to form an idea of the thing. Every image is similar to its object in one respect, and dissimilar in all others, whether it be a painting, a statue, the musical or dramatic representation of a mental mood, etc. Thus the ideas of the external world are images of the regular sequence of natural events, and if they are formed correctly according to the laws of our thinking, and we are able by our actions to translate them back into reality again, the ideas we have are also the *only true* ones for our mental capacity. All others would be false.
>
> In my opinion, it is a mistake, therefore, to try to find pre-established harmony between the laws of thought and those of nature, an identity between nature and mind, or whatever we may call it. A system of signs may be more or less perfect and convenient. Accordingly, it will be more or less easy to employ, more exact in denoting or more inexact, just as is the case with different languages. But otherwise each system can be adapted to the case more or less well. If there were not a number of similar natural objects in the world, our faculty of forming shades of conception would indeed not be of any use to us. Were there no solid bodies, our geo-

metrical faculties would necessarily remain undeveloped and unused, just as the physical eye would not be of any service to us in a world where there was no light. If in this sense anybody wishes to speak of an adaptation of our laws of mind to the laws of nature, there is no objection to it. Evidently, however, such adaptation does not have to be either perfect or exact. The eye is an extremely useful organ practically, although it cannot see distinctly at all distances, or perceive all sorts of aether vibrations, or concentrate exactly in one point all the rays that issue from a point. Our intellectual faculties are connected with the activities of a material organ, namely the brain, just as the faculty of vision is connected with the eye. Human intelligence is wonderfully effective in the world, and brings it under a strict law of causation. Whether it necessarily must be able to control whatever is in the world or can happen—I can see no guarantee for that.[19]

THE FOUNDATIONS OF GEOMETRY

Between 1866 and 1878 Helmholtz wrote six papers on the foundations of geometry. The first two, "Ueber die thatsächlichen Grundlagen der Geometrie" (1866) and "Ueber die Thatsachen, die der Geometrie zum Grunde liegen" (1868), were mathematical and technical in content. Both covered virtually the same material. As the titles indicate, Helmholtz argued that geometry is based upon certain facts and hence is an empirical science. The findings of these two papers correspond in a number of essential points to Riemann's famous *Über die Hypothesen welche der Geometrie zum Grunde liegen* (which was presented as a dissertation in 1854 but not published until 1868). Helmholtz, however, arrived at his conclusions from a different direction and quite independently of Riemann.

In 1870 appeared two more papers, one a review for the first volume of *The Academy* of a number of works in mathematics and mechanics; the other, the famous popular lecture on "The Origin and Meaning of Geometric Axioms." (A version of this lecture was published in the third issue of *Mind* in 1876.) In these papers Helmholtz attempted to summarize and develop in non-mathematical language the results of the growing body of discoveries in non-Euclidean geometry. Here he applied for the first time the findings of metamathematical investigations to some of the traditional philosophical problems concerning the axioms of geometry, the perception of spatial relations, and the Kantian conception of space.

---

19. *Helmholtz's Treatise on Physiological Optics*, III, 24.

The last two papers in this series appeared in 1878: "The Origin and Meaning of Geometric Axioms (II)," a careful, detailed answer to some criticisms of his earlier essays, and "The Facts of Perception," in which he sought to summarize the epistemological implications of his study of physiological optics and geometry.

As Helmholtz himself noted in his first two papers, it was his investigations of visual perception of spatial relations that had led him to consider the general nature of our knowledge of space. The question to which he first directed his attention was: To what extent are the propositions of geometry empirical in reference? To what extent are they simply definitions, deductions from definitions, or propositions whose truth depends upon the way they are expressed? What, in short, are the factual, objective foundations of geometry?

His starting point was the fact that all spatial measurements depend upon the establishment and observation of congruence. But the congruence of two figures or point systems can be determined only if they can be moved toward one another without any alteration in form; that is, the congruence must be independent of any possible movement of the figures. Helmholtz therefore proceeded to investigate the general analytical characteristics that a space must possess in order for such movement to be possible.

This investigation led him to identify four facts upon which he concluded that geometry is based. The first has to do with the continuity and dimensions of space:

> In a space of n dimensions, the location of any point is determined by the measurement of n independent magnitudes, which are continuous functions. Thus—with the exception of certain points, lines, surfaces, or in general certain configurations of less than n dimensions—with any movement of a point, the magnitudes serving as coordinates vary continuously, at least one of them not remaining constant.[20]

The second fact relates to the existence of rigid but movable bodies: "the existence of rigid bodies movable relative to a system of points is presupposed, since this is necessary in order to undertake the comparison of spatial magnitudes by means of congruence." The third concerns the free mobility of these bodies: "it is presupposed that every point can be moved continuously to the

---

20. "Über die thatsächlichen Grundlagen der Geometrie," *Wissenschaftliche Abhandlungen* (Leipzig, 1883), II, 614.

position of any other insofar as it is not fixed by the mathematical relations which exist between it and the other points of the rigid system to which it belongs." The last concerns the independence of the form of a rigid body from rotation: "two congruent bodies remain congruent after one has undergone a complete rotation of any kind." Thus the independence of the congruence of two bodies or rigid point systems from place, orientation, and relative rotation are the facts upon which geometry is based.[21]

As he continued his investigations, Helmholtz quickly learned that these conditions are satisfied, not only by the space of Euclid, but also by the spaces of Lobachevsky and Riemann–indeed, by any space with a constant measure of curvature. Whether real space has a constant measure of curvature or not, however, can only be determined by experience. If it is constant, it is also an empirical question whether the measure of curvature is zero (as in Euclidean "flat" space), greater than zero (as on the surface of a sphere), or less than zero (as in pseudospherical space). As a result, Helmholtz drew the conclusion that Kant was mistaken in claiming that axioms of Euclidean geometry were synthetic, a priori principles–principles which are necessarily true of space but which can be established on the basis of a priori considerations alone. Spaces other than Euclidean are conceivable, and the geometries of these spaces can be formulated.

On the basis of our experience, however, it is impossible to determine conclusively which of these geometries is that of real space. If space does indeed have a constant measure of curvature equal to zero (as experience seemed to indicate), and if the number of dimensions of space is three, and if space is infinitely extended, then no geometry except that of Euclid is possible. Because these other spaces are perfectly conceivable, however, Kant may very well have been right in thinking of space as a necessary form of intuition (as Helmholtz himself was to argue in 1878). But Kant was wrong in claiming Euclidean characteristics for this form of intuition. A powerful blow had been dealt to the Kantian interpretation of space.[22]

At the conclusion of his review in *The Academy*, Helmholtz summarized his position:

---

21. *Ibid.*, pp. 614–15.

22. See Hans Freudenthal, "The Main Trends in the Foundations of Geometry in the 19th Century," in *Logic, Methodology, and Philosophy of Science*, edited by E. Nagel, P. Suppes, and A. Tarski (Stanford, 1962), pp. 613–21, for some interesting material on the reception of non-Euclidean geometry in the latter half of the nineteenth century.

We may resume the results of these investigations by saying that the axioms on which our geometrical system is based are not necessary truths, depending solely on irrefragable laws of our thinking. On the contrary, other systems of geometry may be developed analytically with perfectly logical consistency. Our axioms are, indeed, the scientific expression of a most general fact of experience, the fact, namely, that in our space bodies can move freely without altering their form. From this fact of experience it follows, that our space is a space of constant curvature, but the value of this curvature can be found only by actual measurements. Riemann, indeed, finishes his paper with the somewhat startling conclusion that the axioms of Euclid may be, perhaps, only approximately true. They have been verified by experience to that degree of precision which practical geometry and astronomy have reached hitherto, and, therefore, there is no doubt that the radius of curvature of our space, if it should be spherical or pseudospherical, is infinitely great when compared with the dimensions of our planetary system. But we are not absolutely certain that it would prove to be infinite when compared with the distances of fixed stars, or with the dimensions of space itself.[23]

ESSAYS AND ADDRESSES

In addition to his extremely active career as a teacher, writer, and research scientist, Helmholtz found (or made) time for a number of related activities. By the middle 1850's, only a few years after "The Conservation of Force" and the invention of the ophthalmoscope, he had already become either a regular or an honorary member of many scientific societies. He participated actively in their meetings and was in frequent attendance at scientific congresses. He was also active in the affairs of the universities where he taught, holding the post of rector at both Heidelberg and Berlin.

He felt keenly the gulf between scientists and the general public, and during his career he gave many lectures, some accompanied by carefully planned experimental demonstrations, in an attempt to bring the latest developments in the sciences to the attention of the public. These popular lectures ranged as widely as his own interests—from aesthetics to scientific cosmology, from the movement of glaciers to philosophy. Many were masterful works in their own right, models of their genre; they were referred to widely and translated quickly into English and French. Some were popular presentations of his own or others' researches; some were positive contributions to various fields of science. Frequently they were position papers, sum-

---

23. "The Axioms of Geometry," *The Academy*, I (1870), 130–31.

marizing developments, announcing discoveries, and at the same time providing a broad perspective on the scientific approach to a general area of natural phenomena.

A number of these popular lectures are closely related to his own technical researches in the sciences. "The Physiological Causes of Harmony in Music" (1857), for example, is a careful formulation of the core of the theory he worked out much more fully over the next several years and published in *The Sensations of Tone.* The three lectures on the "Recent Progress in the Theory of Vision" (1868) are a summary of the central theory in the *Handbook of Physiological Optics*, being divided, like that monumental work, into sections devoted to the anatomy, the physiology, and the psychology of vision. The address on "The Application of the Law of the Conservation of Force to Organic Nature" (1861), which he delivered at the request of Faraday on one of his first trips to England, reflects the interests that had led him fifteen years earlier to write "The Conservation of Force."

Other lectures–given typically as rectorial addresses, or upon important anniversaries, or upon such ceremonial occasions as the opening of scientific congresses–were designed to reflect in the broadest terms Helmholtz' conception of science, of the progress of science, and of its relationships to other areas of thought and experience. "The Relation of the Natural Sciences to Science in General" (1862), "The Aim and Progress of Physical Science" (1869), "The Origin of the Planetary System" (1871), and "Thought in Medicine" (1877) all fall into this class.

In "The Relation of the Natural Sciences to Science in General" he presented his criterion for distinguishing between the *Naturwissenschaften* and the *Geisteswissenschaften.* Broadly speaking, he found these two general areas of science or knowledge characterized by different processes of what might best be called psychological induction, that is, by different ways of forming concepts and/or intuitions. The distinction which he emphasized between these two fundamental ways of thinking was also important in his later works in aesthetics and in his general characterizations of aesthetic and scientific understanding.

In "The Aim and Progress of Physical Science" Helmholtz presented his conception of the ultimate aim of scientific investigations and pointed with pride to the triumphant steps–such as the establishment of the principle of the correlation and conservation of force, the extension of the principle of specific nervous energies, and Darwin's theory of natural selection–that had been made toward that goal. Since the general goal of the physical sciences is the comprehension of nature, the discovery of the causal laws of natural phenomena, he wrote, "if we direct our attention to the question of the progress of physical science as a whole, we may judge it by the degree to which the recognition and knowledge of the causal connections embracing all natural phenomena have advanced."

The sentence has a modern sound, and many today would agree with Helmholtz' criterion of scientific progress. His more specific image of the ultimate goal of science, recalling still the position expressed in the Introduction to "The Conservation of Force" and reflecting the dominant tradition in 1869, has of course fared less well. After arguing again that motion is the fundamental change in nature and that the forces of motion are the fundamental forces, he wrote:

> If motion is the fundamental change which lies at the base of all other changes occurring in the world, then every elementary force is a motive force, and the ultimate aim of physical science must be to find the movements which are the real causes of all other phenomena and to determine the motive forces upon which these movements depend. In other words, its aim is to reduce all phenomena to mechanics.

The address "Thought in Medicine" was delivered on the thirty-fifth anniversary of Helmholtz' own doctorate in medicine. It was a triumphant moment for the physicalistic physiology to which the young students of Johannes Müller had dedicated themselves a generation earlier. Their faith in its future had been amply justified. By 1877 a revolution had been successfully carried out in medical research and theory; most of the audience who heard the address had been educated, or were being educated, in the new medicine.

Each of his addresses on Goethe combines an analysis of some aspect of the poet's thought with some of Helmholtz' own investigations. The early "The Scientific Researches of Goethe" (1853) contains a discussion of Goethe's studies in biology and optics along with a presentation of Helmholtz' own recent discoveries concerning the phenomena of color mixtures. It contains, too, his first attempt to distinguish the artistic from the scientific way of comprehending nature, a subject to which he returned many times in later years. The second paper, "Goethe's Anticipation of Subsequent Scientific Ideas" (1892), shows Helmholtz trying to defend his empirical theory of vision both in science and in aesthetics.

Helmholtz wrote on aesthetics several times during his career. It was both a subject of personal interest and one into which his work in physiological optics and physiological acoustics almost forced him to enter. Each of his writings on art may be characterized as an attempt to explain various aesthetic phenomena scientifically or to provide a physiological and psychological basis for them. Each shows a peculiar tension between a desire to reduce some important phenomena to scientific laws and an appreciation of the immense difficulties in honestly carrying out such a reduction. This can readily be seen in his writings in physiological acoustics and the theory of music.

It can also be sensed in such lectures as those on "The Relation of Optics to Painting" (1871), in which he sought to extend his investigations in physiological optics to the analysis of the techniques used in painting to achieve realistic effects. Many points of technique can be codified rather neatly, and it is possible by analysis to account for many of the effects achieved by painters. Helmholtz was aware, however, that painting and music both have their great masters whose works stand irreducibly aloof from easy subsumption under law. One can say some things about the materials and techniques of an art, but there is still much more to be explained in any work of art.[24]

Always sensitive to fundamentals and always possessing a desire to complete lines of analysis once begun, Helmholtz produced in 1887 a work on arithmetic as a companion to his essays on the foundations of geometry. Entitled "An Epistemological Analysis of Counting and Measurement" and published in a *Festschrift* in honor of Eduard Zeller, it is an attempt to provide a foundation in experience for the fundamental laws of arithmetic. The essay also contains an analysis of the conditions which must be satisfied in order to make measurements–that is, to apply the laws of arithmetic to nature–as well as some basic statements on dimensional analysis and the fundamental units of physics. This analysis later became part of the "Introduction to the Lectures on Theoretical Physics," the published version of the cycle which he delivered at the University of Berlin.

PROFESSOR OF PHYSICS AT BERLIN

By 1871, when he was called to succeed Gustav Magnus as professor of physics at the University of Berlin, Helmholtz had clearly become the leading

---

24. Concerning the influence of Helmholtz' optical investigations on painters, see William Innes Homer, *Seurat and the Science of Painting* (Cambridge, Mass., 1964), *passim.*

and most influential scientist in Germany. His only serious competitor for the post, which had been the most important one in Germany for many years, was Gustav Kirchhoff, who at the time was professor of physics at the University of Heidelberg. Kirchhoff did not wish to leave his friends and home, however, and so Helmholtz was free to accept a professorship in the field where his greatest interest and talent lay.

He had already made contributions to all of the fields that then comprised physics. His first paper on electricity and magnetism had appeared in 1851; he had made extensive investigations in physical optics and acoustics in conjunction with his work in physiological optics and acoustics; and he had written a series of papers on the theory of heat. Perhaps his most famous and influential contributions to physics during these early years, apart from "The Conservation of Force," was his essay "On the Integrals of the Hydrodynamic Equations Which Express Vortex Motion" (1858), which has recently been described as the "greatest advance in hydrodynamics since Euler, Lagrange, and Cauchy."[25] The elegant equations contained in this essay were subsequently used by Lord Kelvin in his theory of the vortex atom and, as noted earlier, in such diverse matters as the explanation of cloud formations and patterns in the sands along ocean beaches.

During the decade immediately following his appointment at the University of Berlin, Helmholtz devoted his energies primarily to the study of electricity and magnetism, areas in which once again the inclusiveness of his understanding and the persistence of his drive to penetrate to fundamentals are evident. Between 1870 and 1874, for example, he published a series of three comprehensive papers "Über die Theorie der Elektrodynamik," which, along with a number of more specialized papers, constitute a quite complete survey of the problems and theories within these areas of physics at that time. In his Preface to Heinrich Hertz's *The Principles of Mechanics*, Helmholtz characterizes his own work:

> In Germany at that time the laws of electromagnetics were deduced by most physicists from the hypothesis of W. Weber, who sought to trace back electric and magnetic phenomena to a modification of Newton's assumption of direct forces acting at a distance and in a straight line. With increasing distance these forces diminish in accordance with the same laws as those assigned by Newton to the force of gravitation, and held by Coulomb to apply to the action between pairs of electrified particles. The

---

25. P. Costabel, "The Achievements and Doubts of Classical Mechanics," in *Science in the Nineteenth Century*, edited by René Taton (New York, 1965), p. 92.

> force was directly proportional to the product of the two quantities of electricity, and inversely proportional to the square of their distance apart; like quantities produced repulsion, unlike quantities attraction. Furthermore, in Weber's hypothesis it was assumed that this force was propagated through infinite space instantaneously, and with infinite velocity. The only difference between the views of W. Weber and of Coulomb consisted in this—that Weber assumed that the magnitude of the force between the two quantities of electricity might be affected by the velocity with which the two quantities approached towards or receded from one another, and also by the acceleration of such velocity. Side by side with Weber's theory there existed a number of others, all of which had this in common—that they regarded the magnitude of the force expressed by Coulomb's law as being modified by the influence of some component of the velocity of the electrical quantities in motion. Such theories were advanced by F. E. Neumann, by his son C. Neumann, by Riemann, Grassmann, and subsequently by Clausius. Magnetised molecules were regarded as the axes of circular electric currents, in accordance with an analogy between their external effects previously discovered by Ampère.
>
> This plentiful crop of hypotheses had become very unmanageable, and in dealing with them it was necessary to go through complicated calculations, resolutions of forces into their components in various directions, and so on. So at that time the domain of electromagnetics had become a pathless wilderness. Observed facts and deductions from exceedingly doubtful theories were inextricably mixed up together. With the object of clearing up this confusion I had set myself the task of surveying the region of electromagnetics, and of working out the distinctive consequences of the various theories, in order, wherever that was possible, to decide between them by suitable experiments.[26]

Helmholtz was one of the first continental physicists to appreciate the investigations of Faraday and Maxwell and to see the advantages of Maxwell's theory in accounting for a number of phenomena. Both by critical theoretical analysis and by experimental investigations, in which he was joined by a series of able assistants, he sought to show the inadequacies of the alternative hypotheses, such as that of Wilhelm Weber. He very quickly saw the talent of Heinrich Hertz, who joined him as student and assistant in 1878, and the two men remained close friends down to Hertz's death on January 1, 1894.

In London in 1881, Helmholtz gave his famous Faraday Lecture to the Chemical Society, in which he pointed out that conceiving of electricity as

---

26. Preface to Heinrich Hertz, *The Principles of Mechanics* (London, 1899; New York: Dover [reprint], 1956), unpaged.

consisting of microscopic particles was in no way inconsistent with the work of Faraday and thus with Maxwell's theory. Indeed, he maintained that "if we accept the hypothesis that the elementary substances are composed of atoms, we cannot avoid concluding that electricity also, positive as well as negative, is divided into definite elementary portions, which behave like atoms of electricity." The lecture caused great excitement among British physicists. The suggestion that electricity consists of particles was not, of course, totally new or original with Helmholtz; prior to 1881, however, it was not the dominant view, and in some quarters it was rejected outright. After this lecture there was a consistent interest in the atomic nature of electricity, culminating in the discovery of the electron by J. J. Thomson.

During the last decade of his life Helmholtz' interest centered on investigations in which the principle of least action played the central role. He was trying to show that this principle, which could be treated as one of the fundamental principles of mechanics, was one to which he could reduce—or at least relate—other parts of theoretical physics. (This endeavor reminds one of Einstein's attempts over a period of years to construct a unified field theory.) Helmholtz did make some progress towards his goal in several papers, such as one relating Maxwell's theory to the principle of least action. The work, however, was very difficult, and it was slowed by the increasing burdens of his offices and by the advances of illness and old age. It ended with his death.

It seems to be the verdict of history that this line of investigation, like many another, leads only to a dead end. In any case, the physics that Helmholtz knew and helped to create was shortly to undergo a series of radical transformations. Helmholtz was one of the last, if not the last, of the great classical physicists. Even before his full six-semester cycle of lectures on theoretical physics had been published, Planck and Einstein (both of whom learned a great deal from his writings) and others had set physics on a new course that made Helmholtz' interests, really only a decade old, seem like part of a rather distant past.

Helmholtz was widely honored during his years at the University of Berlin, increasingly so during the last decade of his life. Among many awards, he received in 1873 the Copley Medal, the highest honor bestowed by the Royal Society of London; in 1886 he was the first recipient of the Graefe Medal, given for outstanding work in ophthalmology. His name was changed twice—in 1882, when he was raised to the rank of the hereditary nobility, and again in 1891, when by royal patent granted by Wilhelm II he was made Wirklicher Geheimrath, with the title of Excellency.

In 1887, while continuing in his post as professor of physics at the University of Berlin, he became the first president and director of the Physikalisch Technische Reichanstalt at Charlottenberg. From that time on he held the two posts simultaneously, one venerable, one new, and both of great distinction.

The occasion of his seventieth birthday in 1891 was widely celebrated, the day being treated almost as a national holiday in Germany. Helmholtz himself noted that there were messages all the way from "Tomsk to Melbourne" among the several hundred he received from admirers, scientists, and learned societies.

He continued to attend conferences and scientific congresses, after 1888 in his capacity as president of the Physikalisch Technische Reichanstalt. In 1893, although he was no longer vigorous, at the request of Wilhelm II he accepted an invitation to represent Germany at the World's Fair in Chicago, where he was made honorary president of the International Electrical Congress. On the return voyage he fell aboard ship; while eventually able to continue some of his activities, he never fully regained his health. Death came the following year, on September 8, 1894.

SELECTED WRITINGS OF HERMANN VON HELMHOLTZ

1.

THE CONSERVATION OF FORCE: A PHYSICAL MEMOIR [1847]

Delivered at a meeting of the Physical Society of Berlin on July 23, 1847. (Footnotes and appendices added in 1881 are so designated in brackets.)

INTRODUCTION

As the main contents of the present memoir are intended chiefly for physicists, I have, independently of philosophical foundations, presented it in the form of a physical hypothesis. I have then developed the consequences of this hypothesis in the several branches of physics and, finally, have compared these consequences with the empirical laws of natural phenomena. The derivation of the propositions contained in the memoir may be thought of as starting from one of two basic principles: either from the principle that it is impossible to obtain an unlimited amount of force capable of doing work as the result of any combination whatsoever of natural objects, or from the assumption that all actions in nature are reducible to forces of attraction and repulsion, the intensity of the forces depending solely upon the distances between the points involved. That these two principles are identical will be shown at the beginning of the memoir. Meanwhile, in this Introduction I shall try to indicate in general the fundamental significance which these principles have insofar as the central and ultimate aim of the physical sciences is concerned.

The task of the physical sciences is to discover laws so that individual natural processes can be traced back to, and deduced from, general principles. These principles, such as the laws of refraction and reflection of light and the laws of Mariotte and Gay-Lussac concerning the volumes of gases, are obviously nothing but generic concepts through which the phenomena falling under them are collectively understood. The search for such laws is the task of the experimental part of our sciences. The theoretical part, on the other hand, seeks to ascertain from their visible effects the unknown causes of natural processes; it seeks to comprehend them according to the law of causality. (See Appendix 1.)

We are compelled to and justified in this undertaking by the fundamental principle that every change in nature must have a sufficient cause. The proximate causes, to which we refer natural phenomena, are themselves either invariable or variable; in the latter case, the same fundamental principle compels us to seek still further for the causes of the variation, and so on, until we arrive finally at causes which operate according to invariable law and which consequently produce under the same external conditions the same effect every time. Thus the final goal of the theoretical natural sciences is to discover the ultimate invariable causes of natural phenomena. Whether all processes may actually be traced back to such causes, in which case nature is completely comprehensible, or whether on the contrary there are changes which lie outside the law of necessary causality and thus fall within the region of spontaneity or freedom, will not be considered here. In any case it is clear that science, the goal of which is the comprehension of nature, must begin with the presupposition of its comprehensibility and proceed in accordance with this assumption until, perhaps, it is forced by irrefutable facts to recognize limits beyond which it may not go.

Science treats the objects in the external world according to two different abstractions. In the first place, considering them apart from their effects on other objects and on our sense organs, it regards them merely as existing objects. As such, they are called *matter* or *material.* To us, matter in itself is inert and without effect; the only qualities we distinguish in it are spatial distinctions and quantity (mass), which is assumed always to be constant. We cannot ascribe qualitative differences to matter itself, for when we speak of different kinds of matter we always consider these to be differences in effect, that is, differences in the *forces* of matter. Thus matter in itself can undergo no change other than a spatial one, that is, a movement. Objects in nature are not, however, inert. Indeed, we come to a knowledge of them only through their effects upon our sense organs, since it is from these effects that we infer objects which produce them. Hence, if we wish to make actual application of the concept of matter, we can do so only by ascribing force to it, that is, by adding a second abstraction, *the capacity to produce effects.*

It is evident that the concepts of matter and force cannot be applied separately to nature. Pure matter would be indifferent to the rest of nature, since it would never produce any changes in it or in our sense organs; a pure force would be something which exists and yet does not exist, for that which exists we call matter. It would be equally erroneous to interpret matter as something which produces effects and force as an empty concept to which nothing existing corresponds; both, on the contrary, are abstractions from reality, formed in precisely the same way. We can perceive matter only through its forces, not in and of itself.

We have observed that natural phenomena should be traced back to inalterable ultimate causes. This requirement may now be restated: as ultimate causes, forces which do not vary in time should be found. In science we call bodies with unchangeable forces (indestructible qualities) chemical elements. If we think of the universe as consisting of elements with inalterable qualities, the only possible changes in such a system are spatial ones, that is, movements. Moreover, the external relations by means of which forces effect changes can only be spatial, and thus the forces acting can only be motive forces, which in their action depend only on spatial relations.

To speak more precisely: natural phenomena should be traced back to the movements of material objects which possess inalterable motive forces that are dependent only on spatial relations.

Motion is change of spatial relations. Spatial relations are possible only among extended bodies; they are impossible in homogeneous empty space. Motion, therefore, can be experienced only as a change in the spatial relation of at least two material bodies relative to one another. Further, motive force, as the cause of the change, can be predicated only in cases involving at least two bodies spatially related to one another and is thus to be defined as the effort of the two bodies to change their relative positions. The force, moreover, which two bodies exert upon one another must be resolved into forces of all of their parts upon one another. Thus mechanics considers the forces of material points, that is, the forces of spatial points containing matter, to be basic. (See Appendix 2.) Points, however, have no spatial relations with respect to one another other than their distance apart, for the direction of the line between them can only be determined relative to at least two other points. Thus the motive force which mass points exert upon one another can only be the cause of an alteration in the distance between the points, that is, either a force of attraction or of repulsion.

This conclusion follows equally from the principle of sufficient reason. The forces which two masses exert upon one another must necessarily be determined both in magnitude and direction as soon as the positions of the masses are completely specified. Between two points, however, only one direction is completely specified, namely, that of the line between them. The force, therefore, which they exert upon one another must be directed along this line, and its intensity must be dependent solely upon the distance between them.

Thus we see that the problem of the physical sciences is to trace natural phenomena back to inalterable forces of attraction and repulsion, the intensity of the forces depending upon distance. The solution of this problem would mean the complete comprehensibility of nature. Analytic mechanics has not, up to the present time, accepted this limitation on the concept of motive force, at first because it was not clear concerning the origin of its fundamental principles, and later, when these had become clear, because it was unable to compute the resultant of combinations of motive forces in cases where such computations could not easily be carried out. Nevertheless, a great many of the general principles concerning the movements of systems of combined masses—for example, the principle of virtual velocities, the conservation of the movement of the center of gravity, the conservation of the principal plane of rotation and of the moment of rotation of free systems, and the conservation of vis viva—are valid only for cases[1] where the bodies act on one another with inalterable forces of attraction and repulsion. It is chiefly the first and the last of the principles just mentioned that have any application to terrestrial phenomena, inasmuch as the others apply only to completely free systems. Further, as we shall show, the first is a special case of the last. This must be regarded as the most general and most important consequence of the derivations we have made.

Theoretical natural science, therefore, if it is not to rest content with a partial view of the nature of things, must take a position in harmony with the present conception of simple forces and the consequences of this conception. Its task will be completed when the reduction of phenomena to simple forces has been completed and when, at the same time, it can be proved that the reduction is the only possible one which the phenomena will allow. This will then be established as the conceptual form necessary for understanding nature, and we shall be able to ascribe objective truth to it.

I. THE PRINCIPLE OF THE CONSERVATION OF VIS VIVA

We shall begin with the assumption that it is impossible, by any combination whatsoever of natural bodies, to create force continuously out of nothing. Carnot and Clapeyron[2] have already deduced theoretically from this proposition a series of laws concerning the latent and specific heats of various

1. Better: "can be proved only for cases." [1881]
2. *Poggendorff's Annalen,* LIX, 446, 566.

substances. Some of these laws have been confirmed by experiment; others have not yet been submitted to this test. The purpose of the present memoir is to extend the same principle, in the same manner, to all branches of physics. This will be done partly in order to show the applicability of the principle to all cases where the laws of phenomena have already been sufficiently investigated and partly—with the support of many analogies from what is already known—in order to make additional inferences concerning laws which are as yet only imperfectly known, thus providing a program for subsequent experimental work.

The principle under consideration can be represented in the following manner: Let us imagine that a system of bodies, which stand in certain spatial relations to one another, are acted upon by the forces mutually exerted among them so that they are moved until other positions are reached. We can regard the velocities acquired in this way as a specific quantity of mechanical work and can translate them into it. If we wish the same forces to act a second time, so as to produce the same quantity of work again, we must somehow, by means of other forces placed at our disposal, bring the bodies back to their original positions. In doing this, however, a certain quantity of work of these other forces will be consumed. Our principle requires in this case that the quantity of work gained by the passage of the system from the first position to the second, and the quantity lost by the passage of the system from the second back to the first, always be equal, no matter what the form of the movement, the path, or the velocity at which the change is effected. If this were not the case, that is, if the quantity of work were greater in one direction than in the other, we could use the first for the production of work and the second to carry the bodies back to their original positions (applying part of the work gained) and in this way produce an indefinitely large amount of mechanical force. We should have built a *perpetuum mobile* which both maintains itself in motion and is capable of imparting force to external objects.

If we inquire into the mathematical expression of this principle, we shall find it in the well-known law of the conservation of vis viva. As is known, a quantity of work produced and consumed may be represented by a weight, m, which is raised to a certain height, h; the quantity of work is then mgh, where g is the force of gravity. In order to rise perpendicularly to the height h, the body m must have a velocity $v = \sqrt{2gh}$, and it will reach the same velocity in falling from the same height. Thus $\frac{1}{2}mv^2 = mgh$; consequently, we may substitute half of the product mv^2, which is known in mechanics as the quantity of the vis viva of the body *m*, in place of the quantity of work. Because it agrees better with the manner of measuring the intensity of forces which has become customary, I propose calling $\frac{1}{2}mv^2$ also the quantity of work. This change is of no consequence insofar as existing applications of the concept of vis viva, which are restricted to the principle under consideration, are concerned; we shall, however, gain considerable advantage from it in what is to follow.

Now, as is well known, the principle of the conservation of vis viva states that if any number whatsoever of material points are set in motion solely by the forces which these points exert upon one another, or by forces which are directed towards fixed centers, then the sum total of the vires vivae is the same at all times when the points occupy the same relative positions, whatever their paths or their velocities during intervening times. If we think of the vires vivae as being applied to raise the parts of the system, or masses equivalent to them, to some height, it follows from what has just been shown that the quantities of work, which are represented in a similar manner, must also be equal under the conditions specified.

The principle of the conservation of vis viva, however, is not valid for all possible kinds of forces. In mechanics it is generally related to the principle of virtual velocities, and this can be established only with respect to material points endowed with forces of attraction or repulsion. We shall now show that the principle is valid only where the forces in action can be resolved into the forces of material points[3] acting in the direction of the lines which unite them, the intensity of the forces depending only upon the distances between the points. In mechanics such forces are generally called *central forces*. It follows further and conversely that in all actions of bodies upon one another, wherever the principle given above is capable of general application to the smallest particles of these bodies, such central forces must be regarded as the simplest and most fundamental ones.

Let us consider the case of a material point with the mass *m* that moves under the influence of the forces of several bodies which are joined together in a fixed system *A*; by mechanics we are able to determine the position and velocity of this point at any given time. We shall therefore regard the time, t, as the fundamental variable; dependent upon it are (1) the position (x, y, z) of *m* relative to a system of coordinates fixed with respect to *A*; (2) the tangential velocity, q; (3) the components of the latter parallel to the axes:

---

3. That is, where their resolution into point forces is presupposed. [1881]

$u = dx/dt$, $v = dy/dt$, $w = dz/dt$; and finally (4) the components of the forces acting;

$$X = m\frac{du}{dt}, \quad Y = m\frac{dy}{dt}, \quad Z = m\frac{dw}{dt}.$$

According to our principle, $\frac{1}{2}mq^2$, and hence q^2 also, must always be the same when m occupies the same position relative to A. It is therefore to be regarded, not merely as a function of the fundamental variable t, but also simply as a function of the coordinates x, y, z. Thus

$$d(q^2) = \frac{d(q^2)}{dx}dx + \frac{d(q^2)}{dy}dy + \frac{d(q^2)}{dz}dz. \tag{1}$$

Since $q^2 = u^2 + v^2 + w^2$, we have $d(q^2) = 2udu + 2vdv + 2wdw$. Substituting for u its value dx/dt and for du its value $X(dt/m)$, and using the corresponding values of v and w, we have

$$d(q^2) = \frac{2X}{m}dx + \frac{2Y}{m}dy + \frac{2Z}{m}dz. \tag{2}$$

Since Equations (1) and (2) must hold good simultaneously for all values of dx, dy, dz, it follows[4] that individually

$$\frac{d(q^2)}{dx} = \frac{2X}{m}, \quad \frac{d(q^2)}{dy} = \frac{2Y}{m}, \quad \frac{d(q^2)}{dz} = \frac{2Z}{m}.$$

If, however, q^2 is a function of x, y, and z only, it follows that X, Y, and Z, that is, the directions and magnitudes of the forces acting, are functions solely of the position of m with respect to A.

Let us imagine now, instead of the system A, a single material point a. It follows from what has just been proved that the direction and magnitude of the force exerted by a upon m are determined solely by the position of m with respect to a. Since the position of m with respect to the individual point a is determined only by the distance ma, the law in this case must be modified so that the direction and magnitude of the force are functions of this dis-

4. This conclusion requires a restriction; see Appendix 3. [1881]

tance, which we shall call r. If we think of the coordinates as being referred to any arbitrarily selected system of axes, the origin of which lies in a, we then have

$$md(q^2) = 2Xdx + 2Ydy + 2Zdz = 0 \qquad (3)$$

whenever

$$d(r^2) = 2xdx + 2ydy + 2zdz = 0,$$

that is, whenever

$$dz = \frac{xdx + ydy}{z}.$$

Setting this value in Equation (3), we obtain

$$\left(X - \frac{x}{z}Z\right)dx + \left(Y - \frac{y}{z}Z\right)dy = 0$$

for any value whatsoever of dx and dy; hence also singly,

$$X = \frac{x}{z}Z, \quad Y = \frac{y}{z}Z.$$

That is to say, the resultant must be directed toward the origin of the system of coordinates, that is, toward the point a which exerts the force.

Hence, in systems to which the principle of the conservation of vis viva[5] can be applied with complete generality, the elementary forces of the material points must be central forces.

II. THE PRINCIPLE OF THE CONSERVATION OF FORCE

We shall now give a still more general expression of the law under consideration for cases in which central forces are acting.

---

5. And the equality of action and reaction. [1881]

If ϕ is the intensity of the force which acts in the direction of r, and if it is considered to be positive when it attracts and negative when it repels, then we have

$$X = -\frac{x}{r}\phi, \quad Y = -\frac{y}{r}\phi, \quad Z = -\frac{z}{r}\phi; \tag{1}$$

and from Equation (2) of the preceding section we have

$$md(q^2) = -2\frac{\phi}{r}(xdx + ydy + zdz);$$

hence

$$\tfrac{1}{2}md(q^2) = -\phi dr.$$

Or, if Q and R, q and r represent two related tangential velocities and distances,

$$\tfrac{1}{2}mQ^2 - \tfrac{1}{2}mq^2 = -\int_r^R \phi\, dr. \tag{2}$$

If we consider this equation closely, we find on the left-hand side the difference of the vires vivae possessed by m at two different distances. In order to understand the significance of the quantity $\int_r^R \phi\, dr$, let us imagine the intensities of ϕ, which belong to different points on the line connecting m and a, represented by perpendicular ordinates erected at these points; the expression just given then represents the area of the surface enclosed by the curve between the two ordinates r and R. As this area may be regarded as the sum of the infinite number of ordinates which lie between r and R, it thus represents the sum total of the intensities of all of the forces which act at all distances between R and r. If we call the forces which tend to move the point m, before the motion has actually taken place, tensional forces (*Spannkräfte*), in contrast to what in mechanics is called vis viva, then the quantity $\int_r^R \phi dr$ is the sum of the tensional forces between the distances R and r, and the law presented above may be expressed as follows: The increase of vis viva of a material point during its motion under the influence of a central force is equal to the sum of the tensional forces which correspond to the relative change in distance.

Let us consider the case of two points at a distance R from one another which are under the influence of a force of attraction. By the action of the force they will be drawn to the smaller distance r, and their velocity, and consequently their vis viva, will be increased. If they should, however, be forced to a greater distance r, their vis viva will diminish and finally be totally consumed. With forces of attraction, we can therefore designate the sum of the tensional forces for the distance between $r = 0$ and $r = R$, $\int_0^R \phi dr$, as that which still remains, and that between $r = R$ and $r = \infty$ as that already consumed; the former can act immediately, while the latter can be called into action only by an equivalent loss of vis viva. It is just the opposite with forces of repulsion: if the points are situated at the distance R from one another, as the distance becomes greater, vis viva will be gained, and the tensional forces still existing will be those between $r = R$ and $r = \infty$, those lost between $r = 0$ and $r = R$.

In order to formulate our law in a completely general fashion, let us imagine an arbitrary number of material points with the masses m_1, m_2, m_3, etc., denoted generally by m_a, with coordinates x_a, y_a, z_a. Let the components of the forces acting parallel to the axes be X_a, Y_a, Z_a; the components of the velocities along the same axes, u_a, v_a, w_a; and the tangential velocity, q_a. Let the distance between m_a and m_b be r_{ab} and the central force between them be ϕ_{ab}. For the single point m_b we now have, analogous to Equation (1),

$$X_n = \sum\left[(x_a - x_n)\frac{\phi_{an}}{r_{an}}\right] = m_n \frac{du_n}{dt}$$

$$Y_n = \sum\left[(y_a - y_n)\frac{\phi_{an}}{r_{an}}\right] = m_n \frac{dv_n}{dt}$$

$$Z_n = \sum\left[(z_a - z_n)\frac{\phi_{an}}{r_{an}}\right] = m_n \frac{dw_n}{dt},$$

where the summation sign Σ includes all members which result when all the separate indices (1, 2, 3, etc.), with the exception of n, are substituted for the index letter a.

If we multiply the first equation by $dx_n = u_n dt$, the second by $dy_n = v_n dt$, the third by $dz_n = w_n dt$, and imagine the three equations thus obtained to be formed for every single point of m_b, as has been done here for m_n, and if we add them all together, we obtain

$$\sum\left[(x_a - x_b)dx_b\frac{\phi_{ab}}{r_{ab}}\right] = \sum\left[\tfrac{1}{2}m_a d(u_a^2)\right]$$

$$\sum\left[(y_a - y_b)dy_b\frac{\phi_{ab}}{r_{ab}}\right] = \sum\left[\tfrac{1}{2}m_a d(v_a^2)\right]$$

$$\sum\left[(z_a - z_b)dz_b\frac{\phi_{ab}}{r_{ab}}\right] = \sum\left[\tfrac{1}{2}m_a d(w_a^2)\right].$$

The members of the series on the left-hand side are obtained by substituting for a all the single indices (1, 2, 3, etc.) and for b, in each case, all the values which are greater or smaller than the value a possesses. Thus the summations may be divided into two groups, in one of which a is always greater than b and in the other always smaller. It is clear that for every member of one group,

$$(x_p - x_q)dx_q\frac{\phi_{pq}}{r_{pq}},$$

a member,

$$(x_q - x_p)dx_p\frac{\phi_{pq}}{r_{pq}},$$

must appear in the other. Adding both together, we obtain

$$-(x_p - x_q)(dx_p - dx_q)\frac{\phi_{pq}}{r_{pq}}.$$

If we make the summations in this manner, add all three and set

$$\tfrac{1}{2}d[(x_a - x_b)^2 + (y_a - y_b)^2 + (z_a - z_b)^2] = r_{ab}dr_{ab},$$

we obtain

$$-\Sigma[\phi_{ab}dr_{ab}] = \Sigma[\tfrac{1}{2}m_a d(q_a^2)] \qquad (3)$$

or

$$-\Sigma\left[\int_{r_{ab}}^{R_{ab}} \phi_{ab} dr_{ab}\right] = \Sigma\left[\tfrac{1}{2} m_a Q_a^2\right] - \Sigma\left[\tfrac{1}{2} m_a q_a^2\right] \tag{4}$$

where R and Q, as well as r and q, designate related values. Again, we have on the left-hand side the sum of the tensional forces consumed and on the right the sum of the vires vivae of the entire system.

We can now express the law as follows: In all cases of the motion of free material points under the influence of their attracting and repelling forces, the intensity of which depends solely upon distance, the loss in tensional force is always equal to the gain in vis viva, and the gain in the former is always equal to the loss in the latter. Hence, *the sum of the tensional forces and vires vivae present is always constant.* In this more general form, we can call our law *the principle of the conservation of force.*

In the derivation of the law just presented, nothing is altered if a number of the points, which we shall denote generally by the letter d, are assumed to be fixed, so that q_d is always equal to zero. The form of the law will then be

$$\Sigma\left[\phi_{ab} dr_{ab}\right] + \Sigma\left[\phi_{ad} dr_{ad}\right] = -\Sigma\left[\tfrac{1}{2} m_b d(q_b^2)\right]. \tag{5}$$

It still remains to be noted in what relation the principle of the conservation of force stands to the most general law of statics, the so-called principle of virtual velocities. This follows, of course, immediately from Equations (3) and (5). If equilibrium is the result, given a certain arrangement of the points m_a, that is, if it is the case that these points come to rest (and thus that $q_a = 0$), it follows from Equation (3) that

$$\Sigma\left[\phi_{ab} dr_{ab}\right] = 0; \tag{6}$$

in the case where there are forces acting from points m_d outside the system, it follows from Equation (5) that

$$\Sigma\left[\phi_{ab} dr_{ab}\right] + \Sigma\left[\phi_{ad} dr_{ad}\right] = 0. \tag{7}$$

In these equations dr represents changes in distance which result from the arbitrary small displacements of the points m_a which are permitted by the conditions of the system. We have seen in previous derivations that an increase of vis viva, and thus a change from rest to motion, can be brought about only by an expenditure of tensional force. Corresponding to this, the last two equations state that in cases where no tensional force is consumed in the first moment in any possible direction of motion, the system, once it is at rest, must remain at rest.

It is well known that all the laws of statics may be deduced from the equations just presented. The most important consequence concerning the nature of forces is as follows: Instead of the arbitrarily small displacements of the points *m,* let us imagine the displacements which would occur if the system were quite rigid in itself, so that in Equation (7) every $dr_{ab} = 0$. It follows from this that

$$\Sigma [\phi_{ab} dr_{ab}] = 0$$

and

$$\Sigma [\phi_{ad} dr_{ad}] = 0.$$

Thus the external as well as the internal forces must now satisfy the conditions of equilibrium. Hence, if any system of bodies whatsoever is, through the action of external forces, brought to a condition of equilibrium, the equilibrium will not be destroyed (1) if the individual points of the system are rigidly connected to one another in their given positions, and (2) if the forces which the points exert upon one another are then removed. From this it follows that if the forces which two material points exert upon one another are brought into equilibrium through the action of two external forces on these points, the equilibrium of the system must persist if, instead of the mutual forces of the points, a rigid connection between them is substituted. Forces, however, which are applied to two points that are in a rigid straight line can be in equilibrium only if they lie in the line themselves, are equal to one another, and act in opposite directions. It follows, therefore, for the forces of the points themselves, which are equal and opposed to the external ones, that they must act in the direction of the line which connects them and hence must be forces of either attraction or repulsion.

The preceding propositions may be summarized as follows:

1. Whenever bodies act upon one another by forces of attraction or repulsion which are independent of time and velocity, the sum of their vires vivae and tensional forces must be constant. The maximum quantity of work which can be obtained from them is therefore fixed and finite.

2. If, on the contrary, bodies possess forces which depend upon time and velocity, or which act in directions other than the lines which unite each pair of material points—as is the case, for example, with forces of rotation—then combinations of such bodies are possible in which force may be either lost or gained ad infinitum. (See Appendix 4.)

3. In the case of the equilibrium of a system of bodies under the action of central forces, if we assume that the bodies of the system are rigidly fixed and that only the system as a whole is movable relative to bodies which lie outside it, then the external and internal forces must be in equilibrium. A rigid system of such bodies, therefore, can never be set in motion by the action of its internal forces, but only by the operation of external ones. If, however, there were forces other than central ones, rigid combinations of bodies could be formed which would move of themselves without having any relationship whatever to other bodies.

III. THE APPLICATION OF THE PRINCIPLE TO THE THEOREMS OF MECHANICS

We shall turn now to the special applications of the law of the conservation of force. Let us first consider briefly those cases in which the principle of the conservation of vis viva has already been accepted and applied.

1. *All movements which occur under the influence of the general force of gravitation,* hence the motion of heavenly and ponderable terrestrial bodies. In the case of the heavenly bodies, the law manifests itself in the increase in velocity which is evident when the paths of the planets are nearer the sun, as well as in the inalterability of the major axes of their orbits, their periods of rotation, and their periods of orbital revolution. In the case of terrestrial bodies, it manifests itself in the well-known law that (a) the terminal velocity of a falling body depends only upon the perpendicular distance traversed and is independent of the form of the path and any lateral displacement, and (b) this velocity, if not lessened by friction or by inelastic impact, is just sufficient to carry the body to the same height from which it originally fell. It has already been mentioned that the height of ascent of a certain weight is used as the unit of measure of work in our machines.

2. *The transmission of motion through incompressibly rigid and fluid bodies,* in cases where there is neither friction nor the impact of inelastic materials. For these cases, our general principle is ordinarily presented as a rule to the effect that with a motion transmitted or changed by mechanical power,

there is always a decrease in the intensity of the force proportional to the increase in velocity. Thus, if by means of some machine which produces a uniform mechanical force, the weight m is raised with the velocity c, then the weight nm will be raised with the velocity c/n, so that in both cases the quantity of tensional force developed by the machine in a unit of time will be given by mgc, where g represents the intensity of the force of gravity.

3. *The motions of perfectly elastic solids and fluids.* As a condition of complete elasticity, we must add to the one usually given—that is, that the body which has been altered in form or volume regain completely its original state—that there be no friction among the particles which constitute the body. Our principle was first recognized and most frequently utilized in connection with the laws of movements of this kind. Among the most common cases of its application to solids may be mentioned the impact of elastic bodies, the laws of which may readily be deduced from our principle and that of the conservation of the center of gravity, and the various elastic vibrations which continue without fresh impulse until, through the friction of the constituent parts and the transmission of motion to external bodies, they are destroyed. With fluids, both liquid (evidently elastic, with a very high modulus of elasticity and an equilibrium position of the particles) and gaseous (with a lower modulus of elasticity and without an equilibrium position), motion is in general propagated by undulations. To this class belong the waves on the surface of liquids, the transmission of sound, and probably also the transmission of light and radiant heat.

The vis viva of a single particle Δm in a medium which is traversed by a train of waves obviously is to be determined from the velocity which the particle possesses at its position of equilibrium. As is well known, the general wave equation gives the velocity u, when a^2 is the intensity; λ, the wavelength; α, the velocity of propagation; x, the abscissa; and t, the time, as follows:

$$u = a \cdot \cos \left[\frac{2\pi}{\lambda} (x - \alpha t) \right].$$

For the position of equilibrium, $u = a$, and thus the vis viva of the particle Δm during the undulatory motion is $\frac{1}{2} \Delta m a^2$, that is, it is proportional to the intensity. If the waves spread spherically from a center, continually increasing masses are set in motion, and if the vis viva is to remain constant, the intensity must decrease. Since the masses embraced by the waves increase as the square of the distance, the well-known law that the intensity decreases in the reciprocal ratio follows as a consequence.[6]

As is well known, the laws of reflection, refraction, and polarization of light at the boundary of two media of different wave velocity have already been deduced by Fresnel from the conservation of vis viva and the assumption that the motion of the boundary particles in both media is the same. With the interference of two trains of waves, there is no destruction of vis viva but only another distribution. Two trains of waves of intensities a^2 and b^2, which do not interfere, give the intensity $a^2 + b^2$ to all points upon which they impinge. If they interfere, the maxima have the value $(a + b)^2$, that is, $2ab$ more, and the minima $(a - b)^2$, that is, just as much less, than $a^2 + b^2$.

The vis viva of elastic waves is destroyed by such processes as those in which they are said to be absorbed. We find the absorption of sound waves to result mainly from the impact of the waves against yielding, inelastic substances, such as curtains and bedding; thus absorption may be regarded as the communication of the motion to these substances, in which it is destroyed by friction. Whether motion can also be destroyed by the friction of air particles against one another is a question which cannot be answered as yet.[7] The absorption of heat rays is accompanied by a commensurate development of heat; we shall consider in the next section how far the latter corresponds to a specific equivalent of force. Force will be conserved if the quantity of heat radiated from one body reappears in the body into which it passes, provided, of course, that none is lost by conduction and none of the rays escape elsewhere. This has certainly been assumed in investigations of radiant heat, but I know of no experiments which furnish a proof of it.

Concerning the absorption of light by partially transparent or totally opaque bodies, we are acquainted with three different processes. In the first place, phosphorescent bodies absorb light in such a way that they can yield it again as light. Secondly, most and perhaps all light rays seem to produce heat. From the light, heat, and chemical rays of the spectrum (apparent obstacles to the acceptance of these identities are gradually being removed), the heat equivalent of the chemical and light rays appears to be very small, considering their intense actions upon the eye. Should, however, the similarity of these

---

6. It should be mentioned here that with progressive plane waves, the magnitude of the tensional force of the compressed or displaced elastic medium comprises a second, and much greater, part of the energy than the vis viva. [1881]

7. This has now been clearly established. [1881]

rays, which act in different ways, not be established, we would have to consider the end result of the motion of light unknown. Thirdly, in many cases the light absorbed produces chemical action. In respect to force relationships, we must here distinguish two kinds of action: (1) those in which only the stimulus to chemical reaction is communicated, in a manner similar to the effect of substances which act as catalysts (such as the action of light upon a mixture of chlorine and hydrogen), and (2) those in which it acts in opposition to chemical processes (as in the decomposition of the salts of silver and in its action upon the green parts of plants). In most of these processes, however, the effect of light is so little known that we are unable to form a judgment concerning the magnitudes of the forces involved. The latter appear to be considerable in quantity and intensity only in their actions on the green parts of plants.

IV. THE FORCE EQUIVALENT OF HEAT

The following are mechanical processes in which an absolute loss of force has until now been taken for granted:

1. *The impact of inelastic bodies.* For the most part, the loss is related to a change in shape and a compression of the bodies involved and thus to an increase in tensional forces, for we find a considerable development of heat accompanying the frequent repetition of impacts, as occurs, for example, when we hammer a piece of metal. Part of the motion is also communicated as sound to adjoining solid bodies and gases.

2. *Friction,* both that at the surface of two bodies which move over one another and that arising in the interior of an object due to changes in shape produced by the displacement of particles. With friction, too, certain small changes in the molecular constitution of the bodies occur, especially when they first begin to rub against one another; later the surfaces generally tend to accommodate themselves to each other, so that as the motion continues, these changes may be considered vanishingly small. In many cases, of course, they do not appear at all—for example, when fluids rub against solid bodies or against one another. Thermal and electrical changes, however, always occur.

It is customary in mechanics to represent friction as a force which acts contrary to some existing motion, the intensity of the force being treated as a function of the velocity. This mode of representation is obviously used only for the purpose of making calculations; it is a very incomplete representation of the complicated processes of action and reaction into which the molecular forces enter. From this customary manner of regarding the subject, it was inferred that with friction vis viva was absolutely lost, just as such a loss was taken for granted in the case of inelastic impact. No account was taken of the fact that, apart from the increase of the tensional forces due to the compression of the bodies rubbed or struck, the heat developed also represents a force by which mechanical effects can be achieved. For the most part, the electricity developed—a force directly in its powers of attraction and repulsion, or indirectly in that it can be used to generate heat—was also neglected. It remains to be asked, therefore, whether the sum of these forces always corresponds to the mechanical force which is lost.

In those cases where molecular changes and the development of electricity are virtually absent, the problem can be reduced to two questions: Is a definite quantity of heat always developed for a certain loss of mechanical force? and how far can a quantity of heat correspond to an equivalent mechanical force? As to the answer to the first question, up to the present time only a few experiments have been made. Joule[8] has investigated the quantity of heat developed by the friction of water in narrow tubes and in vessels in which water is set in motion by a paddle wheel. In the first case, he found that the heat which raises one kilogram of water 1°C. was sufficient to raise 452 kilograms through the height of one meter; in the second case, he found the weight to be 521 kilograms. His methods of measurement, however, meet the difficulties of the investigation so imperfectly that the results can lay little claim to accuracy.[9] These figures are probably too high, inasmuch as with his procedure a quantity of heat might readily have escaped unobserved; besides, the necessary loss of mechanical force in other parts of the machine was not taken into account.

Let us turn now to the second question: How far can a quantity of heat correspond to an equivalent mechanical force? The material theory of heat must necessarily assume that the quantity of caloric is constant; caloric can develop mechanical force, therefore, only by its effort to expand. In this theory the force equivalent of heat can thus consist only in the work produced by the heat as it passes from a warmer to a colder temperature. Conceived of

---

8. J. P. Joule, "On the Existence of an Equivalent Relation Between Heat and the Ordinary Forms of Mechanical Power," *Philosophical Magazine,* XXVII, 205.

9. This judgment refers only to the very earliest of Joule's investigations, those which had become known at that time. His later investigations, carried out with complete professional knowledge and indefatigable energy, merit the highest praise. They give a figure of 425 kilograms. [1881]

in this way, the problem has been treated by Carnot and Clapeyron, and all the consequences of the assumption of such a force equivalent, at least with respect to gases and vapors, have been corroborated.

In order to explain the heat developed by friction, the caloric or material theory must assume either that it is communicated by conduction from the external environment (as was supposed by W. Henry) or that it results from the compression of the surfaces and the particles rubbed off (as was believed by Berthollet). The first of these assumptions has thus far no empirical evidence in its favor. If it were true, then in the neighborhood of places that are rubbed, a cold proportionate to the intense heat often developed should be observed. The second assumption, apart from the completely improbable magnitude it must assume for a condensation which is almost imperceptible with a hydrostatic balance, breaks down completely when applied to the friction of fluids and to the experiments in which pieces of iron have been made red hot and soft by hammering and in which pieces of ice have been melted by friction,[10] for the softened iron and the water of the melted ice could not remain in a compressed condition.

Besides this, the development of heat by the motion of electricity proves that there can actually be an absolute increase in the quantity of heat. Ignoring frictional and voltaic electricity (since it might be assumed that by some sort of connection or relation of electricity to caloric, the latter was only transferred from the place where it originated and was deposited in the heated wire), this leaves two ways of producing electric tensions by purely mechanical agencies, namely, by induction and by the motion of magnets. In these processes, heat does not exist to be transferred.

Imagine that we have a completely isolated, positively charged body, a body which cannot lose its electricity. An insulated conductor brought near it will display free $+E$. We can discharge this upon the inner coating of a battery and then remove the conductor, which will show $-E$. This can, of course, be discharged upon the outer surface of the first or upon a second battery. By repeating this process we can obviously charge a battery of any given size as often as we please, and by means of its discharge we can produce heat without its being dissipated away. We shall, of course, have used up a certain amount of mechanical force, for each time we move the negatively charged conductor from the positively charged body, the attraction between them must be overcome.

10. Humphry Davy, *Essay on Heat, Light, and the Combinations of Light.*

This process is actually carried out, of course, when an electrophorus is used to charge a Leyden jar. The same thing occurs in electromagnetic machines; as long as the magnet and the armature move opposite to one another, electric currents are excited which produce heat in the conducting wire, and since the armature moves constantly relative to the magnet, they use up a certain amount of mechanical force. Evidently an indefinitely large amount of heat can be developed by the bodies constituting the machine without disappearing anywhere. Joule has also tried to prove experimentally that the electromagnetic current develops heat rather than cold in the part of the coil of wire directly under the influence of the magnets.

It follows from these facts that the quantity of heat can be increased absolutely by mechanical forces; thus thermal phenomena cannot be explained by the hypothesis of a substance whose mere presence produces the phenomena. These phenomena, on the contrary, must be traced back to changes, to movements, either of a special substance or of the ponderable or imponderable bodies—such as electricity or the luminiferous ether—already known to us. That which has hitherto been called the quantity of heat is, according to this understanding, the expression of (1) the quantity of vis viva of the thermal motion and (2) the quantity of those tensional forces in the atoms which, by changing their arrangements, can develop such a motion. The first corresponds to what has hitherto been called free heat, the second to latent heat.

If it be permitted to try to make the conception of this motion still clearer, an idea derived from the hypothesis of Ampère seems in general best suited to the present state of science. If we think of bodies as being made up of atoms, which are themselves composed of different particles (chemical elements, electricity, etc.), then three kinds of motion may be distinguished in such atoms: (1) displacement of the center of gravity, (2) rotation around the center of gravity, and (3) displacement of the particles of the atom relative to one another. The first two kinds of motion would be balanced by the forces of neighboring atoms and thus transmitted to them in the form of undulations, a form of propagation which corresponds to the radiation of heat but not to its conduction. Movements of the individual particles of the atoms among themselves would be balanced by the forces existing within the atoms, just as one vibrating string slowly sets a second in motion while losing an equal quantity of motion itself. This kind of propagation seems to be similar to the conduction of heat. Moreover, it is clear that such movements in the atoms may cause changes in the molecular forces and thus produce an expansion and alteration in the state of aggregation. What the nature of the motion is we have no way whatsoever of determining, but the possibility of conceiving of the phenomena of heat as due to such motion is sufficient for our present purposes. The principle of the conservation of force will then hold good wherever the conservation of caloric has previously been assumed—that is, with all phenomena of radiation and conduction of heat from one body to another, and with the absorption and release of heat during changes of aggregation.

Of the different ways in which heat is produced, we have considered radiation and its production by mechanical forces. We shall examine the generation of heat by electricity later. It remains here to consider its development in chemical processes. This has been explained hitherto as the freeing of caloric, which was presumed to be latent in some bodies. According to this view, we must ascribe to every simple body, and to every chemical combination which is capable of entering into other combinations of a higher order, a specific quantity of latent heat which is necessary to its chemical constitution. From this we derive the law, which has been partially verified by experience, that when several substances unite to form some chemical compound, the same quantity of heat is always produced, no matter what the order of the combination or the nature of the intermediate products. According to our way of viewing the subject, the quantity of heat developed by chemical processes would be the quantity of vis viva produced by specific amounts of the chemical forces of attraction, and the law just presented would be the expression of the principle of the conservation of force for cases of this kind.

Although the generation of heat undoubtedly occurs, the few investigations that have been made concerning the force equivalent of heat have been made in connection with its disappearance. Up to now we have been acquainted only with cases in which chemical compounds have been broken up or less dense states of aggregation produced, processes by which heat is rendered latent. Nobody has yet bothered to inquire whether heat disappears with the production of mechanical force, as would of course be necessary for the conservation of force. I can therefore only cite an experiment by Joule, which seems to have been rather carefully made. He found that air, passing from a reservoir with a volume of 136.5 cubic inches which has been immersed in water and in which the air was subjected to a pressure of 22 atmospheres, cooled the surrounding water 4.085°F. when the air issued into the atmosphere and thus had to overcome the resistance it provided. When, on the other hand, the air rushed into a vessel of equal size which had been exhausted of air, thus finding no resistance and exerting no mechanical force, there was no change in temperature.

We have still to examine what bearing the experiments of Clapeyron and Holtzmann have upon the force equivalent of heat and upon our own investigations. Clapeyron starts from three assumptions–that heat can be utilized as a means of developing mechanical force only when it passes from warm bodies to colder ones, that the maximum amount of this force is obtained when the passage of the heat occurs between bodies at the same temperature, and that changes of temperature are caused by the compression and expansion of the heated bodies. This maximum, however, must be the same for all bodies which can produce mechanical force by heating and cooling; if it were different, the body in which a certain quantity of heat was capable of producing the greatest effect could be used in the production of mechanical work, and a part of this work could be used to bring the heat back from the colder to the warmer source. In this way an infinite amount of mechanical force might be gained, assuming, of course, that the quantity of heat is not changed in this process.

The following is the general analytical expression given by Clapeyron for this law:

$$\frac{dq}{dv} \cdot \frac{dt}{dp} - \frac{dq}{dp} \cdot \frac{dt}{dv} = C,$$

where q is the quantity of heat contained in a body and t its temperature, both expressed as functions of the volume v and the pressure p. $1/C$ is the mechanical work which the unit of heat (the heat which can raise the temperature of one kilogram of water 1°C.) produces when it drops one degree in temperature. This is asserted to be the same for all bodies but to vary with the temperature. For gases the formula is

$$C = v\frac{dq}{dv} - p\frac{dq}{dp}.$$

Clapeyron's inferences from the general case given by this formula, at least for gases, have many empirical analogies in their favor. His deduction of the law can only be accepted, however, if the absolute quantity of heat is considered to be constant.[11] Further, his more special formula for gases, which alone is supported by comparison with experience, follows also from the for-

11. It is well known that Clausius later (1850) improved on this point of Carnot's theory. [1881]

mula of Holtzmann, as we shall show in a moment. With regard to the general formula, he has only sought to show that the law which follows from it is at least not contradicted by experience. This law states that if the pressure on different bodies, taken at the same temperature, is increased a small amount, quantities of heat will be developed proportional to their capacity to expand when heated. I shall here call attention only to what must be regarded as at least a very improbable consequence of this law: compression of water at the point of maximum density would produce no heat, and between this temperature and the freezing point it would produce cold.

Holtzmann sets out from the assumption that a certain quantity of heat which enters a gas can produce either an increase of temperature or an expansion at the same temperature. The quantity of work produced by the expansion he assumed to be the mechanical equivalent of the heat. From the acoustical experiments of Dulong concerning the relations of the two specific heats of a gas, he calculated that the heat which raises the temperature of one kilogram of water 1°C. would raise a weight of 374 kilograms one meter. This method of calculation, considered from our point of view, is admissible only if the entire vis viva of the heat which enters the gas is actually given up as mechanical force, hence only if the sum of the vis viva and tensional forces—that is, of free and latent heat—is just the same in more tenuous gases as in denser ones at the same temperature. A gas which expands without producing work will then exhibit no change in temperature, which indeed appears to follow from the experiment of Joule mentioned above. Thus the increase and decrease of temperature by compression and expansion would, under ordinary conditions, be due to the production of heat by mechanical force, and vice versa. In support of the correctness of Holtzmann's law, the great number of its consequences which agree with experience may be mentioned; in particular, the deduction of the formula for the elasticity of water vapor at various temperatures.

Joule determined from his own experiments that the force equivalent of the unit of heat produced by mechanical force (which Holtzmann, from the experiments of others, calculated to be 374) is 481, 464, 479; for the unit of heat produced by friction, on the other hand, he found the force equivalent to be 452 and 521.

The formula of Holtzmann is in agreement with that of Clapeyron for gases. In his formula, however, the undetermined function of the temperature, C, is known, and thus the complete determination of the integral is possible. Holtzmann's formula is

$$\frac{pv}{a} = v\frac{dq}{dv} - p\frac{dq}{dp},$$

where a is the force equivalent of the unit of heat. Clapeyron's formula is

$$C = v\frac{dq}{dv} - p\frac{dq}{dp}.$$

Both are therefore the same if $C = pv/a$; or, since $p = k/v(1 + \alpha t)$, where α is the coefficient of expansion and k is a constant, if

$$\frac{1}{C} = \frac{a}{-k(1 + \alpha t)}.$$

The values calculated by Clapeyron for $1/C$ agree fairly well with this formula, as is shown by Table I.

TABLE I

Temperature, C.	Force Equivalents as Calculated by Clapeyron			Force Equivalents According to the Formula
	a	*b*	*c*	
0.0°	1410		1586	1544
35.5		1365	1292	1366
78.8		1208	1142	1198
100.0		1115	1102	1129
156.8		1076	1072	904

The numbers under a are calculated from the velocity of sound in air; the series b from the latent heat of the vapor of ether, alcohol, water, and the oil of turpentine; and those under c from the expansive force of water vapor at various temperatures. Clapeyron's formula for gases, according to this, is identical with Holtzmann's; its applicability to solids and liquids remains doubtful. (See Appendix 5 concerning the work of Robert Mayer.)

V. THE FORCE EQUIVALENT OF ELECTRICAL PROCESSES

Static Electricity

Machine electricity can be the cause of the generation of force in two different ways: when it moves with its conductor because of its force of attraction and repulsion, and when heat is generated because of the motion of electricity in the conductor that carries it. As is well known, the first kind of mechanical phenomenon has been deduced from the hypothesis of two fluids which attract or repel with a force inversely proportional to the square of the distance; and experience, insofar as it can be compared with theory, has been in agreement with calculations. According to our initial analysis, the conservation of force must hold for such forces. We shall therefore go into the more special laws of the mechanical effects of electricity only insofar as is necessary in order to deduce the laws of the electrical development of heat.

Let $e_{\prime}$ and $e_{\prime\prime}$ be two electrical mass elements, whose unit is that of a charge which at one unit of distance repels an equal charge with unit force. If opposite electricities are distinguished by opposite signs, and if the distance between $e_{\prime}$ and $e_{\prime\prime}$ is r, then intensity of their central force is

$$\phi = -\frac{e_{\prime}e_{\prime\prime}}{r^2}.$$

The gain in vis viva which results from the charges moving from the distance R to r is

$$-\int_R^r \phi\, dr = \frac{e_{\prime}e_{\prime\prime}}{R} - \frac{e_{\prime}e_{\prime\prime}}{r}.$$

When they pass from the distance ∞ to r, this is equal to $-(e_{\prime}e_{\prime\prime}/r)$. Let us, in conformity with Gauss in his researches on magnetism, call this last quantity, which is the sum of the tensional forces consumed and of the vis viva produced by the motion from ∞ to r, the *potential* of the two electrical elements for the distance r. The increase of vis viva due to any motion whatsoever is then equal to the excess of the potential at the end of the motion over its value at the beginning.

Let us call the sum of the potentials of an electrical element, relative to the collective elements of an electrified body, the potential of that element relative to the body; and let us call the sum of the potentials of all the elements of one electrified body, relative to all the elements of another body, the potential of the two bodies. Once more, then, the gain in vis viva is given by the difference between the potentials, provided that the distribution of the electricity in the bodies is not changed, that is, that the bodies are idioelectric. If the distribution undergoes a change, the magnitude of the electric tensional forces in the bodies themselves will also be altered, and the vis viva gained must also then be different.

Equal quantities of positive and negative electricity are generated by all methods of electrification. In the neutralization of the electricities of two bodies, one of which, A, contains as much positive electricity as the other, B, does of negative, half of the positive electricity goes from A to B and half of the negative from B to A. If the potentials of the bodies in themselves are W_a and W_b, and if their potential with respect to each other is V, we can find[12] the entire vis viva which has been gained by subtracting the potentials of the moving electric masses in themselves and with respect to each of the other masses at the start of the movement from the same potentials after the movement has taken place. It should be noticed that the potential of two masses changes in sign when the sign of one of the masses is changed. Thus the following potentials are evident:

1. That of the moved $+\frac{1}{2}E$ from A

with respect to itself	$\frac{1}{4}(W_b - W_a)$
with respect to the moved $-\frac{1}{2}E$	$\frac{1}{4}(V - V)$
with respect to the motionless $+\frac{1}{2}E$.	$\frac{1}{4}(-V - W_a)$
with respect to the motionless $-\frac{1}{2}E$	$\frac{1}{4}(-W_b - V)$

2. That of the moved $-\frac{1}{2}E$ from B

with respect to itself	$\frac{1}{4}(W_a - W_b)$
with respect to the moved $+\frac{1}{2}E$.	$\frac{1}{4}(V - V)$
with respect to the motionless $-\frac{1}{2}E$	$\frac{1}{4}(-V - W_b)$
with respect to the motionless $+\frac{1}{2}E$	$\frac{1}{4}(-W_a - V)$
SUM	$-\left(V + \frac{W_a + W_b}{2}\right)$

This quantity, therefore, indicates the maximum of the vis viva generated and the quantity of tensional force which can be gained by electrification.

In order now, instead of these potentials, to introduce more familiar ideas, let us consider the following: Let us imagine surfaces (which we shall call equilibrium surfaces) to be constructed such that the potential of the electrical elements lying in them possesses the same value relative to one or

---

12. That is, under the assumptions made here. See Appendix 6. [1881]

more electrified bodies in the region. Under these conditions the motion of one of the electric particles from any point whatsoever on one surface to any point on another always increases the vis viva by the same amount; motion within a surface, however, will not change the velocity of a particle. Thus the resultant of the collective forces of attraction of electricity for any single point of the region must be perpendicular to the equilibrium surface which passes through the point, and every surface which is at right angles to this resultant must be an equilibrium surface.

Electric equilibrium cannot exist in a conductor unless the resultants of the whole of the forces of attraction of its own electricities, as well as such other electrified bodies as are present, are perpendicular to the surface of the conductor; otherwise the electric particles would be moved along the surface. It follows that the surface of an electrified conductor is itself an equilibrium surface, and the vis viva gained by a vanishingly small electric particle in its passage from the surface of one conductor to that of another is a constant. Let C_a denote the vis viva gained by a unit of positive electricity in its passage from the surface of the conductor A to an infinite distance, so that C_a is positive for positive changes; A_a, the potential of the same quantity of electricity with respect to A when it occupies a certain point upon the surface of A; A_b, the same with respect to B; W_a, the potential of A with respect to itself; W_b, the same for B; V, that of A with respect to B; Q_a, the quantity of electricity in A; and Q_b, that in B. The vis viva gained by the particle e in its passage from an infinite distance to the surface of A is then

$$-eC_a = e(A_a + A_b).$$

If, instead of e, we substitute successively all the electric particles of the surface A, and for A_a and A_b the corresponding potentials, and add them all together, we obtain

$$-Q_aC_a = V + W_a.$$

In like manner, for the conductor B,

$$-Q_bC_b = V + W_b.$$

Now the constant C must have the same value, not only for the entire surface of a conductor, but also for separate conductors if, with the establishment of a connection which does not noticeably change the distribution of their electricities, they do not exchange electricity with each other; that is, it must be the same for all conductors possessing the same free tension. As the unit of measure for the free tension of an electrified body, we can use a quantity of electricity which, distributed over a sphere (of radius equal to one) placed beyond the distance at which induction can take place, is in electric equilibrium with the body. If the electricity is distributed uniformly over the sphere, its external action, as is well known, will be the same as if the electricity were concentrated at its center. Denoting the quantity of electricity by E and the radius of the sphere by R (= 1), for this sphere the constant

$$C = \frac{E}{R} = E .$$

Thus the constant C is equal to the free tension.

In accordance with this, the quantity of the tensional forces of two conductors which contain equal quantities, Q, of positive and negative electricity is

$$-\left(V + \frac{W_a + W_b}{2}\right) = Q\left(\frac{C_a - C_b}{2}\right) .$$

Since C_b is negative, the algebraic difference $C_a - C_b$ is equal to the sum of $C_a + C_b$. If the strength of the current in the conductor B is very large, and consequently if C_b approaches zero, then the quantity of the electric tension is $\frac{1}{2}QC_a = -V + \frac{1}{2}W_a$. If the distance between the two conductors is also very large, the quantity of electric tension, $\frac{1}{2}QC_a$, is $-\frac{1}{2}W_a$.

We have found that the vis viva generated by the motion of two electric masses is equal to the decrease in the sum $\frac{1}{2}(Q_aC_a + Q_bC_b)$. This vis viva is gained as mechanical force if the velocity of the electricity in the bodies is vanishingly small in comparison to the velocity of propagation of the electric motion; if this is not the case, we must obtain it as heat. The heat, Θ, developed by the discharge of equal quantities, Q, of opposed electricities is therefore

$$\Theta = \frac{1}{2a} Q(C_a - C_b),$$

where a is the mechanical equivalent of the unit of heat; or, if $C_b = 0$, as in the case with batteries whose external coating is not insulated and whose capacity is S, so that $CS = Q$,

$$\Theta = \frac{1}{2a} QC = \frac{1}{2a} \frac{Q^2}{S}.$$

Riess has proved by experiment that, with various charges and various numbers of evenly constructed jars, the quantity of heat developed in similar parts of a given closed circuit is proportional to the quantity Q^2/S. He uses S, however, to designate the surface of the coating of the jars; but with uniform jars this must be proportional to the capacity. Vorsselmann de Heer and Knochenhauer have also concluded from their respective experiments that the development of heat with the same charges of the same battery remains the same, no matter how the connecting wire may be changed. The latter has also established this law for cases where the connecting wire is divided and for induction currents. With regard to the quantity a, we still do not possess any experimental data.

It is easy to explain this law if we assume that the discharge of a battery is, not a simple motion of the electricity in one direction, but a motion backward and forward between the coatings in oscillations which get smaller and smaller until finally the entire vis viva is destroyed by the sum of the resistances. The notion that the discharge current consists of alternately opposed currents is supported, first of all, by their alternately opposed magnetic effects, and secondly, by the fact (observed by Wollaston while attempting to decompose water by electric discharges) that both kinds of gases are developed at both electrodes. At the same time, this assumption explains why in these experiments the electrodes must possess the smallest possible surfaces.

Galvanism

With respect to galvanic phenomena, we have to distinguish two classes of conductors: (1) those which conduct like metals and follow the law of the galvanic tension series, and (2) those which do not follow this law. To the latter belong all compound liquids which during conduction undergo a decomposition proportional to the quantity of electricity conducted.

We can classify the experimental facts in accordance with this distinction: (1) those that refer to what occurs only between conductors of the first class, that is, the charging of different metals which are in contact with unequal electricities; and (2) those which refer to what occurs between conductors of both classes, that is, the differences of electric tensions in open circuits and of electric currents in closed circuits.

Electric currents can never be generated by any combination whatsoever of conductors of the first class; all that is produced by such combinations is electric tensions. These tensions, however, unlike those considered earlier, are not equivalent to some quantity of force and do not indicate a disturbance of electric equilibrium. On the contrary, galvanic tensions result from the establishment of electric equilibrium; no motion of electricity can be produced by these tensions other than changes in the position of the conductor itself due to changes in the distribution of the electricity contained in it.

Let us imagine all the metals of the earth brought into contact with one another and the resultant distribution of electricity established. No other combination whatsoever of the same metals can produce a change in the free electrical tensions until contact is established also with a conductor of the second class.

The concept of a contact force, a force which acts at the place where two different metals touch one another and which develops and sustains their different electrical tensions, has previously been left rather undefined in an attempt to include under it the phenomena resulting from the contact of conductors of both the first and the second class. This was done at a time when the invariant, essential feature which distinguishes the two, namely, the chemical processes, was not yet recognized as such. Because of this indefiniteness in the way a contact force was regarded, it appeared to be such that infinite quantities of free electricity—and hence of mechanical force, heat, and light—might be generated by means of it, if a conductor of the second class could be found which was not electrolyzed during the conduction. It was precisely this aspect, of course, which excited such decided opposition to the contact theory, in spite of its simplicity and the precision of the explanations which it did provide.[13]

Thus the principle which we are presenting here directly contradicts the earlier idea of a contact force if the necessity of chemical processes is not also implied by it. If these are admitted, however—that is, if it is assumed that the conductors of the second class do not follow the law of the series of galvanic tensions because they conduct only through electrolysis—then the concept of a contact force can at once be greatly simplified and can be reduced to forces of attraction and repulsion. All phenomena exhibited by conductors of the first class clearly may be derived from the assumption that different chemical substances possess different attractive forces with respect to the two electric-

13. M. Faraday, *Experimental Researches in Electricity.*

ities and that these forces are exerted only at insensible distances, while the electricities themselves act on one another at measurable distances.

According to this, a contact force would consist in the difference of the forces of attraction which the metallic particles at the place of contact exert upon the electricities at that point, and electric equilibrium would exist when an electric particle, in passing from one metal to the other, neither gains nor loses any vis viva. If $c_{,}$ and $c_{,,}$ are the free tensions of the two metals, and if $a_{,}e$ and $a_{,,}e$ are the vires vivae gained by the electric particle e in its passage to one or the other uncharged metal, then the force gained by the passage of e from one charged metal to the other is

$$e(a_{,} - a_{,,}) - e(c_{,} - c_{,,}).$$

With equilibrium this must be equal to zero, and thus

$$a_{,} - a_{,,} = c_{,} - c_{,,};$$

that is, the difference in tension must be constant with different pieces of the same metal and must follow the law of the series of galvanic tensions with different metals.

Since we are concerned here with the conservation of force, we must consider primarily the following effects of galvanic currents: the generation of heat, chemical processes, and polarization. (Electrodynamic effects will be considered in the section on magnetism.) The generation of heat is the same in all currents. With respect to the other two effects, for our purposes we can divide the currents into those which produce chemical decomposition only, those which produce polarization only, and those which produce both.

Let us investigate the conditions of the conservation of force, first, in batteries within which there is no polarization, inasmuch as these are the only ones concerning which specific quantitative laws have been obtained. The intensity of the current J in a battery of n elements is given by Ohm's law as

$$J = \frac{nA}{W},$$

where the constant A is the electromotive force of a single element and W is the resistance of the battery; A and W are independent of the intensity in batteries of this type. Since, during a given period of time, the only things in such a battery that will change are its chemical relations and the quantity of heat, the law of the conservation of force thus requires that the heat gained by the chemical processes which take place be equal to the quantity actually gained. In a simple part of a metallic conductor with the resistance w, the heat developed in a period of time, t, is, according to Lenz,

$$\vartheta = J^2 wt,$$

where the unit for w is a length of a wire in which one unit of current develops one unit of heat in one unit of time. With branching conductors, if the resistance in the individual branches is denoted by w_a, the total resistance w is given by the equation

$$\frac{1}{w} = \sum\left[\frac{1}{w_a}\right]$$

and the intensity J_n in the branch w_n by

$$J_n = \frac{J_w}{w_n};$$

hence the heat ϑ_n in the same branch is

$$\vartheta_n = J^2 w^2 \cdot \frac{1}{w_n} \cdot t,$$

and the total heat developed in all the branches is

$$\vartheta = \Sigma[\vartheta_a] = J^2 w^2 \sum\left[\frac{1}{w_a}\right] t = J^2 wt.$$

Hence, if Lenz's law holds for fluid conductors, the total quantity of heat developed in a battery in which the conduction is effected through any number of branches is, as Joule found,

$$\Theta = J^2 Wt = nAJt.$$

There are two kinds of constant batteries, those constructed according to the system of Daniell and those on Grove's principle. In the first, the chemical process consists in the solution of the positive metal in an acid, while the negative one is precipitated from the same solution. If we take as the unit of intensity a current which in one unit of time decomposes an equivalent of water (producing, say, one gram of oxygen per second), then in the time t the equivalent of the positive metal dissolved will be nJt, and the same quantity of the negative will be precipitated. If we use a_z to designate the heat which an equivalent of the positive metal develops by its oxidation and the solution of its oxide in the acid, and a_c to designate the same for the negative metal, then the quantity of heat developed chemically will be

$$\Theta = nJt(a_z - a_c).$$

The heat developed chemically will thus be equal to that developed electrically if

$$A = a_z - a_c,$$

that is, if the electromotive force of two metals so combined is proportional to the difference of the quantities of heat developed by their combustion and by their combination with acids.

In the batteries constructed according to Grove's principle, polarization is nullified by allowing the hydrogen which separates to reduce the oxygen-rich solution which surrounds the negative metal. To this class belong (1) the batteries of Grove and Bunsen, consisting of amalgamated zinc, dilute sulphuric acid, fuming nitric acid, platinum or coal; and (2) batteries in which chromic acid is used (these have been subjected to careful measurement), consisting of amalgamated zinc, dilute sulphuric acid, and copper or platinum. The chemical processes are the same in the two batteries in which nitric acid is used and also the same in the two with chromic acid. From this it follows, according to the analysis presented above, that the electromotive forces must also be equal, which according to Poggendorff's measurements has proved to be precisely the case. (The battery made with coal and chromic acid is very inconstant and has a considerably higher electromotive force, at least at the beginning. It therefore should not be included here, but belongs among those in which polarization is evident.) In these constant batteries the electromotive force is thus independent of the negative metal. We can include them in the type of Daniell's cell if we regard as the negative element the solution which is immediately in contact with a conductor of the first class and which contains the particles of nitrous acid or the oxide of chromium. Thus regarded, an element of Grove or Bunsen would be a battery made of zinc and nitrous acid, and an element constructed with chromic acid would be a battery made of zinc and the oxide of chromium.

The batteries with polarization may be divided into two classes, those which give rise to polarization but no chemical decomposition, and those which result in both. To the first class, which produces an inconstant, quickly disappearing current, belong (1) the simple batteries of Faraday, with solutions of caustic potash, sulphuret of potassium, and nitrous acid; (2) those made with the more strongly negative metals in common acids when the positive metals are not able to decompose the acid—for example, copper with silver, gold, or platinum; coal in sulphuric acid; etc.; and (3) among the more complicated cells, all those in which are inserted decomposition cells, whose polarization prevails over the electromotive force of the other elements. Up to the present time no precise quantitative experiments on the intensity of these batteries have been carried out because of the great fluctuation of their currents. In general, the intensity of the currents seems to depend upon the nature of the immersed metals; the duration of the current increases with an increase in the size of the surfaces and decreases with an increase in the intensity of the current; and the currents can be renewed, even after they have almost disappeared, by moving the metal in the liquid or by bringing it into contact with the air, which cancels the polarization of the plate at which hydrogen has been liberated. The residual currents which are evident in fine galvanometric instruments are probably due to such effects.

The entire process involves the establishment of an electric equilibrium among the particles of the fluid and the metals; the fluid particles appear to order themselves differently, and, at least in many cases, chemical changes appear to occur at the metallic surfaces. In compound batteries, in which the polarization of the originally equal plates is the effect of the action of currents from other elements, we can regain the lost force of the original current in the form of a secondary current after we have removed these other elements and have formed a circuit out of the polarized cell itself. The absence of special information prevents us from making a more precise application of the principle of the conservation of force here.

The most complicated case is provided by those batteries in which chemical decomposition and polarization take place side by side; to this class belong batteries in which gases are produced. The current in these batteries, as in those involving polarization only, is strongest at the beginning and sinks more or less quickly to a point at which it remains fairly constant. With single elements of this type, or with compound batteries composed only of such elements, the polarization current ceases very, very slowly. It is easier, if one wishes to obtain steady currents quickly, to combine constant with inconstant elements, especially if the plates of the latter are comparatively small. Up to the present time very few measurements have been made on such combinations. From the few which I have been able to find (by Lenz and Poggendorff) it follows that the intensity of such batteries, when different resistances are introduced, cannot be expressed by the simple formula of Ohm; if the constants in the formula are calculated at low intensities, the results for higher intensities are found to be too great. It is therefore necessary to regard the numerator, or the denominator, or both, as functions of the intensity; the facts known at the present time do not enable us to decide which of these is the correct alternative.

If we wish to apply the principle of the conservation of force to these currents, we must separate them into two classes: (1) inconstant, or polarization, currents, with respect to which what we have already said concerning pure polarization is applicable, and (2) constant, or decomposing, currents. The same mode of treatment is applicable to the latter as that applied to constant currents in which no gas is developed; the quantity of heat generated by the current must be equal to that which results from the chemical decomposition. If, for example, in a combination of zinc with a negative metal in dilute sulphuric acid, the quantity of heat liberated by an atom of zinc during its solution and the liberation of hydrogen is $a_z - a_h$, then the quantity of heat developed in the time dt would be

$$J(a_z - a_h)dt.$$

If the development of heat in all parts of such a battery is proportional to the square of the intensity, that is, to $J^2 W dt$, we should have, as before,

$$J = \frac{a_z - a_h}{W},$$

which is the simple formula of Ohm. Since this is not applicable in the present instance, it follows that there are certain transverse sections in the battery in which the development of heat is subject to another law and whose resistance therefore cannot be regarded as constant. If, for example, the heat liberated in any cross-section is directly proportional to the intensity—as it must be in the case, among others, of heat liberated due to a change of aggregation—and thus $\vartheta = Jdt$, we have

$$J(a_z - a_h) = J^2 w + J\mu$$

$$J = \frac{a_z - a_h - \mu}{w}.$$

The quantity μ would thus appear in the numerator of Ohm's formula. The resistance of such a transverse section would be $w = \vartheta/J^2 = \mu/J$. If, however, the development of heat is not exactly proportional to the intensity—in other words, if the quantity μ is not perfectly constant but increases with the intensity—then we have the case which corresponds to the observations of Lenz and Poggendorff.

As soon as the current due to polarization has ceased, the electromotive force of such a battery may, by analogy to constant batteries, be considered that between zinc and hydrogen. In the language of the contact theory, it would be that between zinc and the negative metal, minus the polarization of the latter by hydrogen. We need then only to think of this maximum of polarization as being independent of the intensity of the current and as varying with different metals exactly as the electromotive force varies.

The numerator in Ohm's formula, calculated from measurements of intensity with different resistances, can contain, however, in addition to the electromotive force, a quantity which is the result of the resistance at the points of transition and which perhaps is different for different metals. Since the chemical processes remain the same, the existence of such a transition resistance follows, according to the principle of the conservation of force, from the fact that the intensities of these circuits cannot be calculated from Ohm's law.[14]

I have been unable to find any reliable observations to support the view that in circuits where the polarization current has ceased, the numerator in Ohm's formula is dependent on the nature of the negative metal. In order to stop the polarization current quickly, it is necessary to increase the density of the current on the polarized plate as much as possible, partly by introducing cells with a constant electromotive force and partly by diminishing the

---

14. The occlusion of hydrogen in the metals was not yet known. [1881]

surface of the plate. In the experiments of Lenz and Saweljew which bear on this point, according to their own statement they did not achieve a constant current; thus the electromotive forces calculated from their observations also contain those of the polarization currents. They found for zinc and copper in sulphuric acid, 0.51; for zinc and iron, 0.76; and for zinc and mercury, 0.90.

I may remark, in conclusion, that an experiment has been conducted by Joule to establish the equality of the heat developed chemically and electrically. His method of investigation, however, is open to many objections. For example, he assumed that the law of tangents holds with the highest degree of precision for the tangent compass; he did not work with constant currents, but calculated their intensity from the mean of the first and the last deflections; and he assumed that even with the development of gases, the electromotive force, as well as the resistance of the cells, was constant. Hess has already drawn attention to the divergence between Joule's quantitative determinations of heat and the values found by others. According to a note in *Comptes Rendus,* 1843 (No. 16), A. E. Becquerel is said to have corroborated the same law experimentally.

We have seen that, in order to bring the idea of a contact force into agreement with our principle, we are forced to refer it back to simple forces of attraction and repulsion. Let us now try to trace the electric motions between metals and fluids back to the same forces.

Let us imagine the parts of the compound atoms of a fluid to be endowed with different forces of attraction for the two electricities and, accordingly, to be differently charged. When these parts of the atoms are separated at the metallic electrodes, each atom, according to the law of electrolysis, yields up a quantity of $\pm E$ which is independent of the electromotive forces. We can therefore imagine that in chemical combinations the atoms are already combined with equivalents of $\pm E$, the combinations following the same laws as the stochiometric equivalents of different combinations of the ponderable substances.

If two different metals are immersed in a fluid without a chemical process taking place, the positive components are attracted by the negative metal and the negative components by the positive metal. The consequence is a change in the direction and distribution of the different electric fluid particles, an occurrence which we recognize as a polarization current. The motive force of this current is the electric difference between the metals, and its initial intensity is therefore proportional to that difference. Its duration must, the intensities being equal, be proportional to the number of atoms which spread themselves over the plates and consequently to the surface areas of the plates.

In currents which are accompanied by chemical decomposition, on the other hand, a permanent equilibrium between the fluid particles and the metals is not reached because the positively charged surface of the metal is continually removed, being converted into a part of the fluid, and thus a continuous renewal of the charge must occur in back of it. Once the motion has started, it is accelerated every time an atom of the positive metal, united with an equivalent of positive electricity, enters into the solution and neutralizes an atom of the negative component, as long as the magnitude of the attractive force of the first atoms for the $\pm E$, designated by a_z, is greater than that of the second, a_c. The movement in this way would increase in velocity ad infinitum if the loss of vis viva through the development of heat did not increase at the same time. The velocity, therefore, increases only until this loss, J^2Wdt, is equal to the consumption of the tensional force, $J(a_z - a_c)dt$, or until

$$J = \frac{a_z - a_c}{W}.$$

I believe that this division of galvanic currents into those which result in polarization and those which give rise to chemical decomposition, as required by the principle of the conservation of force, is the only means by which the difficulties of the contact theory and of the chemical theory can both be avoided.

Thermoelectric Currents

We must look for the origin of the force of these currents in the effects at soldered connections, discovered by Peltier, by which a current contrary to a given one is developed.

Let us imagine a constant hydroelectric current, into the conducting wire of which a piece of another metal is soldered, the temperature of the junctions being t' and t''. During the interval of time dt the electric current will generate the heat J^2Wdt in the entire conductor. In addition, at one of the points where the metals are soldered together, the quantity $q_{,}dt$ will be developed, while at the other the quantity $q_{,,}dt$ will be absorbed. If A is the electromotive force of the entire circuit and thus $AJdt$ the heat produced chemically, it follows from the law of the conservation of force that

$$AJ = J^2W + q_{,} - q_{,,}. \tag{1}$$

If B_t is the electromotive force of the thermoelectric cell, when one of the soldered junctions has the temperature t and the other any constant temperature whatsoever (for example, 0°), then for the entire circuit we have

$$J = \frac{A - (B_{t'} - B_{t''})}{W} \tag{2}$$

When $t' = t''$,

$$J = \frac{A}{W}.$$

This, set in Equation (1), gives

$$q_{,} = q_{,,};$$

that is, when the temperature at the soldered junctions of the two metals is the same and the intensity of the current is constant, the heat developed and that absorbed are equal, independently of the cross-section. If we assume that the process is the same at every point of the cross-section, it follows that the quantity of heat developed in equal areas of different cross-sections is proportional to the density of the current, hence that the quantities generated by different currents in all the cross-sections are directly proportional to the intensity of the current.

When the solderings have different temperatures, it follows from Equations (1) and (2) that

$$(B_{t'} - B_{t''})J = q_{,} - q_{,,};$$

that is, with the same current intensity, both the heat developed and the binding force increase with the temperature in the same proportion as the electromotive force.

I am unacquainted with any quantitative experiments with which either of these inferences might be compared.

VI. THE FORCE EQUIVALENT OF MAGNETISM AND ELECTROMAGNETISM

Magnetism

A magnet is capable of generating a certain vis viva by the forces of attraction and repulsion which it exerts upon other magnets and upon unmagnetized iron. Since the phenomena of magnetic attraction may be deduced completely from the assumption of two fluids which attract or repel in the inverse ratio of the square of their distance, it follows, without going beyond the analysis presented at the beginning of this memoir, that the principle of the conservation of force must hold for the movements of magnetic bodies relative to one another. In the interest of the theory of induction, however, we must consider the laws of these movements a little more closely here.

1. If $m_,$ and $m_{,,}$ are two magnetic elements, referred to a unit such that at a distance of one unit an equal quantity of magnetism is repelled with a force of one unit, and if opposite magnetisms are distinguished by opposite signs, and if r is the distance between $m_,$ and $m_{,,}$, then the intensity of their central force is

$$\phi = -\frac{m_, m_{,,}}{r^2}.$$

The gain in vis viva during the passage from an infinite distance to r is $-m_, m_{,,}/r$.

2. If we call this quantity the potential of the two elements, and if we extend the term *potential* to magnetic bodies as we did in the case of electricity, we obtain the increase in vis viva during the movement of two bodies whose magnetism remains constant (for instance, the movement of steel magnets) by subtracting the potential at the beginning of the motion from that at the end. On the other hand, as in the case of electricity, the gain of vis viva during the motion of magnetic bodies subject to changes in distribution is measured by the changes in the expression

$$V + \tfrac{1}{2}(W_a + W_b),$$

where V is the potential of the bodies relative to each other and W_a and W_b those of the bodies in themselves. If B is an invariable steel magnet, the approach of a body with variable magnetism generates a vis viva equal to the increase of the sum $V + \tfrac{1}{2}W_a$.

3. It is known that the external effects of a magnet can always be represented by a certain distribution of the magnetic fluids on its surface. We can thus substitute the potential of such a surface for the potential of the magnet. We then find, as in the case of surfaces which conduct electricity, that for a perfectly soft mass of iron, A, which is magnetized by a magnet, B, the gain C in vis viva for a unit of positive magnetism, during the passage from the surface of the iron to an infinite distance, is given by the equation

$$-QC = V + W_a.$$

Since every magnet possesses equal quantities of north and south magnetism, Q is always equal to zero. For such a piece of iron—or for a piece of steel of the same form, position, and distribution of magnetism, thus one whose magnetism is completely neutralized by the magnet E—it follows that

$$V = -W_a.$$

4. V, however, is the vis viva generated by a steel magnet during its approach to the point where its magnetism is completely neutralized. According to this equation, V must be the same no matter what magnet is approached, provided that the approach is always continued until the magnetism is completely neutralized, for W_a always remains the same. In contrast, the vis viva of a similar piece of iron which has been brought up to the same distribution of magnetism, is, as shown above,

$$V + \tfrac{1}{2}W = -\tfrac{1}{2}W$$

and thus only half as great as that of the pieces already magnetized. It is to be remembered that W itself is negative, hence that $-\tfrac{1}{2}W$ is always positive.

If a piece of unmagnetized steel approaches a magnet, and if it retains the magnetism imparted to it, later, when it is removed to a distance, $-\tfrac{1}{2}W$ will be lost in mechanical work, and the magnet thus created will be capable of producing $-\tfrac{1}{2}W$ more work than could the piece of unmagnetized steel.

Electromagnetism

Electrodynamic phenomena have been referred by Ampère to forces of attraction and repulsion exerted by the elements of a current, with the intensity of the forces depending upon the direction and velocity of the current. His derivations, however, do not incorporate the phenomena of induction. These, together with electrodynamic phenomena, have been traced by W. Weber back to the forces of attraction and repulsion of the electric fluids themselves, the intensity of the forces in this case depending upon the velocity of approach or recession and upon the increase in this velocity. Up to the present time no hypothesis has been found by which these phenomena can be reduced to constant central forces. Neumann has developed the laws of induced currents by extending the experimental law of Lenz for complete circuits to the smallest elements of them. In the case of closed circuits, these laws coincide with those developed by Weber. In like manner, the laws of Ampère and Weber for the electrodynamic effects in closed circuits coincide with Grassmann's derivation of the same laws from forces of rotation. Experience gives us no further information, for experiments have been conducted only on closed or nearly closed circuits. We shall therefore confine the application of our principle to closed circuits and show that the same laws follow from it.

Ampère has already shown that the electrodynamic effects of a closed current can always be re-established by a certain distribution of the magnetic fluids on a surface which is bounded by the current. Because of this, Neumann has extended the concept of potential to closed currents by substituting the potential of such a surface for these effects.

5. If a magnet moves under the influence of a current, the vis viva which it gains must be supplied by the tensional forces consumed in the current. During the interval of time dt, according to the notation used earlier, $AJdt$ in units of heat is produced, or $aAJdt$ in mechanical units, where a is the mechanical equivalent of the unit of heat. The vis viva generated in the circuit is aJ^2Wdt, and that gained by the magnet is $J\frac{dV}{dt}dt$, where V represents its potential with respect to the conductor through which a unit of current passes. Thus

$$aAJdt = aJ^2Wdt + J\frac{dV}{dt}dt,$$

and consequently

$$J = \frac{A - \frac{1}{a} \cdot \frac{dV}{dt}}{W}.$$

We can consider the quantity $\frac{1}{a} \cdot \frac{dV}{dt}$ to be a new electromotive force, the force of the induction current. It acts always in opposition to the force which moves the magnet in the direction it is following, or in opposition to the force which would increase its velocity. Since the induction force is independent of the intensity of the current, it must remain constant, if no current was present prior to the movement of the magnet.

If the intensity is changing, the entire current induced during a given period of time is

$$\int J dt = -\frac{1}{aW}\int \frac{dV}{dt}\,dt = \frac{\frac{1}{a}(V_{\prime} - V_{\prime\prime})}{W},$$

where $V_{\prime}$ is the potential at the beginning and $V_{\prime\prime}$ that at the end of the motion. If the magnet comes from a very great distance, we have

$$\int J dt = -\frac{\frac{1}{a}V_{\prime\prime}}{W},$$

independent of the route or velocity of the magnet.

We can express the law in the following way: The total electromotive force of the induction current which is generated by the movement of a magnet relative to a closed conductor is equal to the change which takes place in the potential of the magnet relative to the conductor when the latter is carrying the current $1/a$. The unit of electromotive force here is that which generated a unit of current against a unit of resistance. The unit of resistance is that in which the unit of current given above develops a unit of heat in a unit of time. The same law was formulated by Neumann, except that instead of $1/a$ he had an undetermined constant ϵ.

6. If a magnet moves under the influence of a conductor, relative to which its potential with one unit of current is ϕ, and also under the influence of a piece of iron magnetized by this conductor, relative to which its potential for the magnetism caused by one unit of current is χ, we then have, as before,

$$aAJ = aJ^2W + J\frac{d\phi}{dt} + J\frac{d\chi}{dt}$$

and hence

$$J = \frac{A - \frac{1}{a}\left(\frac{d\phi}{dt} + \frac{d\chi}{dt}\right)}{W}.$$

The electromotive force of the induction current due to the presence of the piece of iron is thus

$$-\frac{1}{a}\frac{d\chi}{dt}.$$

If the same distribution of magnetism is produced in an electromagnet by the current n as is produced by a magnet which is brought near it, then according to what has been stated in section 4 (above), its potential $n\chi$ with respect to the magnet must be equal to its potential nV with respect to the conducting wire, if V is the potential for one unit of current. Thus $\chi = V$. Hence, if an induction current is excited by means of the magnetization of the piece of iron by the magnet, the electromotive force is

$$-\left(\frac{1}{a}\right)\cdot\left(\frac{d\chi}{dt}\right) = -\left(\frac{1}{a}\right)\cdot\left(\frac{dV}{dt}\right),$$

and as in section 5 (above), the total current is

$$\int Jdt = \frac{\frac{1}{a}(V_{,} - V_{,,})}{W},$$

where $V_{,}$ and $V_{,,}$ are the potentials of the magnetized iron with respect to the conducting wire before and after the magnetization. Neumann deduced this law from its analogy with the preceding case.

7. If an electromagnet is magnetized under the influence of a current, heat is lost through the induction current. If the iron is soft, the same induction current will move in the opposite direction when the inducing circuit is broken, and the heat will be regained. If it is a piece of steel, which retains its magnetism, the heat is permanently lost; in its place we gain a quantity of magnetic force which is capable of doing work and which is equal, as shown in section 4 (above), to one half of the potential of the magnet when the neutralization is complete. From analogy with the preceding case, it does not seem improbable that, as Neumann also concluded, the electromotive force corresponds to its entire potential and that some of the motion of the magnetic fluids, because of its velocity, is lost as heat and thus gained in the magnet.

8. If two closed conductors move relative to one another, the intensity of the currents can be changed in both. If V is their potential relative to each other for a unit of current, we have, as in the preceding cases and for the same reasons,

$$A_{\prime}J_{\prime} + A_{\prime\prime}J_{\prime\prime} = J_{\prime}^2 W_{\prime} + J_{\prime\prime}^2 W_{\prime\prime} + \frac{1}{a} J_{\prime} J_{\prime\prime} \frac{dV}{dt}.$$

If the intensity of the current in the conductor $W_{\prime\prime}$ is a great deal smaller than that in $W_{\prime}$, so that the electromotive force of induction which is excited by $W_{\prime\prime}$ in $W_{\prime}$ vanishes in comparison to the force $A_{\prime}$ and we can set $J = A_{\prime}/W_{\prime}$, we obtain from the equation just given

$$J_{\prime\prime} = \frac{A_{\prime\prime} - \frac{1}{a} J_{\prime} \frac{dV}{dt}}{W_{\prime\prime}}.$$

The electromotive force of induction is thus the same as that generated by a magnet which possesses the same electrodynamic force as the inducing current. This law has been established experimentally by W. Weber.

If, on the other hand, the intensity in $W_{\prime}$ is vanishingly small compared with that in $W_{\prime\prime}$, we have

$$J_{\prime} = \frac{A_{\prime} - \frac{1}{a} J_{\prime\prime} \frac{dV}{dt}}{W_{\prime}}.$$

The electromotive forces of the conductors relative to each other, therefore, are also equal if the intensities of the currents are equal, whatever the form of the conductors may be.

Here again, the total force of induction which, during a given movement of one conductor relative to another one, produces a current which is not changed by the induction is equal to the change in its potential relative to another conductor carrying $-(1/a)$. In this form Neumann deduced the law from the analogy of magnetic and electrodynamic forces; he also extended it to the case where induction is produced in motionless conductors by variations in the intensity of the current. W. Weber has shown the agreement of his assumptions concerning electrodynamic forces with these theorems. We do not obtain any conclusions for this case from the law of the conservation of force except that, by the reaction of the current induced to the inducing current, a decrease in the latter must occur which is equivalent to a loss of heat which in turn is equivalent to the heat gained in the induced current. In the effect of a current upon itself, the same relation must exist between the initial weakening of the current and the extra-current. No further consequences, however, can be deduced here, since the form of the increase in the current is not known. Moreover, Ohm's law is not immediately applicable, since these currents cannot occupy the entire circuit simultaneously.

Of known natural processes, those of organic beings have still to be considered. In plants the processes are mainly chemical, although we also find, at least in some, a slight development of heat occurring. In particular, there is a vast quantity of chemical tensional forces stored up in plants, the equivalent of which we obtain as heat when they are burned. The only vis viva which we know to be absorbed during the growth of plants is that of the chemical rays of sunlight; we are totally at a loss, however, for means of comparing the force equivalents which are thereby lost and gained.

For animals, we already have some fairly specific facts from which we can start. Animals take in oxygen and the complicated oxidizable compounds which are generated in plants, and give them back partly burned as carbonic acid and water, partly reduced to simpler compounds. Thus they use up a certain quantity of chemical tensional forces and, in their place, generate heat and mechanical force. Since the latter represents only a small amount of work compared with the quantity of heat, the question of the conservation of force reduces to the question whether the combustion and transformation of the substances which serve as food generate a quantity of heat equal to that given out by animals. According to the experiments of Dulong and Despretz, this question can, at least approximately, be answered in the affirmative.

In conclusion, I must refer to some remarks of Matteucci which have been directed against the views advocated in this memoir. (They appear in the *Supplément à la Bibliothèque universelle de Genève,* No. 16 [May 15, 1847], p. 375.) He proceeds from the proposition that a chemical process cannot generate as much heat when it simultaneously produces electricity, magnetism, or light as when these are not produced. He then takes pains to show, by a series of measurements which he adduces, that zinc, during its solution in sulphuric acid, generates just as much heat when the solution is effected directly by chemical processes as when it forms a cell with platinum. He also claims that an electric current produces just as great a chemical and thermal effect while it is deflecting a magnet as when no such deflection is produced.

That Matteucci regards these facts as objections is due to a complete misunderstanding of the views he attempts to refute, as will be evident at once from a consideration of our presentation of these relations.

He also adduces two calorimetric experiments, one concerning the heat developed by the combination of caustic baryta with concentrated or dilute sulphuric acid, the other concerning the heat generated by an electric current in a wire surrounded by gases of different cooling capacities, wherein the gases and the wire sometimes were glowing and sometimes were not. He found that the quantity of heat in the former cases was no smaller than in the latter. When we reflect, however, upon the inadequacy of our calorimetric instruments, it will not seem surprising that differences in cooling due to radiation (which passes with more or less difficulty through the surrounding diathermanous bodies, according to its luminous or non-luminous nature) should escape observation. In his first experiment, the union of the baryta with sulphuric acid took place in a non-diathermanous vessel made of lead, from which the light rays could not escape. We need not dwell any further on the imperfections of Matteucci's methods in carrying out these measurements.

I believe, by what has been presented in the preceding pages, that I have proved that the law under consideration does not contradict any known fact within the natural sciences; on the contrary, in a great many cases it is corroborated in a striking manner. I have tried to state, in the most complete manner possible, the inferences which follow from a combination of this law with other known laws of natural phenomena and which still await experimental proof. The purpose of this investigation, in which I have attempted to guard against purely hypothetical considerations, was to lay before physicists as fully as possible the theoretical, practical, and heuristic importance of this law, the complete corroboration of which must be regarded as one of the principle problems of physics in the immediate future.

APPENDICES [1881]

Appendix 1

The philosophical discussion in the Introduction is strongly influenced by Kant's epistemological insights. I still consider these correct. It was only later that it became clear to me that the principle of causality is, indeed, nothing but the presupposition of the lawful regularity or uniformity of all natural phenomena. A law, considered as an objective power, we call *force. Cause,* according to its original meaning, is the unchanging existent (that is, matter) which lies behind the changes of phenomena; the law of its effects is force. The impossibility, alluded to in the memoir, of conceiving of these in isolation from each other thus follows simply from the fact that the law of an effect presupposes certain conditions under which it is realized. A force separated from matter would be the objectification of a law which lacked the conditions for its realization.

Appendix 2

The necessity for analyzing the forces of bodies into forces of material points can, for the masses *upon* which forces act, be derived from the principle of the complete comprehensibility of nature, since complete knowledge of movement is lacking if the motion of each individual material point cannot be given. There does not seem to me, however, to be an equal necessity for such analysis for the masses *from* which the forces arise.

The discussion in Parts I and II of the text is in part acceptable only if this resolution into point forces can be considered well established to begin with. That motive forces, as defined by Newton, are, according to the parallelogram law, the constructed resultant of all individual forces which arise from the collection of all individual mass elements, I still consider a law of nature established through experience. It expresses a fact: the acceleration which a mass point undergoes if several causes are acting together is the resultant (geometrical sum) of the accelerations which the individual causes acting individually would produce. There is, to be sure, a case in which two bodies, namely, two magnets, when they act simultaneously on a third, do not exert a force that is simply the resultant of the forces that each alone would exert. I agree in this case with the assumption that each individual magnet changes the arrangement of an invisible, imponderable substance in the other. I can no longer consider the principle of comprehensibility, however, as sufficient to justify the conclusion that the effect arising out of the combined operation of two or more causes of movement necessarily must be found by the (geometrical) summation of the individual forces.

The empirical content of Newton's second axiom—as well as the principle, expressed a little later, that the forces which two masses exert upon one another must necessarily be determined if the positions of the masses are specified completely—has been abandoned in those electrodynamic theories in which the force between electrical quanta is dependent upon their velocity and acceleration. The investigations made in this direction up to the present time have continued to go contrary to the mechanical principles of the equality of action and reaction and of the constancy of energy, which had been established as exceptionless according to our earlier experience. If, for electricity in conductors, only unstable equilibrium were to exist, the unequivocality and specificity of the solutions of electrical problems would be lost; and to make a force dependent upon an absolute movement—upon a change in the relation of a mass to something which can never possibly be perceived (that is, distinctionless absolute space)—appears to me an assumption which entails the surrender of hope for the complete solution of scientific problems. This position, in my opinion, should be accepted only if all other theoretical possibilities have been exhausted.

Appendix 3

This frequently used proof is insufficient for the case in which forces are dependent upon the velocity or acceleration, as Lipschitz has pointed out to me. For, one can set

$$X = \frac{dU}{dx} + Q \cdot \frac{dz}{dt} - R \cdot \frac{dy}{dt}$$

$$Y = \frac{dU}{dy} + R \cdot \frac{dx}{dt} - P \cdot \frac{dz}{dt}$$

$$Z = \frac{dU}{dz} + P \cdot \frac{dy}{dt} - Q \cdot \frac{dx}{dt},$$

where U is a function of the coordinates. If $P, Q, R,$ on the other hand, are arbitrary functions of the coordinates and their differential quotients, then

$$X \cdot \frac{dx}{dt} + Y \cdot \frac{dy}{dt} + Z \cdot \frac{dz}{dt} = \frac{dU}{dt} = \frac{d}{dt}\left[\frac{1}{2} mq^2\right],$$

that is, the vis viva is a function of the coordinates. The supplement to the values of the force components provided by the factors P, Q, R represents a resultant force which is perpendicular to the resultant velocity of the point moved. Such a force would, as can be seen, change the curvature of the path but not the vis viva.

If one maintains both the validity of the law of action and reaction and the possibility of a resolution into point forces, the general proposition in the text remains correct. For a pair of points this law permits only such forces as are equal in intensity but opposite to each other in the direction of the line that joins them. The forces perpendicular to the velocity could thus affect the movement only in the case where both velocities are perpendicular to this line.

The concluding proposition of the section thus must contain the supplement made in footnote 5.

Appendix 4

This proposition has also been expressed too strongly, since we must restrict the preceding general proposition to cases where the principle of the equality of action and reaction is generally valid. If we allow the latter to lapse, the fundamental electrodynamic law presented recently by Clausius indicates a situation in which forces that are dependent upon the velocities and accelerations still cannot produce an infinite amount of mechanical force.

Appendix 5

Concerning the history of the discovery of the law of the conservation of force, let me supply an omission here by noting that Robert Mayer published his essays "Über die Kräfte der unbelebten Natur" in 1842 and "Die organische Bewegung in ihrem Zusammenhange mit dem Stoffwechsel" in 1845. The conviction that there is an equivalence between heat and work was expressed in the first essay; the equivalence of heat was calculated to be 365 kg/m in the same way in which it was done by Holtzmann. In its general aim the second essay is in reality the same as mine. I only came to know of them later, but since then I have never failed, when I had to discuss the origin of the law publicly, to place Robert Mayer's name first. I have also tried to champion his claim, insofar as I can represent him, against the friends of Joule, who have tended to deny him completely. A letter to P. G. Tait, which I wrote in this connection, was published by him in the Preface to his book *Sketch of Thermodynamics* (Edinburgh, 1868). I reprint it here:

> I must say that to me the discoveries of Kirchhoff in this area (radiation and absorption) appear to be one of the most instructive cases in the history of science, precisely because so many others had previously been so close to making the same discoveries. Kirchhoff's predecessors in this field were related to him in roughly the same way in which, with respect to the conservation of force, Robert Mayer, Colding, and Séguin were related to Joule and William Thomson.
>
> With respect to Robert Mayer, I can, of course, understand the position you have taken in opposition to him; I cannot, however, let this opportunity pass without stating that I am not completely of the same opinion. The progress of the natural sciences depends always upon new inductions being formed out of available facts, and upon the consequence of these inductions, insofar as they refer to new facts, being compared with reality through the use of experiments. There can be no doubt concerning the necessity of this second undertaking. This part of science often requires a large amount of work and great ingenuity, and we are obligated in the highest degree to those who do it well. The fame of discovery, however, remains with those who have found the new idea; the

later experimental verification is quite a mechanical occupation. Further, we cannot demand unconditionally that the person who discovers a new idea also be obligated to carry out the second part of the work. If this were the case, we would have to reject the greatest part of the work of all mathematical physicists. William Thomson, for example, produced a number of theoretical papers concerning Carnot's law and its consequences before he performed a single experiment, and it would not occur to any one of us to treat these lightly.

Robert Mayer was not in a position to conduct experiments; he was repulsed by the physicists with whom he was acquainted (this also happened to me several years later); it was only with difficulty that he could find space for the publication of his first condensed formulation of this principle. You must know that as a result of this rejection he at last became mentally ill. It is now difficult to set oneself back into the modes of thought of that period, and to make clear to oneself how absolutely new the whole idea seemed at that time. I should imagine that Joule too must have fought for a long time in order to gain recognition for his discovery.

Thus, although no one can deny that Joule did much more than Mayer, and although one must admit that in the first publications of Mayer there were many things that were unclear, still I believe that one must accept that Mayer formulated this idea, which determined the most important recent progress in the natural sciences, independently and completely. His reward should not be lessened because at the same time another man in another country and under other conditions made the same discovery and, to be sure, carried it through afterwards better than he did.

In recent years the devotees of metaphysical speculation have sought to stamp the law of the conservation of force as one possessing a priori validity and have come to hail Mayer as a hero in the realm of pure thought. What they consider the high point of his work, however—the metaphysically formulated pseudo-proof of the a priori necessity of this law—still appears to any scientist accustomed to the strict methodology of science the weakest part of his presentation of the principle. This has undoubtedly been the reason why Mayer's work has remained unrecognized for such a long time in scientific circles. It was originally from the other direction, that is, through Joule's masterly work, that the conviction of the validity of the law came to be widespread, at which point attention was attracted to the essays of Mayer.

This law, like all knowledge of the processes of the real world, has been discovered by induction. That man could never build a *perpetuum mobile,* that is, could never gain mechanical force without a corresponding expenditure of force, was an induction made gradually as the result of a great deal of vain searching.

The French Academy had for a long time placed the *perpetuum mobile* in the same category with the squaring of the circle and had ceased to accept any apparent solutions to the problem. This should be understood as the expression of a conviction which was widely shared by the experts. I had expressed it often enough myself during my student days and had heard the inadequacies of the proofs offered in its support discussed. The question concerning the origin of animal heat gave rise to a careful, complete discussion of all the facts related to it. When I began the present memoir, I thought of it only as a piece of critical work, certainly not as an original discovery concerning the priority of which a fight might be waged. I was afterwards somewhat surprised over the opposition which I met with among the experts; publication of my work in *Poggendorff's Annals* was denied me, and among the members of the Berlin Academy only K. G. J. Jacobi, the mathematician, accepted it. Fame and material reward were not to be gained at that time with the new principle; quite the opposite. That I did not in any way strive myself, in the composition of the manuscript, after a priority which did not belong to me (something which my opponents among the metaphysicians have tried to attribute to me) is made completely clear, I believe, by the fact that I mentioned other investigators who, as far as I knew then, had worked in the same direction. Because of these works, especially those of Joule, there could be no priority claim insofar as I was concerned, above all with respect to the general principle.

If my knowledge of the literature in 1847 was incomplete, I beg to be excused, for I wrote the essay in the city of Potsdam, where my literary resources were restricted to the contents of the Gymnasium library. At that time the *Fortschritte der Physik* of the Berlin Physical Society and similar resources, with which now, of course, it would be very easy to orient oneself in the literature of physics, were still lacking.

Appendix 6

The concept of the potential of a body relative to an electric conductor is taken here in a sense somewhat different from the way it has usually appeared later in scientific literature. Since I could find no predecessors in the scanty literature available to me at that time for the use of these concepts, I allowed myself in formulating them to be led by the analogy of the potentials of two different conductors relative to one another (Part V in the text). If these carriers are represented as congruent and the corresponding surface elements are equally charged, this constitutes the potential V of each. If one then thinks of the two bodies as having been transported to congruent positions, V will be that which I have used W to designate here. In this version, each combination of any two electrical elements, e and ϵ, appears twice in the calculations. The W so formed—and as it appears in the text—is not the value of the work, which is $\frac{1}{2}W$. In my later writings I have subscribed to the usage of other authors and have designated $\frac{1}{2}W$ as the potential of the bodies in themselves.

2.

THE SCIENTIFIC RESEARCHES OF GOETHE [1853]

A lecture delivered before the German Society of Königsberg in 1853

IT was inevitable that Goethe, whose comprehensive genius was most strikingly apparent in that sober clearness with which he grasped and reproduced with lifelike freshness the realities of nature and human life in their minutest details, would, by those very qualities of his mind, be drawn toward the study of natural science. And in that department he was not content with acquiring what others could teach him, but soon attempted, as so original a mind was sure to do, to strike out on an independent and very characteristic line of thought. He directed his energies not only to the descriptive but also to the experimental sciences, the chief results being his botanical and osteological treatises, on the one hand, and his theory of color, on the other. The first germs of these researches belong for the most part to the last decade of the eighteenth century, though some of them were not completed or published till later. Since that time, science not only has made great progress but has widely extended its range. It has assumed in some respects an entirely new aspect; it has opened out new fields of research and undergone many changes in its theoretical views. I shall attempt in the following lecture to sketch the relation of Goethe's researches to the present standpoint of science and to bring out the guiding idea that is common to them all.

The peculiar character of the descriptive sciences–botany, zoology, anatomy, and the like–is a necessary result of the work imposed upon them. They undertake to collect and sift an enormous mass of facts and, above all, to bring them into a logical order or system. Up to this point their work is only the dry task of a lexicographer; their system is nothing more than a muniment-room in which the accumulation of papers is so arranged that anyone can find what he wants at any moment. The more intellectual part of their work–and their real interest–only begins when they attempt to feel after the scattered traces of law and order in the disjointed, heterogeneous mass and out of it to construct an orderly system, accessible at a glance, in which every detail has its due place and gains additional interest from its connection with the whole.

In such studies, our poet's organizing capacity and insight found a congenial sphere, and the epoch was propitious to him. He found ready to his hand a sufficient store of logically arranged materials in botany and comparative anatomy, copious and systematic enough to admit of a comprehensive view and to indicate the way to some happy glimpse of an all-pervading law. His contemporaries, if they made any efforts in this direction, wandered without a compass, or else they were so absorbed in the dry registration of facts that they scarcely ventured to think of anything beyond. It was reserved for Goethe to introduce two ideas of infinite fruitfulness.

The first was the conception that the differences in the anatomy of different animals are to be looked upon as variations from a common phase or type, induced by differences of habit, locality, or food. The observation which led him to this fertile conception was by no means a striking one; it is to be found in a monograph on the intermaxillary bone, written as early as 1786. It was known that in most vertebrate animals (that is, mammalia, birds, amphibia, and fishes) the upper jaw consists of two bones, the upper jawbone and the intermaxillary bone. The former always contains in the mammalia the molar and canine teeth, the latter the incisors. Man, who is distinguished from all other animals by the absence of the projecting snout, has, on the contrary, on each side only one bone, the upper jawbone, containing all the teeth.

Goethe discovered in the human skull faint traces of the sutures which in animals unite the upper and middle jawbones; he concluded that man had originally possessed an intermaxillary bone, which had subsequently coalesced with the upper jawbone. This obscure fact opened up to him a source of the most intense interest in the field of osteology, generally decried as the driest of studies. That details of structure should be the same in man and in animals when the parts continue to perform similar functions involved nothing extraordinary. In fact, Camper had already attempted, on this principle, to trace similarities of structure even between man and fishes. But the persistence of this similarity, at least in a rudimentary form, even in a case when it evidently does not correspond to any of the requirements of the complete human structure and so must be adapted to them by the coalescence of two parts originally separate, struck Goethe's far-seeing eye and suggested to him a far more comprehensive view than had hitherto been taken.

Further studies soon convinced him of the universality of his newly discovered principle, so that in 1795 and 1796 he was able to define more clearly the idea that had struck him in 1786 and to commit it to writing in his *Sketch of a General Introduction to Comparative Anatomy.* He there asserts with the utmost confidence and precision that all differences in the structure of animals must be looked upon as variations of a single primitive type, induced by the coalescence, alteration, increase, diminution, or even the complete removal of single parts of the structure—the very principle, in fact, which has become the leading idea of comparative anatomy in its present stage. Nowhere has it been better or more clearly expressed than in Goethe's writings. Subsequent authorities have made but few essential alterations in his theory; the most important of these is that we no longer undertake to construct a common type for the whole animal kingdom but are content with one for each of Cuvier's great divisions. The industry of Goethe's successors has accumulated a well-sifted stock of facts, infinitely more copious than he could command, and has followed up successfully into the minutest details a principle which he could only indicate in a general way.

The second leading conception which science owes to Goethe is the existence of an analogy among the different parts of a single organic being, similar to that which we have just pointed out as existing among corresponding parts of different species. In most organisms we see a great repetition of single parts. This is most striking in the vegetable kingdom: each plant has a great number of similar stem leaves, similar petals, similar stamens, and so on. According to Goethe's own account, the idea first occurred to him while looking at a fan palm at Padua. He was struck by the immense variety of changes of form which the successively developed stem leaves exhibit—by the way the first simple root leaflets are replaced by a series of more and more divided leaves, till we come to the most complicated.

He afterward succeeded in discovering the transformation of stem leaves into sepals and petals, and of sepals and petals into stamens, nectaries, and ovaries. Thus he was led to the doctrine of the metamorphosis of plants, which he published in 1790. Just as the anterior extremity of vertebrate animals takes different forms—becoming in man and in apes an arm; in other animals a paw with claws, or a forefoot with a hoof, or a fin, or a wing—but always retains the same divisions, the same position, and the same connection with the trunk, so the leaf appears as a cotyledon, stem leaf, sepal, petal, stamen, nectary, ovary, etc., all resembling one another to a certain extent in origin and composition, and even capable, under certain unusual conditions, of passing from one form into another (as, for example, may be seen by anyone who looks carefully at a full-blown rose, where some of the stamens are completely, some partially, changed into petals). This view of Goethe's, like the other, is now completely adopted into science and enjoys the universal assent of botanists, though of course some details are still matters of controversy, as, for instance, whether the bud is a single leaf or a branch.

In the animal kingdom, the composition of an individual out of several similar parts is very striking in the great subkingdom of the articulata—for example, in insects and worms. The larva of an insect, or the caterpillar of a butterfly, consists of a number of perfectly similar segments; only the first and last of them differ, and that but slightly, from the others. After their transformation into perfect insects, they furnish clear and simple exemplifications of the view which Goethe had grasped in his doctrine of the metamorphosis of plants, namely, the development of apparently very dissimilar forms from parts originally alike. The posterior segments retain their original simple form; those of the breastplate are drawn close together and develop feet and wings, while those of the head develop jaws and feelers; so that in the perfect insect, the original segments are recognizable only in the posterior part of the body. In the vertebrata, again, a repetition of similar parts is suggested by the vertebral column but has ceased to be observable in the external form. A fortunate glance at a broken sheep's skull, which Goethe found by accident on the sand of the Lido at Venice, suggested to him that the skull itself consists of a series of very much altered vertebræ. At first sight, no two things can be more unlike than the broad, uniform cranial cavity of the mammalia, enclosed by smooth plates, and the narrow cylindrical tube of the spinal marrow, composed of short, massy, jagged bones. It was shrewd to detect the transformation in the skull of a mammal; the similarity is more striking in the amphibia and fishes. It should be added that Goethe left this idea unpublished for a long time, apparently because he was not quite sure how it would be received. Meantime, in 1806, the same idea occurred to Oken, who introduced it to the scientific world and afterwards disputed with Goethe the priority of discovery. In fact, Goethe waited till 1817, when the opinion had begun to find adherents, and then declared that he had had it in his mind for thirty years. The number and composition of the vertebræ of the skull are still a subject of controversy, but the principle has maintained its ground.

Goethe's views, however, on the existence of a common type in the animal kingdom do not seem to have exercised any direct influence on the progress of science. The doctrine of the metamorphosis of plants was introduced into botany as his distinct and recognized property, but his views on osteology were at first disputed by anatomists; they only subsequently attracted attention, when the science had, apparently on independent grounds, found its way to the same discovery. He himself complains that his first ideas of a common type encountered nothing but contradiction and scepticism at the time when he was working them out in his own mind and that even men of the freshest and most original intellect, like the two Von Humboldts, listened to them with something like impatience. But it is almost a matter of course that in any natural or physical science, theoretical ideas attract the attention of its cultivators only when they are advanced in connection with the whole of the evidence on which they rest and thus justify their title to recognition. Be that as it may, Goethe is entitled to the credit of having caught the first glimpse of the guiding ideas toward which the sciences of botany and anatomy were tending and by which their present form is determined.

But great as is the respect which Goethe has secured by his achievements in the descriptive natural sciences, the denunciation heaped by all physicists on his researches in their department, especially on his theory of color, is at least as uncompromising. This is not the place to plunge into the controversy that raged on the subject and so I shall only attempt to state clearly the points at issue and to explain the principle involved and the latent significance of the dispute.

To this end it is of some importance to go back to the history of the origin of the theory, and to its simplest form, because at that stage of the controversy the points at issue are obvious and admit of easy, distinct statement, unencumbered by disputes about the correctness of detached facts and complicated theories.

Goethe himself describes very gracefully, in the confession at the end of his *Theory of Color*, how he came to take up the subject. Finding himself unable to grasp the æsthetic principles involved in effects of color, he resolved to resume the study of the physical theory, which he had been taught at the university, and to repeat for himself the experiments connected with it. With that view he borrowed a prism from Hofrath Büttner, of Jena, but was prevented by other occupations from carrying out his plan and kept it by him for a long time unused. The owner of the prism, a very orderly man, after several times asking in vain, sent a messenger with instructions to bring it back directly. Goethe took it out of the case and thought he would take at least one peep through it. To make certain of seeing something, he turned it toward a long white wall under the impression that, as there was plenty of light there, he could not fail to see a brilliant example of the resolution of light into different colors—a supposition, by the way, which shows how little Newton's theory of the phenomenon was then present to his mind. Of course he was disappointed. On the white wall he saw no colors; they only appeared where it was bounded by darker objects. Accordingly he made the observation (which, it should be added, is fully accounted for by Newton's theory) that color can only be seen through a prism where a dark object and a bright one have the same boundary. Struck by this observation, which was quite new to him, and convinced that it was irreconcilable with Newton's theory, he induced the owner of the prism to relent and devoted himself to the question with the utmost zeal and interest. He prepared sheets of paper with black and white spaces and studied the phenomenon under every variety of condition until he thought he had sufficiently proved his rules. He next attempted to explain his supposed discovery to a neighbor, who was a physicist, and was disagreeably surprised to be assured by him that the experiments were well known and fully accounted for in Newton's theory. Every other natural philosopher whom he consulted told him exactly the same, including even the brilliant Lichtenberg, whom he tried for a long time to convert, but in vain. He studied Newton's writings and fancied he had found some fallacies in them which accounted for the error. Unable to convince any of his acquaintances, he at last resolved to appear before the bar of public opinion and in 1791 and 1792 published the first and second parts of his *Contributions to Physical Optics.*

In that work he describes the appearances presented by white discs on a black ground, black discs on a white ground, and colored discs on a black or white ground, when examined through a prism. As to the results of the experiments, there is no dispute whatever between him and the physicists. He describes the phenomena he saw with great truth to nature; the style is lively, and the arrangement is such as to make a conspectus of them easy and inviting. In short, in this as in all other cases where facts are to be described, he proves himself a master. At the same time he expresses his conviction that the facts he has adduced are calculated to refute Newton's theory. There are two points especially which he considers fatal to it: first, that the center of a broad, white surface remains white when seen through a prism; and secondly, that even a black streak on a white ground can be entirely decomposed into colors.

Newton's theory is based on the hypothesis that there exist light of different kinds, distinguished from one another by the sensation of color which they produce in the eye. Thus there are red, orange, yellow, green, blue, and violet light and light of all intermediate colors. Different kinds of light, or differently colored lights, produce, when mixed, derived colors, which to a certain extent resemble the original colors from which they are derived; to a certain extent they form new tints. White is a mixture of all the before-named colors in certain definite proportions. But the primitive colors can always be reproduced by analysis from derived colors, or from white, while themselves incapable of analysis or change. The cause of the colors of transparent and opaque bodies is that when white light falls upon them, they destroy some of its constituents and send to the eye other constituents, no longer mixed in the right proportions to produce white light. Thus a piece of red glass looks red because it transmits only red rays. Consequently all color is derived solely from a change in the proportions in which light is mixed and is therefore a property of light, not of the colored bodies, which only furnish an occasion for its manifestation.

A prism refracts transmitted light; that is to say, deflects it so that it makes a certain angle with its original direction. The rays of simple light of different colors have, according to Newton, different refrangibilities and therefore, after refraction in the prism, pursue different courses and separate from one another. Accordingly a luminous point of infinitely small dimensions appears, when seen through the prism, to be first displaced and, secondly, extended into a colored line, the so-called prismatic spectrum, which shows what are called the primary colors in the order above-named. If, however, you look at a broader luminous surface, the spectra of the points near the middle are superposed, as may be seen from a simple geometrical investigation, in such proportions as to give white light, except at the edges, where certain of the colors are free. This white surface appears displaced, as the luminous point did; but instead of being colored throughout, it has on one side a margin of blue and violet, on the other a margin of red and yellow. A black patch between two bright surfaces may be entirely covered by their colored edges; and when these spectra meet in the middle, the red of the one and the violet of the other combine to form purple. Thus the colors into which, at first sight, it seems as if the black were analyzed are in reality due, not to the black strip, but to the white on each side of it.

It is evident that at first Goethe did not recollect Newton's theory well enough to be able to find out the physical explanation of the facts I have just glanced at. It was afterward laid before him repeatedly and in a thoroughly intelligible form, for he speaks about it several times in terms that show he understood it quite correctly. But he was still so dissatisfied with it that he persisted in his assertion that the facts just cited are of a nature to convince anyone who observes them of the absolute incorrectness of Newton's theory. Neither here nor in his later controversial writings does he ever clearly state in what he conceives the insufficiency of the explanation to consist. He merely repeats again and again that it is quite absurd. And yet I cannot see how anyone, whatever his views about color, can deny that the theory is perfectly consistent with itself and that if the hypothesis from which it starts be granted, it explains the observed facts completely and even simply. Newton himself mentions these spurious spectra in several passages of his optical works without going into any special elucidation of the point, considering, of course, that the explanation follows at once from his hypothesis. And he seems to have had good reason to think so, for Goethe no sooner began to call the attention of his scientific friends to the phenomena than all with one accord, as he himself tells us, met his difficulties with this explanation from Newton's principles, which, though not actually in his writings, instantly suggested itself to everyone who knew them.

Readers who try to follow attentively and thoroughly every step in this part in the controversy are apt to experience at this point an uncomfortable, almost painful feeling to see a man of extraordinary abilities persistently declaring that there is an obvious absurdity lurking in a few inferences which are apparently quite clear and simple. He searches and searches, and at last—unable, with all his efforts, to find any such absurdity or even the appearance of it—he gets into a state of mind in which his own ideas are, so to speak, crystallized. But it is just this obvious, flat contradiction that makes Goethe's point of view in 1792 so interesting and so important. At this point he has not as yet developed any theory of his own. There is nothing under discussion but a few easily grasped facts, as to the correctness of which both parties are agreed; yet both hold distinctly opposite views, neither party even understanding what his opponent is driving at. On the one side are a number of physicists, who, by a long series of the ablest investigations, the most elaborate calculations, and the most ingenious inventions, have brought optics to such perfection that it, alone among the physical sciences, is beginning almost to rival astronomy in accuracy. Some have made the phenomena the subject of direct investigation; all, thanks to the accuracy with which it is possible to calculate beforehand the result of every alternative in the construction and combination of instruments, have had the opportunity of putting the inferences deduced from Newton's views to the test of experiment—and all, without exception, agree in accepting them. On the other side is a man whose remarkable mental endowments and singular talent for seeing through whatever obscures reality we have had occasion to recognize, not only in poetry, but also in the descriptive parts of the natural sciences—and this man assures us with the utmost zeal that the physicists are wrong. He is so convinced of the correctness of his own view that he cannot explain the contradiction except by assuming narrowness or malice on their part; finally he declares that he cannot help looking upon his own achievement in the theory of color as far more valuable than anything he has accomplished in poetry.[1]

So flat a contradiction leads us to suspect that behind it there must be some deeper antagonism of principle, some difference of organization between his mind and theirs, to prevent them from understanding each other. I shall try to indicate in the following pages what I conceive to be the grounds of this antagonism.

Goethe, though he exercised his powers in many spheres of intellectual activity, is nevertheless *par excellence* a poet. In poetry, as in every other art, the essential thing is to make the material of the art, be it words, or music, or color, the direct vehicle of an idea. In a perfect work of art, the idea must be present and dominate the whole, almost unknown to the poet himself, not as the result of a long intellectual process, but as inspired by a direct intuition of the inner eye, or by an outburst of excited feeling.

An idea thus embodied in a work of art and dressed in the garb of reality does indeed make a vivid impression by appealing directly to the senses, but it loses, of course, that universality and intelligibility which it would have had if presented in the form of an abstract notion. The poet, sensing how the charm of his works is involved in an intellectual process of this type, seeks to apply it to other materials. Instead of trying to arrange the phenomena of nature under definite conceptions, independent of intuition, he sits down to contemplate them as he would a work of art, complete in itself and certain to yield up its central idea, sooner or later, to a sufficiently receptive student. Accordingly, when he sees the skull on the Lido and this suggests to him the vertebral theory of the cranium, he notes that the experience has revived his old belief, already confirmed by experience, that nature has no secrets from the attentive observer. So again in his first conversation with Schiller on the metamorphosis of plants. To Schiller, as a follower of Kant, the idea is the goal, ever to be sought but ever unattainable and therefore never to be exhibited as realized in a phenomenon. Goethe, on the other hand, as a genuine

1. See Eckermann's *Conversations.*

poet, conceives that he finds in the phenomenon the direct expression of the idea. He himself tells us that nothing brought out more sharply the separation between himself and Schiller. This, too, is the secret of his affinity with the natural philosophy of Schelling and Hegel, which likewise proceeds from the assumption that nature shows us by direct intuition the several steps by which a conception is developed. Hence too the ardor with which Hegel and his school defended Goethe's scientific views. Moreover, this view of nature accounts for the war which Goethe continued to wage against complicated experimental researches. Just as a genuine work of art cannot bear retouching by a strange hand, so he would have us believe nature resists the interference of the experimenter who tortures her and disturbs her; in revenge, she misleads the impertinent killjoy by a distorted image of herself.

> Mysterious in the light of day,
> Nature will not be denied her veil.
> And what she does not make manifest to your spirit
> Cannot be forced from her with levers and screws.
>
> *Geheimnisvoll am lichten Tag*
> *Lässt sich Natur des Schleiers nicht berauben,*
> *Und was sie deinem Geist nicht offenbaren mag,*
> *Das zwingst du ihr nicht ab mit Hebeln und mit Schrauben.*

Accordingly, in his attack upon Newton he often sneers at spectra, tortured through a number of narrow slits and glasses, and commends the experiments that can be made in the open air under a bright sun, not merely as particularly easy and particularly enchanting, but also as particularly convincing! The poetic turn of mind is very marked even in his morphological researches. If we examine what has really been accomplished by the help of the ideas which he contributed to science, we shall be struck by the very singular relation which they bear to it. No one will refuse to be convinced if you lay before him the series of transformations by which a leaf passes into a stamen, an arm into a fin or a wing, a vertebra into the occipital bone. The idea that all the parts of a flower are modified leaves reveals a connecting law which surprises us into acquiescence. But now try to define the leaflike organ, to determine its essential characteristics, so as to include all the forms we have named. You will find yourself in a difficulty, for all distinctive marks vanish, and you have nothing left, except that a leaf in the wider sense of the term is a lateral appendage of the axis of a plant. Try then to express the proposition "the parts of the flower are modified leaves" in the language of scientific definition, and it reads, "the parts of the flower are lateral appendages of the axis." To see this does not require a Goethe. So again it has been objected, not unjustly, to the vertebral theory that it must extend the notion of a vertebra so much that nothing is left but the bare fact: a vertebra is a bone. We are equally perplexed if we try to express in clear scientific language what we mean by saying that such and such a part of one animal corresponds to such and such a part of another. We do not mean that their physiological use is the same, for the same piece which in a bird serves as the lower jaw, becomes in mammals a tiny tympanal bone. Nor would the shape, the position, or the connection of the part in question with other parts serve to identify it in all cases. Yet it has been found possible in most cases, by following the intermediate steps, to determine with tolerable certainty which parts correspond to each other. Goethe himself said this very clearly. In speaking of the vertebral theory of the skull, he observed, "Such an *aperçu*, such an intuition, conception, representation, notion, idea, or whatever you choose to call it, always retains something esoteric and indefinable, struggle as you will against it. As a general principle, it may be enunciated but cannot be proved; in detail it may be exhibited but can never be put in a cut-and-dried form." And so, or nearly so, the problem stands to this day.

The difference may be brought out still more clearly if we consider how physiology, which investigates the relations of vital processes as cause and effect, would have to treat this idea of a common type of animal structure. The science might ask: Is it, on the one hand, a correct view that during the geological periods through which the earth has passed, one species has been developed from another, so that, for example, the breast fin of the fish has gradually changed into an arm or a wing? Or shall we say that the different species of animals were created equally perfect—that the points of resemblance among them are to be ascribed to the fact that in all vertebrate animals the first steps in development from the egg can only be effected by nature in one way, almost identical in all cases, and that the later analogies of structure are determined by these features common to all embryos? Probably the majority of observers incline to the latter view,[2] for the agreement among the embryos of different vertebrate animals in the earlier stages is very striking. Thus even young mammals have occasionally rudimentary gills on the side of the neck, like fishes. It seems, in fact, that what are in the mature animals corresponding parts originate in the same way during the process of

2. This was written before the appearance of Darwin's *Origin of Species.*

development, so that scientific men have lately begun to make use of embryology as a sort of check on the theoretical views of comparative anatomy. It is evident that by the application of the physiological views just suggested, the idea of a common type would acquire definiteness and meaning as a distinct scientific conception.

Goethe did much: he saw by a happy intuition that there was a law, and he followed up the indications of it with great shrewdness. But what law it was he did not see, nor did he even try to find it out. That was not in his line. Moreover, even in the present condition of science, a definite view on the question is impossible; the very form in which it should be proposed is scarcely yet settled. And therefore we readily admit that in this department Goethe did all that was possible at the time when he lived. I said just now that he treated nature like a work of art. In his studies on morphology he reminds one of a spectator at a play, with strong artistic sympathies. His delicate instinct makes him feel how all the details fall into their places and work harmoniously together and how some common purpose governs the whole; yet while this exquisite order and symmetry give him intense pleasure, he cannot formulate the dominant idea. That is reserved for the scientific critic of the drama, while the artistic spectator feels perhaps, as Goethe did in the presence of natural phenomena, an antipathy to such dissection, fearing, though without reason, that his pleasure may be spoiled by it.

Goethe's point of view on the theory of color is much the same. We have seen that he rebels against the physical theory just at the point where it gives complete and consistent explanations from principles once accepted. Evidently it is not the insufficiency of the theory to explain individual cases that is a stumbling block to him. He takes offense at the assumption made for the sake of explaining the phenomena, which seem to him so absurd that he looks upon the interpretation as no interpretation at all. Above all, the idea that white light could be composed of colored light seems to have been quite inconceivable to him. At the very beginning of the controversy, he rails at the disgusting Newtonian white of the natural philosophers—an expression which suggests that this was the assumption that most annoyed him.

Again, in his later attacks on Newton, which were not published till after his Theory of Color was completed, he strives to show that Newton's facts might be explained on his own hypothesis and that therefore Newton's hypothesis was not fully proved, rather than attempting to prove that Newton's hypothesis was inconsistent with itself or with the facts. Indeed, he seems to consider the obviousness of his own hypothesis so overwhelming that it need only be brought forward to upset Newton's entirely. There are only a few passages where he disputes the experiments described by Newton. Some of them, apparently, he could not succeed in refuting because the result is not equally easy to observe in all positions of the lenses used and because he was unacquainted with the geometrical relations by which the most favorable positions of them are determined. In other experiments on the separation of simple colored light by means of prisms alone, Goethe's objections are not quite groundless, inasmuch as the isolation of single colors cannot by this means be so effectually carried out that after refraction through another prism, there are no traces of other tints at the edges. A complete isolation of light of one color can only be effected by very carefully arranged apparatus, consisting of combined prisms and lenses—a set of experiments which Goethe postponed to a supplement and finally left unnoticed. When he complains of the complication of these contrivances, we need only think of the laborious, roundabout methods which chemists must often adopt to obtain certain elementary bodies in a pure form, and we need not be surprised to find that it is impossible to solve a similar problem in the case of light in the open air in a garden with a single prism in one's hand.[3] Goethe must, consistently with his theory, deny *in toto* the possibility of isolating pure light of one color. Whether he ever experimented with the proper apparatus to solve the problem remains doubtful, as the supplement in which he promised to detail these experiments was never published.

To give some idea of the passionate way in which Goethe, usually so temperate and even courtier-like, attacks Newton, I quote from a few pages of the controversial part of his work the following expressions, which he applies to the propositions of this consummate thinker in physical and astronomical science: "incredibly impudent," "mere twaddle," "ludicrous explanation," "admirable for schoolchildren in a go-cart," and "but I see nothing will do but lying, and plenty of it."

Thus, in the theory of color, Goethe remains faithful to his principle that nature must reveal her secrets of her own free will, that she is but the transparent representation of the ideal world. Accordingly, he demands, as a pre-

3. I venture to add that I am acquainted with the impossibility of decomposing or changing simple colored light, the two principles which form the basis of Newton's theory, not merely by hearsay, but from actual observation, having been under the necessity in one of my own researches of obtaining light of one color in a state of the greatest possible purity. (See *Poggendorff's Annalen, LXXXVI*, 501, on Sir D. Brewster's *New Analysis of Sunlight.*)

liminary to the investigation of physical phenomena, that the observed facts shall be so arranged that one explains the other; thus we may attain an insight into their connection without having to trust to anything but our senses. This demand looks most attractive but is essentially wrong in principle, for in physical science a natural phenomenon is not considered fully explained until it has been traced back to the ultimate forces which are concerned in its production and maintenance. Since we can never become cognizant of forces *qua* forces, but only of their effects, we are compelled in every explanation of natural phenomena to leave the sphere of sense, to pass to things which are not objects of sense and are defined only by abstract conceptions. When we find a stove warm and then observe that a fire is burning in it, we say, though somewhat inaccurately, that the former sensation is explained by the latter. But in reality this is equivalent to saying: we are always accustomed to find heat where fire is burning, and now a fire is burning in the stove, so we shall find heat there. That is to say, we bring our single fact under a more general, better known fact, rest satisfied with it, and call it falsely an explanation. The generality of the observation, however, does not necessarily imply an insight into causes; such an insight is only obtained when we can make out what forces are at work in the fire and how the effects depend upon them.

But this step into the region of abstract conceptions, which must necessarily be taken if we wish to penetrate to the causes of phenomena, scares the poet away. In writing a poem he has been accustomed to look, as it were, right into the subject and to reproduce his intuition without formulating any of the steps that led him to it. His success is proportionate to the vividness of the intuition. Such is the fashion in which he would have nature attacked. But the natural philosopher insists on transporting him into a world of invisible atoms and movements, of attractive and repulsive forces, whose intricate actions and reactions, though governed by strict laws, can scarcely be taken in at a glance. To him the impressions of sense are not an irrefragable authority. He examines what claim they have to be trusted. He asks whether things which they pronounce alike are really alike and whether things which they pronounce different are really different, and often finds that he must answer, no!

The result of such examination, as at present understood, is that the organs of sense do indeed give us information about external effects produced on them, but they convey those effects to our consciousness in a totally different form, so that the character of a sensuous perception depends not so much on the properties of the object perceived as on those of the organ by which we receive the information. All that the optic nerve conveys to us, it conveys under the form of a sensation of light, whether it be the rays of the sun, or a blow in the eye, or an electric current passing through it. Again, the auditory nerve translates everything into phenomena of sound; the nerves of the skin, into sensations of temperature or touch. The same electric current whose existence is indicated by the optic nerve as a flash of light, or by the organ of taste as an acid flavor, excites in the nerves of the skin the sensation of burning. The same ray of sunshine, which is called light when it falls on the eye, we call heat when it falls on the skin. But despite their different effects upon our organization, the daylight which enters through our windows and the heat radiated by an iron stove do not in reality differ more or less from each other than the red and blue constituents of light. In fact, just as in the undulatory theory the red rays are distinguished from the blue rays only by their longer period of vibration and their smaller refrangibility, so the dark heat rays of the stove have a still longer period and still smaller refrangibility than the red rays of light but are in every other respect exactly similar to them. All these rays, whether luminous or non-luminous, have heating properties, but only a certain number of them, to which for that reason we give the name of light, can penetrate through the transparent part of the eye to the optic nerve and excite a sensation of light. Perhaps the relation between our senses and the external world may be best enunciated as follows: our sensations are for us only *signs* of the objects of the external world and correspond to them only in some such way as written characters or articulate words to the things they denote. They give us, it is true, information respecting the properties of things around us, but no better information than we give a blind man about color by verbal descriptions.

We see that science has arrived at an estimate of the senses very different from that which was present to the poet's mind. And Newton's assertion that white was composed of all the colors of the spectrum was the first germ of the scientific view which has subsequently been developed. For at that time there were none of those galvanic observations which paved the way to a knowledge of the functions of the nerves in the production of sensations. Natural philosophers asserted that white, to the eye the simplest and purest of all our sensations of color, was compounded of less pure and quite complex materials. It seems to have flashed upon the poet's mind that all his principles were unsettled by the results of this assertion, and that is why the hypothesis seemed to him so unthinkable, so ineffably absurd. We must look upon his theory of color as a forlorn hope, a desperate attempt to rescue from the attacks of science the belief in the direct truth of our sensations. This will account for the enthusiasm with which he strives to elaborate and defend his theory, for the passionate irritability with which he attacks his opponent, for the overweening importance he attaches to these researches in comparison with his other achievements, and for his inaccessibility to conviction or compromise.

If we now turn to Goethe's own theories on the subject, we must, on the grounds above stated, expect to find that he cannot, without being untrue to his own principle, give us anything deserving to be called a scientific explanation of the phenomena, and that is exactly what happens. He starts with the proposition that all colors are darker than white, that they have something of shade in them (on the physical theory, white compounded of all colors must necessarily be brighter than any of its constituents). The direct mixture of dark and light, of black and white, gives grey; the colors must therefore owe their existence to some form of co-operation of light and shade. Goethe imagines he has discovered it in the phenomena presented by slightly opaque or hazy media. Such media usually look blue when the light falls on them and they are seen in front of a dark object, but yellow when a bright object is looked at through them. Thus in the daytime the air looks blue against the dark background of the sky, while the sun, when viewed, as is the case at sunset, through a thick and hazy stratum of air, appears yellow. The physical explanation of this phenomenon (which, however, is not exhibited by all such media; for instance, by plates of unpolished glass) would lead us too far from the subject. According to Goethe, the semi-opaque medium imparts to the light something corporeal, something of the nature of shade, such as is requisite, he would say, for the formation of color.

This conception alone is enough to perplex anyone who looks upon it as a physical explanation. Does he mean to say that material particles mingle with the light and fly away with it? But this is Goethe's fundamental experiment; this is the typical phenomenon under which he tries to reduce all the phenomena of color, especially those connected with the prismatic spectrum. He looks upon all transparent bodies as slightly hazy and assumes that the prism imparts to the image which it shows to an observer something of its own opacity. Here again it is hard to get a definite conception of what is meant. Goethe seems to have thought that a prism never gives perfectly defined images, but only indistinct, half-obliterated ones, for he puts them all in the same class with the double images which are exhibited by parallel plates of glass and by Iceland spar. The images formed by a prism are, it is true, indistinct in compound light, but they are perfectly defined when simple light is used. If you examine, he says, a bright surface on a dark ground through a prism, the image is displaced and blurred by the prism. The anterior edge is pushed forward over the dark background, and consequently a hazy light on a dark ground appears blue, while the other edge is covered by the image of the black surface which comes after it and consequently, being a light image behind a hazy dark color, appears yellowish red. But why the anterior edge appears in front of the ground, the posterior edge behind it, and not vice versa, he does not explain. Let us analyze this explanation and try to grasp clearly the conception of an optical image.

When I see a bright object reflected in a mirror, the reason is that the light which proceeds from it is thrown back exactly as if it came from an object of the same kind behind the mirror. The eye of the observer receives the impression accordingly, and therefore he imagines he really sees the object. Everyone knows that there is nothing real behind the mirror to correspond to the image—that no light can penetrate thither—but that what is called the image is simply a geometrical point in which the reflected rays, if produced backwards, would intersect. Accordingly, no one expects the image to produce any real effect behind the mirror. In the same way the prism shows us images of objects which occupy a different position from the objects themselves; that is to say, the light which an object sends to the prism is refracted by it, so that it appears to come from an object lying to one side, called the image. This image is not real; it is, as in the case of reflection, the geometrical point in which the refracted rays intersect when produced backwards. And yet, according to Goethe, this image is to produce real effects by its displacement; the displaced patch of light makes, he says, the dark space behind it appear blue, just as an imperfectly transparent body would, and the displaced dark patch makes the bright space behind it appear reddish yellow. That he really treats the image as an actual object in the place it appears to occupy is obvious, especially as he is compelled to assume, in the course of his explanation, that the blue and red edges of the bright space are respectively before and behind the dark image which, like it, is displaced by the prism.

Goethe remains loyal, in short, to the appearance presented to the senses and treats a geometrical locus as if it were a material object. He does not scruple at one time to make red and blue destroy each other, as in the blue edge of a red surface seen through the prism, and at another to construct out of them a beautiful purple, as when the blue and red edges of two neighboring white surfaces meet in a black ground. And when he comes to Newton's more complicated experiments, he is driven to still more marvelous expedients. As long as you treat his explanations as a pictorial way of representing the physical processes, you may acquiesce in them and even frequently find them vivid and characteristic, but as physical elucidations of the phenomena they are absolutely irrational.

In conclusion, it must be obvious to everyone that the theoretical part of Goethe's theory of color is not natural philosophy at all. At the same time we can, to a certain extent, recognize that the poet wanted to introduce a totally different method into the study of Nature, and more or less understand how he came to do so. Poetry is concerned solely with the "beautiful show" which makes it possible to contemplate the ideal; how that show is produced is a matter of indifference. Even nature is, in the poet's eyes, but the sensible expression of the spiritual. The natural philosopher, on the other hand, tries to discover the levers, the cords, and the pulleys which work behind and shift the scenes. Of course the sight of the machinery spoils the beautiful show, and therefore the poet would gladly talk it out of existence; ignoring cords and pulleys as the chimeras of a pedant's brain, he would have us believe that the scenes shift themselves or are governed by the idea of the drama.

It is precisely characteristic of Goethe that he, alone among poets, must needs break a lance with natural philosophers. Other poets are either so entirely carried away by the fire of their enthusiasm that they do not trouble themselves about the disturbing influences of the outer world, or else they rejoice in the triumphs of mind over matter, even on that unpropitious battlefield. But Goethe, whom no intensity of subjective feeling could blind to the realities around him, cannot rest satisfied until he has stamped reality itself with the image and superscription of poetry. This constitutes the peculiar beauty of his poetry. At the same time, it fully accounts for his resolute hostility to the machinery that every moment threatened to disturb his poetic repose and for his determination to attack the enemy in his own camp.

But we cannot triumph over the machinery of matter by ignoring it; we can triumph over it only by subordinating it to the aims of our moral intelligence. We must familiarize ourselves with its levers and pulleys, fatal though it be to poetic contemplation, in order to be able to govern them after our own will. Therein lies the complete justification of physical investigation and its vast importance for the advance of human civilization.

From what I have said it will be apparent that Goethe followed the same line of thought in all his contributions to science but that the problems he encountered were of diametrically opposite characters. Perhaps, when it is understood how the self-same characteristic of his intellect, which in one branch of science won for him immortal renown, entailed upon him egregious failure in the other, there will be dissipated, in the minds of many worshippers of the great poet, a lingering prejudice against natural philosophers, whom they suspect of being blinded by narrow professional pride to the loftiest inspirations of genius.

3.

THE PHYSIOLOGICAL CAUSES OF HARMONY IN MUSIC [1857]

A lecture delivered in Bonn in 1857

IN the native town of Beethoven, the mightiest among the heroes of harmony, no subject seemed to me better adapted for a popular audience than music itself. Following, therefore, the direction of my researches during the last few years, I shall endeavor to explain to you what physics and physiology have to say regarding the most cherished art of the Rhenish land, music and musical relations. Music has hitherto withdrawn itself from scientific treatment more than any other art. Poetry, painting, and sculpture borrow at least the material for their delineations from the world of experience. They portray nature and man. Not only can their material be critically investigated in respect to its correctness and truth to nature, but scientific art criticism, however much enthusiasts may have disputed its right to do so, has actually succeeded in making some progress in investigating the causes of that aesthetic pleasure which it is the intention of these arts to excite. In music, on the other hand, it seems at first sight as if those were still in the right who reject all "anatomization of pleasurable sensations." This art, borrowing no part of its material from the experience of our senses, not attempting to describe, and only exceptionally to imitate the outer world, necessarily withdraws from scientific consideration the chief points of attack which other arts present, and hence seems to be as incomprehensible and wonderful as it is certainly powerful in its effects.

We are therefore obliged, and we propose, to confine ourselves primarily to a consideration of the material of the art: musical sounds or sensations. It always struck me as a wonderful and peculiarly interesting mystery that in the theory of musical sounds, in the physical and technical foundations of music, which above all other arts seems in its action on the mind the most immaterial, evanescent, and tender creator of incalculable, indescribable states of consciousness—that here especially the science of purest and strictest thought, mathematics, should prove pre-eminently fertile. Thorough bass is a kind of applied mathematics. In considering musical intervals, divisions of time, and so forth, numerical fractions and sometimes even logarithms play a prominent part. Mathematics and music! the most glaring possible opposites of human thought! and yet connected, mutually sustained! It is as if they would demonstrate the hidden consensus of all the actions of our mind, which in the revelations of genius gives us a feeling of unconscious utterances of a mysteriously active intelligence.

When I considered physical acoustics from a physiological point of view and thus more closely followed up the part which the ear plays in the perception of musical sounds, much became clear whose connection had not been previously evident. I shall attempt to inspire you with some of the interest these questions have awakened in my own mind by endeavoring to exhibit a few of the results of physical and physiological acoustics.

The short time at my disposal obliges me to confine my attention to one particular point, but I shall select the most important of all, which will best show you the significance and results of scientific investigation in this field; I mean the foundation of harmony. It is an acknowledged fact that the numbers of the vibrations of concordant tones bear to one another ratios expressible by small whole numbers. But why? What have the ratios of small whole numbers to do with harmony? This is an old riddle, propounded by Pythagoras and hitherto unsolved. Let us see whether the means at the command of modern science will furnish the answer.

First of all, what is a musical tone? Common experience teaches us that all sounding bodies are in a state of vibration. This vibration can be seen and felt; in the case of loud sounds we feel the trembling of the air even without touching the sounding bodies. Physical science has ascertained that any series of impulses which produces a vibration of the air will, if repeated with sufficient rapidity, generate sound. This sound becomes a *musical* tone when such rapid impulses recur with perfect regularity and in precisely equal times. Irregular agitation of the air generates only noise. The *pitch* of a musical tone depends on the number of impulses which take place in a given time; the more there are in the same time, the higher or sharper is the tone. And, as we have observed, there is found to be a close relationship between the well-known harmonious musical intervals and the number of the vibrations of the air. If twice as many vibrations are performed in the same time for one tone as for another, the first is the octave above the second. If the numbers of vibrations in the same time are as 2 to 3, the two tones form a fifth; if they are as 4 to 5, the two tones form a major third.

If you observe that the numbers of the vibrations which generate the tones of the major chord *c e g c* are in the ratio of the numbers 4:5:6:8, you can deduce from these all other relations of musical tones by imagining a new major chord, having the same relations of the numbers of vibrations, to be formed upon each of the above-named tones. The numbers of vibrations within the limits of audible tones which would be obtained by executing the calculation thus indicated are extraordinarily different. Since the octave above any tone has twice as many vibrations as the tone itself, the second octave above will have four times, the third eight times as many. Our modern pianofortes have seven octaves. Their highest tones, therefore, perform 128 vibrations in the time that their lowest tone makes one single vibration.

The deepest *c* which our pianos usually possess answers to the sixteen-foot open pipe of the organ—musicians call it the contra-*C*—and makes thirty-three vibrations in one second of time. This is very nearly the limit of audibility. You will have observed that these tones have a dull, bad quality of sound on the piano and that it is difficult to determine their pitch and the accuracy of their tuning. On the organ the contra-*C* is somewhat more powerful than on the piano, but even here some uncertainty is felt in judging of its pitch. On larger organs there is a whole octave of tones below the contra-*C*, reaching to the next lower *c*, with 16½ vibrations in a second. But the ear can scarcely separate these tones from an obscure drone; and the deeper they are, the more plainly can it distinguish the separate impulses of the air to which they are due. Hence they are used solely in conjunction with the next higher octaves, to strengthen their notes and produce an impression of greater depth.

With the exception of the organ, all musical instruments, however diverse the methods in which their sounds are produced, have their limit of depth at about the same point in the scale as the piano, not because it would be impossible to produce slower impulses of the air of sufficient power, but because the *ear* refuses its office and hears slower impulses separately without gathering them up into single tones. (The oft-repeated assertion of the French physicist Savart that he heard tones of eight vibrations in a second upon a peculiarly constructed instrument seems due to an error.)

Ascending the scale from the contra-*C*, pianofortes usually have a compass of seven octaves, up to the so-called five-accented *c*, which has 4,224 vibrations in a second. Among orchestral instruments only the piccolo flute can reach as high; it will even give one tone higher. The violin usually mounts no higher than the *e* below, which has 2,640 vibrations. Of course we exclude the gymnastics of heaven-scaling virtuosi, who are ever striving to excruciate their audience by some new impossibility. Such performers may aspire to three whole octaves lying above the five-accented *c* and very painful to the ear, for the existence of such tones has been established by Despretz, who, by exciting small tuning forks with a violin bow, obtained and heard the eight-accented *c,* having 32,770 vibrations in a second. Here the sensation of tone seemed to have reached its upper limit, and the intervals were really indistinguishable in the later octaves.

The musical pitch of a tone depends entirely on the number of vibrations of the air in a second, not at all upon the mode in which they are produced. It is quite indifferent whether they are generated by the vibrating strings of a piano or violin, the vocal chords of the human larynx, the metal tongues of the harmonium, the reeds of the clarinet, oboe, or bassoon, the trembling lips of the trumpeter, or the air cut by a sharp edge in organ pipes or flutes. A tone of the same number of vibrations has always the same pitch, by whichever one of these instruments it is produced. That which distinguishes the note *a* of a piano, for example, from the equally high *a* of the violin, flute, clarinet, or trumpet, is called the *quality of the tone*, to which we shall have to return presently.

As an interesting example of these assertions, I should like to show you a peculiar physical instrument for producing musical tones, called the siren (Fig. 1), which is especially adapted to establish the properties resulting from the ratios of the numbers of vibrations.

In order to produce tones upon this instrument, the portvents g_0 and g_1 are connected by means of flexible tubes with a bellows. The air enters into round brass boxes, a_0 and a_1, and escapes by the perforated covers of these boxes at c_0 and c_1. But the holes for the escape of air are not perfectly free. Immediately in front of the covers of both boxes are two other perforated discs, fastened to a perpendicular axis *k*, which turns with great readiness. In the figure, only the perforated disc can be seen at c_0; immediately below it is the similarly perforated cover of the box. In the upper box, c_1, only the edge of the disc is visible. If the holes of the disc are precisely opposite to those of the cover, the air can escape freely; but if the disc is made to revolve, so that some of its unperforated portions stand before the holes of the box, the air cannot escape at all. When the disc is turned rapidly, the ventholes of the box are alternately opened and closed. During the opening, air escapes; during the closure, no air can pass. Hence the continuous stream of air from the bellows is converted into a series of discontinuous puffs, which, when they follow one another with sufficient rapidity, gather themselves together into a tone.

Fig. 1

Each of the revolving discs of this instrument (which is more complicated in its construction than any of the kind hitherto made and hence admits of a much greater number of combinations of tone) has four concentric circles of holes, the lower set having 8, 10, 12, 18, and the upper set 9, 12, 15, and 16 holes, respectively. The series of holes in the covers of the boxes are precisely the same as those in the discs, but under each of them lies a perforated ring, which can be so arranged, by means of the stops *i i i i*, that the corresponding holes of the cover can either communicate freely with the inside of the box or are entirely cut off from it. We are thus enabled to use any one of the eight series of holes singly or to combine them two and two, or three and three, together in any arbitrary manner.

The round boxes, h_0 h_0 and h_1 h_1, of which halves only are drawn in the figure, serve by their resonance to soften the harshness of the tone.

The holes in the boxes and discs are cut obliquely, so that when the air enters the boxes through one or more of the series of holes, the wind itself drives the discs around with a perpetually increasing velocity.

On beginning to blow the instrument, we first hear separate impulses of the air, escaping as puffs, as often as the holes of the disc pass in front of those of the box. These puffs of air follow one another more and more quickly, as the velocity of the revolving discs increases, just like the puffs of steam of a locomotive on beginning to move with the train. They next produce a whirring and whizzing, which constantly becomes more rapid. At last we hear a dull drone, which, as the velocity further increases, gradually gains in pitch and strength.

Suppose that the discs have been brought to a velocity of 33 revolutions in a second, and that the series with 8 holes has been opened. At each revolution of the disc all these 8 holes will pass before each separate hole of the cover. Hence there will be 8 puffs for each revolution of the disc, or 8 times 33, that is, 264 puffs in a second. This gives us the once-accented c' of our musical scale.[1] But on opening the series of 16 holes instead, we have twice as many, or 16 times 33, that is, 528 vibrations in a second. We hear exactly the octave above the first c', that is, the twice-accented c''.[2] By opening both the series of 8 and 16 holes at once, we have both c' and c'' at once and can convince ourselves that we have the absolutely pure harmony of the octave. By taking 8 and 12 holes, which give numbers of vibrations in the

1. [That is, "middle C," written on the ledger line between the bass and treble staves.—A. J. Ellis, translator.]
2. [Or *c* on the third space of the treble staff.—A. J. E.]

ratio of 2 to 3, we have the harmony of a perfect fifth. Similarly 12 and 16, or 9 and 12, give fourths; 12 and 15 give a major third, and so on.

The upper box is furnished with a contrivance for slightly sharping or flatting the tones which it produces. This box is movable upon an axis and connected with a toothed wheel, which is worked by the driver attached to the handle *d*. By turning the handle slowly while one of the series of holes in the upper box is in use, we can make the tone sharper or flatter, according as the box moves in the opposite direction to the disc or in the same direction as the disc. When the motion is in the opposite direction, the holes meet those of the disc a little sooner than they otherwise would, so the time of vibration of the tone is shortened, and the tone becomes sharper. The contrary ensues in the other case.

Now, on blowing through 8 holes below and 16 above, we have a perfect octave, as long as the upper box is still; but when it is in motion, the pitch of the upper tone is slightly altered, and the octave becomes false. On blowing through 12 holes above and 18 below, we have a perfect fifth as long as the upper box is at rest, but if it moves, the harmony is perceptibly marred.

These experiments with the siren show us, therefore:

1. That a series of puffs following one another with sufficient rapidity produce a musical tone.

2. That the more rapidly they follow one another, the higher is the tone.

3. That when the ratio of the number of vibrations is exactly 1 to 2, the result is a perfect octave; when it is 2 to 3, a perfect fifth; when it is 3 to 4, a pure fourth, and so on. The slightest alteration in these ratios destroys the purity of the harmony.

You will perceive from these data that the human ear is affected by vibrations of the air within certain degrees of rapidity—from about 20 to about 32,000 in a second—and that the sensation of musical tone arises from this effect. That the sensation thus excited is a sensation of musical tone does not depend in any way upon the peculiar manner in which the air is agitated, but solely on the peculiar powers of sensation possessed by our ears and auditory nerves. I remarked a little while ago that when the tones are loud, the agitation of the air is perceptible to the skin. In this way deaf mutes can perceive the motion of the air which we call sound. But they do not hear, that is, they have no sensation of tone in the ear. They feel the motion by the nerves of the skin, producing that peculiar description of sensation called whirring.

The limits of the rapidity of vibration within which the ear feels an agitation of the air to be sound depend also wholly upon the peculiar constitution of the ear. When the siren is turned slowly and the puffs of air succeed each other slowly, you hear no musical sound. By the continually increasing rapidity of its revolution, no essential change is produced in the kind of vibration of the air. Nothing new happens externally to the ear. The only new result is the sensation experienced by the ear, which then for the first time begins to be affected by the agitation of the air. Hence the more rapid vibrations receive a new name and are called *sound.* If you admire paradoxes, you may say that aerial vibrations do not become sound until they fall upon a hearing ear.

I must now describe the propagation of sound through the atmosphere. The motion of a mass of air through which a tone passes belongs to the so-called wave motions—a class of motions of great importance in physics. Light, as well as sound, is one of these motions.

The name is derived from the analogy of waves on the surface of water, and these will best illustrate the peculiarity of this description of motion. When a point in a surface of still water is agitated—as by throwing in a stone—the motion thus caused is propagated in the form of waves, which spread in rings over the surface of the water. The circles of waves continue to increase even after rest has been restored at the point first affected. At the same time, the waves become continually lower, the farther they are removed from the center of motion, and gradually disappear. On each wave ring we distinguish ridges, or crests, and hollows, or troughs. Crest and trough together form a wave, and we measure its length from one crest to the next.

While the wave passes over the surface of the fluid, the particles of the water which form it do not move on with it. This is easily seen by floating a chip of straw on the water. When the waves reach the chip, they raise or depress it, but when they have passed over it, the position of the chip is not perceptibly changed. Now a light, floating chip has no motion different from that of the adjacent particles of water. Hence we conclude that these particles do not follow the wave but, after some pitching up and down, remain in their original positions. That which really advances as a wave is, consequently, not the particles of water themselves but only a superficial form, which continues to be built up by fresh particles of water. The paths of the separate particles of water are more nearly vertical circles, in which they revolve with a tolerably uniform velocity as long as the waves pass over them.

In Fig. 2 the dark wave line *A B C* represents a section of the surface of the water over which waves are running in the direction of the arrows above *a* and *c*. The three circles *a, b,* and *c* represent the paths of particular particles of water at the surface of the wave. The particle which revolves in the circle *b* is supposed, at the time that the surface of the water presents the form *A B C*, to be at its highest point *B* and the particles revolving in the circles *a* and *c* to be simultaneously in their lowest positions.

The respective particles of water revolve in these circles in the direction marked by the arrows. The dotted curves represent other positions of the passing waves, at equal intervals of time, partly before the assumption of the *A B C* position (as for the crests between *a* and *b*), and partly after the same (for the crests between *b* and *c*). The positions of the crests are marked with figures. The same figures in the three circles show where the respective revolving particles would be at the moment the wave assumed the corresponding form. It will be noticed that the particles advance by equal arcs of the circles as the crest of the wave advances by equal distances parallel to the water level.

In the circle *b* it will be seen further that the particle of water in its positions 1, 2, 3 hastens to meet the approaching wave-crests 1, 2, 3. It rises on its left-hand side and is carried on by the crest from 4 to 7 in the direction of its advance. Afterward, the particle falls behind the crest, sinks down again on the right-hand side, and finally reaches its original position at 13. (In the lecture itself, Fig. 2 was replaced by a working model in which the movable particles, connected by threads, really revolved in circles, while connecting elastic threads represented the surface of the water.)

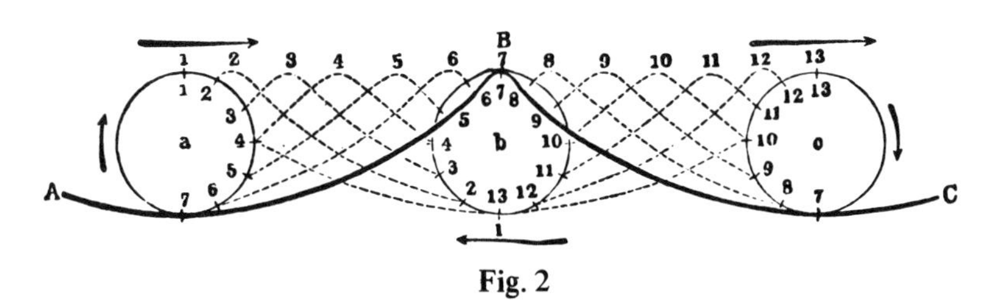

Fig. 2

All particles at the surface of the water, as you see by this drawing, describe equal circles. The particles of water at different depths move in the same way, but as the depths increase, the diameters of their circles of revolution rapidly diminish.

In this way, then, arises the appearance of a progressive motion along the surface of the water, while in reality the moving particles of water do not advance with the wave but perpetually revolve in their small circular orbits.

To return from waves of water to waves of sound. Imagine that an elastic fluid like air replaces the water and that the waves of this replaced water are compressed by an inflexible plate laid on their surface, the fluid being prevented from escaping laterally from the pressure. On the waves being thus flattened out, the ridges where the fluid had been heaped up will produce much greater density than the hollows, from which the fluid had been removed to form the ridges. Hence the ridges are replaced by condensed strata of air and the hollows by rarefied strata. Now further imagine that these compressed waves are propagated by the same law as before and also that the vertical circular orbits of the several particles of water are compressed into horizontal straight lines. Then the waves of sound will retain the peculiarity of having the particles of air oscillating backwards and forwards in a straight line, while the wave itself remains merely a progressive form of motion, continually composed of fresh particles of air. The immediate result then would be waves of sound spreading out horizontally from their origin.

But the expansion of waves of sound is not limited, like those of water, to a horizontal surface. They can spread out in any direction whatsoever. Suppose the circles generated by a stone thrown into the water to extend in all directions of space, and you will have the spherical waves of air by which sound is propagated. Hence we can continue to illustrate the peculiarities of the motion of sound by the well-known visible motions of waves of water.

The lengths of various waves of water, measured from crest to crest, are extremely different. A falling drop or a breath of air curls the surface of the water gently. The waves in the wake of a steamboat toss the swimmer or skiff severely. The waves of a stormy ocean can find room in their hollows for the keel of a ship of the line, and their ridges can scarcely be looked down upon from a masthead. The waves of sound present similar differences. The little curls of water with short lengths of wave correspond to high tones, the giant ocean billows to deep tones. Thus the contrabass *C* has a wave thirty-five feet long and its higher octave a wave of half the length, while the highest tones of a piano have waves only three inches in length.[3]

You perceive that the pitch of the tone corresponds to the length of the wave. To this we should add that the height of the ridges, or (transferred to air) the degree of alternate condensation and rarefaction, corresponds to the

3. [The exact lengths of waves corresponding to certain notes, or symbols of tone, depend upon the standard pitch assigned to one particular note, and this differs in different countries. Hence the figures of the author have been left unreduced. They are sufficiently near to those usually adopted in England to occasion no difficulty to the reader in these general remarks.—A. J. E.]

loudness and intensity of the tone. But waves of the same height may have different forms; the crest of the ridge, for example, may be rounded or pointed. Corresponding varieties occur in waves of sound of the same pitch and loudness, and the so-called *timbre* or quality of tone corresponds to the *form* of the waves of water.

The conception of form is transferred from waves of water to waves of sound. Supposing waves of water of different forms to be pressed flat as before, the surface, having been leveled, will of course display no differences of

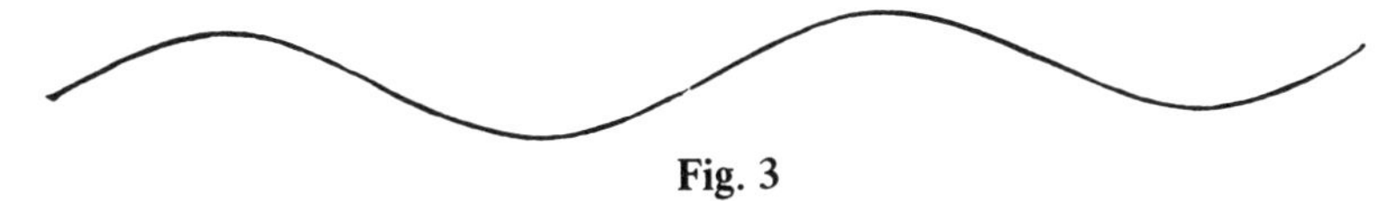

Fig. 3

form; but in the interior of the mass of water we shall have different distributions of pressure, and hence of density, which exactly correspond to the differences of form in the uncompressed surface. In this sense we can continue to speak of the form of waves of sound and can represent it geometrically. We make the curve rise where the pressure, and hence density, increases and fall where it diminishes—just as if we had a compressed fluid beneath the curve, which would expand to the height of the curve in order to regain its natural density.

Unfortunately, the form of waves of sound, on which depends the quality of the tones produced by various sounding bodies, can at present be assigned in only a very few cases. Among those which we are able to determine with some exactness is one of great importance, here termed the *simple* or *pure* wave form and represented in Fig. 3. It can be seen in waves of water only when their height is small in comparison with their length and when they run over a smooth surface without external disturbance and without any action of wind. Ridge and hollow are gently rounded, equally broad and symmetrical, so that, if we inverted the curve, the ridges would exactly fit into the hollows, and conversely. This form of wave would be more precisely defined by saying that the particles of water describe exactly circular orbits of small diameters with exactly uniform velocities. To this simple wave form corresponds a peculiar species of tone which, for reasons to be hereafter assigned, depending upon its relation to quality, we shall term a *simple* tone. Such a tone is produced by striking a tuning fork and holding it before the opening of a properly tuned resonance tube. The tone of tuneful human voices, singing the vowel *oo* in *too*, in the middle positions of their register, appears not to differ materially from this form of wave.

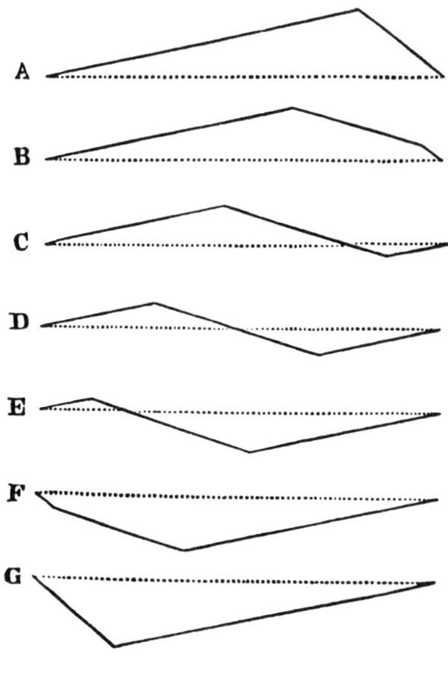

Fig. 4

We also know the laws of the motion of strings with sufficient accuracy to assign in some cases the form of motion which they impart to the air. Thus Fig. 4 represents the forms successively assumed by a string struck, as in the German zither, by a pointed stylus.[4] *A* represents the form assumed by the string at the moment of percussion. Then, at equal intervals of time, follow the forms *B, C, D, E, F, G*; and then, in inverse order, *F, E, D, C, B, A*, and so on in perpetual repetition. The form of motion which such a string, by means of an attached sounding board, imparts to the surrounding air probably corresponds to the broken line in Fig. 5, where *h h* indicates the position of equilibrium, and the letters *a b c d e f g* show the line of the wave which is produced by the action of several forms of string marked by the corresponding capital letters in Fig. 4. It is easily seen how greatly this form of wave (which

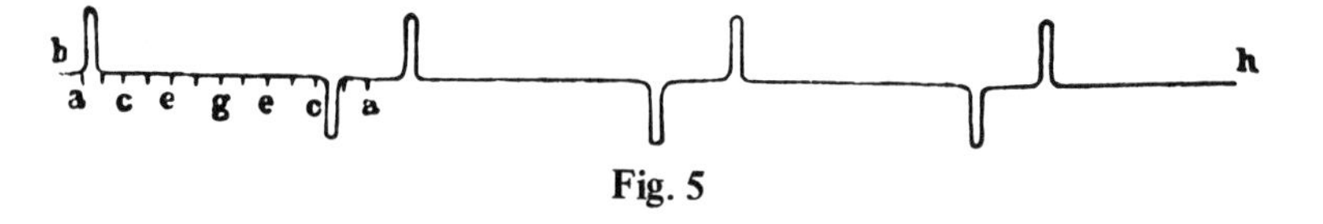

Fig. 5

4. [The plectrum of the ancient *lyra*, or the quill of the old harpsichord, which may be easily imitated on a guitar.—A. J. E.]

of course could not occur in water) differs from that of Fig. 3 (independently of magnitude), as the string only imparts to the air a series of short impulses, alternately directed to opposite sides.[5]

The waves of air produced by the tone of a violin would, on the same

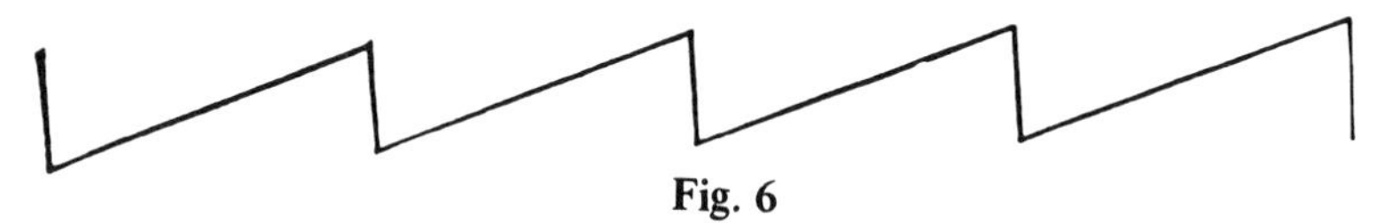

Fig. 6

principle, be represented by Fig. 6. During each period of vibration the pressure increases uniformly; at the end it falls back suddenly to its minimum.

It is to such differences in the forms of the waves of sound that the variety of quality in musical tones is due. We may carry the analogy even further. The more uniformly rounded the form of wave, the softer and milder is the quality of tone. The more jerking and angular the wave form, the more piercing the quality. Tuning forks, with their rounded forms of wave (Fig. 3), have an extraordinarily soft quality, and the qualities of tone generated by the zither and violin resemble in harshness the angularity of their wave forms (Figs. 5 and 6).

Finally, I would direct your attention to an instructive spectacle which I have never been able to view without a certain degree of physico-scientific delight because it displays to the bodily eye, on the surface of water, what otherwise could only be recognized by the mind's eye of the mathematical thinker in a mass of air traversed in all directions by waves of sound. I allude to the composition of many different systems of waves, as they pass over one another, each undisturbedly pursuing its own path. We can watch it from the parapet of any bridge spanning a river, but it is most complete and sublime when viewed from a cliff beside the sea. It is then rare not to see innumerable systems of waves of various length propagated in various directions. The longest come from the deep sea and dash against the shore. Where the boiling breakers burst, shorter waves arise and run back again towards the sea. Perhaps a bird of prey darting after a fish gives rise to a system of circular waves, which, rocking over the undulating surface, are propagated with the same regularity as on the mirror of an inland lake. And thus from the distant horizon, where white lines of foam on the steel blue surface betray the coming trains of wave, down to the sand beneath our feet, where the impression of

5. It is here assumed that the sounding board and the air in contact with it immediately obey the impulse given by the end of the string without exercising a perceptible reaction on the motion of the string.

their arcs remains, there is unfolded before our eyes a sublime image of immeasurable power and unceasing variety, which, as the eye at once recognizes its pervading order and law, enchains and exalts without confusing the mind.

Just in the same way you must conceive the air of a concert hall or ballroom traversed in every direction, not merely on the surface, by a variegated crowd of intersecting wave systems. From the mouths of the male singers proceed waves of six to twelve feet in length; from the lips of the female singers dart shorter waves, from eighteen to thirty-six inches long. The rustling of silken skirts excites little curls in the air, each instrument in the orchestra emits its peculiar waves, and all these systems expand spherically from their respective centers, dart through one another, are reflected from the walls of the room and thus rush backwards and forwards, until they succumb to the greater force of newly generated tones.

Although this spectacle is veiled from the material eye, we have another bodily organ, the ear, specially adapted to reveal it to us. This analyzes the interdigitation of the waves (which in such cases would be far more confused than the intersection of the water undulations), separates the several tones which compose it, and distinguishes the voices of men and women–even of individuals–the peculiar qualities of tone given out by each instrument, the rustling of the dresses, the footfalls of the walkers, and so on.

It is necessary to examine the circumstances with greater minuteness. When a bird of prey dips into the sea, rings of waves arise, which are propagated as slowly and regularly upon the moving surface as upon a surface at rest. These rings are cut into the curved surface of the waves in precisely the same way they would have been cut into the still surface of a lake. The form of the external surface of the water is determined in this, as in other more complicated cases, by taking the height of each point to be the height of all the ridges of the waves which coincide at this point at one time, after deducting the sum of all similarly simultaneously coincident hollows. Such a sum of positive magnitudes (the ridges) and negative magnitudes (the hollows), where the latter have to be subtracted instead of being added, is called an algebraic sum. Using this term, we may say that *the height of every point of the surface of the water is equal to the algebraic sum of all the portions of the waves which at that moment there concur.*

It is the same with the waves of sound. They too are added together at every point of the mass of air, as well as in contact with the listener's ear. For them also the degree of condensation and the velocity of the particles of air in the passages of the organ of hearing are equal to the algebraic sums of the separate degrees of condensation and of the velocities of the waves of sound, considered apart. This single motion of the air produced by the simultaneous action of various sounding bodies has now to be analyzed by the ear into the separate parts which correspond to their separate effects. For doing this the ear is much more unfavorably situated than the eye. The latter surveys the whole undulating surface at a glance; the ear can, of course, only perceive the motion of the particles of air which impinge upon it. Yet the ear solves its problem with the greatest accuracy, certainty, and specificity. This power of the ear is of supreme importance for hearing. Were it not present, it would be impossible to distinguish different tones.

Some recent anatomical discoveries appear to give a clue to the explanation of this important power of the ear.

You will all have observed the phenomenon of the sympathetic production of tones in musical instruments, especially stringed instruments. The string of a pianoforte, when the damper is raised, begins to vibrate as soon as its proper tone is produced in its neighborhood with sufficient force by some other means. When this foreign tone ceases, the tone of the string will be heard to continue some little time longer. If we put little paper riders on the string, they will be jerked off when its tone is produced in the neighborhood. This sympathetic action of the string depends on the impact of the vibrating particles of air against the string and its sounding board.

Each separate wave crest (or condensation) of air which passes by the string is, of course, too weak to produce a sensible motion in it. But when a long series of wave crests (or condensations) strike the string in such a manner that each succeeding one increases the slight tremor which resulted from the action of its predecessors, the effect finally becomes sensible. It is a process of exactly the same nature as the swinging of a heavy bell. A powerful man can scarcely move it sensibly by a single impulse. A boy, by pulling the rope at regular intervals corresponding to the time of its oscillations, can gradually bring it into violent motion.

This peculiar reinforcement of vibration depends entirely on the rhythmical application of the impulse. When the bell has been once made to vibrate as a pendulum in a very small arc, if the boy always pulls the rope as it falls and at a time when his pull augments the existing velocity of the bell, this velocity, increasing slightly at each pull, will gradually become considerable. But if the boy applies his power at irregular intervals, sometimes increasing and sometimes diminishing the motion of the bell, he will produce no sensible effect.

In the same way that a mere boy is thus enabled to swing a heavy bell, the tremors of light and mobile air suffice to set in motion the heavy, solid mass of steel contained in a tuning fork, provided that the tone which is excited in the air is exactly in unison with that of the fork. In this case also, every impact of a wave of air against the fork increases the motions excited by the like previous blows.

This experiment is most conveniently performed on a fork which is fastened to a sounding board (see Fig. 7), the air being excited by a similar fork of precisely the same pitch. If one is struck, the other will be found after a few seconds to be sounding also. Damp the first fork, by touching it for a moment with a finger, and the second will continue the tone. The second will then bring the first into vibration, and so on. But if a very small piece of wax be attached to the end of one of the forks, rendering its pitch scarcely perceptibly lower than the other, the sympathetic vibration of the second fork ceases, because the times of oscillation are no longer the same. The blows which the waves of air excited by the first fork inflict upon the sounding board of the second fork are indeed, for a time, in the same direction as the motions of the second fork and consequently increase the latter, but after a very short time they cease to be so and consequently destroy the slight motion which they had excited.

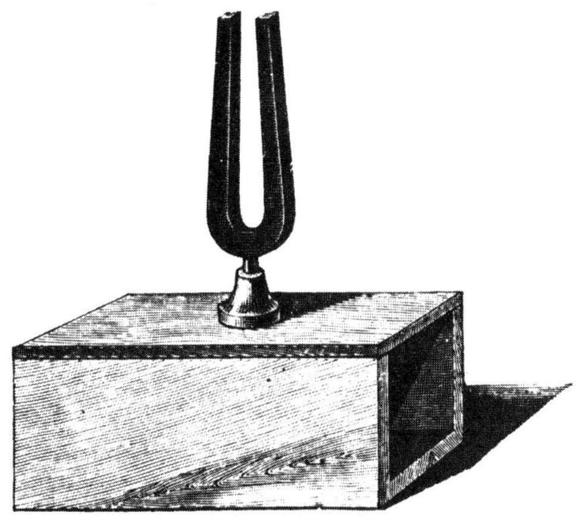

Fig. 7

Lighter and more mobile elastic bodies, such as strings, can be set in motion by a much smaller number of aerial impulses. Hence they can be set in sympathetic motion much more easily than tuning forks and by means of musical tones which are far less accurately in unison with themselves. If several tones are sounded in the neighborhood of a pianoforte, no string can be set in sympathetic vibration unless it is in unison with one of those tones. To test this, depress the forte pedal (thus raising the dampers), and put paper riders on all the strings. They will of course leap off when their strings are put in vibration. Then let several voices or instruments sound tones simultaneously in the neighborhood. All those riders, and only those, will leap off which are placed upon strings that correspond to tones of the same pitch as those sounded. You perceive that a pianoforte is also capable of analyzing the wave confusion of the air into its elementary constituents.

The process which actually goes on in our ear is probably very like that just described. Deep in the petrous bone out of which the internal ear is hollowed lies a peculiar organ, the cochlea or snail shell–a cavity filled with water and so called from its resemblance to the shell of a common garden snail. This spiral passage is divided throughout its length into three sections, upper, middle, and lower, by two membranes stretched across it. The Marchese Corti discovered some very remarkable formations in the middle section, consisting of innumerable plates, microscopically small and arranged side by side in an orderly manner, like the keys of a piano. They are connected at one end with the fibers of the auditory nerve and at the other with the stretched membrane.

Fig. 8 shows this extraordinarily complicated arrangement for a small part of the partition at the cochlea. The arches which leave the membrane at *d* and are reinserted at *e*, reaching their greatest height between *m* and *o*, are probably the parts which are suited for vibration. They are spun round with innumerable fibrils, among which some nerve fibers can be recognized, coming to them through the holes near *c*. The transverse fibers *g, h, i, k,* and the cells *o*, also appear to belong to the nervous system. There are about three thousand arches similar to *d e* lying orderly beside each other, like the keys of a piano, in the whole length of the partition of the cochlea.

In the so-called vestibule also, where the nerves expand upon little membranous bags swimming in water, elastic appendages, similar to stiff hairs, have been lately discovered at the ends of the nerves. The anatomical arrangement of these appendages leaves scarcely any room to doubt that they are set into sympathetic vibration by the waves of sound which are conducted through the ear. Now if we venture to conjecture–it is at present only a conjecture, but after careful consideration I am led to think it very probable–that every such appendage is tuned to a certain tone like the strings of a piano,

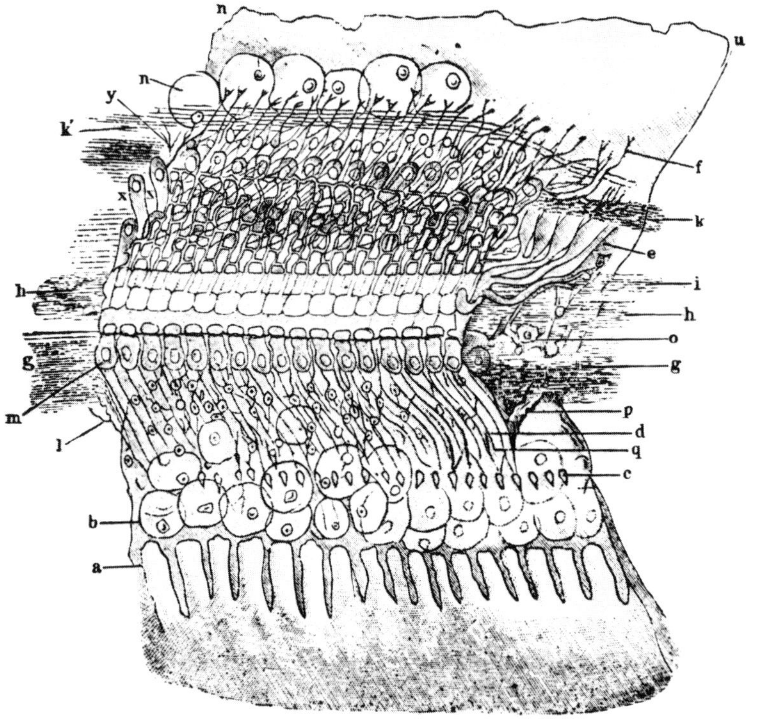

Fig. 8

then the recent experiment with a piano shows you that when (and only when) that tone is sounded, the corresponding hairlike appendage may vibrate and the corresponding nerve fiber experience a sensation, so that the presence of each single such tone in the midst of a whole confusion of tones must be indicated by the corresponding sensation.

Experience shows us that the ear really possesses the power of analyzing waves of air into their elementary forms. Up to this point we have been considering those compound motions of the air which have been caused by the simultaneous vibration of several elastic bodies. Since the forms of the waves of sound of different musical instruments are different, there is room to suppose that the kind of vibration excited in the passages of the ear by one such tone will be exactly the same as the kind of vibration which in another case is there excited by two or more instruments sounded together. If the ear analyzes the motion into its elements in the latter case, it cannot well avoid doing so in the former, where the tone is due to a single source. And this is found to be really the case.

I have previously mentioned the form of wave with gently rounded crests and hollows, and termed it simple or pure. In reference to this form the French mathematician Fourier has established a celebrated and important theorem which may be translated from mathematical into ordinary language thus: *Any form of wave whatever can be compounded of a number of simple waves of different lengths.* The longest of these simple waves has the same length as that of the given form of wave, the others have lengths one-half, one-third, one-fourth, etc., as great.

By the different modes of uniting the crests and hollows of these simple waves, an endless multiplicity of wave forms may be produced.

For example, the wave curves *A* and *B* (Fig. 9) represent waves of simple tones, *B* making twice as many vibrations as *A* in a second of time and being consequently an octave higher in pitch. *C* and *D*, on the other hand, represent the waves which result from the superposition of *B* on *A*. The dotted curves in the first halves of *C* and *D* are repetitions of so much of the figure *A*. In *C*, the initial point *e* of the curve *B* coincides with the initial point d_0 of *A*. But in *D*, the deepest point b_2 of the first hollow in *B* is placed under the initial point of *A*. The result is two different compound curves, the first, *C*, having

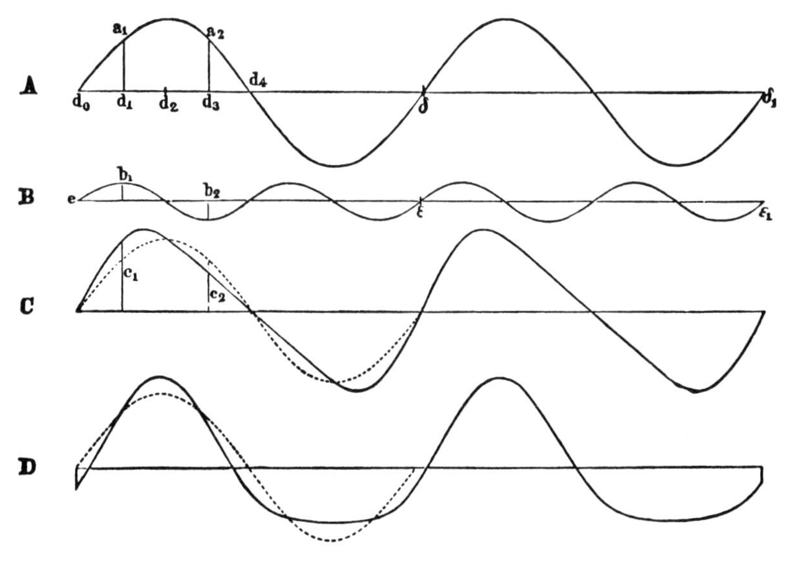

Fig. 9

steeply ascending and more gently descending crests, but so related that if the figure were reversed, the elevations would exactly fit into the depressions. But in *D* we have pointed crests and flattened hollows, which are, however, symmetrical with respect to right and left.

Other forms are shown in Fig. 10, which are also compounded of two simple waves, *A* and *B*, of which *B* makes three times as many vibrations in a second as *A* and consequently is the twelfth higher in pitch. The dotted curves in *C* and *D* are, as before, repetitions of *A*. *C* has flat crests and flat hollows, *D* has pointed crests and pointed hollows.

These extremely simple examples will suffice to give a conception of the great multiplicity of forms resulting from this method of composition. Supposing that, instead of two, several simple waves were selected, with heights and initial points arbitrarily chosen, an endless variety of changes could be effected. In point of fact, any given form of wave could be reproduced.[6]

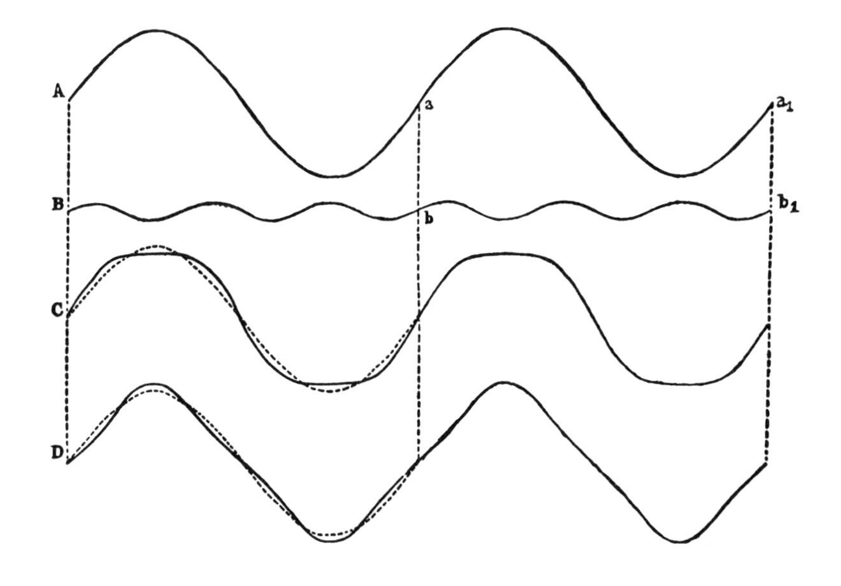

Fig. 10

6. Of course the waves could not overhang, but waves of such a form would have no possible analogue in waves of sound [which, the reader will recollect, are not actually in the forms here drawn but have only condensations and rarefactions, conveniently replaced by these forms.—A. J. E.].

When various simple waves concur on the surface of water, the compound wave form has only a momentary existence. The longer waves move faster than the shorter, and consequently the two kinds of wave immediately separate, giving the eye an opportunity to recognize the presence of several systems of waves. But when waves of sound are similarly compounded, they never separate, because long and short waves traverse air with the same velocity. The compound wave is permanent and continues its course unchanged, so that when it strikes the ear there is nothing to indicate whether it originally left a musical instrument in this form or was compounded on the way out of two or more undulations.

What does the ear do? Does it analyze this compound wave, or does it grasp it as a whole? The answer to these questions depends upon the sense in which we take them. We must distinguish two different points—the audible *sensation*, as it is developed without any intellectual interference, and the idea, which we form in consequence of that sensation. We have, as it were, to distinguish between the material ear of the body and the spiritual ear of the mind. The material ear does precisely what the mathematician effects by means of Fourier's theorem and what the pianoforte accomplishes when a confused mass of tones is presented to it. It analyzes those wave forms which were not originally due to simple undulations, such as those furnished by tuning forks, into a sum of simple tones and feels the tone due to each separate simple wave separately, whether the compound wave originally proceeded from a source capable of generating it or became compounded on the way.

For example, when a string is struck, it gives a tone corresponding, as we have seen, to a wave form widely different from that of a simple tone. When the ear analyzes this wave form into a sum of simple waves, it hears at the same time a series of simple tones corresponding to these waves.

Strings are peculiarly favorable for such an investigation because they are themselves capable of assuming extremely different forms in the course of their vibration and these forms may also be considered, like those of aerial undulations, as compounded of simple waves. Fig. 4 showed the consecutive forms of a string struck by a simple rod. Fig. 11 gives a number of other forms of vibration of a string, corresponding to simple tones. The continuous line shows the extreme displacement of the string in one direction, and the dotted line in the other. At *a* the string produces its fundamental tone, the deepest simple tone it can produce, vibrating in its whole length, first on one side and then on the other. At *b* it falls into two vibrating sections, separated

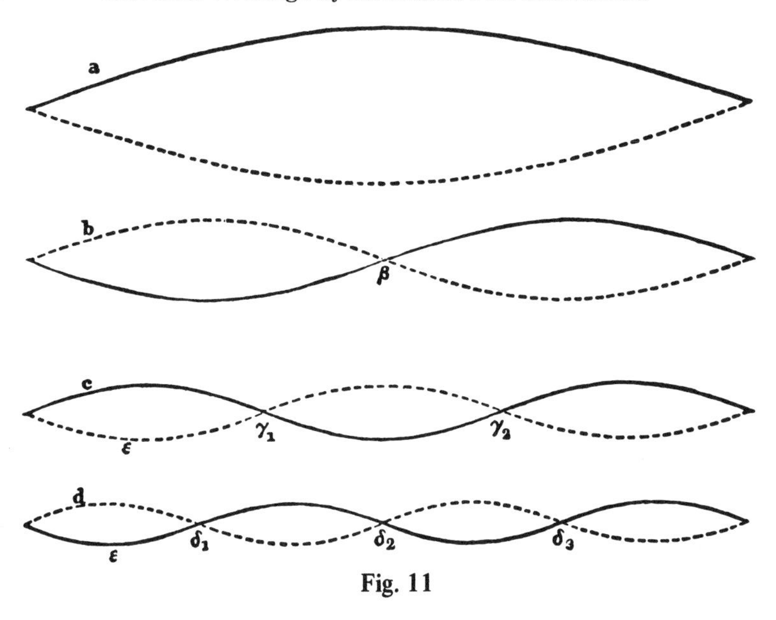

Fig. 11

by a single stationary point β, called a *node* (knot). The tone is an octave higher, the same as each of the two sections would separately produce, and it performs twice as many vibrations in a second as the fundamental tone. At *c* we have two nodes, γ_1 and γ_2, and three vibrating sections, each vibrating three times as fast as the fundamental tone and hence giving its twelfth. At *d* there are three nodes, δ_1, δ_2, δ_3, and four vibrating sections, each vibrating four times as quickly as the fundamental tone and giving the second octave above it. In the same way, forms of vibration may occur with 5, 6, 7, etc., vibrating sections, each performing respectively 5, 6, 7, etc., times as many vibrations in a second as the fundamental tone. All other vibrational forms of the string may be conceived as compounded of a sum of such simple vibrational forms.

The vibrational forms with stationary points or nodes may be produced by gently touching the string at one of these points either with the finger or a rod and then rubbing the string with a violin bow, plucking it with the finger, or striking it with a pianoforte hammer. The bell-like harmonics or flageolet tones of strings, so much used in violin playing, are thus produced.

Now suppose that a string has been excited and, after its tone has been allowed to continue for a moment, it is touched gently at its middle point β (Fig. 11*b*) or δ_2 (Fig. 11*d*). The vibrational forms *a* and *c*, for which this point is in motion, will be immediately checked and destroyed; but the vibrational forms *b* and *d*, for which this point is at rest, will not be disturbed, and the tones due to them will continue to be heard. In this way we can readily discover whether certain members of the series of simple tones are contained in the compound tone of a string when excited in any given way, and the ear can be rendered sensible of their existence. Once these simple tones in the sound of a string have been thus rendered audible, the ear will be able to observe them readily in the untouched string after a little accurate attention.

The series of tones which are thus made to combine with a given fundamental tone is perfectly determinate. They are tones which perform twice, thrice, four times, etc., as many vibrations in a second as the fundamental tone. They are called the upper partials, or harmonic overtones, of the fundamental tone. If this last be *c*, the series may be written as follows in musical notation:[7]

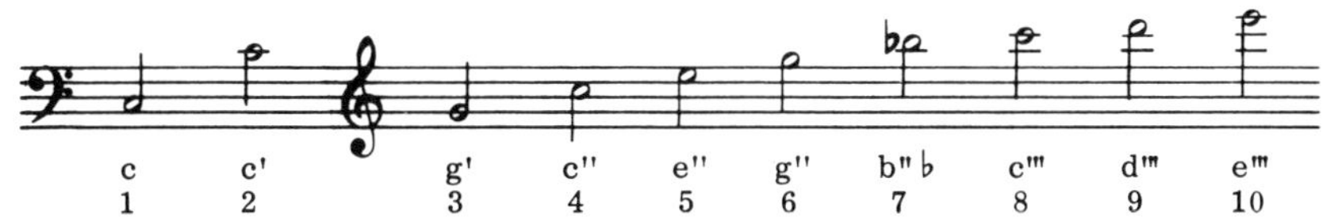

Not only strings, but almost all kinds of musical instruments, produce waves of sound which are more or less different from those of simple tones and are therefore capable of being compounded out of a greater or less number of simple waves. The ear analyzes them all by means of Fourier's theorem better than the best mathematician, and on paying sufficient attention can distinguish the separate simple tones due to the corresponding simple waves. This corresponds precisely to our theory of the sympathetic vibration of the organs described by Corti. Experiments with the piano, as well as the mathematical theory of sympathetic vibrations, show that any upper partials which may be present will also produce sympathetic vibrations. It follows, therefore, that in the cochlea of the ear every external tone will set in sympathetic vibration, not merely the little plates with their accompanying nerve fibers, corresponding to its fundamental tone, but also those corresponding to all the upper partials, and consequently that the latter must be heard as well as the former.

A simple tone is one excited by a succession of simple wave forms. All

7. [It being understood that, on account of the temperament of a piano, these are not precisely the fundamental tones of the corresponding strings on that instrument; in particular the upper partial, *b″* ♭, is necessarily much flatter than the fundamental tone of the corresponding note on the piano.—A. J. E.]

other wave forms, such as those produced by the greater number of musical instruments, excite sensations of a variety of simple tones. Consequently, every tone of every musical instrument must be strictly regarded, so far as the sensation of musical tone is concerned, as a chord with a predominant fundamental tone.

This whole theory of upper partials, or harmonic overtones, will perhaps seem new and singular. Probably few or none of those present, however frequently they may have heard or performed music and however fine may be their musical ear, have perceived the existence of any such tones, although, according to my representations, they must be always and continuously present. A special act of attention is requisite in order to hear them, and unless we know how to perform this act, the tones remain concealed. As you are aware, no perceptions obtained by the senses are merely sensations impressed on our nervous systems. A distinct intellectual activity is required to pass from a nervous sensation to the conception of an external object, which the sensation has aroused. The sensations of our nerves of sense are mere signs indicating certain external objects, and it is usually only after considerable practice that we acquire the power of drawing correct conclusions from our sensations respecting the corresponding objects.

It is a universal law of the perceptions obtained through the senses that we pay only so much attention to the sensations actually experienced as is sufficient for us to recognize external objects. In this respect we are very one-sided, inconsiderate partisans of practical utility—far more so, indeed, than we suspect. All sensations which have no direct reference to external objects we are accustomed, as a matter of course, entirely to ignore. We do not become aware of them until we make a scientific investigation of the action of the senses or have our attention directed by illness to the phenomena of our own bodies. Thus we often find patients, when suffering from a slight inflammation of the eyes, become aware for the first time of those beads and fibers known as muscae volitantes (*mouches volantes*) swimming about within the vitreous humor of the eye; then they often hypochondriacally imagine all sorts of coming evils because they fancy that these appearances are new, whereas they have generally existed all their lives.

Who can easily discover that there is an absolutely blind point, the so-called punctum cecum, within the retina of every healthy eye? How many people know that the only objects they see single are those at which they are looking and that all other objects behind or before these appear double? I could adduce a long list of similar examples, which have not been brought to light until the actions of the senses were scientifically investigated and which remain obstinately concealed until attention is drawn to them by appropriate means, often an extremely difficult task to accomplish.

To this class of phenomena belong the upper partial tones. It is not enough for the auditory nerve to have a sensation. The intellect must reflect upon it. Hence my former distinction of a material and a spiritual ear.

We always hear the tone of a string accompanied by a certain combination of upper partial tones. A different combination of such tones belongs to the tone of a flute, or of the human voice, or of a dog's howl. Whether a violin or a flute, a man or a dog, is close by us is a matter of interest for us to know, and our ear takes care to distinguish the peculiarities of their tones with accuracy. The *means* by which we can distinguish them, however, is a matter of perfect indifference. Whether the cry of the dog contains the higher octave or the twelfth of the fundamental tone has no practical interest for us and never occupies our attention. The upper partials are consequently thrown into that unanalyzed mass of peculiarities of a tone which we call its *quality*. As the existence of upper partial tones depends on the wave form, we see, as I stated previously, that the quality of tone corresponds to the form of wave.

The upper partial tones are most easily heard when they are not in harmony with the fundamental tone, as in the case of bells. The art of the bell-founder consists precisely in giving bells such a form that the deeper and stronger partial tones shall be in harmony with the fundamental tone, as otherwise the bell would be unmusical, tinkling like a kettle. But the higher partials are always out of harmony, and hence bells are unfitted for artistic music. On the other hand, it follows, from what has been said, that the upper partial tones are all the more difficult to hear, the more accustomed we are to the compound tones of which they form a part. This is especially the case with the human voice, and many skillful observers have consequently failed to discover them there.

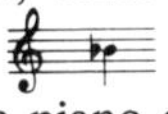

The preceding theory was wonderfully corroborated by a method which enabled not only myself but other persons to hear the upper partial tones of the human voice. No particularly fine musical ear is required for this purpose, as was formerly supposed, but only proper means for directing the observer's attention. Let a powerful male voice sing the note e♭  to the vowel O in *ore*, close to a good piano. Then lightly touch on the piano the note *b′* in the next octave above, and listen attentively to the sound of the piano as it dies away. If this *b′*♭ is a real upper partial in the compound tone uttered by the singer, the sound of the piano will apparently not die away at all, but the corresponding upper partial of the voice will be heard as if the note of the piano continued.[8] By properly varying the experiment, it will be found possible to distinguish the vowels from one another by their upper partial tones.

The investigation is rendered much easier by aiding the ear with small globes of glass or metal, as in Fig. 12. The larger opening *a* is directed to the source of sound, and the smaller, funnel-shaped end is applied to the drum of the ear. The enclosed mass of air, which is almost entirely separated from that without, has its own proper tone or keynote, which will be heard, for example, on blowing across the edge of the opening *a*. If this proper tone of the globe is excited in the external air, either as a fundamental or as an upper partial tone, the encluded mass of air is brought into violent sympathetic vibration, and the ear thus connected with it hears the corresponding tone with much increased intensity. By this means it is extremely easy to determine whether the proper tone of the globe is or is not contained in a compound tone or mass of tones.

Fig. 12

On examining the vowels of the human voice, it is easy to recognize, with the help of such resonators as have just been described, that the upper partial tones of each vowel are peculiarly strong in certain parts of the scale. Thus O in *ore* has its upper partials in the neighborhood of *b′*♭, A in *father* in the neighborhood of *b″*♭ (an octave higher). The following gives a general view

8. [In repeating this experiment the observer must remember that the e♭ of the piano is not a true twelfth below the *b′*♭. Hence the singer should first be given b′♭ from the piano, which he will naturally sing as *b*♭, an octave lower, and then take a true fifth below it. A skillful singer will thus hit the true twelfth and produce the required upper partial *b′*♭. On the other hand, if he sings *e*♭ from the piano, his upper partial *b′*♭ will probably beat with that of the piano.—A. J. E.]

of those portions of the scale where the upper partials of the vowels, as pronounced in the north of Germany,[9] are particularly strong:

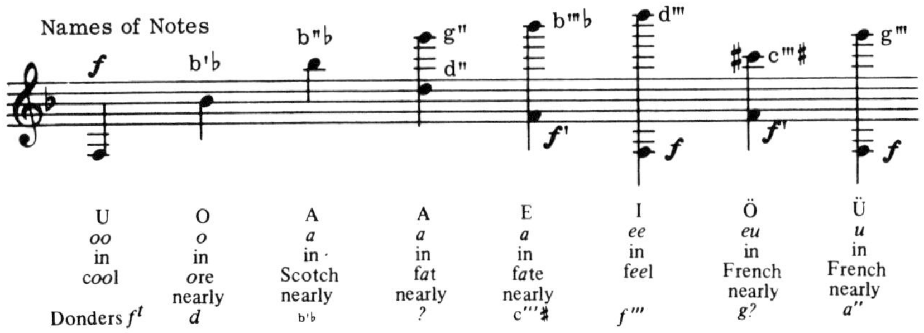

The following easy experiment clearly shows that it is indifferent whether the several simple tones contained in a compound tone like a vowel uttered by the human voice come from one source or several. If the dampers of a pianoforte are raised, not only do the sympathetic vibrations of the strings

9. [The corresponding English vowel sounds are probably none of them precisely the same as those pronounced by the author. It is necessary to note this, for a very slight variation in pronunciation would produce a change in the fundamental tone and consequently a more considerable change in the position of the upper partials. The tones given by Donders, which are written below the English equivalents, are cited on the authority of Helmholtz' *Tonempfindungen* (3rd Edition, 1870), where Helmholtz says: "Donders' results differ somewhat from mine, partly because his refer to a Dutch, and mine to a North German, pronunciation, and partly because Donders, not having had the assistance of tuning forks, could not always correctly determine the octave to which the sounds belong" (p. 171). The author also remarks that $b''\flat$ answers only to the deep German A (which is the broad Scotch *a'*, or *aw* without labialization) and that if the brighter Italian A (English *a* in f*a*ther) be used, the resonance rises a third, to d''' (p. 167). Dr. C. L. Merkel, of Leipzig, in his *Physiologie der menschlichen Sprache* (1866), after citing Helmholtz' experiments as detailed in his *Tonempfindungen*, gives the following as "the pitches of the vowels according to his most recent examination of of his own habits of speech, as accurately as he is able to note them" (p. 109):

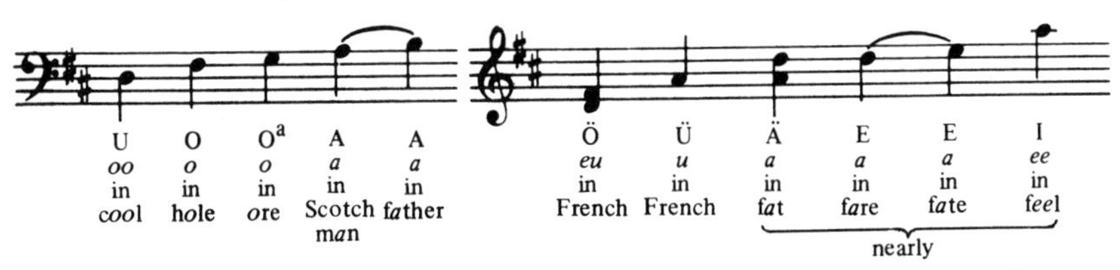

"Here the note *a* applies to the *timbre obscur* of A with low larynx, and *b* to the *timbre clair* of A with high larynx, and similarly the vowel E may pass from *d''* to *e''* by narrowing the channel in the mouth. The intermediate vowels Ö, Ä, have also two different timbres; hence their pitch is not fixed. The most frequent are consequently written over one another; the lower note is for the obscure, and the higher for the bright timbre. But the vowel Ü seems to be tolerably fixed as *a'*, just as its parent U and I are upon *d* and *a''*, and it has consequently the pitch of the ordinary *a'* tuning fork." —A. J. E.]

furnish tones of the same *pitch* as those uttered beside it; but if we sing A (*a* in *father*) to any note of the piano, we hear an A quite clearly returned from the strings; and if E (*a* in *fare* or *fate*), O (*o* in *hole* or *ore*), and U (*oo* in *cool*), be similarly sung to the note, E, O, and U will be echoed back. It is only necessary to hit the note of the piano with great exactness.[10] The sound of the vowel is produced solely by the sympathetic vibration of the higher strings, which correspond with the upper partial tones of the tone sung. In this experiment the tones of numerous strings are excited by a tone proceeding from a single source, the human voice, which produces a motion of the air, equivalent in form, and therefore in quality, to that of this single tone itself.

We have hitherto spoken only of compositions of waves of different lengths. We shall now compound waves of the same lengths which are moving in the same direction. The result will be entirely different, according as the elevations of one coincide with those of the other (in which case elevations of double the height and depressions of double the depth are produced), or the elevations of one fall on the depressions of the other. If both waves have the same height, so that the elevations of one exactly fit into the depressions of the other, both elevations and depressions will vanish in the second case, and the two waves will mutually destroy each other. Similarly two waves of sound, as well as two waves of water, may mutually destroy each other, when the condensations of one coincide with the rarefactions of the other. This remarkable phenomenon, wherein sound is silenced by a precisely similar sound, is called the interference of sounds.

This is easily proved by means of the siren already described. When the upper box is placed so that the puffs of air may proceed simultaneously from the rows of twelve holes in each wind chest, their effect is reinforced, and we obtain the fundamental tone of the corresponding tone of the siren very full and strong. But if the boxes are arranged so that the upper puffs es-

10. [My own experience shows that if any vowel at any pitch be loudly and sharply sung, or called out, beside a piano of which the dampers have been raised, that vowel will be echoed back. There is generally a sensible pause before the echo is heard. Before repeating the experiment with a new vowel, whether at the same or a different pitch, damp all the strings and then again raise the dampers. The result can easily be made audible to a hundred persons at once, and it is extremely interesting and instructive. It is peculiarly so if different vowels be sung to the same pitch, so that they have all the same fundamental tone, and the upper partials only differ in intensity. For female voices the pitches [musical notation] *a'* to *c''* are favorable for all vowels. This is a fundamental experiment for the theory of vowel sounds and should be repeated by all who are interested in speech.—A. J. E.]

cape when the lower series of holes is covered and the fundamental tone vanishes, we only hear a faint sound of the first upper partial, which is an octave higher and which is not destroyed by interference under these circumstances.

Interference leads us to the so-called musical beats. If two tones of exactly the same pitch are produced simultaneously and their elevations coincide at first, they will never cease to coincide; and if they did not coincide at first, they never will coincide. The two tones will either perpetually reinforce, or perpetually destroy each other. But if the two tones have only approximately equal pitches and their elevations at first coincide, so that they mutually reinforce each other, the elevations of one will gradually outstrip the elevations of the other. Times will come when the elevations of the one fall upon the depressions of the other, and then other times when the more rapidly advancing elevations of the one will have again reached the elevations of the other. These alternations become sensible by that alternate increase and decrease of loudness which we call a beat.

These beats may often be heard when two instruments which are not exactly in unison play a note of the same name. When the two or three strings which are struck by the same hammer on a piano are out of tune, the beats may be distinctly heard. Very slow and regular beats often produce a fine effect in sostenuto passages, as in sacred part-songs, by pealing through the lofty aisles like majestic waves, or by a gentle tremor giving the tone a character of enthusiasm and emotion. The greater the difference of the pitches, the quicker the beats. As long as no more than four to six beats occur in a second, the ear readily distinguishes the alternate reinforcements of the tone. If the beats are more rapid, the tone grates on the ear or, if it is high, becomes cutting. A grating tone is one interrupted by rapid breaks, like that of the letter R, which is produced by interrupting the tone of the voice by a tremor of the tongue or uvula.[11] When the beats become more rapid, the ear finds a continually increasing difficulty when attempting to hear them separately, even though there is a sensible roughness of the tone. At last they become entirely undistinguishable and, like the separate puffs which compose a tone, dissolve as it were into a continuous sensation of tone.[12]

11. [The trill of the uvula is called the Northumbrian burr and is not known out of Northumberland, in England. In France it is called the *r grasseyé* or *provençal*, and is the commonest Parisian sound of *r*. The uvula trill is also very common in Germany, but it is quite unknown in Italy.—A. J. E.]

12. The transition of beats into a harsh dissonance was displayed by means of two organ pipes, of which one was gradually put more and more out of tune with the other.

Hence, while every separate musical tone excites in the auditory nerve a uniform sustained sensation, two tones of different pitches mutually disturb each other and split up into separable beats, which excite a feeling of discontinuity as disagreeable to the ear as similar intermittent but rapidly repeated sources of excitement are unpleasant to the other organs of sense—for example, flickering and glittering light to the eye, scratching with a brush to the skin. This roughness of tone is the essential character of dissonance. It is most unpleasant to the ear when the two tones differ by about a semitone, in which case, in the middle portions of the scale, from twenty to forty beats ensue in a second. When the difference is a whole tone, the roughness is less; and when it reaches a third, it usually disappears, at least in the higher parts of the scale. The (minor or major) third may in consequence pass as a consonance.

Even when the fundamental tones have such widely different pitches that they cannot produce audible beats, the upper partial tones may beat and make the tone rough. Thus, if two tones form a fifth (that is, one makes two vibrations in the same time as the other makes three), there is one upper partial in each tone which makes six vibrations in the same time. If the ratio of the pitches of the fundamental tones is exactly as 2 to 3, the two upper partial tones of six vibrations are precisely alike and do not destroy the harmony of the fundamental tones. But if this ratio is only approximately as 2 to 3, then these two upper partials are not exactly alike and hence will beat and roughen the tone.

It is very easy to hear the beats of such imperfect fifths because, as our pianos and organs are now tuned, all the fifths are impure, although the beats are very slow. By properly directed attention, or still better with the help of a properly tuned resonator, it is easy to hear that it is the particular upper partials here spoken of that are beating together. The beats are necessarily weaker than those of the fundamental tones because the beating upper partials are themselves weaker. Although we are not usually clearly conscious of these beating upper partials, the ear feels their effect as a want of uniformity or a roughness in the mass of tone, whereas a perfectly pure fifth, with pitches precisely in the ratio of 2 to 3, continues to sound with perfect smoothness, without any alterations, reinforcements, diminutions, or roughnesses of tone. As has already been mentioned, the siren proves in the simplest manner that the most perfect consonance of the fifth precisely corresponds to this ratio between the pitches. We have now learned the reason of the roughness experienced when any deviation from that ratio has been produced.

In the same way, two tones which have their pitches exactly in the ratios of 3 to 4, or 4 to 5, and consequently form a perfect fourth or a perfect major third, sound much better when sounded together, than two others whose pitches slightly deviate from this exact ratio. In this manner, any given tone being assumed as fundamental, there is a precisely determinate number of other degrees of tone which can be sounded at the same time without producing any want of uniformity or any roughness of tone, or which will at least produce less roughness than any slightly greater or smaller intervals of tone under the same circumstances.

This is the reason why modern music, which is essentially based on the harmonious consonance of tones, has been compelled to limit its scale to certain determinate degrees. But even in ancient music, which allowed only one part to be sung at a time and hence had no harmony in the modern sense of the word, it can be shown that the upper partial tones contained in all musical tones sufficed to determine a preference in favor of progressions through certain determinate intervals. When an upper partial tone is common to two successive tones in a melody, the ear recognizes a certain relationship between them, serving as an artistic bond of union. Time is, however, too short for me to enlarge on this topic, as we should be obliged to go far back into the history of music.

I shall but mention that there exists another kind of secondary tone, which is only heard when two or more loudish tones of different pitch are sounded together and is hence termed *combinational.*[13] These secondary tones are likewise capable of beating and hence producing roughness in the chords. Suppose a perfectly just major third, *c′ e′* (ratio of pitches, 4 to 5), is sounded on the siren, or with properly tuned organ pipes, or on a violin;[14] then a faint *c* two octaves deeper than the *c′* will be heard as a combinational tone. The same *c* is also heard when the tones *e′ g′* (ratio of pitches 5 to 6) are sounded together.[15]

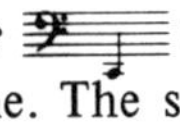

13. [These are of two kinds, *differential* and *summational*, according as their pitch is the difference or sum of the pitches of the two generating tones. The former are the only combinational tones here spoken of. The discovery of the latter was entirely due to the theoretical investigations of the author.—A. J. E.]

14. [In the ordinary tuning of the English concertina this major third is just, and generally this instrument shows the differential tones very well. The major third is very false on the harmonium and piano.—A. J. E.]

15. [This minor third is very false on the English concertina, harmonium, or piano, and the combinational tone heard is consequently very different from the true *c*.—A. J. E.]

16. [The combinational tone *c*, an octave higher, is also produced once from the fifth *c′ g′*.—A. J. E.]

If the three tones *c′*, *e′*, *g′*, having their pitches precisely in the ratios 4, 5, 6, are struck together, the combinational tone *c* is produced twice[16] in perfect unison and without beats. But if the three notes are not exactly thus tuned,[17] the two *c* combinational tones will have different pitches and produce faint beats.

The combinational tones are usually much weaker than the upper partial tones, hence their beats are much less rough and sensible than those of the latter. They are consequently but little observable, except in tones which have scarcely any upper partials, as those produced by flutes or the closed pipes of organs. But it is indisputable that on such instruments part music scarcely presents any line of demarcation between harmony and disharmony and is consequently deficient both in strength and character. On the contrary, all good musical qualities of tones are comparatively rich in upper partials, possessing the five first, which form the octaves, fifths, and major thirds of the fundamental tone. Hence, in the mixture stops of the organ, additional pipes are used, giving the series of upper partial tones corresponding to the pipe producing the fundamental tone, in order to generate a penetrating, powerful quality of tone to accompany congregational singing. The important part played by the upper partial tones in all artistic musical effects is here also indisputable.

We have now reached the heart of the theory of harmony. Harmony and disharmony are distinguished by the undisturbed current of the tones in the former, which are as flowing as when produced separately, and by the disturbances created in the latter, in which the tones split up into separate beats. All that we have considered tends to this end. In the first place the phehomenon of beats depends on the interference of waves. Hence they could only occur if sound were due to undulations. Next, the determination of consonant intervals necessitated a capability in the ear of feeling the upper partial tones and analyzing the compound systems of waves into simple undulations, according to Fourier's theorem. It is entirely in accord with this theorem that the pitches of the upper partial tones of all serviceable musical tones must stand to the pitch of their fundamental tones in the ratios of the whole numbers to one, and that consequently the ratios of the pitches of concordant intervals must correspond with the smallest possible whole numbers.

---

17. [As on the English concertina or harmonium, on both of which the consequent effect may be clearly heard.—A. J. E.]

How essential is the physiological constitution of the ear which we have just considered, becomes clear by comparing it with that of the eye. Light is also an undulation of a peculiar medium, the luminous ether, diffused through the universe; and light, as well as sound, exhibits phenomena of interference. Light, too, has waves of various periodic times of vibration, which produce in the eye the sensation of color, red having the greatest periodic time, then orange, yellow, green, blue, violet—the periodic time of violet being about half that of the outermost red. But the eye is unable to decompose compound systems of luminous waves, that is, to distinguish compound colors from one another. It experiences from them a single, unanalyzable, simple sensation, that of a mixed color. It is indifferent to the eye whether this mixed color results from a union of fundamental colors with simple or with non-simple ratios of periodic times. The eye has no sense of harmony in the same meaning as the ear. There is no music to the eye.

Aesthetics endeavors to find the principle of artistic beauty in its unconscious conformity to law. Today I have endeavored to lay bare the hidden law on which depends the agreeableness of consonant combinations. It is in the truest sense of the word unconsciously obeyed, so far as it depends on the upper partial tones, which, though felt by the nerves, are not usually consciously present to the mind. Their compatibility or incompatibility, however, is felt without the hearer knowing the cause of the feeling he experiences.

These phenomena of agreeableness of tone, as determined solely by the senses, are of course merely the first step toward the beautiful in music. For the attainment of that higher beauty which appeals to the intellect, harmony and disharmony are only means, although essential and powerful means. In disharmony the auditory nerve feels hurt by the beats of incompatible tones. It longs for the pure efflux of the tones into harmony. It hastens toward that harmony for satisfaction and rest. Thus both harmony and disharmony alternately urge and moderate the flow of tones, while the mind sees in their immaterial motion an image of its own perpetually streaming thoughts and moods. Just as in the rolling ocean, this movement, rhythmically repeated and yet every varying, rivets our attention and hurries us along. But whereas in the sea, blind physical forces alone are at work and the final impression on the spectator's mind is nothing but solitude, in a musical work of art the movement follows the outflow of the artist's own emotions. Now gently gliding, now gracefully leaping, now violently stirred, penetrated by or laboriously contending with the natural expression of passion, the stream of sound, in primitive vivacity, bears over into the hearer's soul unimagined moods which the artist has overheard from his own, and finally raises him to that repose of everlasting beauty of which God has allowed but few of his elect favorites to be the heralds.

But I have reached the confines of physical science and must close.

4.

THE APPLICATION OF THE LAW OF THE CONSERVATION OF FORCE TO ORGANIC NATURE [1861]

An address to the Royal Society of London on April 12, 1861

THE most important progress in natural philosophy by which the present century is distinguished has been the discovery of a general law which embraces and rules all the various branches of physics and chemistry. This law is of as much importance for the highest speculations on the nature of forces as for immediate and practical questions in the construction of machines. This law at present is commonly known by the name of the principle of conservation of force. It might be better perhaps to call it, with Mr. Rankine, the conservation of energy, because it does not relate to that which we commonly call *intensity* of force. It does not mean that the intensity of the natural forces is constant, but it relates more to the whole amount of power which can be gained by any natural process and by which a certain amount of work can be done. For example, if we apply this law to gravity, it does not mean (what is strictly and undoubtedly true) that the intensity of the gravity of any given body is the same as often as the body is brought back to the same distance from the center of the earth. Or with regard to the other elementary forces of nature–for example, chemical force: when two chemical elements come together, so that they influence each other, either from a distance or by immediate contact, they will always exert the same force upon each other, the same force both in intensity and in its direction and in its quantity. This other law indeed is true, but it is not the same as the principle of conservation of force.

We may express the meaning of the law of conservation of force by saying that every force of nature, when it effects any alteration, loses and exhausts its faculty to effect the same alteration a second time. But while, by every alteration in nature, that force which has been the cause of this alteration is exhausted, there is always another force which gains as much power of producing new alterations in nature as the first has lost. Although, therefore, it is the nature of all inorganic forces to become exhausted by their own working, the power of the whole system in which these alterations take place is neither exhausted nor increased in quantity, but only changed in form.

Some special examples will enable you better to understand this law than any general theories. We will begin with gravity, that most general force, which not only exerts its influence over the whole universe but which at the same time gives the means of moving to a great number of our machines. Clocks and smaller machines, you know, are often set in motion by a weight. The same is really the case with water mills. Water mills are driven by falling water, and it is the gravity, the weight of the falling water, which moves the mill. Now you know that by water mills, or by a falling weight, every machine can be put in motion and that by these motive powers every sort of work can be done which can be done at all by any machine. You see, therefore, that the weight of a heavy body, either solid or fluid, which descends from a higher place to a lower place is a motive power and can do every sort of mechanical work. Now if the weight has fallen down to the earth, then it has the same amount of gravity, the same intensity of gravity; but its power to move, its power to work, is exhausted; it must become again raised before it can work anew. In this sense, therefore, I say that the faculty of producing new work is exhausted, is lost; and this is true of every power of nature when this power has produced alteration.

Hence, therefore, the faculty of producing work, of doing work, does not depend upon the intensity of gravity. The intensity of gravity may be the same, the weight may be in a higher position or in a lower position, but the power to work may be quite different. The power of a weight to work, or the amount of work which can be produced by a weight, is measured by the product of the height to which it is raised and the weight itself. Therefore our common measure is foot-pound[s], that is, the product of the number of feet and the number of pounds. Now we can, by the force of a falling weight, raise another weight; as, for example, the falling water in a water mill may raise the weight of a hammer. Therefore it can be shown that the work of the raised hammer, expressed in foot-pounds, that is, the weight of the hammer multiplied by the height expressed in feet to which it is raised–that this amount of work cannot be greater than the product of the weight of water which is falling down and the height from which it fell down.

Now, we have another form of motive power, of mechanical motive power; that is, velocity. The velocity of any body in this sense, if it is producing work, is called [the] vis viva, or living force, of that body. You will find many examples of it. Take the ball of a gun: if it is shot off and has a great velocity, it has an immense power of destroying; and if it has lost its velocity, it is quite a harmless thing. The great power it has depends only on its velocity. In the same sense, the velocity of the air, the velocity of the wind, is motive power; for it can drive windmills, and by the machinery of the windmills it can do every kind of mechanical work. Therefore you see that also velocity in itself is a motive force.

Take a pendulum which swings to and fro. If the pendulum is raised to the side, the weight is raised up; it is a little higher than when it hangs straightly down, perpendicular. Now, if you let it fall and it comes to its position of equilibrium, it has gained a certain velocity. Therefore, at first, you had motive power in the form of a raised weight. If the pendulum comes again to the position of equilibrium, you have motive power in the form of vis viva, in the form of velocity, and then the pendulum goes again to the other side, and it ascends again till it loses its velocity; then again vis viva, or velocity, is changed into elevation of the weight. So you see in every pendulum that the power of a raised weight can be changed into velocity, and the velocity into the power of a raised weight. These two are equivalent.

Then take the elasticity of a bent spring. It can do work, it can move machines or watches. The crossbow contains such springs. These springs of the watch and crossbow are bent by the force of the human arm, and they become in that way reservoirs of mechanical power. The mechanical power which is communicated to them by the force of the human arm, afterwards is given out by a watch during the next day. It is spent by degrees to overpower the friction of the wheels. By the crossbow the power is spent suddenly. If the instrument is shot off, the whole amount of force which is communicated to the spring is then again communicated to the shaft and gives it a great vis viva.

Now the elasticity of air can be a motive power in the same way as the elasticity of solid bodies; if air is compressed, it can move other bodies. Let us take the air gun; there the case is quite the same as with the crossbow. The air is compressed by the force of the human arm; it becomes a reservoir of mechanical power; and if it is shot off, the power is communicated to the ball in the form of vis viva, and the ball has afterwards the same mechanical power as is communicated to the ball of a gun loaded with powder.

The elasticity of compressed gases is also the motive power of the mightiest of our engines, the steam engine, but there the case is different. The machinery is moved by the force of the compressed vapors, but the vapors are not compressed by the force of the human arm, as in the case of the compressed air gun. The compressed vapors are produced immediately in the interior of the boiler by the heat which is communicated to the boiler from the fuel. You see, therefore, that in this case the heat comes in the place of the force of the human arm, so that we learn by this example that heat is also a motive power. This part of the subject, the equivalence of heat as a motive power with mechanical power, has been that branch of this subject which has excited the greatest interest and has been the subject of deep research.

It may be considered as proved at present that if heat produces mechanical power, that is, mechanical work, a certain amount of heat is always lost. On the other hand, heat can be also produced by mechanical power, namely, by friction and the concussion of unelastic bodies. You can bring a piece of iron into a high temperature, so that it becomes glowing and luminous, by only beating it continuously with a hammer. Now, if mechanical power is produced by heat, we always find that a certain amount of heat is lost; and this is proportional to the quantity of mechanical work produced by that heat. We measure mechanical work by foot-pounds, and the amount of heat we measure by the quantity of heat which is necessary to raise the temperature of one pound of water by one degree, taking the centigrade scale. The equivalent of heat has been determined by Mr. Joule, of Manchester. He found that one unit of heat, or that quantity of heat which is necessary for raising the temperature of a pound of water 1° C., is equivalent to the mechanical work by which the same mass of water is raised to 423½ meters, or 1389 English feet. This is the mechanical equivalent of heat. Hence, if we produce so much heat as is necessary for raising the temperature of one pound of water by one degree, then we must apply an amount of mechanical work equal to raising one pound of water 1389 English feet, and lose it for gaining again that heat.

By these considerations, it is proved that heat cannot be a ponderable matter, but that it must be a motive power, because it is converted into motion or into mechanical power and can be either produced by motion or mechanical power. Now, in the steam engine we find that heat is the origin of the motive power, but the heat is produced by burning fuel, and therefore the origin of the motive power is to be found in the fuel, that is, in the chemical forces of the fuel and in the oxygen with which the fuel combines. You see from this that the chemical forces can produce mechanical work and can be measured by the same units and by the same measures as any other mechanical force. We may consider the chemical forces as attractions–in this instance, as attraction of the carbon of the fuel for the oxygen of the air–and if this attraction unite[s] the two bodies, it produces mechanical work just in the same way as the earth produces work, if it attract[s] a heavy body.

Now the conservation of force, of chemical force, is of great importance for our subject today, and it may be expressed in this way. If you have any quantity of chemical materials, and if you cause them to pass from one state into a second state in any way, so that the amount of the materials at the beginning and the amount of the materials at the end of this process be the same, then you will have always the same amount of work, of mechanical work or its equivalent, done during this process. Neither more nor less work can be done by the process. Commonly, no mechanical work in the common sense is done by chemical force, but usually it produces only heat; hence the amount of heat produced by any chemical process must be independent of the way in which that chemical process goes on. The way may be determined by the will of the experimenter as he likes.

We see, therefore, that the energy of every force in nature can be measured by the same measure, by foot-pounds, and that the energy of the whole system of bodies which are not under the influence of any exterior body must be constant, that it cannot be lessened or increased by any change. Now the whole universe represents such a system of bodies endowed with different sorts of forces and of energy, and therefore we conclude from the facts I have brought before you that the amount of working power, or the amount of energy in the whole system of the universe, must remain the same, quite steady and unalterable, whatever changes may go on in the universe.

If we accept the hypothesis of Laplace that in the first state the universe was formed by a chaos of nebulous matter, spread out through infinite space, then we must conclude that at this time the only form of energy existing in this system was the attraction of gravitation, and it was therefore the same sort of energy as is possessed by a raised weight. Afterwards, astronomers suppose, this nebulous matter was conglomerated and aggregated to solid masses. Great quantities of this nebulous matter, possibly from a great distance, fell together; and thus their attraction, or the energy of their attraction, was destroyed; and hence heat must have been produced; and the facts we know at present are sufficient to enable us to calculate the amount of this heat, that is, of the whole heat which must have been produced during the whole process of conglomeration. This amount of heat is immensely great, so that it surpasses all our ideas and all the limits of our imagination. If we calculate this quantity of heat, and suppose that the sun contained at the same time the whole heat, and that the sun had the same specific heat as water, the sun would be heated to twenty-eight millions of degrees, that is, to a temperature surpassing all temperatures we know on earth. However, this temperature could not exist at any time in the sun because the heat which was produced by the aggregation of the masses must also be spent partially by radiation into space. I give only the result of these calculations, in order that you may see from it what a great amount of heat could be produced in this way. The same process goes on also at present in the falling stars and meteors which come down to the earth from planetary spaces. Their velocity is destroyed by the friction of the air and by the concussion with the surface of the earth, and we see how they become luminous, and if they are found on the earth, we find them hot.

The sun also at present is hotter than any heated body here on the earth. That is shown by the latest experiments made by Professors Kirchhoff and Bunsen, of Heidelberg, on the spectrum of the sun, by which it is proved that in the atmosphere of the sun, iron and other metals are contained as vapors which cannot be changed into vapors by any amount of heat on the earth.

Our earth contains a great amount of energy in the form of its interior heat. This part of its energy produces the volcanic phenomena, but it is without great influence upon the phenomena of the surface because only a very small amount of this heat comes through. It can be calculated that the amount of heat which goes from the interior to the surface cannot raise the temperature of the surface any higher than the thirteenth part of a degree.

We have another power which produces motion on the surface of the earth. I mean the attraction of the sun and of the moon producing the tides.

All the other phenomena on the surface of the earth are produced by the radiation of the sun, by the sunbeams; and the greater part of those changes which occur on the surface of our earth are caused by the heat of the sun. As the heat of the sun is distributed unequally over the surface, some parts of the atmosphere become heated more than other parts; the heated parts of the atmosphere rise up, and so winds and vapors are produced. They come down at first as clouds in the higher parts of the atmosphere and then as rain upon the surface of the earth; they are collected as rivers and go again down into the sea. So you see that all the meteorological phenomena of our earth are produced by the effect of the solar beams, by the heat of the sun.

The light of the sun is the cause of another series of phenomena, and the principal products of the light of the sun are plants because plants can only grow with the help of the sunlight. It is only by the help of the sunlight that they can produce the inflammable matter which is deposited in the bodies of plants and which is extracted from the carbonic acid and the water contained in the atmosphere and in the earth itself.

This may give you an idea of the sense and bearing of the general principle on which I propose to speak. As many English philosophers have been occupied with working out the consequences of this most general and important principle for the theory of heat, for the energy of the solar system, for the construction of machines, you will hear these results better explained by your own countrymen; I shall abstain from entering further into this part of the subject. At the same time that Mr. Grove showed that every force of nature is capable of bringing into action every other force of nature, Mr. Joule, of Manchester, began to search for the value of the mechanical equivalent of heat and to prove its constancy, principally guided by the more practical interests of engineering. The first exposition of the general principle was published in Germany by Mr. Mayer, of Heilbronn, in the year 1842. Mr. Mayer was a medical man and much interested in the solution of physiological questions, and he found out the principle of the conservation of force guided by these physiological questions. At the same time also, I myself began to work on this subject. I published my researches a little later than Mr. Mayer, in 1847.

Now, at first sight it seems very remarkable and curious that even physiologists should come to such a law. It appears more natural that it should be detected by natural philosophers or engineers, as it was in England; but there is, indeed, a close connection between both the fundamental questions of engineering and the fundamental questions of physiology with the conservation of force. For getting machines into motion, it is always necessary to have motive power, either in water, fuel, or living animal matter. The constructors of machines, instruments, watches, within the last century, who did not know the conservation of force, were induced to try if they could not keep a machine in motion without any expenditure for getting the motive power. Many of them worked for a long time very industriously to find out such a machine which would give perpetual motion and produce any mechanical work which they liked. They called such a machine a perpetual mover. They thought they had an example of such a machine in the body of every animal. There, indeed, motive power seemed to be produced every day without the help of any external mechanical force. They were not aware that eating could be connected with the production of mechanical power. Food, they believed, was wanted only to restore the little damages in the machine, or to keep off friction, like the oil which made the axles of wheels to run smoothly. Now, at first by the mathematicians of the last century, the so-called principle of the conservation of vis viva was detected, and it was shown that by the action of the purely mechanical powers it was not possible to construct a perpetual mover; but it remained still doubtful if it would not be possible to do so by the interposition of heat, or electricity, or chemical force. At last, the general law of conservation of force was discovered, and stated, and established; and this law shows that also by the connection of mechanical powers with heat, with electricity, or with chemical force, no such machine can be constructed to give a perpetual motion and to produce work from nothing.

We must consider the living bodies under the same point of view and see how it stands with them. Now if you compare the living body with a steam engine, then you have the completest analogy. The living animals take in food that consists of inflammable substances, fat and the so-called hydrocarbons, as starch and sugar, and nitrogenous substances, as albumen, flesh, cheese, and so on. Living animals take in these inflammable substances and oxygen—the oxygen of the air, by respiration. Therefore, if you take (in the place of fat, starch, and sugar) coals or wood and the oxygen of the air, you have the substances in the steam engine. The living bodies give out carbonic acid and water; and then if we neglect very small quantities of more complicated matters which are too small to be reckoned here, they give up their nitrogen in the form of urea.

Now let us suppose that we take an animal on one day, and on any day afterwards; and let us suppose that this animal is of the same weight the first day and the second day and that its body is composed quite in the same way on both days. During the time—the interval of time—between these two days the animal has taken in food and oxygen and has given out carbonic acid, water, and urea. Therefore a certain quantity of inflammable substance, of nutriment, has combined with oxygen and has produced nearly the same substances, the same combinations, which would be produced by burning the food in an open fire, at least—fat, sugar, starch, and so on. And those substances which contained no nitrogen would give us quite in the same way carbonic acid and water if they [were] in the open fire, as if they burned in the living body; only the oxidation in the living body goes on more slowly. The albuminous substances would give us the same substances, and also nitrogen, as if they were burned in the fire.

You may suppose, for making both cases equal, that the amount of urea which is produced in the body of the animal may be changed without any very great development of heat into carbonate of ammonia, and carbonate of ammonia may be burned and gives nitrogen, water, and carbonic acid. The amount of heat which would be produced by burning urea into carbonic acid and nitrogen would be of no great value when compared with the great quantity of heat which is produced by burning the fat, the sugar, and the starch. Therefore we can change a certain amount of food into carbonic acid, water, and nitrogen, either by burning the whole in the open fire, or by giving it to living animals as food and burning afterwards only the urea. In both cases we come to the same result.

Now I have said that the conservation of force for chemical processes requires a fixed amount of mechanical work, or its equivalent, to be given out during this process; and the amount is exactly the same in whatever way the process may go on. And therefore we must conclude that by the animal as much work must be done, must be given out—the same equivalent of mechanical work—as by the chemical process of burning. Now let us remark that the mechanical work which is spent by an animal and which is given to the external world consists, firstly, in heat, and secondly, in real mechanical work. We have no other forms of work, or of equivalent of work, given out by living animals. If the animal is reposing, then the whole work must be given out in the form of heat; and therefore we must conclude that a reposing animal must produce as much heat as would be produced by burning its food. A small difference would remain for the urea; we must suppose that the urea produced by the animal is also burned and taken together with the heat immediately produced by the animal itself. Now we have experiments made upon this subject by the French philosophers Dulong and Despretz. They found that these two quantities of heat—the one emitted by burning, the other by the living animal—are nearly identical, at least so far as could be established at that time and with those previous researches which existed at that time. The heat which is produced by burning the materials of the food is not quite known even now. We want to have researches on the heat produced by the more complicated combinations which are used as food. Dulong and Despretz have calculated the heat according to the theoretical supposition of Lavoisier, which supposition is nearly right, but not quite right; therefore there is a little doubt as to the amount of the heat. But experiments show that at least to the tenth part of that heat the quantities are really equal; and we may hope, if we have better researches on the heat produced by burning the food, that these quantities will also be more equal than they were found to be by Dulong and Despretz.

Now if the body be not reposing, but if muscular exertion take place, then also mechanical work is done. The mechanical work is very different according to the different kinds of muscular exertion. If we walk only on a plane surface, we must overpower the resistance of friction and the resistance of the air, but these resistances are not so great that the work which we do by walking on a plane is of great amount. Our muscles can do work in very different ways. By the researches of Mr. Redtenbacher, the director of the Polytechnic School of Karlsruhe, it is proved that the best method of getting the greatest amount of work from a human body is by the treadmill, that is, by going up a declivity. If we go up the declivity of a hill, we raise the weight of our own body. In the treadmill the same work is done, only the mill goes always down, and the man on the mill remains in his place.

Now we have researches on the amount of air which is taken in and of carbonic acid given out during such work in the treadmill, made by Dr. Edward Smith. He found that a most astonishing increase of respiration takes place during such work. Now you all know that if you go up a hill, you are hindered in going too fast by the great frequency and the great difficulty of respiration. This, then, becomes far greater than by the greatest exertion of walking on a plain, and really the difficulty is produced by the great mechanical work which is done in the same time. Now, partly from the experiments of Dulong and Despretz, and partly from the experiments of Dr. Edward Smith, we can calculate that the human body, if it be in a reposing state, but not sleeping, consumes so much oxygen and burns so much carbon and hydrogen that during one hour so much heat is produced that the whole body, or a weight of water equal to the weight of the body, would be raised in temperature 1.2° C. (2.2° F.). Now Dr. Edward Smith found that by going in the treadmill at such a rate that if he went up a hill at the same rate, he would have risen during one hour 1712 feet, that during such a motion he exhaled five times as much carbonic acid as in the quiet state and ten times as much as in sleeping. Therefore the amount of respiration was increased in a most remarkable way. If we now calculate these numbers, we find that the quantity of heat which is produced during one hour of repose is 1.2° C. and that these are nearly equivalent to rising 1712 feet, so that therefore the amount of mechanical work done in a treadmill, or done in ascending a hill at a good rate, is equivalent to the whole amount of heat which is produced in a quiescent state. The whole amount of the decomposition in the living body is five times as great as in a reposing and wakeful state. Of these five quantities, one quantity is spent for mechanical work, and four fifths remain in the form of heat. Always in ascending a hill, or in doing great mechanical work, you become hot, and the production of heat is extremely great, as you well know, without making particular experiments. Hence you see how much the decomposition in the body is increased by doing really mechanical work.

Now these measurements give us another analogy. We see that in ascending a mountain we produce heat and mechanical work, and that the fifth part of the equivalent of the work which is produced by the chemical process is really gained as mechanical work. Now if we take our steam engine, or a hot-air engine, or any other engine which is driven by heat in such a way that one body is heated and expands and by the expansion other bodies are moved—I say, if we take any thermodynamic engine, we find that the greatest amount of mechanical work which can be gained by chemical decomposition or chemical combination is only an eighth part of the equivalent of the chemical force and seven eighths of the whole are lost in the form of heat. And this amount of mechanical work can only be gained if we have the greatest difference of temperature which can be produced in such a machine. In the living body we have no great difference of temperature; and in the living body the amount of mechanical work which could be gained if the living body were a thermodynamic engine, like the steam engine or the hot-air engine, would be much smaller than one eighth. Really, we find from the great amount of work done, that the human body is in this way a better machine than the steam engine, only its fuel is more expensive than the fuel of steam engines.

There is another machine which changes chemical force into mechanical power; that is, the magnetoelectric machine. By these magnetoelectric machines a greater amount of electrical power can be changed into mechanical work than in our artificial thermodynamic machines. We produce an electric current by dissolving zinc in sulphuric acid and liberating another oxidizable matter. Generally it is only the difference of the attraction of zinc for oxygen compared with the attraction of copper or nitrous acid for oxygen. In the human body we burn substances which contain carbon and hydrogen, and therefore the whole amount of attraction of carbon and hydrogen for oxygen is put into action to move the machine; and in this way the power of the living body is greater and more advantageous than the power of the magnetoelectric machine.

Let us now consider what consequences must be drawn when we find that the laws of animal life agree with the law of the conservation of force, at least as far as we can judge at present regarding this subject. As yet we cannot prove that the work produced by living bodies is an exact equivalent of the chemical forces which have been set into action. It is not yet possible to determine the exact value of either of these quantities so accurately as will be done ultimately, but we may hope that at no distant time it may be possible to determine this with greater accuracy. There is no difficulty opposed to this task. Even at present I think we may consider it as extremely probable that the law of the conservation of force holds good for living bodies.

Now we may ask, what follows from this fact as regards the nature of the forces which act in the living body? The majority of the physiologists in the last century, and in the beginning of this century, were of opinion that the processes in living bodies were determined by one principal agent, which they chose to call the *vital principle.* The physical forces in the living body, they supposed, could be suspended or again set free at any moment by the influence of the vital principle, and [they surmised] that by this means this agent could produce changes in the interior of the body, so that the health of the body would be thereby preserved or restored. Now the conservation of force can exist only in those systems in which the forces in action (like all forces of inorganic nature) have always the same intensity and direction if the circumstances under which they act are the same. If it were possible to deprive any body of its gravity, and afterwards to restore its gravity, then indeed we should have the perpetual motion. Let the weight come down as long as it is heavy; let it rise if its gravity is lost; then you have produced mechanical work from nothing. Therefore this opinion that the chemical or mechanical power of the elements can be suspended, or changed, or removed in the interior of the living body, must be given up if there is complete conservation of force.

There may be other agents acting in the living body than those agents which act in the inorganic world; but those forces, as far as they cause chemical and mechanical influences in the body, must be quite of the same character as inorganic forces in this, at least—that their effects must be ruled by necessity and must be always the same, when acting in the same conditions, and that there cannot exist any arbitrary choice in the direction of their actions. This is that fundamental principle of physiology which I mentioned in the beginning of this discourse.

Still at the beginning of this century, physiologists believed that it was the vital principle which caused the processes of life, and that it detracted from the dignity and nature of life if anybody expressed his belief that the blood was driven through the vessels by the mechanical action of the heart, or that respiration took place according to the common laws of the diffusion of gases. The present generation, on the contrary, is hard at work to find out the real causes of the processes which go on in the living body. They do not suppose that there is any other difference between the chemical and the mechanical actions in the living body and out of it than can be explained by the more complicated circumstances and conditions under which these actions take place, and we have seen that the law of the conservation of force legitimizes this supposition. This law, moreover, shows the way in which this fundamental question, which has excited so many theoretical speculations, can be really and completely solved by experiment.

5.

THE RELATION OF THE NATURAL SCIENCES TO SCIENCE IN GENERAL [1862]

An academic discourse delivered at Heidelberg on November 22, 1862

TODAY we are gathered, as is our annual custom, in grateful memory of an enlightened sovereign of this kingdom, Karl Friedrich, who, in an age when the ancient fabric of European society seemed about to be torn apart, strove with lofty purpose and untiring zeal to promote the welfare of his subjects and, above all, their moral and intellectual development. Rightly did he judge that he could not realize this benevolent goal more effectively than by the revival and encouragement of this university. Speaking, as I do now, on such an occasion and at once in the name and in the presence of the whole university, I have thought it well to try to undertake—insofar as it is possible from the narrow point of view of a single student—a general view of the interrelationships of the several sciences and of their study.

At the present day the relations among the different sciences, which have led us to combine them under the name *Universitas Literarum,* may appear to have become looser than ever. We see scholars and scientists absorbed in specialties of such vast extent that even the most universal genius can hope to master no more than a small part of our present knowledge. The philologists of the last three centuries, for example, found ample occupation in the study of Greek and Latin; at best they learned in addition—for immediate practical purposes—two or three European languages. Now comparative philology aims at nothing less than a knowledge of all the languages of all the branches of the human race, in order to ascertain from them the laws by which language itself has been formed. It has already made progress towards this goal. Even classical philology is no longer restricted to the study of those works which, because of their artistic perfection and precision of thought, or because of the importance of their contents, have become models of prose and poetry for all ages. On the contrary, we have learned that every lost fragment of an ancient author, every gloss of a pedantic grammarian, every allusion of a Byzantine court-poet, every broken tombstone of a Roman official found in some corner of Hungary or Spain or Africa, may contribute a fresh fact or fresh evidence and thus serve to increase our knowledge of the past. Still another group of scholars is busy with the vast task of collecting and cataloguing, in a useful fashion, every available bit of information about classical antiquity. Add to this, in history, the study of original documents; the critical examination of papers accumulated in the archives of states and towns; the combination of details scattered in memoirs, in correspondence, and in biographies; and the deciphering of hieroglyphics and cuneiform inscriptions—and add in natural history the increasingly comprehensive classification of minerals, plants, and animals, living as well as extinct—and there opens out an expanse of knowledge the contemplation of which may well bewilder us.

In all these sciences the range of investigation widens as fast as the means of observation improve. The zoologists of past times were satisfied for the most part to have described the teeth, hair, feet, and other external characteristics of an animal. The anatomist confined himself to human anatomy insofar as he could make it out with the help of the knife, saw, and scalpel, along with the occasional aid of various injections—and human anatomy passed for an unusually extensive, difficult study. Now we are no longer satisfied with the science called gross human anatomy, a science which was once thought, quite unreasonably, to be almost completed. We have added comparative anatomy (that is, the anatomy of all animals) and microscopic anatomy, both of which, being sciences of infinitely wider range, now absorb completely the interest of students.

The four elements of the ancient world and of medieval alchemy have increased in the chemistry of the present day to sixty-four, the last three of which are known due to a method invented in our own university.[1] Not only is the number of the elements far greater than before, but the methods for producing complicated compounds of them have been so vastly improved that what is called organic chemistry—which embraces only the study of compounds of carbon with oxygen, hydrogen, nitrogen, and a few other elements—has already taken its place as an independent science.

"As the stars in the heavens" was, in ancient times, the natural expression for a number beyond our comprehension. (Pliny even thought it presumptu-

---

1. The method of spectrum analysis developed by Bunsen and Kirchhoff, both of Heidelberg.

ous ["rem etiam Deo improbam"] on the part of Hipparchus to have undertaken to count the stars and to determine their relative positions.) Yet none of the catalogues before the seventeenth century, which were constructed without the aid of telescopes, gave more than 1,000 to 1,500 stars of the first to the fifth magnitudes. At the present time several observatories are engaged in extending these catalogues to include stars down to the tenth magnitude. Thus upwards of 200,000 fixed stars are to be catalogued and their places accurately determined. The most immediate consequence of these observations has been the discovery of a great number of new planets, so that instead of the six known in 1781, there are now seventy-five.[2]

Contemplation of this enormous activity in all branches of science may well make us stand amazed at the audacity of man and exclaim with the Chorus in *Antigone:* "Many are the wonders of the world, but none so wonderful as man!" Who can grasp all this? Who can hold all these threads in his hand or correctly find his way among them? One obvious consequence of this vast extension of the limits of science is that every student is forced to choose a narrower and narrower field for his own studies and can maintain only an imperfect acquaintance with allied fields of research. We are inclined to laugh when we hear that in the seventeenth century Kepler was appointed professor of both mathematics and moral philosophy at Gratz, and that at Leiden, at the beginning of the eighteenth, Boerhaave was simultaneously professor of botany, chemistry, and clinical medicine, and therefore of pharmacy as well. At the present time we require at least four professors–in a university with its full complement of teachers, seven or eight–to represent all these branches of science. The same is true in other disciplines.

One of my strongest motives for discussing today the relations of the various sciences to one another is that I am myself a student of natural philosophy and that recently natural philosophy has been criticized for striking out on a path of its own, for separating itself more and more sharply from the other sciences which are joined by common philological and historical ties. This separation has, in fact, long been apparent, but it seems to me to have developed mainly under the influence of Hegelian philosophy or at least to have been brought into more distinct relief by that philosophy. Certainly at the end of the last century, when Kantian philosophy reigned supreme, such a schism never developed. On the contrary, Kant's philosophy rested upon ex-

---

2. At the end of November 1864, the 82nd of the small planets, Alcmene, was discovered. On May 11, 1883, the 233rd was discovered, and the number grows yearly.

actly the same foundations as the physical sciences, as is evident from his own scientific works and especially from his cosmological hypothesis, which was based upon Newton's law of gravitation and which later came to be widely accepted under the name of Laplace's nebular hypothesis. The sole purpose of Kant's critical philosophy was to examine the sources and the validity of our knowledge and to establish standards or criteria for the other sciences. According to his teaching, a principle established a priori by pure thought was a rule applicable only to the methods of pure thought and to nothing else; it had no real, positive content.

Hegel's philosophy of identity was bolder. It began with the hypothesis that not only mental phenomena but even the actual world (that is, nature and man) is the result of the thought of a creative mind—a mind, it was supposed, similar in kind to the human mind. It followed from this hypothesis that the human mind is able, without any guidance from external experience, to think again the thoughts of the creator and to rediscover them by its own inner activity. Such was the view with which the philosophy of identity set to work to construct a priori the results of the various sciences. This procedure might possibly be more or less successful in matters of theology, law, politics, language, art, history—in short, in all sciences (properly classed together under the general heading of moral sciences) whose subject matter grows out of our moral and intellectual nature. The state, the church, art, and language exist in order to satisfy certain moral or mental needs of man. Even if natural forces, or chance, or the rivalry of other men, interpose obstacles, the efforts of the human mind to satisfy its needs, being systematically directed to one end, must eventually triumph over all such fortuitous hindrances. It would not be absolutely impossible for a philosopher, starting from an exact knowledge of the mind, to trace out the general course of human development in these areas, provided of course that he had a broad knowledge of facts available to which he could attach his abstractions and speculations.

Hegel was materially assisted in his attempt to carry out his program by the profound philosophical views on historical and scientific subjects with which the writings of his immediate predecessors, both poets and philosophers, abound. He had, for the most part, only to collect and combine these insights in order to produce a system calculated to impress people by a number of acute and original observations. He thus succeeded in gaining an enthusiastic reception from most of the educated men of his time and in raising extravagantly sanguine hopes of solving the deepest enigmas of human life—all the more sanguine, surely, because the structure of his system was disguised by a strange, abstract phraseology, which was perhaps really understood by but a few of his worshippers.

Still, even granting that Hegel was more or less successful in constructing a priori the leading results of the moral sciences, this in itself was no proof of the correctness of the hypothesis of identity from which he began. On the contrary, the facts of nature were the crucial test. That in the moral sciences traces of the activity of the human intellect and of the several stages of its development should present themselves was a matter of course; but surely, if nature really resulted from the thought of a creative mind, the philosophic system ought without difficulty to accommodate at least comparatively simple natural processes and phenomena. We venture to say that it was at this point that Hegel's philosophy broke down completely. His philosophy of nature seemed, at least to natural philosophers, absolutely meaningless. Of all the distinguished scientists who were his contemporaries, not one was found to stand up for his ideas. Accordingly, Hegel himself, convinced of the importance of winning for his philosophy in the field of physical science that recognition which had been so freely granted to it elsewhere, launched out with unusual vehemence and acrimony against the natural philosophers, especially against Sir Isaac Newton, the first and greatest representative of scientific investigation. The philosophers, led by Hegel, accused the scientific men of narrowness; the scientists retorted that the philosophers were crazy. And so it came about that men of science began to lay some stress on the banishment of all philosophic influences from their work. Some of them, indeed, including men of the greatest acuteness, went so far as to condemn philosophy altogether, not merely as useless but as mischievous dreaming. Thus, it must be confessed, not only were the illegitimate pretensions of the Hegelian system to subordinate to itself all other studies rejected, but no attention was paid to the legitimate aims of philosophy: the critical analysis of the sources of knowledge, and the establishment of standards for intellectual endeavors.

In the moral sciences the course of things was different, although ultimately the result was almost the same. In all branches of the moral sciences—in theology, politics, jurisprudence, aesthetics, philology—there arose enthusiastic Hegelians who tried to reform their several disciplines in accordance with the doctrines of their master and, by the royal road of speculation, to reach the promised land instantly and gather in the harvest, something which hitherto had been accomplished only by long, laborious study. And so, for some time, a hard and fast line was drawn between the moral and the physical sciences. Indeed, the latter was often denied the very name of science.

The strained relationship did not last long at its original intensity. By a brilliant series of discoveries and practical applications the physical sciences proved to the eyes of all that they contained a healthy germ of extraordinary fertility; it was impossible to withhold recognition and respect from them any longer. In other departments of science, too, conscientious, factually oriented investigators soon protested against the overbold flights of speculation. In the final analysis, therefore, it cannot be denied that the philosophies of Hegel and Schelling did exercise some beneficial influence. Since their time the attention of investigators in the moral sciences has been more clearly and more consistently directed toward defining the scope of those sciences and toward their intellectual content. The great amount of labor bestowed on these speculative systems has not been entirely wasted.

As empirical investigation of facts has again come to the fore in the moral sciences, the opposition between them and the physical sciences has become less and less marked. Yet we must not forget that although this opposition was brought out in an unnecessarily exaggerated form by the Hegelian philosophy, it has its roots in the nature of things and will sooner or later make itself felt. It depends partly upon the kinds of intellectual processes characteristic of the two groups of sciences and partly, as their very names indicate, on the subjects of which they treat. It is not easy for a physicist to convey to a linguist or to a jurist a clear idea of some complicated process of nature; he must demand of them certain abstractions from phenomena, as well as some skill in the use of geometrical and mechanical conceptions. In this it is frequently difficult for them to follow him. On the other hand, an artist or a theologian will perhaps find the natural philosopher too much inclined to mechanical and materialistic explanations, which seem commonplace and chilling to their feelings and inspirations. Nor will the linguist or the historian, who are bound to the jurist and the theologian by common philological and historical studies, fare better with the natural philosopher. They will find him shockingly indifferent to literary treasures and perhaps more indifferent than he should be to the history of his own science. In short, there is no denying that while the moral sciences deal directly with the most precious interests of the human mind and with the institutions it has brought into being, the natural sciences are concerned with external, indifferent matter, obviously indispensable because of the uses to which it may be put but apparently without any immediate bearing on the cultivation of the intellect.

It has been shown, then, that the sciences have split into countless branches, that a real and deeply felt opposition has grown up between different groups of them, and finally that no single individual can comprehend all or even a considerable part of them. Under these circumstances is it still reasonable to keep them together in one educational institution? Is the union of the four faculties to form one university a mere relic of the Middle Ages? Many arguments have been adduced for separating them—for establishing the medical faculty in the hospitals of our large cities, establishing polytechnic institutes for the natural sciences, and forming special seminaries for the theologians and jurists. Long may the German universities be preserved from such a fate! Then, indeed, would the connection among the different sciences be finally broken. How essential that connection is, not only in formal ways necessary for the continuance of scientific work, but also in material ways necessary for the successful application of that work, may be shown by a few brief considerations.

First, I would say that the union of the different faculties is necessary in order to maintain a healthy equilibrium among our various intellectual energies. Each discipline tries certain of our intellectual faculties more than others and strengthens them by constant exercise. Any sort of one-sided development, however, is attended by dangers; it disqualifies us from using those faculties that are less exercised and so renders us less capable of taking a general view. Above all, it leads us to overvalue ourselves. Anyone who has proved himself much more successful than others in one kind of intellectual work is apt to forget that there are many things that others can do better than he. Self-adulation is a mistake—I would have every student remember—which is the worst enemy of intellectual activity. How many men of ability have forgotten to practice that criticism of themselves which is so essential to the student and so hard to exercise! How many have been completely crippled in their progress because they have thought dry, laborious drudgery beneath them and have devoted all their energies to a quest for brilliant theories and wonder-working discoveries! How many such men have become bitter misanthropes and have put an end to a melancholy existence because they failed to obtain among their colleagues that recognition which must be won by labor and results and which is withheld from mere would-be genius! The more isolated a man is, the more liable he is to this danger. On the other hand, nothing is better for the spirit than to feel yourself forced to strain every nerve to win the admiration of the men whom you, in your turn, most admire.

If we compare the intellectual activities involved in the pursuit of the various branches of science, we are struck by certain generic differences dividing one group of sciences from another. In addition to these differences, of course, every man of conspicuous ability has his own special mental constitution, which fits him for one line of thought rather than for another. If you compare, for example, the work of two contemporary investigators, even in closely allied branches of science, you can generally see that the more distinguished the men are, the more clearly their individuality comes out and the less qualified either of them would be to carry on the researches of the other. But today I must and shall restrict myself to a discussion of some of the more general differences in the intellectual work in the various branches of science.

I have already mentioned the enormous mass of information accumulated in the sciences. It is obvious that the organization and arrangement of this knowledge must be proportionately developed, if we are not to be hopelessly lost in a maze of erudition. The better the order and systemization are, the more extensive the accumulation of knowledge can be without the system suffering. One of the reasons why we can clearly surpass our predecessors in each individual discipline is that they have shown us how to organize our knowledge.

This organization consists, first of all, in rather mechanical arrangements of materials, such as are to be found in our catalogues, lexicons, registers, indexes, digests, scientific and literary annuals, law codes, systems of natural history, and the like. By means of these any information that is not readily available in the memory is immediately accessible to anyone who wishes it. With a good lexicon a schoolboy of the present day can achieve results in the interpretation of the classics which an Erasmus, with the erudition of a lifetime, could hardly attain. Works of this kind form, so to speak, our intellectual principal, with the interest of which we trade; it is like capital invested in land. The learning buried in catalogues, lexicons, and indexes looks as bare and uninviting as the soil of a farm; the uninitiated cannot see or appreciate the labor and capital invested there; to them the work of the plowman seems eternally dull, weary, and monotonous. True, the compiler of a lexicon or of a system of natural history must be prepared to undertake labor as wearying and as stubborn as the plowman's, but one must not suppose that his work is of secondary importance or that it is in any way as dry and mechanical as it looks when we have it before us in black and white. In this, as in any other sort of scientific work, it is necessary to establish every fact by careful observation, to verify and collate each one, and to separate what is important from what is not. All this requires a man with a thorough understanding both of the purpose of the compilation and of the methods and content of the science. For such a man every detail has its relation to the whole and its special interest. Otherwise such an activity would be the vilest drudgery imaginable. That the influence of the progressive development of scientific ideas extends to these works is obvious from the constant demand for new lexicons, new systems of natural history, new law codes, new star catalogues, etc., all indicating the advancements in the art of ordering and organizing scientific knowledge.

Our knowledge, however, cannot lie dormant in the form of catalogues. The very fact that we must carry it about in black and white shows that our intellectual mastery of it is incomplete. It is not enough to be acquainted with facts; scientific knowledge begins only when their laws and their causes are unveiled. Logical ordering of the facts begins with the relating of similar things and the formulation of a general concept embracing them. Such a concept, as the name implies, encompasses a number of individual facts and stands for them in our thought. We call such a concept *generic* when it embraces a number of existing objects; we call it a *law* when it embraces a number of events or occurrences. When, for example, I have ascertained that all mammals—that is, all warm-blooded, viviparous animals—breathe by the use of lungs and have two chambers in their hearts and at least three tympanic bones, I need no longer remember these anatomical peculiarities in the individual cases of the monkey, the dog, the horse, and the whale: the general rule includes a vast number of single instances and stands for them in my memory. When I express the law of refraction, not only does this law embrace all cases of light rays falling at all possible angles on the plane surface of water and inform me of the consequence; it includes all cases of rays of any color incident on the surface of any transparent substance whatsoever. The law thus includes an infinite number of cases, which it would have been absolutely impossible to carry in one's memory. Moreover, not only does this law include those cases which we ourselves or other men have already observed, but we should not hesitate to apply it to new, unobserved cases with absolute confidence in the reliability of our predictions. In the same way, if we were to discover a new species of mammal, one not yet examined, we are entitled to assume, with a confidence bordering on certainty, that it has lungs, two chambers in its heart, and three or more tympanic bones.

Thus, when we combine intellectually the facts of experience and form general ideas, whether they be generic concepts or laws, we not only bring our knowledge into a form in which it can be easily used and easily retained, but we actually extend it, inasmuch as we are justified in extending the rules and the laws we have discovered to all similar cases that may afterwards be presented to us.

The examples mentioned above—that is, the examples of mammals and refraction phenomena—are such that there is no great difficulty in combining individual cases intellectually so as to form general ideas. The phenomena lie clearly before our eyes and can be followed in all their stages. In complicated cases, however, it is not so easy to distinguish completely between related and unrelated phenomena and to embrace the related phenomena in a clear, well-defined concept. Assume, for example, that we know a man to be ambitious; we can perhaps predict with some assurance that if he were to act under certain conditions, he would follow the dictates of his ambition and decide on a certain line of action. We cannot, however, define with absolute precision what constitutes an ambitious man, nor can we say by what standard the intensity of his ambition is to be measured; nor, further, can we say precisely what degree of ambition must be present in order to ensure some given direction to the man's action under some particular set of circumstances. Accordingly, we make comparisons among previously observed actions of the man in question and among the actions of other men who have behaved in the same way in similar situations, and we make our inferences concerning future action without being able to express either the major or the minor premise in a definite, clearly defined form—perhaps even without being aware that our prediction rests on such an analogy as I have described. In such cases our conclusion is based on a certain psychological instinct, not on conscious reasoning, although in reality we have gone through an intellectual process identical with that which leads us to assume that a newly discovered mammal has lungs.

Now this kind of induction, which cannot be entirely assimilated to any logical form, nor expressed as an exceptionless universal law, plays a most important role in human life. The whole process by which we translate our sensations into perceptions depends upon it, as may clearly be shown by an investigation of what are called optical illusions. For example, when the nervous system in the eye is excited by a blow, we imagine that we see light in our field of vision, since throughout our lives we have felt excitations in the optic nerves only when light was there. We have become habituated to identifying the sensations of those nerves with the presence of light in the field of vision, and we continue to do this even when there is no such light there. The same kind of induction plays a major role in psychological processes in general, due to the extraordinary complexity of the influences involved in the formation of character and in the excitation of momentary moods. In ascribing to ourselves free will—that is, the power to act as we please without being subject to a strict, exceptionless law of causality—we deny completely the possibility of referring at least part of the way in which our mental activity expresses itself to a rigorous law.

In opposition to *logical induction,* which operates only with clearly defined universal propositions, we shall call this kind of reasoning *aesthetic induction,* since it is most conspicuous in the most exceptional works of art. It is an important part of artistic talent to be able to reproduce—by words, forms, colors, or musical tones—the external indications of character, mood, or emotion and to grasp by a kind of instinctive intuition, uncontrolled by any definable rule, the steps by which we pass from one mood to another. If we find that an artist has consciously followed general rules and abstract ideas, we begin to think his work poor and commonplace, and we cease to admire it. The work of great artists, on the other hand, brings before us character and mood with such lifelikeness, such a wealth of individual traits, and such an overwhelming conviction of truth that it almost seems—because all disturbing influences have been eliminated—more real than reality itself.

If now, after these reflections, we proceed to review the various sciences and to classify them according to the method by which they arrive at their conclusions, we are struck by a fundamental difference between the natural and the moral sciences. The natural sciences are, for the most part, in a position to reduce their inductive inferences to clearly defined general rules and principles. The moral sciences, on the other hand, in the great majority of cases, have to do with judgments arrived at by psychological instinct. History, for example, must first establish the reliability of the witnesses or reporters who supply basic information: after the facts they relate have been verified, the important but difficult task of ascertaining their motives begins, and this can be accomplished only by psychological intuition. Philology—insofar as it is concerned with the history of literature and art and with the interpretation and emendation of the texts handed down to us—must seek, as it were, to feel out both the author's literal meaning and the accessory notions which he wished his words to suggest. For such purposes it is necessary to start with a correct insight into both the personality of the author and the genius of the language in which he wrote.

All of this affords scope for aesthetic but not for strictly logical induction. It is possible to reach a conclusion only when one has a large number of individual facts already in one's memory, ready to be applied instantly to some particular question. Accordingly, one of the first requirements in studies of this kind is a ready, accurate memory. Many celebrated historians and philologists have, indeed, astonished their contemporaries with their extraordinary strength of memory. Of course, memory alone is insufficient; there must also be a facility for recognizing real resemblances and a delicate, fully trained insight into the springs of human action. This insight in turn is unattainable without a certain sympathetic warmth and an interest in observing the working of other men's minds. While intercourse with our fellow-men in daily life must provide the foundation of these psychological insights, it is the study of history and art that makes them richer and more complete, for in these we study men acting under unusual conditions, and through this we come to appreciate the full range of the forces which lie hidden in our breasts.

None of this group of sciences, with the exception of grammar, leads us to the formulation of rigorous, universally valid, general laws. Even the laws of grammar are a product of the human will, though they can hardly be said to have been framed deliberately and according to some grand plan. Rather, they have grown up gradually, as they have been required, and they are presented to students who are learning a language in the form of commands, that is, laws imposed by external authority.

Theology and jurisprudence are related to historical and philological studies; the latter two, in fact, serve as studies preparatory and auxiliary to the former. The laws of both theology and jurisprudence are commands or injunctions. They are laws imposed by external authority to regulate, from a moral and legal standpoint, the actions of mankind, as distinguished from natural laws, which are generalizations based on a large number of facts. As with natural law, however, the subsumption of an individual under grammatical, legal, moral, or theological rules takes a well-known logical form: a rule forms the major premise of such an inference, while the minor premise states that the case in question satisfies the conditions necessary for the application of the rule. Whether the minor premise really is true—as in grammatical analyses, where the meanings of sentences are to be made clear, or in legal judgments concerning the truth of alleged facts, the intentions of parties involved in litigation, and the meaning of the documents they have introduced into court—will in most cases be determined by psychological intuition. It should not be forgotten, of course, that, like the syntax of a fully developed language, a system of jurisprudence gradually elaborated, as ours has been, by the practice of more than two thousand years has reached a high degree of logical completeness and consistency. Cases which do not fall obviously under one or another of the laws already laid down are quite exceptional. Such exceptions there will always be, however, for the legislation of men can never reach the absolute consistency and completeness of the laws of nature. In cases of exceptions there is no course open but to try to guess the intentions of the legislators and, if necessary, to base our judgment on analogies to similar cases.

Grammar and jurisprudence have certain advantages as means for training the intellect inasmuch as they tax all the intellectual powers to about the same degree. For this reason secondary education in modern European nations is built primarily upon the study of the grammar of foreign languages. One's mother tongue and modern foreign languages, however, since they are acquired solely by practice, do not provide for the exercise of any conscious logical thinking, although by the study of these languages we may cultivate an appreciation for the beauty of artistic expression. The two classical languages, Latin and Greek, on the other hand, besides their exquisite logical subtlety and aesthetic beauty, have an additional advantage which they seem to possess in common with most ancient languages: they show by numerous, distinct inflections the precise relations of words and sentences to one another. For languages are, as it were, worn away by long use; in the interest of brevity and rapidity of expression, grammatical distinctions are cut to a minimum and thus made much less specific. This is obvious when one compares any modern European language—especially English, where the process has gone further than in any other—with Latin. This seems to me the real reason why the modern languages are far less suitable than the ancient as instruments of education. And as grammar is the staple of school education, so legal studies are rightly used as a means of training persons of maturer age, even when such studies are not specifically required for professional purposes.

At the opposite extreme from philology and history, with regard to the kinds of intellectual labor they require, stand the natural and physical sciences. I do not mean to say that in many branches of these sciences an instinctive appreciation of analogies and a certain artistic sense have no part to play. On the contrary, in natural history, decisions concerning which characteristics are to be considered important for classification and which unimportant, or what divisions of the animal and vegetable kingdoms are to be considered more natural than others, are really left to an instinct of this kind, acting not in accordance with any strictly definable rule. Moreover, it is very significant that it was an artist, Goethe, who gave the first impetus to the researches of comparative anatomy into the analogies of corresponding organs in different species of animals (and to the parallel botanical theory of the metamorphosis of plants), thus pointing the direction which the science of comparative anatomy has followed ever since. But even in those branches of natural science where we must deal with the least understood vital processes, it is generally far easier to discover comprehensive principles and to express them in definitive language than it is in cases where we must base our judgment on an analysis of the human mind. It is only when we come to the experimental and mathematical sciences, however, and especially when we come to pure mathematics, that we see the peculiar characteristics of the natural and physical sciences fully brought out.

The essential differentia of these sciences seems to me to consist in the comparative ease with which individual observations and experiences are brought together under general laws of extraordinary comprehensiveness and unconditional validity. (In the moral sciences, it is precisely here that insuperable difficulties are encountered.) Indeed, in mathematics the general propositions which stand as axioms at the head of a system are so few in number, so comprehensive in scope, and so immediately obvious that no proof whatsoever is required for them. Let me remind you that the whole of pure mathematics (arithmetic) is developed from three axioms: *Things which are equal to the same thing are equal to each other. If equals are added to equals, the sums are equal. If unequals are added to equals, the sums are unequal.* And the axioms of geometry and theoretical mechanics are no more numerous.

These three sciences are developed out of a few axioms by the continuous process of deducing more and more complicated propositions. Moreover, algebra is not confined to finding the sums of a finite number of quantities; in higher analysis we learn to add infinite series, whose terms increase or diminish according to the most varied laws, and in fact to solve problems which could never be completed by direct addition. An instance of this kind shows us the conscious logical activity of the mind in its purest, most perfect form. Further, it enables us to see the extreme care that must be taken, the extreme caution with which it is necessary to advance, the accuracy required in order to determine exactly the scope of such universal principles as have been attained, and the difficulty in forming and understanding abstract concepts. On the other hand, it enables us to see the certainty, range, and fertility of this kind of intellectual work.

The fertility of this method comes out most strikingly in applied mathematics, especially in mathematical physics, which of course includes physical astronomy. Newton showed, through a mechanical analysis of the motion of the planets, that every particle of matter in the universe attracts every other particle with a force which varies inversely as the square of the distance between them. By virtue of that one law of gravitation, astronomers have been able, given only the position, velocity, and mass of each planet of the solar system at any one time, to calculate with the greatest accuracy the movements of the planets at the remotest past and the most distant future. Still further, we recognize the operation of this law in the movements of double stars whose distances from us are so great that their light takes years to reach us—in some cases, indeed, so great that all attempts to measure the distances have failed.

The discovery of the law of gravitation and its implications are the most impressive accomplishments that the logical powers of the human mind have so far achieved. I do not mean to say that there have not been other men with equal or greater powers of abstraction than Newton and those astronomers who either paved the way for his discovery or worked out its implications. It is only that there has never been presented to the human mind such an awesome subject as the complex movements of the planets—movements which once served merely to nourish the astrological superstitions of ignorant stargazers, but which now are reduced to a single law, capable of rendering the most exact account of the minutest detail of their motions. In addition to these great examples and following patterns set by them, a number of other areas have developed within physics, among which physical optics and the theory of electricity and magnetism should be mentioned.

The experimental sciences have one great advantage over others in the search for general laws of nature: they can change at pleasure the conditions under which something occurs and thus, in looking for a law, can confine themselves to a small number of characteristic instances. The law must, of course, then stand the test of application to more complex cases. The physical sciences, therefore, once the correct methods have been found, have made comparatively rapid progress. Not only have they enabled us to look back into primeval periods, when nebulous masses were forming into suns and planets and were being heated by the energy of their own contraction; not only have they enabled us to investigate the chemical constituents of the solar atmosphere and of the remotest fixed stars; they have enabled us to turn the forces of nature to our own use and to make them the servants of our will.

Enough has been said to indicate how the intellectual processes characteristic of this group of sciences differ from those required in the moral sciences. The mathematician requires no memory whatsoever for individual facts, the physicist hardly any. Hypotheses based on the recollection of similar cases may be useful to guide one onto a correct track, but they have no real value until they have led to a strict, precisely formulated law. Nature does not allow us to doubt for a moment that we have to do with a strict chain of cause and effect, admitting of no exceptions. To us as students, therefore, goes forth the mandate to continue working until we have discovered invariant laws; only then may we rest satisfied, for only then can our knowledge grapple victoriously with time and space and the forces of nature.

The difficult work of conscious logical reasoning demands great perseverance and great caution; as a rule it moves very slowly and is rarely illuminated by brilliant flashes of genius. It knows little of that facility with which the most varied experiences come thronging into the memory of the philologist or historian. Rather, it is an essential condition of the methodical progress of thought that the mind remain concentrated on a single issue, undistracted by secondary points, undisturbed by wishes and hopes, and move steadily in a direction deliberately chosen. A celebrated logician, John Stuart Mill, expresses the conviction that the inductive sciences have recently done more for the advancement of logical methods than the labors of philosophers properly so called. One basic reason for such an assertion must undoubtedly be that in no other department of knowledge can an error in a chain of reasoning be so easily detected by the incorrectness of the results as in those sciences where the results of reasoning can be directly compared with the facts of nature.

Although I have maintained that it is in the physical sciences, especially in such branches of them as are developed mathematically, that the solution of scientific problems has been most successfully achieved, you must not imagine that I wish to disparage other studies in comparison with them. If the natural and physical sciences have the advantage in their greater formal perfection, the moral sciences deal with a richer subject matter—with questions which lie closer to the interests and feelings of men, with the human mind itself, its motives and its different dispositions and activities. The moral sciences clearly have the loftier and more difficult task. They cannot afford, however, to lose sight of the example set by those branches of science which, because they deal with more manageable materials, have—at least in form—made greater progress. Not only have they something to learn from them concerning methods, but they may also draw encouragement from the richness of their results. For I believe that our age has learned many things from the natural sciences. The absolute, unconditional attention to facts, the great care with which they are collected, the distrust of appearances, the effort to find everywhere relations of cause and effect and to proceed on the assumption of their existence—these qualities which distinguish our century from preceding ones seem to me to point to such an influence.

I do not intend to go deeply into the question of how large a place the study of mathematics, as the representative of conscious logical reasoning, should have in the schools. In any case it is only a question of time. As the scope of science is extended, its systemization and organization must be improved, and inevitably students will find themselves compelled to go through a stricter course of training than grammar can provide. In my own experience, what strikes me about students who move from our secondary schools into scientific and medical studies is, first of all, a certain laxity in the application of strict, universal laws. The grammatical rules on which they have been trained are for the most part followed by long lists of exceptions, so the students are not in the habit of relying unconditionally on the certainty of legitimate deductions from strict, universal laws. Secondly, I find them on the whole too inclined to trust to authority, even in cases where they might form an independent judgment. (In philological studies, of course, because it is rarely possible to survey all of a subject and because decisions concerning disputed questions often depend upon an aesthetic feeling for beauty of expression and for the genius of a language, even the best instructor must often consult authorities.) Both faults are traceable to a certain laziness and uncertainty of thought, the sad effects of which are not confined to subsequent scientific studies. The best remedy for both is surely to be found in the study of mathematics, where there is absolute certainty in the reasoning and where no authority is recognized other than one's own intelligence.

So much for the several branches of science considered as different but supplementary exercises for the intellect. Knowledge alone is not, however, the sole purpose of man on earth. Although the sciences arouse and educate the subtlest powers of the mind, anyone who studied purely for the sake of knowing would surely not be fulfilling the purpose of his existence. Often we see men of considerable gifts, to whom fortune has given a comfortable situation without at the same time giving them ambition or energy enough to produce anything, dragging out a weary, unsatisfied existence, while imagining all the time that they are following the noblest aim of life in constantly devoting themselves to the increase of their knowledge and the cultivation of their minds. Action alone gives a man a life worth living. He must aim either at the practical application of his knowledge or at the extension of the limits of science itself. To extend the limits of science is surely to work for the progress of humanity. And so we pass to the second of the links uniting the labors in the different sciences.

Knowledge is power. Our age, more than any other, is in a position to demonstrate the truth of this maxim. We have taught the forces of inanimate nature to minister to the wants of human life and to the designs of the human intellect. The application of steam has multiplied our physical strength a thousand- and a millionfold; weaving and spinning machines have relieved us of unrewarding, monotonous labor. Communication among men, with its far-reaching influence on material and intellectual progress, has increased far beyond what anyone could have dreamed when the oldest among us was born. But it is not only by machines, not only by rifled cannons and armor-plated ships, not only by accumulated stores of money and other goods necessary to the welfare of a nation that our powers have multiplied, although these things have exercised so unmistakable an influence that even the proudest, most obstinate despotisms of our time have been forced to consider removing restrictions on industry and conceding to the industrial middle classes a due voice in national councils.

It is the political and legal organization of states and the moral discipline of individual citizens which determine the superiority of the developed states over the undeveloped. As surely as a nation remains indifferent to the influences of civilization, so surely is it on the road to destruction. The several conditions of national prosperity act and react upon one another. Wherever the administration of justice is uncertain, wherever the interests of the majority cannot be made known by legitimate means, there the development of national resources and the power dependent on them is impossible. Nor is it possible to make good soldiers except out of men who have learned under just laws to educate the sense of honor that characterizes an independent man. Certainly they cannot be made out of men who have lived like the submissive slaves of a capricious tyrant.

Accordingly—and here we ignore any consideration of ideals—all nations are interested in the progress of knowledge for the simple reason of self-preservation. Moreover, they are interested not only in the development of the physical sciences and their technical application, but also in the progress of the legal, political, and moral sciences and of accessory historical and philological studies. No nation which desires to be independent and influential can afford to remain behind. And this has not escaped the notice of the cultivated peoples of Europe. Never before has so large a share of public resources been devoted to universities, schools, and scientific institutions. Indeed, we in Heidelberg have this year occasion to congratulate ourselves on another rich endowment granted by our government and our parliament.

I spoke at the beginning of my address about the increased division of labor in science and about the improved organization among scientific investigators. In fact, scientists form a kind of organized army, laboring on behalf of the whole nation, and generally under its direction and at its expense, to augment the stock of such knowledge as may serve to promote industrial enterprise, increase wealth, adorn life, improve political and social relations, and further the moral development of individual citizens. But we shall not consider the immediate practical results of this work; that we leave to those who know no better arguments.

We are convinced that whatever contributes to the knowledge of the forces of nature and to the powers of the human mind is worthwhile and may, in its own proper time, bear practical fruit, very often where we least expect it. Who would have dreamed, when Galvani first touched the muscles of a frog with different metals and noticed their contraction, that eighty years later, by virtue of a process that had its first manifestations in anatomical researches, all Europe would be spanned by wires flashing information from Madrid to St. Petersburg with the speed of lightning? In the hands of Galvani, and at first even in Volta's, electric currents were capable of exerting only the feeblest forces and could only be detected by the most delicate equipment. Had they been neglected on the ground that the investigation promised no immediate practical result, we should now be ignorant of the most important and most interesting of the links among the various forces of nature. When young Galileo, then a student at Pisa, noticed one day during divine services a chandelier swinging backward and forward and satisfied himself, by counting his pulse, that the period of the oscillations was independent of the arc through which the chandelier moved, who could possibly have known that this discovery would eventually enable us, by means of the pendulum, to attain an accuracy in the measurement of time until then deemed impossible and, further, would enable the storm-tossed seaman in the most distant oceans to determine in what degree of longitude he was sailing?

Whoever in scientific investigations looks only for immediate practical results may generally be assured that he will look in vain. The most that science can achieve is complete knowledge and understanding of the actions of natural and moral forces. Each individual scientist must be content to find his reward in his pleasure in new discoveries, in new victories of mind over matter, or in the aesthetic beauty of a well-ordered field of knowledge, where the interconnections and relations of all the elements are clear to the mind and where everything indicates the presence of a ruling intellect. He must rest satisfied with the realization that he too has contributed something to the increasing fund of knowledge on which the dominion of man over all the forces hostile to intelligence rests. He should not expect always to receive recognition and reward commensurate with the value of his work. It is only too true that many a man to whom a monument has been erected after his death would have been delighted to receive during his lifetime a tenth part of the money spent in doing honor to his memory. We must acknowledge, however, that the value of scientific discoveries is now far more fully appreciated than formerly by the general public and that instances of the authors of great advances in science living in obscurity and poverty have become rarer and rarer. Indeed, the governments and peoples of Europe have on the whole admitted it to be their duty to acknowledge distinguished achievements in science by appropriate appointments or special rewards.

The sciences all have one common aim: to establish the rule of intellect over the world. While the moral sciences aim directly at making the content and resources of the intellectual life more abundant and more interesting, and while they seek to separate what is worthwhile in thought from what is worthless, the physical sciences strive indirectly toward the same goal, inasmuch as they labor to make men less and less slaves to the insistent basic necessities of life. Each investigator works in his own area, choosing for himself those tasks for which he is best fitted by his abilities and his training. But each must realize that it is only in conjunction with others that he can further the common work and that it is therefore his duty to make the results of his investigations easily and completely accessible to others. If he does this, he will get assistance from others and will again in his turn be able to render aid to them.

The annals of science are rich in cases indicating how such mutual services have been exchanged, even between areas of science apparently very remote from one another. Historical chronology is based upon astronomical calculations of eclipses of the sun and moon, the accounts of which are preserved in ancient histories. Conversely, many of the important data of astronomy—for instance, the invariability of the length of the day, and the periods of several comets—rest upon ancient historical reports. Again, in recent years physiologists, in particular Brücke, have begun to draw up a complete system of all the vocables that can be produced by the organs of speech and to use this system as the basis of a universal alphabet, adapted to all human languages. Thus physiology has entered the service of comparative philology and has already succeeded in accounting for many apparently anomalous substitutions by showing that they are governed, not by the laws of euphony, as previously believed, but by similarities among the movements of the mouth that produce them. Again, comparative philology gives us information about the relationships among and the migrations of different tribes in prehistoric times, including the degree of civilization that various groups had reached at the time when they separated from others. We can determine these facts because the names which these early peoples learned to give to objects reappeared later as words common to several languages. Thus the study of languages actually gives us historical data for periods for which no other historical evidence exists. Still further, notice the help which, not only the sculptor, but the archaeologist concerned with the investigation of ancient statues gets from anatomy. And (if I may be permitted to refer to my own most recent studies) it is possible, by reference to physical acoustics and to the physiological theory of the sensations of tone, to establish the basic principles upon which our musical system is constructed, something which is essentially within the sphere of aesthetics. In fact, the physiology of the sense organs is most intimately related to psychology, inasmuch as physiology proves that our perceptions are the results of mental processes which do not fall within the sphere of consciousness and which must therefore remain inaccessible to any conscious introspection.

I have been able to mention only some of the most striking instances of this interdependence among the various sciences and only those that could be mentioned in a few words. Naturally, I have tried to choose examples from the most widely separated sciences. Far greater, of course, is the influence which related sciences exert upon one another, but of that I need not speak, for each of you is aware of it from his own experience.

In conclusion, I would say: Let each of us think of himself, not as a man seeking to gratify his own thirst for knowledge, or to promote his own private advantage, or to shine by his own abilities, but rather as a fellow laborer on one great common work, upon which the highest interests of humanity rest. Then surely we shall not fail of our reward in the approval of our own conscience and in the esteem of our fellow citizens. To keep up these relations among all investigators and all branches of knowledge, to animate them all to vigorous cooperation toward their common end—this is the great office of the universities. It is essential that the four faculties shall always go hand in hand, and with this conviction we must strive, insofar as we are able, to press onward toward the fulfillment of our great mission.

6.

RECENT PROGRESS IN THE THEORY OF VISION [1868]

A course of lectures delivered in Frankfurt and Heidelberg and published in the Preussische Jahrbücher, *1868**

I. THE EYE AS AN OPTICAL INSTRUMENT

The physiology of the senses is a borderland in which the two great divisions of human knowledge, the natural and the mental sciences, encroach on one another's domain. Here problems arise which are important for both and which only the combined labor of both can solve.

No doubt the first concern of physiology is with material changes in material organs, and that of the special physiology of the senses is with the nerves and their sensations, insofar as these are excitations of the nerves. But in the course of investigation into the functions of the sensory organs, natural science cannot avoid considering the perception of external objects, which is the result of these excitations of the nerves, for the simple reason that the existence of a particular perception often reveals a nervous excitation which would otherwise have escaped our notice. On the other hand, perception of external objects must always be an act of our powers of comprehension and must therefore be accompanied by consciousness. It is a mental activity. The further exact investigation of this process has been pushed, the more it has revealed to us an ever-widening field of such mental activities, which are involved even in those perceptions which at first sight appear to be most simple and immediate. These concealed activities have been but little discussed because we are so accustomed to regard the perception of an external object as a single, instantaneous act, which does not admit of analysis.

It is scarcely necessary for me to remind my present readers of the funda-

*A number of Helmholtz' brief biographical footnotes, usually stating only the individual's dates and places of birth and death, are omitted. All his substantive footnotes are retained, as are most of those by the original translator, Philip H. Pye-Smith.

mental importance of this field of inquiry to almost every other department of science. For perception by the senses supplies, after all, directly or indirectly, the material of all human knowledge, or at least the stimulus necessary to develop every inborn faculty of the mind. It supplies the basis for man's every action in relation to the outer world. And if this level of mental processes is admitted to be the simplest and lowest of its kind, it is none the less important and interesting, for there is little hope that he who does not begin at the beginning of knowledge will ever arrive at its end.

It is by this path that the art of experiment, which has become so important in natural science, found entrance into the hitherto inaccessible field of mental processes—at first, to be sure, only insofar as it has enabled us to establish, through an investigation of various kinds of sensations, which ones give rise to what images or ideas. But from this beginning already flow many inferences as to the nature of the mental processes underlying perception. It is in this broad sense, therefore, that I shall attempt to give here some account of recent physiological investigations.

I am the more desirous of doing so because I have lately published a complete survey of the field of physiological optics,[1] and I welcome this opportunity to draw together in a compendious form my views and conclusions on the subject, which might escape notice among the numerous details of a book devoted to the specific purposes of natural science. In that work I took great pains to convince myself of the truth of every fact of even the slightest importance by personal observation and experiment. Indeed, there is no longer much controversy on the more important facts. (The chief difference of opinion now concerns the extent of certain individual differences in sense perception.) During the last few years a great number of distinguished investigators have, under the influence of the rapid progress of ophthalmic medicine, studied the physiology of vision; and in proportion as the number of observed facts has increased, they have become more capable of scientific arrangement and explanation. I need not remind those of my readers who are conversant with the subject how much labor must be expended to establish many facts which appear comparatively simple and almost self-evident.

So that what follows may be understood in all its complexity, I shall first discuss the *physical* characteristics of the eye as an optical instrument; next, the *physiological* processes of excitation and conduction in the parts of the

1. *The Handbook of Physiological Optics* was published at Leipzig in 1867.

nervous system which belong to it; and lastly, the *psychological* question—how perceptions are produced by the changes which take place in the optic nerve.

The first part of our inquiry, which cannot be passed over because it is the foundation of what follows, will involve a considerable repetition of what is already generally known so that what is new can be shown in its proper context. It is precisely this aspect of the subject which excites so much interest, for it is the real starting point of that remarkable progress which ophthalmic medicine has made during the last twenty years—a progress which, for its rapidity and scientific character, is perhaps without parallel in the history of the healing art.

Every lover of mankind must rejoice in these achievements, which ward off or cure so much misery which until now we were powerless to treat, but a man of science has special reason to look on them with pride. For this wonderful advance has been achieved, not by groping and lucky finding, but by rigorously pursued investigations, and thus it carries with it the pledge of still further successes. Astronomy was once the pattern from which the other sciences learned how the right method will lead to success; now ophthalmic medicine demonstrates how much may be accomplished in the treatment of disease by extended application of well-understood methods of investigation and by accurate insight into the causal connection of phenomena. It is no wonder that the right sort of men were drawn to an arena which offered to the true scientific spirit—the spirit of patient, cheerful work—a prospect of new and noble victories over the opposing powers of nature. It was because there were so many of these men that the success was so brilliant. Let me name, out of the whole number, a representative of each of the three nations of common origin which have contributed most to the result: Von Graefe in Germany, Donders in Holland, and Bowman in England.

There is another point of view from which this advance in ophthalmology may be regarded—and with equal satisfaction. Schiller says of science:

> Who woos the goddess must not hope the wife.
>
> *Wer um die Göttin freit, suche in ihr nicht das Weib.*[2]

History teaches us—and we shall have an opportunity to observe in the present inquiry—that the most important practical results have sprung unexpectedly

---

2. From Schiller's *Sprüche.* More freely, "Let not him who seeks the love of a goddess expect to find in her the woman."

from investigations which to the ignorant might seem mere busy trifling and which even those better able to judge could regard only with the intellectual interest which pure theoretical inquiry excites.

Of all our senses the eye has always been held the choicest gift of nature, the most marvelous product of her creative force. Poets and orators have celebrated its praises; philosophers have extolled it as a crowning instance of perfection in an organism; and opticians have tried to imitate it as an unsurpassed model. The most enthusiastic admiration of this wonderful organ is only natural when we consider the functions it performs—when we dwell on its power to transcend space, on the swiftness of succession of its brilliant images, and on the wealth of images it provides us. It is by the eye alone that we know the countless shining worlds which fill immeasurable space, the landscapes of our own earth, with all the varieties of sunlight which grace them, the wealth of form and color among flowers, and the strong and happy life that moves in animals. Next to the loss of life itself, that of eyesight is the heaviest.

But even more important than the delight in beauty and the admiration of the majesty of Creation which we owe to the eye, are the certainty and exactness with which we can judge by sight the position, distance, and size of the objects which surround us. For this knowledge is the necessary foundation for all our actions, from moving a needle through a tangled skein of thread to leaping from one cliff to another when life itself depends on the right measurement of the distance. The very success of the movements and actions dependent on the accuracy of the images which the eye gives us is a continual confirmation of that accuracy. If sight were to deceive us as to the position and distance of external objects, we would become aware of the deception immediately when we tried to grasp or approach them. This daily verification by our other senses of the impressions we receive by sight produces so firm a conviction of their absolute and complete truth that the exceptions noted by philosophy or physiology, however well grounded they may seem, have no power to shake it. No wonder, then, that there exists a widespread conviction that the eye is an optical instrument so perfect that none formed by human hands can ever be compared with it, and that its exact, complicated construction should be regarded as the full explanation of the accuracy and variety of its functions.

Actual examinations of the performance of the eye as an optical instrument, carried on chiefly during the last ten years, have brought about a remarkable change in these views, just as the test of facts has disabused our minds of so many similar fancies. But as, in similar cases, reasonable admiration actually increases when the really important functions are more clearly understood and their purpose better estimated, so it may well prove with our more exact knowledge of the eye. For the remarkable performance of this little organ can never be denied, and while we might feel compelled to withdraw our admiration in certain respects, we must experience it afresh in others.

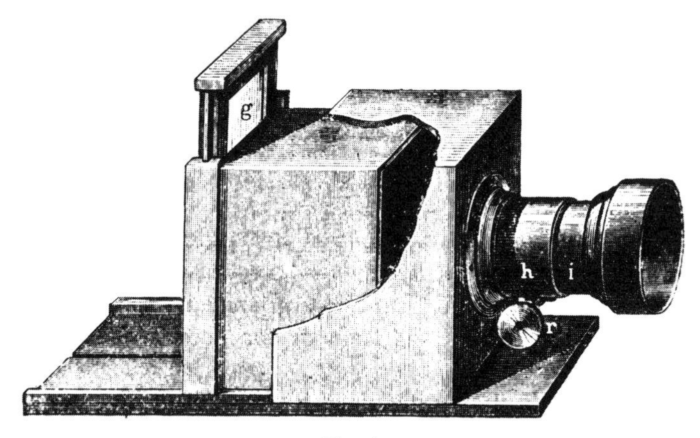

Fig. 1

Regarded as an optical instrument, the eye is a camera obscura. This apparatus is well known in the form used by photographers (Fig. 1). A blackened box constructed of two parts, one of which slides in the other, has in front a combination of lenses (fixed inside the tube *hi*) which refract the incident rays of light and unite them at the back of the instrument into an optical image of the objects which lie in front of the camera. When the photographer first arranges his instrument, he receives the image upon a plate of ground glass, *g*. It is there seen as a small, elaborate picture in its natural colors, more clear and beautiful than the most skillful painter could imitate; it is, however, upside down. The next step is to substitute for this glass a prepared plate upon which the light exerts a permanent chemical effect, stronger on the more brightly illuminated parts, weaker on those which are darker. These chemical changes, having once taken place, are permanent: by their means the image is fixed upon the plate.

The natural camera obscura of the eye (seen in a diagrammatic section in Fig. 2) has its blackened chamber globular instead of cubical and made, not of wood, but of a thick, strong, white substance known as the sclerotic coat.

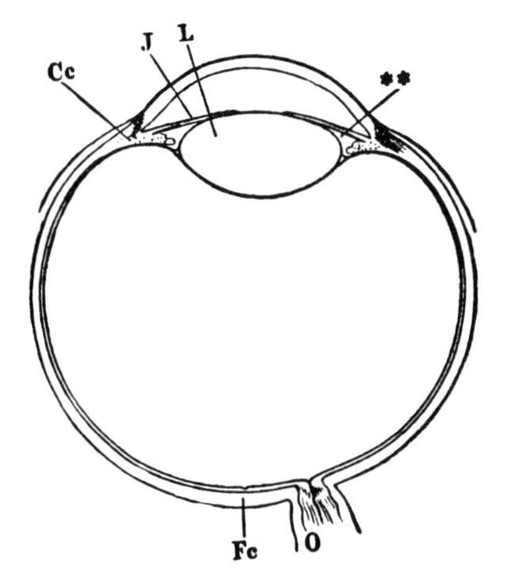

Scale: 2/1

Fig. 2

It is this which is partly seen between the eyelids as the so-called white of the eye. This globular chamber is lined with a delicate coat of winding blood vessels covered inside by black pigment. But the interior of the eye is not empty like the camera; it is filled with a transparent jelly as clear as water.

The lenses of the camera obscura are represented, first, by a convex transparent window like a pane of horn (the cornea), which is fixed in front of the sclerotic coat like a watch glass in front of its metal case. Its fusion with the sclerotic coat and its own firm texture make the cornea's position and curvature constant. But the glass lenses of the photographer are not fixed; they are movable by means of a sliding tube which can be adjusted by a screw (Fig. 1, *r*) so as to bring the objects in front of the camera into focus. The nearer the objects are, the farther the lenses are pushed forward; the farther away they are, the more the lenses are drawn back. The eye has the same task of bringing objects–at one time near, at another distant–into focus at the back of its dark chamber, so some power of adjustment or *accommodation* is necessary. This is accomplished by the movements of the crystalline lens (Fig. 2, *L*), which is placed a short distance behind the cornea. It is covered by a curtain of varying color, the iris (*J*), which is perforated in the center by a round hole, the pupil; the edges of the iris are in contact with the front of the crystalline lens. Through this opening we see through the transparent and, of course, invisible lens the black chamber within.

The crystalline lens is circular, biconvex, and elastic. It is attached at its edge to the inside of the eye by means of a circular band of folded membrane which surrounds it like a plaited ruff; this is called the ciliary body, or zonule of Zinn (Fig. 2, **). The tension of this ring (and so of the lens itself) is regulated by a series of muscular fibers known as the ciliary muscle (*Cc*). When this muscle contracts, the tension of the lens is diminished, and through its physical property of elasticity, its surfaces—chiefly the front one—become more convex than when the eye is at rest. Its refractive power is thus increased, and the images of near objects are brought to a focus on the back of the dark chamber of the eye.

The healthy eye when at rest sees distant objects distinctly; by the contraction of the ciliary muscle it is accommodated for those which are near. This mechanism has been one of the greatest riddles of the physiology of the eye since the time of Kepler; and our present knowledge of its mode of action is of the greatest practical importance, due to the frequency of defects in the power of accommodation. No other problem in optics has given rise to so many contradictory theories as this. The key to its solution was found when the English surgeon Sanson first observed very faint reflections of light through the pupil from the two surfaces of the crystalline lens. He thus acquired a reputation as an unusually careful observer, for this phenomenon was anything but obvious. It can be seen only by strong side illumination, in darkness otherwise complete, and only when the observer takes a certain position—and then all he sees is a faint misty reflection. But this faint reflection was destined to become a shining light in a dark corner of science. It was in fact the first appearance observed in the living eye which came directly from the lens. Sanson immediately applied his discovery to ascertain whether the lens was in its place in cases of impaired vision. Max Langenbeck made the next step by observing that the reflections from the lens alter during accommodation. These alterations were employed by Cramer of Utrecht—and also independently by the present writer—to arrive at an exact knowledge of all the changes which the lens undergoes during the process of accommodation. I succeeded in applying to the movable eye in a modified form the principle of the heliometer, an instrument by which astronomers are able to measure so accurately the small distances between stars, in spite of their constant apparent motion in the heavens, that they can thus sound the depths of the region of the fixed stars. An instrument constructed for the purpose, the ophthalmometer, enables us to measure in the living eye the curvature of the cornea and of the two surfaces of the lens, the distance of these from each other, etc., with greater precision than had been possible even after death. By this means we can ascertain the entire range of changes of the optical apparatus of the eye insofar as it affects accommodation.

The physiological problem was thus solved. Ophthalmologists, especially Donders, next investigated the individual defects of accommodation which give rise to the conditions known as near- and farsightedness. It was necessary to devise trustworthy methods to ascertain the precise limits of the power of accommodation even with inexperienced and uninstructed patients. It became apparent that very different conditions had been confused with near- and farsightedness and that this confusion had made the choice of suitable glasses uncertain. It was also discovered that some of the most obstinate and obscure affections of the sight, formerly reputed to be "nervous," were attributable simply to defects of accommodation and could be readily corrected by using suitable glasses. Moreover, Donders proved that the same defects of accommodation are the most frequent cause of squinting, and Von Graefe showed that neglected, progressive shortsightedness tends to produce a severely dangerous expansion and deformity of the back of the globe of the eye. Thus relations were recognized, where least expected, between the newly discovered optical mechanism and important diseases. The result was no less beneficial to the patient than interesting to the physiologist.

We must now speak of the part which receives the optical image when brought to a focus in the eye. This is the retina, a thin membranous expansion of the optic nerve which forms the innermost of the coats of the eye. The optic nerve (Fig. 2, *O*) is a cylindrical cord which contains a multitude of minute fibers protected by a strong tendinous sheath. The nerve enters the interior of the eye from behind, in the middle of its posterior hemisphere, but rather to the inner (nasal) side. Its fibers then spread out in all directions over the surface of the retina. They end by becoming connected with ganglion cells and nuclei, like those found in the brain, and through them with structures not found elsewhere, called rods and cones. The rods are slender cylinders; the cones, or bulbs, are somewhat thicker, flask-shaped structures. All are ranged perpendicular to the surface of the retina and closely packed together, so as to form a regular mosaic layer behind it. Each rod is connected with one of the minutest nerve fibers, each cone with one somewhat thicker. This layer of rods and cones (also known as *membrana Jacobi*) has been proved by direct experiments to be the really sensitive layer of the retina, the only structure in which the action of light is capable of producing a nervous excitation.

There is in the retina a remarkable spot which is near its center a little to the outer (temporal) side and which from its color is called the yellow spot. The retina is here somewhat thickened, but in the middle of the yellow spot is found a depression, the fovea centralis, where the retina is reduced to those elements which are absolutely necessary for exact vision. Fig. 3 (from Henle) shows a thin transverse section of this central depression taken from a retina which had been hardened in alcohol. *Lh* (Lamina hyalina, membrana limitans) is an elastic membrane which separates the retina from the vitreous. The cones (*b*) are here smaller than elsewhere, measuring only 1/400th of a millimeter in diameter, and form a close, regular mosaic. The other, more or less opaque elements of the retina are seen to be wanting, except the corpuscles (*g*), which belong to the cones. At *f* are the fibers which unite these with the rest of the retina. This consists of a layer of fibers of the optic nerve (*n*) in front and then two layers of nerve cells known as the internal and external ganglion layers (*gli, gle*), with a stratum of fine granules (*gri*) between them. All these parts of the retina are absent at the bottom of the fovea centralis; their gradual thinning away at its borders is seen in the diagram. Nor do the blood vessels of the retina enter the fovea; they end in a circle of delicate capillaries around it.

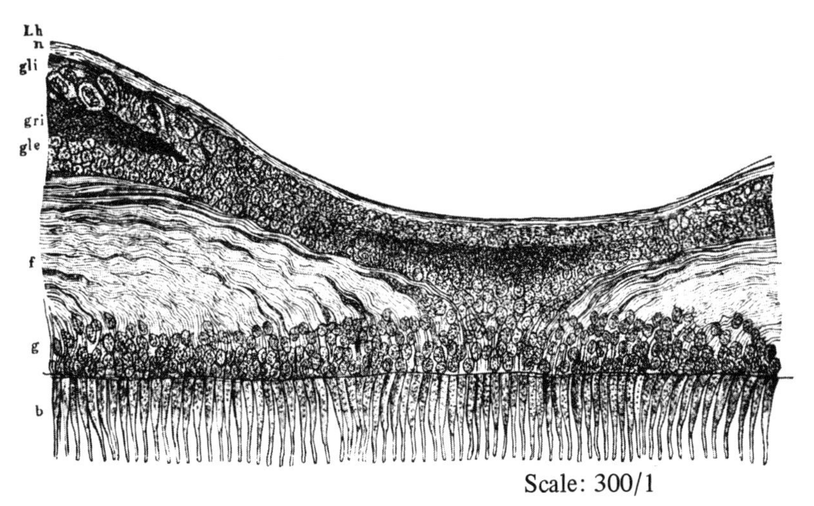

Fig. 3

This fovea, or pit of the retina, is of great importance for vision, since it is the spot where the most exact discrimination of distances is made. The cones are here packed most closely together and receive light which has not been impeded by other, semitransparent parts of the retina. We may assume that a single nervous fibril runs from each of these cones through the trunk of the optic nerve to the brain, without touching its neighbors, and there produces its special impression, so that the excitation of each individual cone produces a distinct and separate effect upon the sense.

The production of optical images in a camera obscura depends on the well-known fact that the rays of light from an illuminated object are broken or refracted, as they pass through the lenses of the instrument, into new directions which bring them all to a single point, the focus, at the back of the camera. An ordinary burning glass has the same property. If we allow the rays of the sun to pass through it and hold a sheet of white paper at the proper distance beyond it, we may notice two effects. In the first place (and this is often disregarded), the burning lens, although made of transparent glass, throws a shadow like any opaque body. Next we see in the middle of this shadow a spot of dazzling brilliance, the image of the sun. The rays which, if the lens had not been there, would have illuminated the whole space occupied by the shadow, are concentrated by the refracting power of the burning glass upon the bright spot in the middle, and so both light and heat are more intense there than where the unrefracted solar rays fall. If, instead of the disc of the sun, we choose a star or any other point as the source of light, its light will be similarly drawn into a point at the focus of the lens, and the image of the star will appear upon the white paper. If there is another fixed star nearby, its light will be collected at a second illuminated point on the paper; and if the star happens to send out red rays, its image on the paper will appear red. The same will be true of any number of neighboring stars, the image of each corresponding to it in brilliance, color, and relative position. If, instead of a multitude of separate luminous points, we have a continuous series of them in a bright line or surface, a similar line or surface will be produced upon the paper. But here also, if the piece of paper is at the proper distance, all the light that proceeds from any one point will be brought to a focus at a point which corresponds to it in strength and color of illumination. As a corollary, no point of the paper will receive light from more than a single point of the object.

If we replace our sheet of white paper with a prepared photographic plate, each point of its surface will be altered by the light which is concentrated on it. This light is derived from the corresponding point in the object, and the surface point matches it in relative intensity. Hence the changes which occur on the plate will correspond to the chemical intensity of the rays which fall upon it.

This is exactly what takes place in the eye. Instead of the burning glass we have the cornea and crystalline lens; and instead of the piece of paper, the retina. If a well focused image is thrown upon the retina, each cone will be reached by as much light as proceeds from the corresponding point in the field of vision. The nerve fiber which arises from each cone will thus be excited only by the light proceeding from the corresponding point in the field, while other nerve fibers will be excited by the light proceeding from other points of the field. Fig. 4 illustrates this effect: the rays which come from the point *A* in the object of vision are so bent that they all unite at *a* on the retina, while those from *B* unite at *b*. As a result, the light from each separate bright point of the field of vision excites a separate sensation; the variation in brightness among the points of the field of vision can be detected by the sense; and the separate sensations each arrive separately at the seat of consciousness.

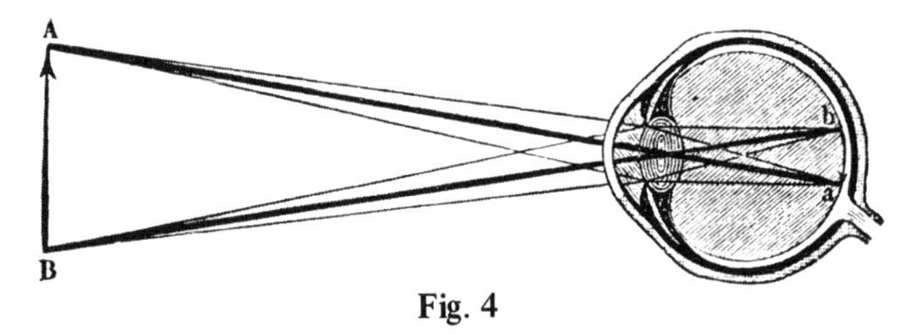

Fig. 4

If we compare the eye with other optical instruments, we observe the advantage it has over them. This is its very large field of vision—for each eye separately, 160° (nearly two right angles) laterally and 120° vertically; and for both together, somewhat more than two right angles from left to right. The field of view of artificial instruments is usually very small, and it becomes smaller as the size of the image increases.

But we must also admit that we are accustomed to expect in these instruments complete precision of the image in its entire extent, while it is only necessary for the image on the retina to be exact over a very small surface, namely, that of the yellow spot. The diameter of the central pit corresponds in the field of vision to an angular magnitude which can be covered by the nail of one's forefinger when the hand is stretched out as far as possible. In this small part of the field our power of vision is so accurate that it can distinguish the distance between two points of only 1′ angular magnitude, that is, a distance equal to one sixtieth of the diameter of the fingernail. This distance corresponds to the width of one of the cones of the retina. All the other parts of the retinal image are seen imperfectly, and the more so, the nearer they fall to the edge of the retina. Thus the image we receive by the eye is like a picture which is minutely and elaborately finished in the center but only roughly sketched in at the borders. But even though at each instant we only see a very small part of the field of vision accurately, we see this in the context of what surrounds it, and we see enough of this outer, larger part of the field to notice any striking object and particularly any change that takes place in it. (All of this is unattainable in a telescope.)

But if the objects are too small, we cannot discern them at all with the greater part of the retina.

> When, lost in boundless blue on high,
> The lark pours forth his thrilling song,
>
> *Wenn über uns, im blauen Raum verloren*
> *Ihr schmetternd Lied die Lerche singt.*[3]

the "ethereal minstrel" is "lost" until we can bring her image to a focus upon the central pit of our retina. Only then are we able to see her. For to *look* at anything means to place the eye in such a position that the image of the object falls on that small region of perfectly clear vision. This we may call direct vision, while indirect vision is that exercised with the lateral parts of the retina—indeed, with all except the yellow spot.

The deficiencies which result from the inexactness of vision and the smaller number of cones in most of the retina are compensated for by the rapidity with which we can turn the eye to one point after another of the field of vision. It is this rapidity of movement which really constitutes the chief advantage of the eye over other optical instruments. Indeed, the peculiar way in which we are accustomed to give our attention to external objects—turning it to one thing at a time, and as soon as that has been taken in, hastening to another—enables the sense of vision to accomplish as much as is necessary, so we have practically the same advantage as if we enjoyed an accurate view of the whole field of vision at once.

It is not until we begin to examine our sensations closely that we even

3. The lines in the well-known passage of *Faust*.

become aware of the imperfections of indirect vision. Whatever we want to see, we look at and see it accurately. What we do not look at, we do not as a rule care for at the moment, and so we do not notice how imperfectly we see it. Indeed, it is only after long practice that we are able to turn our attention to an object in the field of indirect vision (as is necessary for some physiological observations) without looking at it and thus bringing it into direct view. It is equally difficult to fix the eye on an object for the number of seconds required to produce the phenomenon of an afterimage. To get a well-defined one requires a good deal of practice.

The importance of the eye as an organ of expression depends largely on the same fact. For the movements of the eyeball–its glances–are among the most direct signs of the movement of the attention, that is, the movements of the mind, of the person who is looking at us.

Just as quickly as the eye turns upward, downward, and from side to side, the accommodation changes, so as to bring into focus the object to which our attention is at the moment directed. Thus near and distant objects pass in rapid succession into accurate view. All these changes of direction and accommodation take place far more slowly in artificial instruments. A photographic camera can never show near and distant objects clearly at once, nor can the eye; but the eye shows them so rapidly one after another that most people, who have not thought about how they see, do not realize that there is any change at all.

Let us now examine some defects in the optical properties of the eye. We shall pass over the individual defects of accommodation, which have already been mentioned as the cause of near- and farsightedness. These defects appear to be partly the result of our artificial way of life, partly of the changes of old age. Elderly persons lose their power of accommodation, and their range of clear vision becomes confined within more or less narrow limits. To exceed these they must resort to the aid of glasses.

But there is another quality which we expect of optical instruments, namely, that they shall be free from dispersion, or achromatic. Dispersion of light depends on the fact that the colored rays, which united make up the white light of the sun, are not refracted in exactly the same degree by any transparent substance known. Hence the size and position of the optical images thrown by these differently colored rays are not quite the same: they do not perfectly overlap each other in the field of vision. The white surface of the image appears fringed with violet or orange, according as the red or blue rays are broader, and this of course detracts somewhat from the sharpness of the outline.

Many of my readers know what a curious part the inquiry into the eye's chromatic dispersion has played in the invention of achromatic telescopes. It is a celebrated instance of how a right conclusion may sometimes be drawn from two false premises. Newton thought he had discovered a relation between the refractive and dispersive powers of various transparent materials, from which it followed that achromatic refraction was impossible. Euler, on the other hand, concluded that since the eye is achromatic, the relation discovered by Newton could not be correct. Reasoning from this assumption, he constructed theoretical rules for making achromatic instruments, and Dolland carried them out. But Dolland himself observed that the eye could not be achromatic because its construction did not conform to Euler's rules. At last Fraunhofer actually measured the degree of chromatic aberration of the eye.

An eye constructed to bring red light from an infinite distance to a focus on the retina can do the same with violet rays only from a distance of two feet. With ordinary light this is not noticed, for these extreme colors are the least luminous of all and so the images they produce are scarcely observed beside the more intense images of the intermediate yellow, green, and blue rays. But the effect is very striking when we isolate the extreme rays of the spectrum by means of violet glass. Glass colored with cobalt oxide allows the red and blue rays to pass but stops the green and yellow rays, that is, the brightest rays of the spectrum. If those of my readers who have eyes of ordinary focal distance will look at a lighted streetlamp from a distance through a violet glass, they will see a red flame surrounded by a broad, bluish-violet halo. This is the dispersive image of the flame thrown by its blue and violet light. The phenomenon is a simple, complete proof of the fact of chromatic aberration in the eye. The reason why this defect is so little noticed under ordinary circumstances, and why it is in fact somewhat less than a glass instrument of the same construction would have, is that the chief refractive medium of the eye is water, which possesses a less dispersive power than glass.[4] In any case, the chromatic aberration of the eye, though present, does not materially affect vision with ordinary white illumination.

4. The diffraction in the eye is, however, rather greater than an instrument made with water would produce under the same conditions.

A second defect which is of great importance in optical instruments of high magnifying power is what is known as spherical aberration. Spherical refracting surfaces unite the rays which proceed from a luminous point into approximately a single focus, only when each ray falls nearly perpendicularly upon the corresponding part of the refracting surface. If all those rays which form the center of the image are to be exactly united, a lens with other than a spherical surface must be used, and this cannot be made with sufficient mechanical perfection. Now the eye has refracting surfaces which are partly elliptical, and here again the natural prejudice in its favor led to an erroneous belief that spherical aberration was thus prevented. But this was a still greater blunder.

More accurate investigation showed that much greater defects than spherical aberration are present in the eye—defects which are easily avoided in optical instruments by a little care in the manufacture, and compared with which the amount of spherical aberration is very unimportant. The careful measurements of the curvature of the cornea—first made by Senff of Dorpat; next, by myself with a better adapted instrument (the writer's ophthalmometer already referred to); and afterwards carried out in numerous cases by Donders, Knapp, and others—have proved that the cornea of most human eyes is not a perfectly symmetrical curve but is variously bent in different directions. I have also devised a method of testing the centering of a living eye, that is, ascertaining whether the cornea and the crystalline lens are symmetrically placed with regard to their common axis. By this means I discovered in the eyes which I examined slight but distinct deviations from accurate centering. The result of these two defects of construction is the condition called astigmatism, which is found more or less in most human eyes and prevents our seeing vertical and horizontal lines at the same distance perfectly clearly at the same time. If the degree of astigmatism is excessive, it can be obviated by the use of glasses with cylindrical surfaces, a circumstance which has lately much attracted the attention of ophthalmologists.

Nor is this all. A refracting surface which is imperfectly elliptical, such as an ill-centered telescope, does not give a single illuminated point as the image of a star. According to the surface and arrangement of the refracting media, it may give elliptical, circular, or linear images. The images of an illuminated point, as the human eye brings them to focus, are even more inaccurate: they are irregularly radiated. The reason lies in the construction of the crystalline lens, the fibers of which are arranged around six diverging axes (Fig. 5). The rays we see around stars and other distant lights are images of the radiated

Fig. 5

structure of our own lens; and the universality of this optical defect is proved by the fact that any figure with diverging rays is called star-shaped. It is from the same cause that the moon, while her crescent is still narrow, appears to many persons as a double or triple image.

It is not too much to say that if an optician wanted to sell me an instrument which had all these defects, I should think myself quite justified in blaming his carelessness in the strongest terms and giving him back his instrument. Of course I shall not do this with my eyes. I shall be only too glad to keep them as long as I can, defects and all. Still, the fact that however bad they may be, I can get no others, does not at all diminish their defects, so long as I maintain the narrow but indisputable position of a critic on purely optical grounds.

We are not yet done, however, with the list of the defects of the eye.

We expect that the optician will use good, clear, perfectly transparent glass for his lenses. Otherwise a bright halo will appear around each illuminated surface in the image: what should be black will look grey, and what should be white will be dull. But this is precisely what occurs in the image our eyes give us of the outer world. (The obscurity of dark objects when seen near very bright ones depends essentially on this defect.) If we throw a strong light[5] through the cornea and crystalline lens, they appear dingy white, less transparent than the aqueous humor which lies between them. This defect

5. E.g., from a lamp, concentrated by a bull's-eye condenser.

is most apparent in the blue and violet rays of the solar spectrum, for there the phenomenon of fluorescence[6] comes in to increase it.

Although the crystalline lens looks so beautifully clear when taken out of the eye of an animal just killed, it is in fact far from optically uniform in structure. One can see the shadows and dark spots within one's own eye (the so-called entoptic objects) by looking at an extensive bright surface—the clear sky, for instance—through a very narrow opening. These shadows are chiefly due to the fibers and spots in the lens.

There are also a number of minute fibers, corpuscles, and folds of membrane which float in the vitreous humor and are seen when they come close in front of the retina, even under the ordinary conditions of vision. They are then called muscae volitantes, because when the observer tries to look at them, they naturally move with the movement of the eye and, seeming continually to flit away from the point of vision, look like flying insects. These objects are present in everyone's eyes and usually float in the highest part of the globe of the eye, out of the field of vision, whence on any sudden movement of the eye they are dislodged and swim freely in the vitreous humor. They may occasionally pass in front of the central pit and so impair sight. It is a remarkable proof of the way in which we observe, or fail to observe, the impressions made on our senses that these muscae volitantes often appear something quite new and disquieting to persons whose sight is beginning to suffer from any cause, although these objects have doubtless been dislodged many times before.

A knowledge of the way in which the eye is developed in man and other vertebrates explains these irregularities in the structure of the lens and the vitreous humor. Both are produced by an invagination of the integument of the embryo. A dimple is first formed, then deepens to a round pit, and then expands until its orifice becomes relatively minute; it is finally closed, and the pit becomes completely shut off. The cells of the scarfskin which line this hollow form the crystalline lens, the true skin beneath them becomes its capsule, and the loose tissue which underlies the skin is developed into the

---

6. This term is given to the property which certain substances possess of becoming for a time faintly luminous as long as they receive violet or blue light. The bluish tint of a solution of quinine and the green color of uranium glass depend on this property. The fluorescence of the cornea and crystalline lens appears to depend upon the presence in their tissue of a very small quantity of a substance like quinine. For the physiologist this property is most valuable, for by its aid he can see the lens in a living eye by throwing on it a concentrated beam of blue light; thus he can ascertain that it is placed close behind the iris, not separated by a large "posterior chamber," as was long supposed. For seeing, however, the fluorescence of the cornea and lens is simply disadvantageous.

vitreous humor. The mark where the neck of the fossa was sealed can still be recognized as one of the entoptic images of many adult eyes.

The last defect of the human eye which must be noticed is the existence of certain inequalities of the surface which receives the optical image. Not far from the center of the field of vision is a break in the retina where the optic nerve enters. Here there is nothing but nerve fibers and blood vessels; since the cones are absent, any rays of light which fall on the optic nerve itself are unperceived. This blind spot produces a corresponding gap in the field of vision, where nothing is visible. (Fig. 6 shows the posterior half of the globe of a right eye which has been cut across. *R* is the retina with its branching blood vessels. The point from which these diverge is that at which the optic nerve enters. To the reader's left is the yellow spot.)

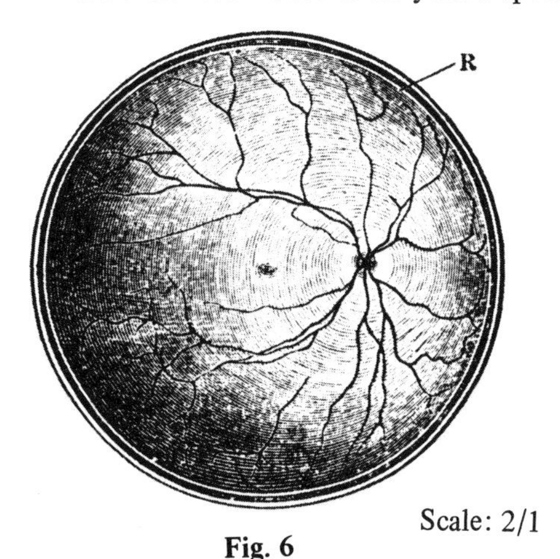

Scale: 2/1

Fig. 6

The gap caused by the presence of the optic nerve is no slight one. It is about 6° in horizontal and 8° in vertical dimension. Its inner border is about 12° horizontally distant from the temporal, or external, side of the center of distinct vision. The way to recognize this blind spot most readily is doubtless known to many of my readers. Take a sheet of white paper, and mark on it a little cross; then to the right of this, on the same level and about three inches off, draw a round black spot half an inch in diameter. Now, holding the paper at arm's length, shut the left eye, fix the right upon the cross, and bring the paper gradually nearer. When it is about eleven inches from the eye, the black spot will suddenly disappear; it will again come into sight as the paper is moved nearer. This blind spot is so large that it might prevent our seeing eleven full moons if placed side by side, or a man's face at a distance of only six or seven feet. Mariotte, who discovered the phenomenon, amused Charles II of England and his courtiers by showing them how they might see each other with their heads cut off.

There are, in addition, a number of smaller gaps in the field of vision in which a small bright point—a fixed star, for example—may be lost. These are caused by the blood vessels of the retina. The vessels run in the front layers and so cast their shadow on the part of the sensitive mosaic which lies behind them. The larger ones shut off the light from reaching the rods and cones altogether; the more slender at least limit its amount. These splits in the image presented by the eye may be recognized by making a hole in a card with a fine needle and looking through it at the sky, moving the card a little from side to side all the time. A still better experiment is to throw sunlight through a small lens upon the white of the eye at the outer angle (lateral canthus), while the globe is turned as much as possible inwards. The shadow of the blood vessels is then thrown across onto the inner wall of the retina, and we see them as gigantic branching lines, like Fig. 6 magnified.

These vessels lie in the front layer of the retina itself, and of course their shadow can be seen only when it falls on the proper sensitive layer. Thus this phenomenon furnishes a proof that the hindmost layer is that which is sensitive to light. By its help it has become possible actually to measure the distance between the sensitive and the vascular layers of the retina. It is done as follows: If the focus of the light thrown onto the white of the eye (the sclerotic) is moved slightly backward and forward, the shadow of the blood vessels and its image in the field of vision will, of course, move also. The extent of these movements can be easily measured. From these data Heinrich Müller, of Würzburg (whose too early loss to science we still deplore), determined the distance between the two foci and found it to equal exactly the thickness which actually separates the layer of rods and cones from the vascular layer of the retina.

The condition of the point of clearest vision (the yellow spot, or macula lutea) is disadvantageous in another way. It is less sensitive to weak light than the other parts of the retina. It has long been known that many stars of inferior magnitude—for example, the Coma Berenicæ and the Pleiades—are seen more brightly if looked at somewhat obliquely than when their rays fall full upon the eye. This can be proved to depend partly on the yellow color of the macula, which weakens blue more than other rays. It may also be partly the result of the absence of vessels at this yellow spot (as noted above), which interferes with its free communication with the life-giving blood.

All these imperfections would be exceedingly troublesome in an artificial camera obscura and in the photographic picture it produced. But they are not so in the eye—so little, indeed, that it was very difficult to discover some of them. The reason for their not interfering with our perception of external objects is not simply that we have two eyes, so that one makes up for the defects of the other. For even when we do not use both, as in the case of persons blind in one eye, the impression we receive from the field of vision is free from the defects which the irregularity of the retina would be expected to occasion. The chief reason is that we are continually moving the eye. Furthermore, the imperfections almost always affect those parts of the field to which we are not at the moment directing our attention.

But it remains after all a wonderful paradox that we are so slow to observe these and other peculiarities of vision (such as the afterimages of bright objects), so long as they are not strong enough to prevent our seeing external objects. It is a fact which we meet constantly, not only in optics, but in studying the perceptions produced by other senses on the consciousness. The difficulty with which we perceive the defect of the blind spot is well shown by the history of its discovery. Its existence was first demonstrated by theoretical arguments. While the long controversy whether the perception of light resided in the retina or the choroid was still undecided, Mariotte asked himself what perception there is where the choroid is deficient. He made experiments to ascertain this point and in the course of them discovered the blind spot. Millions of men had used their eyes for ages, thousands had thought over the nature and cause of their functions, and yet it was only by a remarkable combination of circumstances that a simple phenomenon was noticed which would apparently have revealed itself to the slightest observation. Even now, anyone who tries for the first time to repeat the experiment which demonstrates the existence of the blind spot, finds it difficult to divert his attention from the fixed point of clear vision without losing sight of it in the attempt. It is only by long practice in optical experiments that even an experienced observer is able, as soon as he shuts one eye, to recognize the blank space in the field of vision which corresponds to the blind spot.

Other phenomena of this kind have only been discovered by accident and usually by persons whose senses were peculiarly acute and whose power of observation was unusually stimulated. Among these may be mentioned Goethe, Purkinje, and Johannes Müller. When a subsequent observer tries to repeat with his own eyes these experiments as he finds them described, it is of course easier for him than for the discoverer; but even now there are many phenomena described by Purkinje which have never been seen by anyone else, although it cannot be certainly held that they depended on individual peculiarities of this acute observer's eyes.

The phenomena of which we have spoken, as well as a number of others, may be explained by the general rule that it is much easier to recognize any change in the condition of a nerve than a constant, equable impression on it. In accordance with this rule, peculiarities in the excitation of separate nerve fibers which are equally present during our whole lifetime (such as the shadow of the blood vessels of the eye, the yellow color of the central pit of the retina, and most of the fixed entoptic images) are never noticed at all. If we want to observe them, we must employ unusual modes of illumination and, particularly, constant change of its direction. According to our present knowledge of the conditions of nervous excitation, it seems to me very unlikely that we have here to do with a simple property of sensation. It must, I think, be explained rather as a phenomenon belonging to our power of attention, and I now only touch upon the question in passing, since its full discussion will come afterwards in its proper connection.

So much for the physical properties of the eye. If I were asked why I have spent so much time explaining its imperfection, I would answer, as I said at first, that my object was not to depreciate the performance of this wonderful organ or to diminish our admiration of its construction. My object was to make the reader understand, at the first step of our inquiry, that it is not any mechanical perfection of our sensory organs which secures for us such wonderfully true and exact impressions of the outer world. The next section of this inquiry will introduce much bolder and more paradoxical conclusions than any I have yet stated.

We have now seen that the eye in itself is not by any means so complete an optical instrument as it at first appears. Its extraordinary value depends upon the way we use it. Its perfection is practical, not absolute, consisting not in the avoidance of every error, but in the fact that all its defects do not prevent its rendering us the most important and varied services. From this point of view, the study of the eye gives us a deep insight into the true character of organic adaptation generally. This consideration becomes still more interesting when brought into relation with the great and daring conceptions which Darwin has introduced into science, as to the means by which the progressive perfection of the races of animals and plants has been carried on.

Wherever we scrutinize the construction of physiological organs, we find the same character of practical adaptation to the wants of the organism, although there is perhaps no other instance which we can follow in such detail as that of the eye. For the eye has every possible defect that can be found in an optical instrument, as well as some which are peculiar to itself. They are all so counteracted that the inexactness of the image which results from their presence very little exceeds, under ordinary conditions of illumination, the limits which are set to the delicacy of sensation by the dimensions of the retinal cones. But as soon as we make our observations under other than ordinary conditions, we become aware of the chromatic aberration, the astigmatism, the blind spots, the venous shadows, the imperfect transparency of the media, and all the other defects of which I have spoken.

The adaptation of the eye to its function is therefore most complete, as is seen in the very limits which are set to its defects. Here the result which may be reached by innumerable generations working under the Darwinian law of inheritance coincides with what the wisest Wisdom may have devised beforehand. A sensible man will not cut firewood with a razor, and so we may assume that analogously each step in the refinement of the eye must have made the organ more vulnerable or slower in its development. We must also bear in mind that soft, watery animal textures are always unfavorable and difficult material for an instrument of the mind. One result of this mode of construction of the eye, whose importance we shall see by and by, is that clear, complete perceptions of external objects by the sense of sight are possible only when we direct our attention to one part after another of the field of vision in the manner partly described above. We shall also take notice of other conditions which tend to work in the same direction.

But we are apparently not yet much nearer to understanding vision. We have made only one step: we have learned how the optical arrangement of the eye enables it to separate the rays of light which come in from all parts of the field of vision and then to bring together again all those that have proceeded from each single point, so that they may produce their effect upon a single fiber of the optic nerve. Let us now see how much we know of the *sensations* of the eye and how far this will bring us toward the solution of the riddle of vision.

II. THE SENSATION OF SIGHT

In the first section we followed the course of the rays of light as far as the retina and saw the result produced by the peculiar arrangement of the optical apparatus. The light which is reflected from the separate illuminated points of external objects is again united in the sensitive terminal structures of separate nerve fibers, each of which is activated without affecting its neighbors. At this point the older physiologists thought they had solved the problem, insofar as it appeared to them capable of solution. External light fell directly upon a sensitive nervous structure in the retina and was, it seemed, directly felt there.

But during the last century, and still more during the first quarter of this, our knowledge of the processes which take place in the nervous system became so far advanced that as early as 1826, Johannes Müller (in his great work on the *Comparative Physiology of Vision,* which marks an epoch in science), was able to lay down the most important principles of the theory of the sensations derived from the senses. These principles have not only been confirmed in all important respects by subsequent investigation, but have proved capable of even more extensive application than this eminent physiologist could have suspected.

The conclusions at which he arrived are generally designated the theory of the specific energies of the senses. They are no longer so novel that they can be reckoned among the latest advances of the theory of vision, which form the subject of the present essay. Moreover, they have been frequently expounded in a popular form by others, as well as by myself.[7] But that aspect of the theory of vision with which we are now occupied is little more than a further development of the theory of the specific energies of the senses. I must therefore beg my reader to forgive me if, in order to give him a comprehensive view of the whole subject, I bring before him much which he already knows, while I also introduce the more recent additions to our knowledge in their appropriate places.

All that we perceive in the external world is brought to consciousness by means of certain changes which are produced in our sensory organs by external impressions and transmitted to the brain by the nerves. It is in the brain that these impressions first become conscious sensations and are combined so as to produce our conceptions of the objects surrounding us. If the nerves which convey these impressions to the brain are cut through, the sensations—and the perception due to the impressions—immediately cease. In the case of the eye, the visual perception is produced, not directly in each retina, but in the brain itself by means of the impressions transmitted to it from both eyes. The proof of this lies in the fact (which I shall later explain more fully) that

---

7. Helmholtz, "The Nature of Human Sensations," *Königsberger Naturwissenschaftliche Unterhaltungen,* iii (1852); and "Human Vision," a popular scientific lecture (Leipzig, 1855).

the visual image of a solid object of three dimensions is produced only by a combination of the impressions derived from both eyes.

What we directly apprehend is, not the immediate action of the external exciting cause upon the ends of our nerves, but only the changed condition of the nervous fibers, which we call the state of excitation, or functional activity. All the nerves of the body, so far as we now know, have the same structure, and the change which wc call excitation is in each of them a process of precisely the same kind, regardless of the function it subserves. For while the task of some nerves is that (already mentioned) of carrying sensitive impressions from the external organs to the brain, others convey voluntary impulses in the opposite direction, from the brain to the muscles, causing them to contract and so moving the limbs. Still other nerves carry an impression from the brain to certain glands and call forth their secretion, or to the heart and blood vessels and regulate the circulation. But the fibers of all these nerves are the same clear cylindrical threads of microscopic minuteness, containing the same oily, albuminous material. It is true that there are some differences in the diameter of the fibers, but so far as we know, this is due to such subsidiary things as the necessity for a certain strength or for room to accommodate a certain number of independent conducting fibers. It appears to have no relation to their peculiarities of function.

Moreover, all nerves have the same electromotor actions, as the researches of du Bois-Reymond prove. In all nerves the condition of excitation is called forth by the same mechanical, electrical, chemical, or thermal changes. It is propagated with the same speed (about one hundred feet per second) to each end of the fibers and produces the same changes in the electromotor properties of the nerves. Lastly, all nerves die when submitted to the same conditions and, with a slight apparent difference according to their thickness, undergo the same coagulation of their contents. In short, all that we can ascertain of nervous structure and function, apart from the action of the other organs with which they are united and in which during life we see the evidence of their activity, is precisely the same for all the different kinds of nerves. Very recently the French physiologists Philippeau and Vulpian, after dividing the motor and sensitive nerves of the tongue, succeeded in getting the upper half of the sensitive nerve to unite with the lower half of the motor. After the wound had healed, they found that irritation of the upper half, which in normal conditions would be felt as a sensation, now excited the motor branches below and caused the muscles of the tongue to move. We conclude from these facts that all of the differences in the excitation of different nerves depend upon the difference of the organs to which the nerve is united and to which it transmits the state of excitation.

The nerve fibers have been often compared with telegraphic wires traversing a country, and the comparison is well suited to illustrate this striking and important peculiarity of their mode of action. In the telegraphic network we find everywhere the same copper or iron wires carrying the same kind of movement, a stream of electricity, but producing very different results in the various stations according to the auxiliary apparatus with which they are connected. At one station the effect is the ringing of a bell, at another a signal is moved, and at a third a recording instrument is set to work. Chemical decompositions may be produced which will serve to spell out the messages, and even the human arm may be moved by electricity so as to convey telegraphic signals. When the Atlantic cable was being laid, Sir William Thomson found that the slightest signals could be recognized by the sense of taste, if the wire were laid upon the tongue. Again, a strong electric current may be transmitted by telegraphic wires in order to ignite gunpowder for blasting rocks. In short, every one of the hundred different actions which electricity is capable of producing may be called forth by a telegraphic wire laid to whatever spot we please, and it is always the same process in the wire itself which leads to these diverse consequences.

Nerve fibers and telegraphic wires are equally striking examples to illustrate the doctrine that the same cause may, under different conditions, produce different results. However commonplace this may now sound, mankind had to work long and hard before it was understood and before this doctrine replaced the belief previously held in the constant, exact correspondence between cause and effect. We can scarcely say that the truth is even now universally recognized, since in our present subject its consequences have been till lately disputed.

As motor nerves, when irritated, produce movement because they are connected with muscles, and as glandular nerves produce secretion because they lead to glands, so sensitive nerves, when they are irritated, produce sensation because they are connected with sensitive organs. But we have very different kinds of sensation. In the first place, the impressions derived from external objects fall into five groups, entirely distinct from one another, corresponding to the five senses. Their difference is so great that it is not possible to compare in quality a sensation of, say, light with one of sound or of smell. We shall name this difference, so much deeper than that between comparable qualities of the same sense, a difference in the *mode,* or *kind,* of sensation. The differences among impressions belonging to the same sense (for example, the differences among the various sensations of color) we shall call a difference of *quality*.

Whether by the irritation of a nerve we produce a muscular movement, a secretion, or a sensation depends upon whether we are dealing with a motor, a glandular, or a sensitive nerve and not at all upon the means of irritation we may use. It may be giving an electric shock, or tearing the nerve, or cutting it through, or moistening it with a solution of salt, or touching it with a hot wire. In the same way (this great step forward was made by Johannes Müller), the kind of sensation which will ensue when we irritate a sensitive nerve—whether an impression of light, or of sound, or of feeling, or of smell, or of taste will be produced—depends entirely upon which sense the excited nerve subserves and not at all upon the method of excitation we adopt.

Let us now apply this to the optic nerve, which is the object of our present inquiry. In the first place, we know that no kind of action upon any part of the body except the eye and the nerve which belongs to it can ever produce the sensation of light. The stories of somnambulists, which are the only arguments that can be adduced against this belief, we may be allowed to dismiss. On the other hand, it is not light alone which can produce the sensation of light upon the eye, but any power which can excite the optic nerve. If the weakest electric currents are passed through the eye, they produce flashes of light. A blow, or even a slight pressure made upon the side of the eyeball with the finger, makes an impression of light in the darkest room, and under favorable circumstances this may become intense. It is important to remember that in these cases no objective light is produced in the retina, as some of the older physiologists assumed. The sensation of light may be so strong that, if it were really produced by an actual development of light within the eye, a second observer could not fail to see through the pupil the illumination of the retina which would follow. But nothing of the sort has ever been seen. Pressure or the electric current excites the optic nerve, and therefore, according to Müller's law, a sensation of light results; but under these circumstances, at least, there is not the smallest spark of actual light.

In the same way, increased blood pressure, its abnormal constitution in fevers, or its contamination with intoxicating or narcotic drugs can produce sensations of light to which no actual light corresponds. Even in cases in which an eye is entirely lost by accident or by an operation, the irritation of the stump of the optic nerve while it is healing can produce similar subjective effects. It follows from these facts that the peculiarity in kind which distinguishes the sensation of light from all others does not depend upon any peculiar qualities of light itself. Every action which is capable of exciting the optic nerve is capable of producing the impression of light; and the purely subjective sensation thus produced is so precisely similar to that caused by external light that persons unacquainted with these phenomena readily suppose that the rays they see are real, objective beams.

External light produces no distinctive effects in the optic nerve, none that are not produced by other agents of an entirely different nature. In one respect only does light differ from the other causes which are capable of exciting this nerve: namely, that the retina, being placed at the back of the firm globe of the eye and further protected by the bony orbit, is almost entirely withdrawn from other exciting agents and is thus only exceptionally affected by them, while it is continually receiving the rays of light which stream in upon it through the transparent media of the eye.

On the other hand, the optic nerve, by reason of the peculiar structures connected with the ends of its fibers—the rods and cones of the retina—is incomparably more sensitive to rays of light than any other nervous apparatus of the body, since the rest can only be affected by rays which are concentrated enough to produce noticeable elevation of temperature. This explains why the sensations of the optic nerve are for us the ordinary sensible sign of the presence of light in the field of vision, and why we always connect the sensation of light with light itself, even where they are really unconnected. But we must never forget that a survey of all the facts in their natural connection puts it beyond doubt that external light is only one of the exciting causes capable of bringing the optic nerve into functional activity and that there is no exclusive relation between the sensation of light and light itself.

Now that we have considered the action of excitants upon the optic nerve in general, we shall proceed to the qualitative differences of the sensation of light, that is to say, to the various sensations of color. We shall try to ascertain how far these differences of sensation correspond to actual differences in external objects.

Light is known in physics as a movement which is propagated by successive waves in the elastic ether distributed through the universe, a movement of the same kind as the circles which spread upon the smooth surface of a pond when a stone falls on it, or the vibrations which are transmitted through our atmosphere as sound. The chief difference is that the rate with which light spreads, and the rapidity of movement of the minute particles which form the waves of ether, are both enormously greater than that of the waves of water or of air. The waves of light sent forth from the sun differ exceedingly in size, just as the little ripples whose summits are a few inches distant from each other differ from the ocean waves, between whose foaming crests lie valleys of sixty or a hundred feet. But just as high and low, or short and long, waves on the surface of water differ not in kind but only in size, so the various waves of light which stream from the sun differ in their height and length but move all in the same manner and show (with certain differences depending upon the length of the waves) the same remarkable properties of reflection, refraction, interference, diffraction, and polarization. Hence we conclude that the undulating movement of the ether is in all of them the same. We must particularly note that the phenomena of interference (in which light is either strengthened or obscured by light of the same kind, according to the distance it has traversed) prove that all the rays of light depend upon oscillations of waves, and that the phenomena of polarization (which differ according to different lateral directions of the rays) show that the particles of ether vibrate at right angles to the direction in which the ray is propagated.

All the different sorts of rays which I have mentioned produce one effect in common. They raise the temperature of the objects on which they fall, and so are felt by our skin as rays of heat. But the eye perceives only part of these vibrations of the ether as light. It is not at all cognizant of the waves of great length, which I have compared with those of the ocean; these are therefore named the dark heat rays. Such are the rays which proceed from a warm but not red-hot stove and which we recognize as heat, but not as light. Again, the waves of shortest length, which correspond with the very smallest ripples produced by a gentle breeze, are so slightly appreciated by the eye that such rays are also generally regarded as invisible and are known as the dark chemical rays.

Between the very long and the very short waves of ether are waves of intermediate length, which strongly affect the eye but do not essentially differ in any other physical property from the dark heat rays and the dark chemical rays. The distinction between the visible and invisible rays depends only on the different length of their waves and the different physical relations which result. We call these middle rays *light* because they alone illuminate our eyes.

When we consider the heating property of these rays, we also call them luminous heat. Because they produce such very different impressions on our skin and on our eyes, heat was universally considered an entirely different kind of radiation from light until about thirty years ago. But both kinds of radiation are inseparable in the illuminating rays of the sun; indeed, the most careful recent investigations prove that they are precisely identical. To whatever optical processes they may be subjected, it is impossible to weaken their illuminating power without at the same time, and in the same degree, diminishing their heating and their chemical action. Whatever produces an undulatory movement of ether produces thereby all the effects of the undulation, whether light, or heat, or fluorescence, or chemical change.

Those undulations which strongly affect our eyes and which we call light excite the impression of different colors, according to the length of the waves. The undulations with the longest waves appear to us red; and as the length of the waves gradually diminishes, they seem to be golden yellow, yellow, green, blue, and violet (the last color being that of the light rays which have the shortest wave length). This series of colors is universally known in the rainbow. We also see it if we look into the light through a glass prism, and a diamond sparkles with hues which follow in the same order. In passing through transparent prisms, a beam of white light, which consists of a multitude of rays of various colors—that is, various wave lengths—is decomposed by the different degrees of refraction of its several parts (as explained in the first section), so that each of its component hues appears separately. These colors of the several primary kinds of light are best seen in the spectrum produced by a narrow streak of light passing through a glass prism; they are at once the fullest and the most brilliant which the external world can show.

When several of these colors are mixed together, they give the impression of a new color, which generally seems more or less white. If they were all mingled in precisely the same proportions in which they are combined in sunlight, they would give the impression of perfect white. According as the rays of longest, middle, or shortest wave length predominate in such a mixture, it appears as reddish white, greenish white, bluish white, and so on.

Everyone who has watched a painter at work knows that two colors mixed together give a new one. Although the results of the mixture of colored light differ in many particulars from those of the mixture of pigments, on the whole the appearance to the eye is similar. If we allow two different colored lights to fall at the same time upon a white screen, or upon the same part of our retina, we see only a single compound color, more or less different from the two original ones. The most striking difference between the mixture of pigments and that of colored light is that, while painters make green by mixing blue and yellow pigments, the union of blue and yellow rays of light produces white.

The simplest way of mixing colored light is shown in Fig. 7, in which *p* is a small flat piece of glass and *b* and *g* are two colored wafers. The observer looks down at *b* through the glass plate, in which *g* is reflected. If *g* is placed in the correct position so that its image exactly coincides with that of *b,* there will appear to be only a single wafer at *b,* of a color produced by the mixture of the two real ones. In this experiment the light from *b,* which traverses the glass pane, actually unites with that from *g,* which is reflected from it, and the two combined pass on to the retina at *o.*

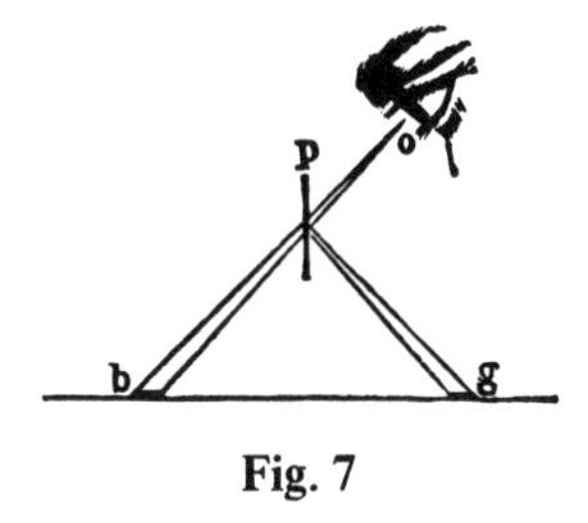

Fig. 7

In general, then, light, which consists of undulations of different wave lengths, produces different impressions upon our eye, namely, those of different colors. But the number of hues which we can recognize is much smaller than that of the various possible combinations of rays with different wave lengths which external objects can convey to our eyes. The retina cannot distinguish between the white which is produced by the union of scarlet and bluish green light and that which is composed of yellowish green and violet, or of yellow and ultramarine blue, or of red, green, and violet, or of all the colors of the spectrum united. All these combinations appear identically as white, yet from a physical point of view they are very different. In fact, the only resemblance among these several combinations is that they are indistinguishable to the human eye. For instance, a surface illuminated with red and bluish green light would come out black in a photograph, and another lighted with yellowish green and violet would appear very bright, while to the eye both surfaces alike seem simply white. Again, if we successively illuminate colored objects with white beams of light of various composition, they appear differently colored; and if we decompose two such beams by a prism, or look at them through a colored glass, the difference between them at once becomes evident.

Other colors also, especially when they are not strongly pronounced, may be composed of very different mixtures and yet appear indistinguishable to the eye, while in every other property, physical or chemical, they are entirely distinct.

Newton first showed how to represent the system of colors distinguishable to the eye in a simple diagrammatic form, and by the same means it is comparatively easy to demonstrate the law of the combination of colors. The primary colors of the spectrum are arranged in a series around the circumference of a circle, beginning with red and passing by imperceptible degrees through the various hues of the rainbow to violet. The red and violet are united by shades of purple, which on the one side change to indigo and blue tints, and on the other through crimson and scarlet to orange. The middle of the circle is left white, and on lines which run from the center to the circumference are represented the various tints which can be produced by diluting the full colors of the circumference until they pass into white. A color wheel of this kind shows all the varieties of hue which can be produced with the same amount of light.

It is possible to arrange the places of the several colors in this diagram and the quantity of light which each reflects, so that when we have determined the resultant of two colors of different known strengths of light (in the same way as we might ascertain the center of gravity of two bodies of different known weights), we shall find their combination color at the "center of gravity" of the two amounts of light. That is to say, in a properly constructed color wheel the combination color of any two colors will be found upon a straight line drawn between them; and compound colors which contain more of one than of the other component hue will be found proportionally nearer to the former and farther from the latter.

We find, however, when we have drawn our diagram, that those colors of the spectrum which are most saturated in nature and which must therefore be placed at the greatest distance from the central white will not arrange themselves in the form of a circle. The circumference of the diagram presents three projections corresponding to red, green, and violet, so that the color circle is more properly a triangle with one corner rounded off, as seen in Fig. 8. The continuous line represents the curve of the colors of the spectrum, and the small circle in the middle the white. At the corners are the three colors I have mentioned, while the sides of the triangle show the transitions from red

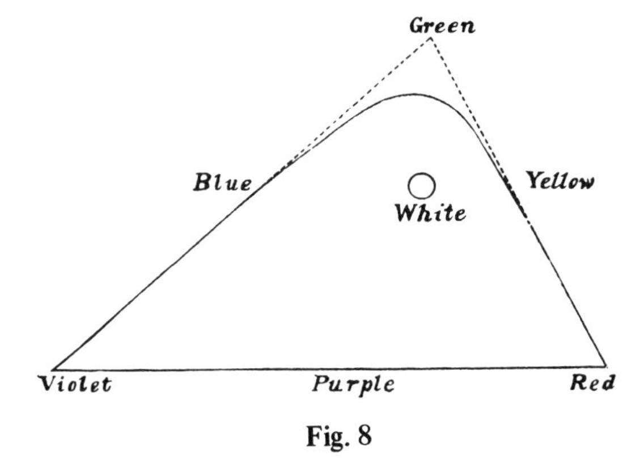

Fig. 8

through yellow to green, from green through bluish green and ultramarine to violet, and from violet through purple to scarlet.

Newton used the diagram of the colors of the spectrum (in a somewhat different form from that just given) only as a convenient way of representing the facts to the eye; but recently Maxwell has succeeded in demonstrating the strict, even quantitative accuracy of the principles involved in its construction. His method is to produce combinations of colors on swiftly rotating discs painted in sectors of various tints. When such a disc is spun rapidly, so that the eye can no longer follow the separate hues, they blend into a uniform combination color, and the quantity of light which belongs to each component hue is directly commensurate with the breadth of the sector of the circle it occupies. The combination colors which are produced in this manner are exactly those which would result if the same quantities of colored light illuminated the same surface simultaneously, as can be experimentally proved. Thus have the relations of size and number been introduced into the apparently inaccessible region of colors, and their differences in quality have been reduced to measurable relations of quantity.

All differences among colors may be reduced to three, which may be described as differences of tone, of fullness (or, as it is technically called, saturation), and of brightness. The differences of *tone* are those which exist among the several colors of the spectrum, to which we give the names red, yellow, blue, violet, and purple. With regard to tone, colors form a series which returns upon itself, a series which we complete when we allow the terminal colors of the rainbow to shade into one another through purple and crimson. It is the same series which we describe as arranged around the circumference of the color wheel.

The *fullness* or saturation of colors is greatest in the pure tints of the spectrum and becomes less in proportion as they are mixed with white light. This is true, at least, for colors produced by external light. (For our sensations it is possible to increase still further the apparent saturation of color, as we shall presently see.) Just as pink is a whitish crimson and flesh color a whitish scarlet, so pale green, straw color, light blue, etc., are all produced by diluting the corresponding colors with white. All compound colors are, as a rule, less saturated than the simple tints of the spectrum.

Lastly, we have the differences of *brightness,* or strength of light, which are not represented in the color wheels. So long as we restrict our observation to colored rays of light, difference in brightness appears to be only one of quantity, not of quality. Black is only darkness–that is, simple absence of light. But when we examine the colors of external objects, black corresponds just as much to a peculiarity of surface reflection as does white, and therefore it has as good a right to be called a color. As a matter of fact, we find in common language a series of terms to express colors with a small amount of light. We call them *dark* (in English, *deep*) when they have little light but are full in tint and *gray* when they are pale. Thus dark blue conveys the idea of depth in tint, of a full blue with a small amount of light; while gray blue is a pale blue with a small amount of light. In the same way, the colors known as maroon, brown, and olive are dark, more or less saturated tints of red, yellow, and green, respectively.

In this way we may reduce all actual objective differences in color, insofar as they are appreciated by the eye, to three kinds: differences of hue (tone), differences of fullness (saturation), and differences of amount of illumination (brightness). It is in this way that we describe the system of colors in ordinary language. But we are able to express this threefold difference in another way.

I said above that a properly constructed color wheel approaches a triangle in its outline. Let us suppose for a moment that it is an exact rectilinear triangle, as made by the dotted line in Fig. 8. (How far this differs from the actual condition we shall have to indicate later.) Let the colors red, green, and violet be placed at the corners, and we see expressed the law which was mentioned above: namely, that all the colors in the interior and on the sides of the triangle are compounds of the three at its corners. It follows that *all differences of hue depend upon combinations in different proportions of the three primary colors.* (It is best to consider the three just named as primary. The old ones–red, yellow, and blue–are inconvenient and were only chosen from experience of painters' colors; it is impossible to make a green out of blue and yellow light.)

We shall better understand the remarkable fact that all the varieties in the composition of external light are mixtures of three primitive colors, if in this respect we compare the eye with the ear. Sound, as I mentioned before, is, like light, an undulating movement, spreading by waves. In the case of sound also, we have to distinguish waves of various lengths which produce upon our ear impressions of different quality. We recognize the long waves as low notes, the short as high-pitched.

The ear may receive many waves of sound—that is to say, many notes—at once, but they do not form compound notes in the same way that colors, when perceived at the same time and place, form compound colors. The eye cannot tell the difference if we substitute orange for red and yellow; but if we hear the notes *c* and *e* sounded at the same time, we cannot then sound *d* in their stead without entirely changing the impression upon the ear. The most complicated harmony of a full orchestra is changed to our perception if we alter any one of its notes. No accord (or consonance of several tones) is, at least for the practiced ear, completely like another composed of different tones; whereas if the ear perceived musical tones as the eye does colors, every accord might be completely formed by combining only three constant notes—one very low, one very high, and one intermediate—simply changing the relative strength of these three primary notes to produce all possible musical effects.

In reality we find that an accord remains unchanged to the ear only so long as the strength of each separate tone which it contains remains unchanged. If we wish to describe it exactly and completely, the strength of each of its component tones must be exactly stated. In the same way, the physical nature of a particular kind of light can only be fully ascertained by measuring and noting the amount of light of each of the simple colors which it contains. But in sunlight, in the light of most stars, and in flames we find a continuous transition of colors into one another through numberless intermediate gradations. Accordingly, we must ascertain the amount of light of an infinite number of compound rays if we wish to arrive at an exact physical knowledge of sunlight or starlight. In the sensations of the eye, however, we need distinguish for this purpose only the varying intensities of three components.

The practiced musician is able to catch the separate notes of the various instruments among the complicated harmonies of an entire orchestra, but the optician cannot directly ascertain the composition of light by means of his eye; he must make use of the prism to decompose the light for him. As soon as this is done, however, the composite character of light becomes apparent. He can then distinguish the light of separate fixed stars from one another by the dark and bright lines which the spectrum shows him, and he can recognize the chemical elements contained in flames on the earth, or even in the intense heat of the sun's atmosphere, in the fixed stars, or in the nebulae. The fact that light derived from each separate source carries with it certain permanent physical peculiarities is the foundation of spectrum analysis—that most brilliant discovery of recent years, which has opened the extreme limits of celestial space to chemical analysis.

There is an extremely interesting and not very uncommon defect of sight which is known as color blindness. In this condition the differences of color are reduced to a still more simple system than that described above; namely, to combinations of only two primary colors. Persons so affected are called color-blind because they confound certain hues which appear very different to ordinary eyes. At the same time, they distinguish other colors quite as accurately—or even, it seems, rather more accurately—than ordinary people. They are usually red-blind; that is to say, there is no red in their system of colors, so they see no difference which is produced by the addition of red. All tints are for them varieties of blue and green (or, as they call it, yellow). Scarlet, flesh color, white, and bluish green appear to them to be identical or, at the utmost, to differ in brightness. The same applies to crimson, violet, and blue and to red, orange, yellow, and green. The scarlet flowers of the geranium have for them exactly the same color as its leaves. They cannot distinguish between the red and green signals of trains. They cannot see the red end of the spectrum at all. Very full scarlet appears to them almost black, so that a red-blind Scotch clergyman went to buy scarlet cloth for his gown, thinking it was black.[8]

In this particular of discrimination of colors, we find remarkable inequalities in different parts of the retina. In the first place, all of us are red-blind in the outermost part of our field of vision. A geranium blossom when moved backward and forward just within the field of sight is recognized only as a moving object. Its color is not seen; if it is waved in front of a mass of leaves

8. [A similar story is told of Dalton, the author of the atomic theory. He was a Quaker and went to the Friends' Meeting at Manchester in a pair of scarlet stockings, which some wag had put in place of his ordinary dark grey ones.—Philip H. Pye-Smith, translator.]

of the same plant, it cannot be distinguished from them in hue. In fact, all red colors appear much darker when viewed indirectly. This red-blind part of the retina is most extensive on the inner or nasal side of the field of vision; and according to recent researches of Woinow, there is at the farthest limit of the field of vision a narrow zone in which all distinction of colors ceases and there remain only differences of brightness. In this outermost circle everything appears white, gray, or black. Probably only those nervous fibers which convey impressions of green light are present in this part of the retina.

In the second place, as I have already mentioned, the spot near the middle of the retina just around the central pit is colored yellow. This makes all blue light appear somewhat darker in the center of the field of sight. The effect is particularly striking with mixtures of red and greenish blue, which appear white when looked at directly but acquire a blue tint when viewed at a slight distance from the middle of the field; on the other hand, when they appear white indirectly, they are red to direct vision. These inequalities of the retina, like the others mentioned in the first section, are rectified by the constant movements of the eye. We know what impressions of indirect vision—the pale, indistinct colors of the external world as usually seen—correspond to those of direct vision; and we learn to judge the colors of objects according to the impression which they would make on us if seen directly. As a result, only unusual or special concentration of attention enables us to recognize the difference of which I have been speaking.

The theory of colors, with all these marvelous and complicated relations, was a riddle which Goethe in vain attempted to solve; nor were we physicists and physiologists more successful. I include myself in the number, for I long toiled at the task without getting any nearer my goal, until I at last discovered that a wonderfully simple solution had been discovered at the beginning of this century and had been in print ever since for anyone to read who chose. This solution was found and published by the same Thomas Young who first showed the right method of arriving at the interpretation of Egyptian hieroglyphics. He was one of the most acute men who ever lived, but he had the misfortune to be too far in advance of his contemporaries. They looked on him with astonishment but could not follow his bold speculations, and so a mass of his most important thoughts remained buried and forgotten in the *Transactions of the Royal Society,* until a later generation managed by slow degrees to rediscover his discoveries and came to appreciate the force of his arguments and the accuracy of his conclusions. (In proceeding to explain the theory of colors proposed by him, I beg the reader to notice that conclusions afterward to be drawn about the nature of the sensations of sight are quite independent of what is hypothetical in this theory.)

Dr. Young supposes that there are in the eye three kinds of nerve fibers, the first of which, when irritated in any way, produces the sensation of red; the second, the sensation of green; and the third, that of violet. He further assumes that the first are excited most strongly by the waves of ether of greatest length and the second (which are sensitive to green light) by the waves of middle length, while the last (which convey impressions of violet) are acted upon only by the shortest vibrations of ether. At the red end of the spectrum the excitation of those fibers which are sensitive to that color predominates; hence this part appears red. Further along the spectrum there is added an impression upon the fibers sensitive to green light, resulting in the mixed sensation of yellow. In the middle of the spectrum, the nerves sensitive to green become much more excited than the other two kinds, so green is the predominant impression. As soon as this becomes mixed with violet, the result is the color known as blue; while at the most highly refracted end of the spectrum, the impression produced on the fibers which are sensitive to violet light overcomes every other.[9]

It is apparent that this hypothesis is merely an extension of Johannes Müller's law of specific energies. Just as the difference between the sensations of light and warmth depends demonstrably upon whether the rays of the sun fall upon nerves of sight or nerves of feeling, so it is assumed in Young's hypothesis that the difference among the sensations of color depends simply upon which kind of nervous fibers is most strongly affected. When all three kinds are equally excited, the result is the sensation of white light. The phenomena that occur in red blindness must be ascribed to a condition in which one kind of nerves, those sensitive to red rays, are incapable of excitation. It is possible that this class of fibers is absent, or at least very sparingly distributed, along the edge of the retina, even in the normal human eye.

It must be confessed that we have at present no anatomical basis in either men or quadrupeds for this theory of colors; but Max Schultze has discovered a structure in birds and reptiles which manifestly corresponds with what we

---

9. The precise tint of the three primary colors cannot yet be ascertained by experiment. The red, it is certain from the experience of the color-blind, belongs to the extreme red of the spectrum. But at the other end Young took violet for the primitive color, while Maxwell considers that it is more properly blue. The question is still an open one: according to J. J. Müller's experiments (*Archiv für Ophthalmologie,* XV, No. 2. p. 208), violet is more probable. The fluorescence of the retina is here a source of difficulty.

should expect to find. In the eyes of many of this group of animals, there are found among the rods of the retina a number which contain a red drop of oil in their anterior end (that which is turned toward the light), while other rods contain a yellow drop, and others none at all. There can be no doubt that red light will reach the rods with a red drop much better than light of any other color, while yellow and green light will find easiest entrance to the rods with the yellow drop. Blue light would be shut off almost completely from both, but it would affect the colorless rods all the more effectually. We may therefore with great probability regard these rods as the terminal organs of nervous fibers which convey impressions of red, yellow, and blue light, respectively.

I have myself subsequently found a similar hypothesis very convenient and well suited to explain in a most simple manner certain peculiarities which have been observed in the perception of musical tones, peculiarities as enigmatic as those we have been considering in the eye. In the cochlea of the internal ear, the ends of the nerve fibers, which lie spread out regularly side by side, are provided with minute elastic appendages (the rods of Corti) arranged like the keys and hammers of a piano. My hypothesis is that each of these separate nerve fibers is constructed so as to be sensitive to a definite tone, to which its elastic fiber vibrates in perfect consonance. This is not the place to describe the special characteristics of our sensations of musical tones which led me to frame this hypothesis. Its analogy with Young's theory of colors is obvious, and it explains the origin of overtones, the perception of the quality of sounds, the difference between consonance and dissonance, the formation of the musical scale, and other acoustic phenomena by as simple a principle as that of Young. But in the case of the ear I could point to a much more distinct anatomical foundation for such a hypothesis, and since that time it has been possible to demonstrate the relation supposed–not, it is true, in man or any other vertebrate animals, whose labyrinth lies too deep for experiment, but in some of the marine Crustacea. These animals have external appendages to their organs of hearing which may be observed in the living animal–jointed filaments to which the fibers of the auditory nerve are distributed–and Hensen, of Kiel, has satisfied himself that some of these filaments are set in motion by certain tones, and others by different ones.

It remains to answer an objection against Young's theory of color. I mentioned above that the outline of the color wheel which marks the position of the most saturated colors (those of the spectrum) approaches a triangle in form; but our conclusions upon the theory of the three primary colors require a perfect rectilinear triangle enclosing the complete color system, for only in that case is it possible to produce all possible tints by various combinations of the three primary colors at the angles. It must be remembered, however, that the color wheel includes the entire series of colors which actually occur in nature, while our theory concerns the analysis of our subjective sensations of color. We need only assume that colored light does not produce sensations of absolutely pure color; that red, for instance, even when completely freed from all admixture of white light, excites not only those nervous fibers which are sensitive to impressions of red but also, to a very slight degree, those which are sensitive to green, and perhaps to a still smaller extent those which are sensitive to violet rays. If so, the sensation which the purest red light produces in the eye is still not the purest sensation of red which we can conceive of as possible. This sensation could only be called forth by a fuller, purer, more saturated red than has ever been seen in this world.

It is possible to verify this conclusion. We are able to produce artificially a sensation of the kind I have described. This fact is not only important as a complete answer to a possible objection to Young's theory; it is also, as will readily be seen, of the greatest importance for understanding the real significance of our sensations of color. In order to describe the experiment I must first give an account of a new series of phenomena.

The result of nervous action is fatigue, proportionate to the intensity of the function performed and its duration. The blood, on the other hand, which flows in through the arteries, is constantly performing its function of replacing used material by fresh, thus carrying away the chemical results of functional activity; that is to say, removing the fatigue.

The occurrence of fatigue as the result of nervous action takes place in the eye as well as other organs. When the entire retina becomes tired, as when we spend some time in the open air in brilliant sunshine, it becomes insensible to weaker light, so that if we pass immediately into a dimly lighted room, we see nothing at first; we are blinded, as we call it, by the previous brightness. After a time the eye recovers itself, and at last we are able to see, and even to read, by the same dim light which at first appeared complete darkness.

It is thus that fatigue of the entire retina shows itself. But it is possible for separate parts of that membrane to become exhausted, if they alone have received a strong light. If we look steadily for some time at any bright object surrounded by a dark background—it is necessary to look steadily so that the image will remain stationary upon the retina and thus fatigue a sharply defined portion of its surface—and afterwards turn our eyes upon a uniform, dark gray surface, we see projected upon it an afterimage of the bright object with the same outline but with reversed illumination. What was dark now appears bright, and what was bright dark, like a photographer's negative. By carefully fixing our attention, it is possible to produce very elaborate afterimages, so much so that occasionally even printing can be distinguished in them. This phenomenon is the result of a local fatigue of the retina. Those parts of the membrane upon which the bright light fell are less sensitive to the light of the dark gray background than are the neighboring regions, so there appears a dark spot upon the really uniform surface, corresponding in extent to the surface of the retina which previously received the bright light.

(Illuminated sheets of white paper are sufficiently bright to produce this afterimage. If we look at much brighter objects–at flames, or at the sun itself–the effect becomes complicated. The strong excitement of the retina does not pass away immediately, but produces a direct or positive afterimage, which at first unites with the negative or indirect one produced by the retinal fatigue. Furthermore, the effects of the different colors of white light differ in both duration and intensity, so that the afterimages become colored and the whole phenomenon much more complicated.)

By means of these afterimages it is easy to convince oneself that the impression produced by a bright surface begins to diminish after the first second and that by the end of a single minute it has lost from a quarter to half of its intensity. The simplest form of experiment for this purpose is as follows: Cover the lower half of a white sheet of paper with a black one, fix the eye upon some point of the white sheet near the margin of the black, and after thirty to sixty seconds draw the black sheet away quickly, without losing sight of the point. The half of the white sheet which is suddenly exposed appears of the most brilliant brightness; and thus it becomes apparent how much the first impression produced by the upper half of the sheet had become blunted and weakened, even in the short time taken by the experiment. And yet (it is also important to remark) the observer had not noticed this blunting at all until the contrast forced it to his attention.

It is possible to produce a partial fatigue of the retina in still another way. We may tire it for certain colors only, by exposing either the entire retina or a portion of it to a single color–say, red–for a certain time (from half a minute to five minutes). According to Young's theory, only one or two kinds of the optic nerve fibers will become fatigued, namely, those which are sensitive to impressions of the color in question. All the rest will remain unaffected. The result is that when the afterimage appears upon a gray background, the uniformly mixed light of the latter can only produce sensations of green and violet in the part of the retina which has become fatigued by red light and thus made red-blind for the time. The afterimage, accordingly, appears bluish green, the complementary color to red.

It is by this means that we are able to produce in the retina the pure, fundamental sensations of saturated colors. If, for instance, we wish to see pure red, we fatigue a part of our retina by exposing it to the bluish green of the spectrum. We thus make this part at once green-blind and violet-blind. We then throw the afterimage upon a red as perfect as the prismatic spectrum allows. The afterimage immediately appears in full and burning red. In contrast, the purest red that the external world can afford appears to the unfatigued part of the retina less saturated than the red of the afterimage and as if it were covered with a whitish mist.

These facts are perhaps enough. I shall not accumulate further details, whose explication would require lengthy descriptions of many separate experiments.

We have already seen enough to answer the question whether it is possible to maintain the natural, innate conviction that the qualities of our sensations, especially our sensations of sight, give us a true copy of corresponding qualities in the outer world. It is clear that they do not. The question was really decided by Johannes Müller's inference, from well ascertained facts, of the law of specific nervous energy. Whether the rays of the sun appear to us as color or as warmth does not depend at all upon their own properties, but simply upon whether they excite the fibers of the optic nerve or those of the skin. Pressure upon the eyeball, a feeble current of electricity passed through it, a narcotic drug carried to the retina by the blood—all are capable of exciting the sensation of light just as well as the sunbeams. The most complete difference which characterizes our several sensations, that is, the distinctive qualities of sight, hearing, taste, smell, and touch—this deepest of all distinctions, so deep that it is impossible to draw any comparison of likeness, or unlikeness, between the sensations of, say, colors and musical tones—depends, as we now see, not at all upon the nature of the external object, but solely upon the central connections of the nerves which are affected.

We now see that the question whether within the special range of each particular sense it is possible to discover a correspondence between its objects and the sensations they produce, is of only secondary importance. What color the waves of ether shall appear to us when they excite the optic nerve depends upon their length. The system of naturally visible colors offers us a series of combinations in the composition of light, but the number of those combinations is wonderfully reduced from an unlimited number to only three. Inasmuch as the most important property of the eye is its minute appreciation of locality, and as it is so much more perfectly organized for this purpose than the ear, we may be well content that it is capable of recognizing comparatively few differences in quality of light. (The ear, which is so well able to distinguish qualities of sound, has scarcely any power to appreciate differences of locality.) But it is certainly matter for astonishment to anyone who trusts unconditionally the impressions of his senses, that neither the limits within which the spectrum affects our eyes nor the humanly perceived differences of color (which are no more than a simplified impression of all the actual differences of light per se) have any demonstrable significance apart from the sense of sight. Rays of light which are precisely the same to our eyes may in all other physical and chemical effects be completely different. Lastly, we find that even the pure fundamental elements of all our sensations of color—the simple primary tints of our perception—cannot be produced by just any kind of external light in the natural unfatigued condition of the eye. These elementary sensations of color can be called forth only after an artificial preparation of the organ; they only exist, in fact, as subjective phenomena.

We conclude, therefore, that as to any correspondence between external light and the sensations it produces, there is only one possible connection—a connection which at first may seem slender, but which is in fact quite sufficient to lead to a countless number of most useful applications. This law of correspondence between what is subjective and objective in vision is as follows:

Similar light produces under like conditions a like sensation of color. Light which under like conditions excites unlike sensations of color is dissimilar.

When two objects or properties correspond to one another in this manner, one is a *sign* for the other. Hitherto the notions of sign and image, or representation, have not been carefully enough distinguished in the theory of perception; and this seems to me to have been the source of countless mistakes and false hypotheses. In an *image* the representation must be of the same kind as that which is represented. Indeed, it is only an image to the extent that it is like in kind. A statue is an image of man insofar as its form reproduces his; even if it is executed on a smaller scale, every dimension must be represented in proportion. A picture is an image, or representation, of the original, first, because it represents the colors of the latter by similar colors, and secondly, because it represents a part of its relations in space (those which belong to perspective) by corresponding relations in space.

The excitation of the nerves in the brain and the ideas in our consciousness can be considered images of processes in the external world insofar as the former parallel the latter, that is, insofar as they represent the similarity of objects by a similarity of signs and thus represent a lawful order by a lawful order.

This is obviously sufficient to enable the understanding to infer what is constant from the varied changes of the external world and to formulate it as a concept or a law. That it is also sufficient for all practical purposes we shall see in the next section of this essay. However, not only uneducated persons, who are accustomed to trust blindly to their senses, but even the educated, who know that their senses may be deceived, are inclined to demur at so complete a lack of any closer correspondence between actual objects and the sensations they produce. Natural philosophers long hesitated, for instance, to admit the identity of the rays of light and of heat, and exhausted all possible means of escaping a conclusion which seemed to contradict the evidence of their senses.

Another example is that of Goethe, as I have endeavored to show elsewhere. He was led to contradict Newton's theory of colors because he could not persuade himself that white, which appears in sensation as the purest manifestation of the brightest light, could be composed of darker colors. Yet Newton's discovery of the composition of light was the first germ of the modern doctrine of the true functions of the senses; and in the writings of his contemporary Locke were correctly laid down the most important principles on which the right interpretation of sensible qualities depends.

However clearly we may feel that this is the inherent difficulty for many people, I have never found their conviction of certainty derived from the senses so distinctly expressed that it is possible to lay hold of the point of error. The reason, it seems to me, is that beneath the popular notions on the subject lie other, more fundamental conceptual errors.

We must not be led astray by confounding the notions of *phenomenon* and *appearance.* The different colors of objects are phenomena caused by certain real differences in their constitution. They are, according to the scientific as well as to the uninstructed view, no mere appearance, even though the way they appear depends chiefly upon the constitution of our nervous system. A deceptive appearance occurs when the normal phenomena of one object are confounded with those of another, but the sensation of color is by no means a deceptive appearance. There is no other way in which color can appear, so there is nothing which we could describe as the normal phenomenon in distinction from the impressions of color received by means of the eye.

Here the principal difficulty seems to me to lie in the notion of *quality*. All difficulty vanishes as soon as we clearly understand that each quality or property of a thing is, in reality, nothing but its capability of exercising certain effects upon other things. These actions either go on between adjacent parts of the same body and so produce alterations in its aggregate condition; or they proceed from one body to another, as in the case of chemical reactions; or they produce their effect on our sense organs and are there recognized as sensations, such as those of sight. Any of these actions is called a *property* when the object it acts on is understood without being expressly mentioned. Thus, when we speak of the solubility of a substance, we mean its behavior toward water; when we speak of its weight, we mean its attraction to the earth; and in the same way we may correctly call a substance blue, understanding as a tacit assumption that we are speaking of its action upon a normal eye.

But if what we call a property, or quality, always implies an action of one thing on another, it can never depend upon the nature of one agent alone. It can exist only in relation to, and dependent on, the nature of some second object which is acted upon. Hence it is really meaningless to talk as if there were properties of light which belong to it absolutely, independent of all other objects, and which we may expect to find exhibited in the sensations of the human eye. The notion of such properties is a contradiction in itself. They cannot possibly exist, and therefore we cannot expect to find any correspondence of our sensations of color with qualities of light.

These considerations naturally suggested themselves long ago to thoughtful minds; they may be found clearly expressed in the writings of Locke and Herbart, and they are completely in accordance with Kant's philosophy. But in former times they demanded a more than usual power of abstraction for their truth to be understood. Now the facts which we have laid before the reader illustrate them in the clearest manner.

After this excursion into the world of abstract ideas, we return once more to the subject of color. Let us examine it as a sensible sign of certain external qualities, either of light itself or of the objects which reflect it.

It is essential for a good sign to be constant—that is, the same object must always produce the same sign. We have already seen that in this particular our sensations of color are imperfect; they are not quite uniform over the entire field of the retina. But the constant movement of the eye makes up for this imperfection, just as it makes up for the unequal sensitiveness of the different parts of the retina to form.

We have also seen that when the retina becomes tired, the intensity of the impression produced on it rapidly diminishes, but here again the usual effect of the constant movements of the eye is to equalize the fatigue of the various parts. Hence we rarely see afterimages; if they appear at all, it is only in cases of brilliant objects, such as very bright flames or the sun itself. And as long as the fatigue of the entire retina is uniform, the relative brightness and color of the different objects in sight remain almost unchanged. So the effect of fatigue is gradually to weaken the apparent illumination of the entire field of vision.

This brings us to consider those differences in the images presented by the eye which depend on different degrees of illumination. Here again we meet with instructive facts. We look at external objects under light of very different intensity, varying from the most dazzling sunshine to the pale beams of the moon—and the light of the full moon is 150,000 times less than that of the sun. Moreover, the color of the illumination may vary greatly. We sometimes employ artificial light, which is always more or less orange in color; or the natural daylight may be altered, as in the green shade of an arbor or in a room with colored carpets and curtains.

As the brightness and color of the illumination change, so of course will the brightness and color of the light which the illuminated objects reflect to our eyes, since all differences in local color depend upon different bodies reflecting and absorbing various proportions of the several rays of the sun. (Cinnabar, for example, reflects the rays of great length without any obvious loss, while it absorbs almost the whole of the other rays. Accordingly, this substance appears of the same red color as the beams which it reflects toward the eye. If it is illuminated with light of some other color, without any mixture of red, it appears almost black.)

These observations teach what we find confirmed by daily experience in a hundred ways—that the apparent color and brightness of illuminated objects vary with the color and brightness of the illumination. This is a fact of the first importance for the painter, for many of his finest effects depend on it.

But what is most important practically is for us to be able to recognize surrounding objects when we see them; it is only seldom that, for some artistic or scientific purpose, we turn our attention to the way they are illuminated. What is constant in the color of an object is, not the brightness and color of the light which it reflects, but the relation between the intensity of the different colored constituents of this reflected light, on the one hand, and that of the corresponding constituents of the light which illuminates it, on the other. This proportion alone is the expression of a constant property of the object in question.

Considered theoretically, the task of judging of the color of a body under changing illumination would seem to be impossible; but in practice we soon find that we are able to judge local color without the least uncertainty or hesitation and under the most different conditions. For instance, white paper in full moonlight is darker than black satin in daylight, but we never find any difficulty in recognizing the paper as white and the satin as black. Indeed, it is much more difficult to satisfy ourselves that a dark object in the sunshine reflects light of exactly the same color, and perhaps the same brightness, as a white object in shadow, than to satisfy ourselves that the proper color of a white paper in shadow is the same as that of a sheet of the same kind lying close to it in the sunlight.

Gray seems to us something altogether different from white, and so it is, regarded as a proper color,[10] for anything which only reflects half the light it receives must have a different surface from one which reflects it all. Yet the impression upon the retina of a gray surface under illumination may be absolutely identical with that of a white surface in the shade. Every painter represents a white object in shadow by means of gray pigment, and if he has correctly imitated nature, it appears pure white. In order to convince oneself of the identity of gray and white in this respect—that is, as illumination colors—the following experiment may be tried: Cut out a circle in gray paper, and concentrate a strong beam of light upon it with a lens, so that the limits of the illumination exactly correspond with those of the gray circle. It will then be impossible to tell that there is any artificial illumination at all. The gray looks white.[11]

---

10. [The local, or proper, color of an object (*Körperfarbe*) is that which it shows in common white light, while the illumination color (as I have translated *Lichtfarbe*) is that which is produced by colored light. Thus the red of some sandstone rocks seen by common white light is their proper color, while that of a snow-covered mountain in the rays of the setting sun is an illumination color.—P. H. P.-S.]

11. [The demonstration is more striking if the gray disk is placed on a sheet of white paper in diffused light.—P. H. P.-S.]

We may assume, and the assumption is justified by certain phenomena of contrast, that illumination by the brightest white we can produce gives a true criterion for judging the darker objects in the neighborhood, since under ordinary circumstances the brightness of any proper color diminishes in proportion as the illumination is diminished or the retinal fatigue increased. This relation holds even for extreme degrees of illumination, insofar as the objective intensity of the light is concerned, but not for our sensation. Under illumination so brilliant as to approach what would be blinding, the degrees of brightness of light objects become less and less distinguishable; in the same way, when the illumination is very feeble, we are unable to appreciate slight differences in the amount of light reflected by dark objects. The result is that in sunshine local colors of moderate brightness approach the brightest, whereas in moonlight they approach the darkest. The painter utilizes this difference to represent noonday or midnight scenes, although pictures, which are usually seen in uniform daylight, do not really admit of any difference of brightness approaching that between sunshine and moonlight. To represent the former, he paints the objects of moderate brightness almost as bright as the brightest; for the latter, he makes them almost as dark as the darkest.

The effect is assisted by another difference in the sensation produced by the same actual conditions of light and color. If the brightness of various colors is equally increased, that of red and yellow becomes apparently stronger than that of blue. Thus, if we select a red and a blue paper which appear of the same brightness in ordinary daylight, the red will seem much brighter in full sunlight, the blue in moonlight or starlight. This peculiarity in our perception is also made use of by painters: they make yellow tints predominate when representing landscapes in full sunshine, while every object of a moonlight scene is given a shade of blue. But it is not only local color which is thus affected; the same is true of the colors of the spectrum.

These examples show very plainly how independent our judgment of colors is of their actual amount of illumination. In the same way, it is scarcely affected by the color of the illumination. We know, of course, in a general way that candlelight is yellowish compared with daylight, but we only learn to appreciate how much the two kinds of illumination differ in color when we bring them together at the same intensity–as, for example, in the experiment of colored shadows. If we admit light from a cloudy sky[12] through a narrow opening into a dark room so that it falls sideways on a horizontal sheet of white paper, while candlelight falls on it from the other side, and if we then

12. This experiment with diffused white daylight may also be made with moonlight.

hold a pencil vertically upon the paper, it will of course throw two shadows. The one made by the daylight is orange and looks so; the other, made by the candlelight, is really white but appears blue by contrast. The blue and the orange of the two shadows are both colors which we call white, when we see them by daylight and candlelight, respectively. Seen together, they appear as two very different, tolerably saturated colors, yet we do not hesitate a moment in recognizing white paper by candlelight as white, and very different from orange.

The most remarkable of this series of facts is that we can separate the color of any transparent medium from that of objects seen through it. This is proved by a number of experiments contrived to illustrate the effects of contrast. If we look through a green veil at a field of snow, the light reflected from it must really have a greenish tint when it reaches our eyes, yet it appears of a reddish tint from the effect of the indirect afterimage of green. So completely are we able to separate the light which belongs to the transparent medium from that of the objects seen through it.[13]

The changes of color in the last two experiments are known as phenomena of contrast. They consist of mistakes as to local color, which for the most part depend upon imperfectly defined afterimages.[14] This effect is known as *successive contrast* and is experienced when the eye passes over a series of colored objects. But a similar mistake may result from our custom of judging local color according to the brightness and color of the various objects seen at the same time. If these relations happen to be different from what is usual, contrast phenomena ensue. When, for example, objects are seen under two different colored illuminations, or through two different colored media (whether real or apparent), these conditions produce what is called *simultaneous contrast.* Thus in the experiment described above (of colored shadows thrown by daylight and candlelight), the doubly illuminated surface of the paper, being the brightest object seen, gives a false criterion for white. Compared with it, the really white but less bright light of the shadow thrown by the candle looks blue. Moreover, in these curious effects of contrast, we must take into account that differences in sensation which are easily apprehended appear to us greater than those which are less obvious. Differences of color which are actually before our eyes are more easily apprehended than

13. A number of similar experiments will be found described in the author's *Handbook of Physiological Optics.*

14. [These afterimages have been described as accidental images, positive when of the same color as the original color, negative when of the complementary color.–P. H. P.-S.]

those which we only keep in memory, and contrasts between objects which are close to one another in the field of vision are more easily recognized than when they are at a distance. All this contributes to the effect. Indeed, there are a number of subordinate circumstances affecting the result which it would be very interesting to follow out in detail, for they throw great light upon the way in which we judge local color: but we must not pursue the inquiry further here. I shall only remark that all these effects of contrast are not less interesting for the scientific painter than for the physiologist, since he must often exaggerate the natural phenomena of contrast in order to produce the impression of greater varieties of light and greater fullness of color than can be actually produced by artificial pigments.

Here we must leave the theory of the sensations of sight. This part of our inquiry has shown us that the qualities of these sensations can be regarded only as *signs* of certain other qualities, which belong sometimes to light itself, sometimes to the bodies it illuminates, and that there is not a single actual quality of the objects seen which precisely corresponds to our sensations of sight. Nay, we have seen that, even regarded as signs of real phenomena in the outer world, they do not possess the one essential requisite of a complete system of signs–constancy–with anything like completeness. All we can say of any given sensation of sight is that under similar conditions, the qualities of this sensation appear in the same way for the same objects.

Yet, in spite of all this imperfection, we have also found that by means of so inconstant a system of signs, we are able to handle the most important part of our problem–to recognize the same proper colors wherever they occur–and considering the difficulties in our way, it is surprising how well we succeed. Out of this inconstant system of brightness and of colors, varying according to the illumination, varying according to the retinal fatigue, and varying according to the part of the retina affected, we are able to determine the proper color of any object, the one constant phenomenon which corresponds to a constant quality of its surface. And we can do this, not after long consideration, but with an instantaneous, involuntary decisiveness.

The inaccuracies and imperfections of the eye as an optical instrument, and the deficiencies of the image on the retina, now appear insignificant in comparison with the incongruities we have met with in the field of sensation. One might almost believe that Nature had here contradicted herself on purpose in order to destroy any dream of a pre-existing harmony between the outer and the inner world.

And what progress have we made in our task of explaining sight? It might seem that we are further off than ever, with the riddle only more complicated and with less hope than ever of finding the answer. The reader may perhaps feel inclined to reproach science as knowing only how to shatter with adverse criticism the beautiful world presented to us by our senses, in order to annihilate the fragments:

> Woe! woe!
> Thou hast destroyed
> The beautiful world
> With powerful fist;
> In ruin 'tis hurled,
> By the blow of a demigod shattered.
> The scattered
> Fragments into the void we carry,
> Deploring
> The beauty perished beyond restoring.[15]
>
> *Du hast sie zerstört*
> *Die schöne Welt*
> *Mit mächtiger Faust;*
> *Sie stürzt, sie zerfällt,*
> *Ein Halbgott hat sie zerschlagen.*
> *Wir tragen*
> *Die Trümmern ins Nichts hinüber,*
> *Und klagen*
> *Über die verlorne Schöne.*

–and may feel determined to stick fast to the "sound common sense" of mankind and believe his own senses more than physiology.

But there is still a part of our investigation which we have not touched–that into our conceptions of space. Let us see whether, after all, our natural reliance upon the accuracy of what our senses teach us will not be justified even before the tribunal of science.

III. THE PERCEPTION OF SIGHT

The colors which have been the subject of the last section are not simply an ornament we should be sorry to lose; they are also a means of assisting us to distinguish and recognize external objects. But for this purpose color is far less important than the means which the eye's rapid, far-reaching power gives

15. [Bayard Taylor's translation of the passage in *Faust.*–P. H. P.-S.]

us of distinguishing the various relations of *locality*. No other sense can be compared with the eye in this respect. The sense of touch, it is true, can also distinguish relations of space and has priority in judging all matter within reach as regards resistance, volume, and weight. But the range of touch is limited, and the distinction it can make between small distances is not nearly so accurate as that of sight. Yet the sense of touch is sufficient, as experiments upon persons born blind have proved, to develop complete conceptions of space. This proves that the possession of sight is not necessary for the formation of these conceptions. As we shall soon see, we are continually controlling and correcting the notions of spatial location derived from the eye by the help of the sense of touch and we always accept the impressions of touch as decisive. The two senses, which really have the same task, though with very different means of accomplishing it, happily make up for each other's deficiencies. Touch is a trustworthy and experienced servant but enjoys only a limited range, while sight rivals the boldest flights of fancy in penetrating to illimitable distances.

This combination of the two senses is of great importance in simplifying our present task. Since we are here concerned only with vision, and since touch is sufficient to produce complete conceptions of spatial location, we may assume these conceptions to be already complete, at least in their general outline, and confine our investigation to ascertaining the point of agreement between the visual and tactile perceptions of space. The question of how it is possible for any conception of spatial location to arise from either or both of these sensations, we shall leave till last.

It is obvious, from a consideration of well-known facts, that even though our sensations are distributed among separate nervous structures, we do not necessarily conceive of the causes of these sensations as locally separate. For example, we may have in one room sensations of light, of warmth, of various notes of music, and also perhaps of an odor, and we may recognize that all these agents are diffused through the air of the room at the same time, without any difference of locality. When a compound color falls upon the retina, we receive three separate elementary impressions, probably conveyed by separate nerves, without having any power of distinguishing them. In a note struck on a stringed instrument or in the human voice, we hear different tones at the same time—one fundamental and a series of harmonic overtones—which are probably received by different nerves, yet we are unable to separate them in space.

Many articles of food produce different impressions of taste upon different parts of the tongue and also produce sensations of odor by their volatile particles ascending into the nostrils from behind. But these different sensations, originating in different parts of the nervous system, are usually fused completely and inseparably in the compound sensation we call taste. No doubt it is possible, with a little attention, to ascertain the parts of the body which receive these sensations, but even when we recognize these as locally separate, it does not follow that we must conceive of the sources of these sensations as separated in the same way.

We find a corresponding fact in the physiology of sight—namely, that we see only a single object with our two eyes, although the impression is conveyed by two distinct nerves. In fact, both kinds of phenomena (taste and binocular vision) are examples of a more universal law.

Hence the fact that a plane optical image of the objects in the field of vision is produced on the retina and that the different parts of this image excite different fibers of the optic nerve, is not sufficient to explain why we refer the sensations thus produced to locally distinct regions of our field of vision. Something else is clearly being added to produce the conception of the separation of these impressions in space.

The sense of touch offers precisely the same problem. When two different parts of the skin are touched at the same time, two different sensitive nerves are excited, but the local separation between these nerves is not a sufficient ground for our recognition that the two parts which have been touched are distinct and for our consequent conception of two different external objects. Indeed, this conception varies according to circumstances. If we touch a table with two fingers and feel under each a grain of sand, we suppose that there are two separate grains of sand. If we place the two fingers one against the other with a single grain of sand between them, we may have the same sensations of touch in the same two nerves as before, yet under these circumstances we suppose that there is only a single grain. Our consciousness of the fingers' position obviously influences the result at which the mind arrives. This is further proved by the experiment of crossing two fingers and putting a marble between them; the single object then produces in the mind the conception of two.

What is the faculty which supplements the anatomical distinction in locality between the different sensitive nerves and which, in cases like those I have mentioned, produces the idea of separation in space? In attempting to answer this question, we cannot avoid a controversy which has not yet been decided.

Some physiologists, following the lead of Johannes Müller, answer that the retina or skin, being itself an organ which is extended in space, receives impressions which carry with them this quality of extension in space; that the conception of locality is innate; and that impressions derived from external objects are transmitted of themselves to corresponding local positions in the image produced in the sensitive organ. We may describe this as the nativistic, or intuitive, theory of spatial intuition. It obviously cuts short all further inquiry into the origin of these intuitions, since it regards them as original, inborn, and incapable of further explanation.

The opposing view was put forth in a more general form by the early English philosophers of the sensation school, Molyneux, Locke, and Jurin. Its application to special physiological problems has only become possible in very modern times, particularly since we have gained more accurate knowledge of the movements of the eye. The invention of the stereoscope by Wheatstone made the difficulties and imperfections of the nativistic theory of vision much more obvious than before and led to another solution, which approached much nearer to the older view and which we shall call the empirical theory of vision. This assumes that none of our sensations give us anything more than signs for external objects and movements and that we can learn how to interpret these signs only through experience and practice. For example, the perception of differences in spatial location can be attained only through movement; in the field of vision it depends upon our experience of the movements of the eye.

This empirical theory must of course also assume a difference among the sensations of various parts of the retina, depending on their location. Otherwise it would be impossible to distinguish any local differences in the field of vision. The sensation of red, when it falls upon the right side of the retina, must in some way be different from the sensation of the same red when it affects the left side. Moreover, this difference between the two sensations must be of another kind from that which we recognize when the same spot in the retina is successively affected by two different shades of red. Lotze has named this difference between the sensations which the same color excites when it affects different parts of the retina the *local sign* of the sensation. We are for the present ignorant of the nature of this difference, but I adopt the name given by Lotze as a convenient expression. While it would be premature to form any further hypothesis as to the nature of these local signs, there can be no doubt of their existence, for it follows from the fact that we are able to distinguish local differences in the field of vision.

The difference, therefore, between the two opposing views is as follows: The empirical theory regards the local signs (whatever they really may be) as signs whose signification must be learned, and is actually learned, in order to arrive at a knowledge of the external world. It is not necessary to suppose any kind of correspondence between these local signs and the actual differences of spatial location which they signify. The nativistic theory, on the other hand, supposes that the local signs are direct perceptions of actual differences in space, both in their nature and in their magnitude.

It is apparent that our present inquiry requires us to consider the far-reaching opposition between these two systems of philosophy: one which assumes that the laws of mental operations are in pre-existing harmony with those of the outer world, and the other which attempts to explain all correspondence between mind and matter as the result of experience.

So long as we confine ourselves to observing a field of two dimensions whose individual parts offer no (or, at any rate, no recognizable) difference in their distances from the eye—so long, for instance, as we look only at the sky and distant parts of the landscape—both theories offer equally good explanations of the way we form perceptions of local relations in the field of vision. The extension of the retinal image corresponds to the extension of the actual image presented by the objects before us; or, at least, any incongruities can be reconciled with the nativistic theory of sight without any very difficult assumptions or explanations.

The first of these incongruities is that in the retinal picture the top and bottom, and the right and left, of the actual image are inverted. This we saw in Fig. 4 (page 154) to result from the rays of light crossing as they enter the pupil; the point *a* is the retinal image of *A*, *b* of *B*. This has always been a difficulty in the theory of vision, and many hypotheses have been invented to explain it, two of which have survived. We may, with Johannes Müller, consider the conceptions of upper and lower in spatial intuitions to be purely relative—that is, to concern one relative only to the other—and assume that the correspondence between what is upper according to our sense of sight and what is upper according to our sense of touch is acquired only through experience, when our hands, as they feel, appear in our field of vision. Or, secondly, we may assume with Fick that since all impressions upon the retina must be conveyed to the brain in order to be perceived there, the nerves of sight and of feeling are so arranged in the brain as to produce a correspondence between the conceptions they produce of upper and under, right and left.

This supposition has, however, no pretense of any anatomical facts to support it.

The second difficulty for the nativistic theory is that while we have two retinal pictures, we do not see double. This difficulty was met by the assumption that both retinas, when they are excited, produce only a single sensation in the brain because the individual points of each retina correspond with those of the other and each pair of corresponding or "identical" points produces the sensation of a single one. There is an actual anatomical arrangement which might support this hypothesis. The two optic nerves cross before entering the brain and at that point become united. Pathological observations make it probable that the nerve fibers from the right-hand halves of both retinas pass to the right cerebral hemisphere, those from the left halves to the left hemisphere. But although corresponding nerve fibers would thus be brought close together, it has not yet been shown that they actually unite in the brain.

These two difficulties do not apply to the empirical theory, since it only supposes that the actual sensible sign, whether simple or complex, is recognized as the sign of that which it signifies. An uninstructed person is as sure as possible of his visual perceptions without ever knowing that he has two retinas, that there is an inverted image on each, or that there is such a thing as an optic nerve to be excited or a brain to receive the impression. He is not troubled by his retinal images being inverted and double. He knows what impression a certain object in a certain position makes on him through his eyesight, and he governs himself accordingly.

But we must learn the signification of the local signs which belong to our sensations of sight, so as to be able to recognize the actual relations which they denote. The potential for this learning depends, first, on our having movable parts of our own body within sight. Once we know, by means of touch, what relation in space is and what movement is, we can learn what changes in the impressions on the eye correspond to the voluntary movements of a hand which we can see. In the second place, when we move our eyes while looking at a field of vision filled with objects at rest, the retina changes its relation to the almost unchanged position of the retinal image. We thus learn what impression the same object makes upon different parts of the retina. An unchanged retinal image, passing over the retina as the eye turns, is like a pair of compasses which we move over a drawing in order to measure its parts.

Even if the local signs of sensation were quite arbitrary, thrown together without any systematic arrangement (a supposition which I regard as improbable), it would still be possible by moving the hand and the eye, as just described, to ascertain which signs go together and which correspond in different regions of the retina to points at similar distances in the two dimensions of the field of vision. This is in accordance with experiments by Fechner, Volkmann, and myself which prove that even the fully developed eye of an adult can only compare accurately the size of those lines or angles in the field of vision whose images can be thrown one after another upon precisely the same spot of the retina by the ordinary movements of the eye.

We may convince ourselves by a simple experiment that the correspondence of the perceptions of touch and sight depend, even in an adult, upon a continuous comparison of the two by means of the retinal images of our hands as they move. If we put on a pair of spectacles with prismatic glasses whose two flat surfaces converge toward the right, all objects appear to be shifted to the right. If we then try to touch anything we see, taking care to shut our eyes before our hand appears in sight, our hand passes to the right of the object. If we follow the movement of the hand with the eye, however, we are able to touch what we intend by bringing the retinal image of the hand up to that of the object. Again, if we handle the object for one or two minutes, watching it all the time, a fresh correspondence is formed between the eye and the hand in spite of the deceptive glasses, and we are able to touch the object with perfect certainty even when our eyes are shut. We can even touch it with our other hand with our eyes shut. This proves that it is not the perception of touch which has been altered by comparison with the false retinal images; on the contrary, it is the perception of sight which has been corrected by that of touch.

Again, if, after trying this experiment several times, we take off the spectacles and then look at any object, taking care not to bring our hands into the field of vision, and if we then try to touch the object with our eyes shut, our hand will pass beyond it on the opposite side–that is, to the left. The new harmony which has been established between the perceptions of sight and touch continues its effects and leads to fresh mistakes when the normal conditions are restored. In preparing objects with needles under a compound microscope, we must learn to harmonize the inverted microscopic image with our muscular sense; and we have to get over a similar difficulty in shaving before a looking glass, which changes right to left.

These instances, in which the image presented in the two dimensions of the field of vision is essentially of the same kind as the retinal images and resembles them, can be equally well explained (or nearly so) by the two opposite theories of vision to which I have referred. But it is quite another matter when we pass to the observation of nearby objects of three dimensions. In this case there is a thorough and complete incongruity between our retinal images, on the one hand, and the actual state of the objects, as well as our correct impression of them, on the other. Here we are compelled to choose between the two opposite theories, and this aspect of our subject—the explanation of our perception of solidity, or *depth,* in the field of vision and of binocular vision, on which that perception chiefly depends—has for many years been the field of much investigation and no little controversy. And no wonder, for we have already learned enough to see that the questions to be decided are of fundamental importance, not only for the physiology of sight, but for a correct understanding of the true nature and limits of human knowledge generally.

Each of our eyes projects a plane image upon its own retina. However we may suppose the conducting nerves to be arranged, the two retinal images when united in the brain can only reappear as a plane image. Yet the actual impression on our mind is a solid image of three dimensions. Here again, as in the system of colors, the outer world is richer than our sensation, but in this case the conception formed by the mind completely represents the reality of the outer world. It is important to remember that this perception of depth is fully as vivid, direct, and exact as that of the plane dimensions of the field of vision. If a man takes a leap from one rock to another, his life depends just as much upon his rightly estimating the distance of the rock on which he is to alight as upon his not misjudging its position to the right or left. As a matter of experience, we find that we can do the one just as quickly and surely as the other.

In what way can this appreciation of what we call depth, solidity, and distance come about? Let us first ascertain the facts.

At the outset of the inquiry we must bear in mind that perception of the solid form of objects and of their relative distance from us is not quite absent, even when we look at them with only one eye and without changing our position. Our means in this case are just the same as those which the painter can employ to give the figures on his canvas the appearance of being solid objects and of standing at different distances from the spectator. It is part of a painter's merit for his figures to stand out boldly, but how does he produce the illusion? In the first place, in painting a landscape he likes to have the sun near the horizon, which gives him strong shadows, for these throw objects in the foreground into bold relief. Next he prefers an atmosphere which is not quite clear, because slight obscurity makes the distance appear far off. Then he is fond of bringing in figures of men and cattle, because by the help of these objects of known size we can easily measure the size and distance of other parts of the scene. Lastly, houses and other regular productions of art are also useful to him for giving a clue to the meaning of the picture, since they enable us easily to recognize the position of horizontal surfaces.

The representation of solid forms by drawings in correct perspective is most successful with objects of regular, symmetrical shape, such as buildings, machines, and implements of various kinds. For we know that all of these are chiefly bounded either by planes which meet at a right angle or by spherical and cylindrical surfaces; and this is sufficient to supply what the drawing does not directly show. Moreover, in the case of figures of men or animals, our knowledge that the two sides are symmetrical further assists the impression conveyed. But objects of unknown and irregular shape, such as rocks or masses of ice, baffle the skill of the most consummate artist; and even their representation in the most complete, perfect manner possible—that is, in a photograph—often shows nothing but a confused mass of black and white. Yet when we have these objects in reality before our eyes, a single glance is enough for us to recognize their form.

The first man who clearly defined the aspects in which it is impossible for any picture to represent actual objects was the great master of painting, Leonardo da Vinci, who was almost as distinguished in natural philosophy as in art. He pointed out in his *Trattato della pittura* that the separate views of the outer world presented by our eyes are not precisely the same. Each eye sees in its retinal image a perspective view of the objects which lie before it, but inasmuch as it occupies a somewhat different position in space from the other, its point of view, and so its whole perspective image, is different. If I hold up my finger and look at it first with the right eye and then with the left eye, it covers slightly different parts of the opposite wall, shifting from left to right. If I hold up my right hand with the thumb toward me, I see with the right eye more of the back of the hand, with the left more of the palm; and the same effect is produced whenever we look at bodies whose parts are at different distances from our eyes. But when I look at a hand represented in the same position in a painting, the right eye sees exactly the same figure as the left. Thus we see that actual solid objects present different images to the two eyes, while a painting shows only the same. Hence follows a difference in the impression made upon the sight which the utmost perfection in a representation on a flat surface cannot supply.

The clearest proof that seeing with two eyes—and the difference between the images they present—constitutes the most important cause of our perception of a third dimension in the field of vision has been furnished by Wheatstone's invention of the stereoscope. I assume that this instrument and the peculiar illusion which it produces are well known. By its help we see the solid shape of the objects represented on the stereoscopic slide, with the same complete evidence of the senses with which we should look at the real objects themselves. This illusion is produced by presenting somewhat different pictures to the two eyes—to the right, one which represents the object in perspective as it would appear to that eye; to the left, one as it would appear to the left. If the pictures are otherwise exact and represent precisely these two different points of view (as can easily be done by photography), we receive on looking into the stereoscope the same impression in black and white as the object itself would give.

Anyone who has sufficient control over the movements of his eyes can combine the two pictures on a stereoscopic slide into a single solid image without the help of an instrument. It is only necessary to direct the eyes so that they focus at the same time upon the corresponding points in the two pictures; but it is easier to do this by the help of an instrument which apparently brings the two pictures to the same place.

In Wheatstone's original stereoscope, represented in Fig. 9, the observer looked with his right eye into the mirror *b* and with his left into the mirror *a*. Both mirrors were placed at an angle to the observer's line of sight, and the two pictures were so placed at *k* and *g* that their reflected images appeared at

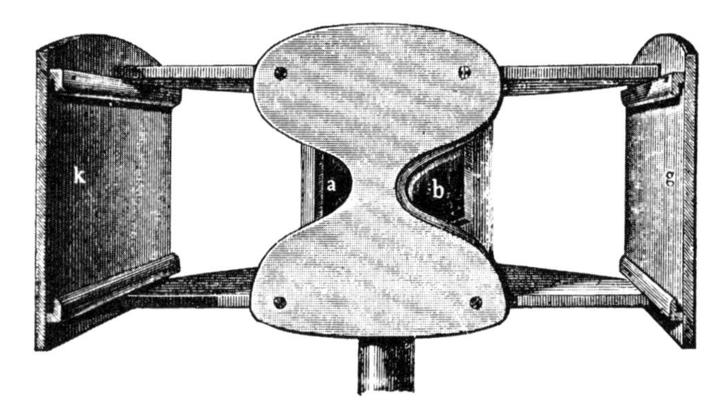

Fig. 9

the same place behind the two mirrors; but the right eye saw the picture *g* in the mirror *b*, while the left saw the picture *k* in the mirror *a*.

A more convenient instrument, though it does not give such sharply defined effects, is the ordinary stereoscope of Brewster, shown in Fig. 10. Here the two pictures are placed on the same slide and laid in the lower part of the stereoscope, which is divided by a partition *S*. Two slightly prismatic glasses with convex surfaces (Fig. 11) are fixed at the top of the instrument. These show the pictures somewhat farther off, somewhat magnified, and at the same time overlapping each other, so that both appear to be in the middle of the instrument. Both pictures are apparently brought to the same spot, though each eye sees only the one which belongs to it.

The illusion produced by the stereoscope is most obvious and striking when other means of recognizing the form of an object fail. This is the case with geometrical outlines of solid figures, such as diagrams of crystals, and also with representations of irregular objects, especially when they are transparent and the shadows do not fall as we are accustomed to see them in opaque objects. Thus glaciers in stereoscopic photographs often appear to the unassisted eye an incomprehensible chaos of black and white; but when seen through a stereoscope, the clear transparent ice, with its fissures and polished

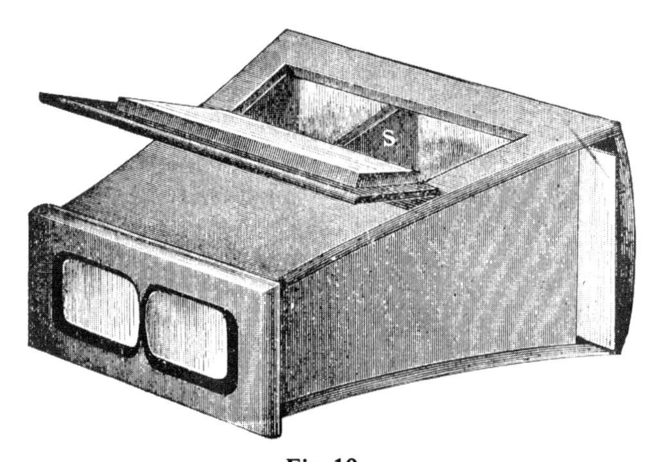

Fig. 10

Fig. 11

surfaces, stands out as if it were real. When I have seen for the first time buildings, cities, or landscapes with which I was familiar from stereoscopic pictures, they have often seemed familiar to me; but I have never experienced this impression after seeing any number of ordinary pictures because these so imperfectly represent the real effect upon the senses.

The accuracy of the stereoscope is no less wonderful. Dove has contrived an ingenious illustration of this: Take two pieces of paper printed with the same type or from the same copperplate, hence exactly alike, and put them in the stereoscope in place of the two ordinary photographs. They will unite into a single completely flat image because, as we have seen above, the two retinal images of a flat picture are identical. But no human skill is able to copy the letters of one copperplate onto another so perfectly that there shall not be some difference between them. If, therefore, we print off the same sentence from the original plate and a copy of it, or the same letters with different specimens of the same type, and put the two pieces of paper into the stereoscope, some lines will appear nearer and some farther off than the rest. This is the easiest way of detecting spurious bank notes: a suspected one is put in a stereoscope along with a genuine specimen of the same kind, and it is instantly apparent whether all the marks in the combined image coincide. This experiment is also important for the theory of vision, since it teaches us in a most striking manner how vivid, sure, and minute is our judgment as to depth derived from the combination of the two retinal images.

We now come to the question: how is it possible for two flat images upon the retina, each from a different perspective and each representing only two dimensions, to combine so as to present a single, solid image of three dimensions?

We must first make sure that we are really able to distinguish between the two flat images offered us by our eyes. If I hold my finger up and look toward the opposite wall, it covers a different part of the wall to each eye, as I mentioned above. I see the finger twice, in front of two different places on the wall; and if, with both eyes open, I see a single image of the wall, I must see a double image of the finger.

In ordinary vision we try to recognize the solid form of surrounding objects, and we either do not notice this double image at all, or notice it only when it is unusually striking. If we desire to see it, we must look at the field of vision in another way—the way an artist does who intends to draw it. He tries to forget the actual shape, size, and distance of the objects he represents.

One might suppose that this is the more simple, original way of seeing things; and hitherto most physiologists have regarded it as the kind of vision which results most directly from sensation, while they have looked on ordinary solid vision as a secondary way of seeing things, which has to be learned through experience. But every draftsman knows how much harder it is to appreciate the apparent form in which objects appear in the field of vision and to measure the angular distance between them than to recognize their actual form and comparative size. In fact, the knowledge of the true relations of surrounding objects, knowledge of which the artist cannot divest himself, is his greatest obstacle in drawing from nature.

If we look at the field of vision with both eyes, as an artist does, fixing our attention upon the outlines as they would appear if projected on a pane of glass between us and them, we become instantly aware of the difference between the two retinal images. We see those objects double which lie farther off or nearer than the point at which we are looking but which are close enough to it laterally to be sufficiently seen. At first we can only recognize double images of objects at very different distances from the eye, but by practice we can learn to see them with objects at nearly the same distance.

All these phenomena (and others like them) of double images of objects seen with both eyes may be reduced to a simple rule which was laid down by Johannes Müller: "For each point of one retina there is on the other a *corresponding* point." In the ordinary flat field of vision presented by the two eyes, the images received by corresponding points as a rule coincide, while images received by those which do not correspond do not coincide. The corresponding points in each retina (ignoring slight deviations) are those which are situated at the same lateral and vertical distance from the point of the retina at which rays of light come to a focus when we fix the eye for exact vision, namely, the yellow spot.

The reader will remember that the nativistic theory of vision necessarily assumes a complete combination of those sensations which are excited by impressions upon corresponding—or, as Müller calls them, identical—points. This supposition was most fully expressed in the anatomical hypothesis that the two nerve fibers which arise from corresponding points of the two retinas actually unite so as to form a single fiber, either at the commissure of the optic nerves or in the brain itself. It is worth noting that Johannes Müller did not definitely commit himself to this mechanical explanation, although he suggested its possibility. He wished his law of identical points to be regarded simply as an expression of facts, and he insisted only that the position in the field of vision of the images they receive is always the same.

But a difficulty arises in that we always distinguish rather imprecisely between double images if it is possible to fuse them into a single intuition–in very striking contrast to the extraordinary precision with which, as Dove has established, we can judge stereoscopic relief. Yet the latter ability depends upon the same differences between the two retinal images as lie at the basis of the phenomenon of double images. The slight difference of distance between the objects represented in the right and left half of a stereoscopic photograph, which suffices to produce the most striking effect of solidity, must be increased twenty- or thirty-fold before it can be recognized in the production of a double image, even if we suppose the most careful observation by one who is well practiced in the experiment.

There are a number of other circumstances which make the recognition of double images either easy or difficult. The most striking instance of the latter is the effect of relief. The more vivid the impression of solidity, the more difficult the double images are to see, so that it is easier to see them in stereoscopic pictures than in the actual objects they represent. On the other hand, the observation of double images is facilitated by varying the color and brightness of the lines in the two stereoscopic pictures, or by putting lines in both which exactly correspond and so will make more evident by contrast the imperfect coalescence of the other lines. All these circumstances ought to have no influence if the combination of the two images in our sensation depended upon an anatomical arrangement of the conducting nerves.

Again, after the invention of the stereoscope, a fresh difficulty arose in explaining our perceptions of solidity by the differences between the two retinal images. First, Brücke called attention to a series of facts which apparently made it possible to reconcile the newly discovered phenomena with the theory of the innate identity of the sensations conveyed by the two retinas. If we carefully observe the way in which we look at stereoscopic pictures or at real objects, we notice that the eye follows the different outlines one after another. At any given moment we see the fixed point single, while the other points appear double. But our attention is usually concentrated upon the fixed point, and we take such little note of the double images that to many people they are a new and surprising phenomenon when first pointed out. In following the outlines of these pictures, or of an actual image, we move our eyes unequally this way and that. Sometimes they converge and sometimes diverge, according as we look at points of the outline which are apparently nearer or farther off; and these differences in movement may give rise to the impression of different degrees of distance of the several lines.

Now it is quite true that by this movement of the eye while looking at stereoscopic outlines, we gain a much more clear and exact image of the raised surface they represent, than we could by fixing our attention upon a single point. Perhaps the simple reason is that when we move our eyes, we look at every point of the figure in succession directly and therefore see it much more sharply defined than when we see only one point directly and the others indirectly. But Brücke's hypothesis that the perception of solidity is produced solely by this movement of the eyes was disproved by experiments made by Dove, which showed that the peculiar illusion of stereoscopic pictures is also produced when they are illuminated with an electric spark. The light then lasts for less than 1/4,000th of a second, in which time heavy bodies move so little, even at great velocities, that they seem to be at rest. While the spark lasts, there cannot be the slightest movement of the eye which can possibly be recognized, yet we receive the complete impression of stereoscopic relief.

Secondly, a combination of the sensations of the two eyes, such as the anatomical hypothesis assumes, is proved not to exist by the phenomenon of stereoscopic luster, which was also discovered by Dove. If the same surface is made white in one stereoscopic picture and black in another, the combined image appears to shine, though the paper itself is quite dull. Stereoscopic drawings of crystals are made so that one shows white lines on a black ground and the other, black lines on a white ground. When looked at through a stereoscope, they give the impression of a solid crystal of shining graphite. By the same means it is possible to produce in stereoscopic photographs the still more beautiful effect of the sheen of water or of leaves.

The explanation of this curious phenomenon is as follows: A dull surface, like unglazed white paper, reflects the light which falls on it equally in all directions and therefore always looks equally bright from whatever point it is seen; hence, of course, it appears equally bright to both eyes. On the other hand, a polished surface, in addition to the reflected light which it scatters equally in all directions, throws back other beams by *regular* reflection, which only pass in definite directions. One eye may receive this regularly reflected light and the other not, in which case the surface appears much brighter to the one than to the other. As this can only happen with shining bodies, the effect of the black and white stereoscopic pictures is like that of a polished surface.

If there were a complete combination of the impressions produced upon both retinas, the union of white and black would give gray. The fact that when they are actually combined in the stereoscope, they produce the effect of luster–an effect which cannot be produced by any kind of uniform gray surface–proves that the impressions on the two retinas are not combined into one sensation. Nor does this effect of stereoscopic luster depend upon an alternation between the impressions of the two eyes, on what is called the rivalry of the retinas, as is proved by illuminating stereoscopic pictures for an instant with an electric spark. With no time for alternation, the same effect is perfectly produced.

In the third place, it can be established, not only that the images received by the two eyes do not coalesce into one sensation, but that the sensations transmitted from the two eyes to the brain are not exactly similar. They can, in fact, be readily distinguished. For, if the sensation given by the right eye were indistinguishably the same as that given by the left, it would make no difference (at least by the light of an electric spark, when no movements of the eye can help us in distinguishing the two images) whether we saw the right-hand stereoscopic picture with the right eye and the left with the left, or put the two pictures into the stereoscope reversed. But in practice we find that it makes all the difference, for if we switch the two pictures, the relief appears inverted: what should be farther off seems nearer, and what should stand out seems to fall back. Since, when we look at objects by the momentary light of an electric spark, they always appear in their true relief and never reversed, it follows that the impression produced on the right eye is not indistinguishable from that on the left.

Lastly, some very curious and interesting phenomena are seen when two dissimilar pictures are put before the two eyes at the same time, pictures which cannot be combined so as to present the appearance of a single object. If, for example, we look with one eye at a page of print and with the other at an engraving,[16] there follows what is called the rivalry of the two fields of vision. The two images are not seen at the same time, one covering the other; rather, at some points one prevails and at others, the other. If they are equally distinct, they usually change places every few seconds. But if the engraving presents anywhere in the field of vision a uniform white or black surface, the printed letters which occupy the same position in the image presented to the other eye will usually prevail exclusively over the uniform surface of the engraving.

16. [The practiced observer is able to do this without any apparatus, but most persons will find it necessary to put the two objects in a stereoscope, or at least to hold a book, or a sheet of paper, or the hand in front of the face to serve for the partition in the stereoscope.–P. H. P.-S.]

In spite of what former observers have said to the contrary, I maintain that it is possible for the observer at any moment to control this rivalry by voluntary direction of his attention. If he tries to read the printed sheet, the letters remain visible, at least at the spot where he is reading. If, on the other hand, he tries to follow the outline and shadows of the engraving, these prevail. I find, moreover, that it is possible to fix the attention upon a very feebly illuminated object and make it prevail over a much brighter one which coincides with it in the retinal image of the other eye. I can, for instance, follow the watermark of a white piece of paper and cease to see strongly marked black figures in the other field. Hence the retinal rivalry is not a trial of strength between two sensations, but depends upon our fixing or failing to fix the attention.

Indeed, there is scarcely any other phenomenon so well fitted for the study of the causes which are capable of determining the attention. It is not enough to form a conscious intention of seeing first with one eye and then with the other. We must form as clear a notion as possible of what we expect to see, and then it will actually appear. If, on the other hand, we leave the mind at liberty without a fixed intention to observe a definite object, the alternation between the two pictures ensues which is called retinal rivalry. In this case we find that, as a rule, bright and strongly marked objects in one field of vision prevail, either completely or at least for a time, over those which are darker and less distinct in the other.

We may vary this experiment by using a pair of spectacles with different colored glasses. We shall find, on looking at the same object with both eyes at once, that there ensues a similar rivalry between the two colors. Everything appears covered over, first with one and then with the other. After a time, however, the vividness of both colors becomes weakened, partly by fatigue of the retinal elements which each affects, partly by the complementary after-images which result. The alternation then ceases, and there ensues a kind of mixture of the two original colors.

It is much more difficult to fix the attention upon a color than upon an object such as an engraving, for the attention—upon which, as we have seen, the whole phenomenon of rivalry depends—fixes itself with constancy only upon an object which continually offers something new for the eye to follow. But we may facilitate this by reflecting, on the surface of each glass nearer the eye, letters or other lines upon which the attention can fix. These reflected images themselves are not colored, but as soon as the attention is fixed upon one of them, we become conscious of the color of the corresponding glass.

These experiments on the rivalry of colors have given rise to a singular controversy among the best observers, and the possibility of such difference of opinion is an instructive hint as to the nature of the phenomenon itself. One party (including Dove, Regnault, Brücke, Ludwig, Panum, and Hering) maintains that the result of a binocular view of two colors is the true combination color. Other observers (among them Heinrich Meyer, of Zürich, Volkmann, Meissner, and Funke) declare quite as positively that under these conditions they have never seen the combination color. I myself agree entirely with the latter, and a careful examination of the cases in which I imagined that I saw the combination color has always proved to me that it was the result of phenomena of contrast. Each time that I brought the true combination color side by side with the binocular mixture of colors, the difference was very apparent. On the other hand, there can of course be no doubt that the observers I first named really saw what they profess, so there must here be great individual difference. Indeed, with certain experiments which Dove recommends as particularly well suited to prove the correctness of his conclusion, such as the binocular combination of complementary polarization colors into white, I could not myself see the slightest trace of a combination color.

This striking difference in a comparatively simple observation seems to me of great interest. It is a remarkable confirmation of the supposition made above, in accordance with the empirical theory of vision, that in general only those sensations are perceived as separated in space which can be separated one from another by voluntary movements. Even when we look at a compound color with one eye, three separate sensations are, according to Young's theory, produced together; but since it is impossible to separate these by any movement of the eye, they always remain locally united. Yet we have seen that even in this case we may become conscious of a separation under certain circumstances; namely, when part of the color obviously belongs to a transparent covering. When two corresponding points of the retinas are illuminated with different colors, it is rare for any separation between them to appear in ordinary vision; if it does, it usually takes place in the part of the field of sight outside the region of exact vision. But there is always a possibility of separating the compound impression thus produced into its two parts, which will appear to some extent independent of each other and will move with the movements of the eye. Depending upon the degree of attention which the observer is accustomed to give to the region of indirect vision and to double images, he may or may not be able to separate the colors which fall on both retinas at the same time.

Mixed hues, whether looked at with one eye or with both, excite many simple sensations of color at the same time, each having exactly the same position in the field of vision. The difference in the way such a compound color is regarded by different people depends upon whether this compound sensation is accepted immediately as a coherent whole, without any attempt at analysis, or whether the observer is able by practice to recognize the parts of which it is composed and to separate them from one another. The former is our usual (though not constant) habit when looking with one eye, while we are more inclined to the latter when using both. But inasmuch as this inclination must depend chiefly upon practice in observing distinctions, gained by preceding observation, it is easy to understand how great individual peculiarities may arise.

If we carefully observe the rivalry which ensues when we try to combine two stereoscopic drawings, one drawn in black lines on a white ground and the other in white lines on black, we shall see that the white and black lines which affect nearly corresponding points of the retinas always remain visible side by side—an effect which of course implies that the white and black grounds are also visible. By this means the brilliant surface, which seems to shine like black lead, makes a much more stable impression than that produced under the operation of retinal rivalry by entirely different drawings. If we then cover the lower half of the white figure on a black ground with a sheet of printed paper, the upper half of the combined stereoscopic image continues to show the phenomenon of luster, while in the lower we see retinal rivalry between the black lines of the other figure and the black marks of the type. As long as the observer attends to the solid form of the object represented, the black and white outlines of the two stereoscopic drawings carry forward the point of exact vision as it moves along them, but the effect can only be kept up by continuing to follow both. So long as he keeps his attention steadily upon both drawings, their impressions will be equally combined. There is no better way of preserving the combined effect of two stereoscopic pictures than this. Indeed, it is possible to combine (at least, partially and for a short time) two entirely different drawings placed in the stereoscope by fixing the attention upon the way in which they cover each other—watching, for instance, the angles at which their lines cross. But as soon as the attention turns from the angle to follow one of the lines which makes it, the picture to which the other line belongs vanishes.

Let us now gather the results to which our inquiry into binocular vision has led us:

1. *The excitation of corresponding points of the two retinas is not indistinguishably combined into a single impression.* If it were, it would be impossible to see stereoscopic luster. And we have found reason to believe that this effect is not a consequence of retinal rivalry, even if we admit the latter phenomenon to belong to sensation at all, rather than to the degree of attention. On the contrary, the appearance of luster is associated with the restriction of this rivalry.

2. *The sensations which are produced by the excitation of corresponding points of each retina are not indistinguishably the same.* Otherwise we should not be able to distinguish the true from the inverted, or pseudoscopic, relief when two stereoscopic pictures are illuminated by an electric spark.

3. *The combination of the two different sensations received from corresponding retinal points is not produced by suppressing one of them for a time.* In the first place, the perception of solidity given by the two eyes depends upon our being conscious of the two different images at the same time. Furthermore, this perception of solidity is independent of any movement of the retinal images, since it is possible under momentary illumination.

We therefore learn that two distinct sensations are transmitted from the two eyes and reach the consciousness at the same time without coalescing. The combination of these two sensations into the single perceptual image of the external world of which we are conscious in ordinary vision must therefore be produced, not by any anatomical mechanism of sensation, but by a mental act.

4. Further, we find that *on the whole, or at least in large part, the coincidence in localization of visual impressions at corresponding points on the retinas does occur at the same place in the field of vision,* but that *the idea which refers the two impressions to the same object can equally well disturb this coincidence considerably.* If each coincidence in localization were given by a direct act of sensation, this sensation could not be overcome by a conflicting idea; but the case is altogether different if the coincidence is a judgment of distance based on experience, resting on acquired knowledge of the significance of local signs. In the latter case, one experience competes only with another; and it is comprehensible that since our idea that two visual images belong to the same object is based on our judgment (by means of our eyesight) of their mutual positions, we shall perceive their distances from the points of fixation in the plane of the visual field as equal, even though they are not precisely so.

It follows further that if the equality of localization of corresponding places in both fields of vision does not rest on experience, the primary comparison of different distances in each individual visual field cannot rest upon immediate experience. For if such were given, the correspondence of both fields in immediate experience would, necessarily and completely, also be given, as soon as the identity of the two points of fixation and the correspondence of the meridian of one eye with the corresponding meridian of the other are determined.

The reader can see how this series of facts has driven us by force to the empirical theory of vision. It is right to mention that fresh attempts have recently been made to explain the origin of our perception of solidity and the phenomena of single and double binocular vision by the assumption of some ready-made anatomical mechanism. We cannot criticize these attempts here; it would lead us too far into details. Although many of these hypotheses are very ingenious (and at the same time very indefinite and elastic), they have always proved insufficient because the actual world offers us far more numerous relations than the authors of these attempts could provide for. As soon as they have arranged one of their systems to explain some particular phenomenon of vision, it is found not to answer for some others. Then, in order to help out the hypothesis, the very doubtful assumption has to be made that in these other cases, sensation is overcome and obliterated by conflicting experience. But what possible confidence could we put in any of our perceptions if we were able to disregard sensations at will, whenever they refer to an object of our attention, in favor of other ideas to which they are opposed? At any rate, it is clear that in every case where experience must finally decide, we shall succeed much better in forming a correct notion of what we see if we have no conflicting sensations to overcome than if a correct judgment must be formed in spite of them.

It follows that the hypotheses which have been successively framed by the various supporters of intuitive theories of vision, in order to suit one phenomenon after another, are really quite unnecessary. No fact has yet been discovered inconsistent with the empirical theory, which does not assume any peculiar modes of physiological action in the nervous system or any hypothetical anatomical structures. It assumes nothing more than the well-known association between the impressions we receive and the conclusions we draw from them, according to the fundamental laws of daily experience. True, we cannot at present offer a complete scientific explanation of the mental processes involved, and there is no immediate prospect of our doing so. But since these processes actually exist, and since every form of the intuitive theory has been obliged to revert to them when all other explanations failed, these mysteries concerning mental operations cannot be regarded from a scientific point of view as constituting a deficiency in the empirical theory of vision.

It is impossible to draw any line in the study of our perceptions of space that will sharply separate those which belong to immediate awareness from those which are the result of extended experience. If we attempt to set such a boundary, we find that experience proves more exact, more direct, and more specific than immediate awareness and, in fact, proves its superiority by overcoming the latter. The only supposition which does not lead to any contradiction is that of the empirical theory, which regards all our perceptions of space as depending upon experience and takes, not only the qualities, but even the local signs of the sense of sight as nothing more than signs, the meaning of which we have to learn by experience.

We become acquainted with their meaning by comparing them with the results of our own movements, the changes we produce in the outer world. The infant first begins to play with its hands. There is a time when it does not know how to turn its eyes or its hands to an object which attracts its attention by its brightness or color. When a little older, a child seizes whatever is presented to it, turns it over and over, looks at it, touches it, and puts it in his mouth. The simplest objects are what a child likes best; he always prefers the most primitive toy to the elaborate inventions of modern ingenuity. After he has looked at such a toy every day for weeks at a time, he learns at last all the perspective images which it presents; then he throws it away and wants a fresh toy to handle. By this means the child learns to recognize the different views which the same object can afford in correlation with the movements which he is constantly giving it.

The conception of the shape of any object, gained in this manner, is the result of associating all such visual images, which are combined in the judgment we form as to its dimensions and shape. And once we have acquired an accurate conception of the form of any object, we can deduce from it the various images it would present to the sight if seen from various points of view, as well as the various movements we should have to impress upon it in order to obtain these successive images.

I have often noticed a striking instance of this process in looking at stereoscopic pictures. If, for example, they consist of elaborate outlines of complicated crystalline forms, it is often at first difficult to combine them. When this is the case, I look for two points in the diagram which correspond and make them overlap by a voluntary movement of my eyes. But as long as I have not recognized the kind of form the drawings are intended to represent, I find that my eyes begin to diverge again, and the two points no longer coincide. Then I try to follow the various lines of the figure, and suddenly I see what form is represented. From that moment on, my eyes pass over the outlines of the apparently solid body with the utmost ease and without ever separating. As soon as we have gained a correct notion of the shape of an object, we have the rule for the movements of the eyes which are necessary for seeing it. In carrying out these movements and thus receiving the visual impressions we expect, we retranslate the notion we have formed into reality; and finding that this retranslation agrees with the original, we become convinced of the accuracy of our conception.

This last point is, I believe, of great importance. The meaning we assign to our sensations depends upon experiment, not upon mere observation of what takes place around us. We learn by experiment that the correspondence between two processes takes place at any moment which we choose and under conditions which we can alter as we choose. Mere observation would not give us the same certainty, even though often repeated under different conditions, for we should thus learn only that the processes in question appear together frequently (or even always, as far as our experience goes). Mere observation would not teach us that they appear together at any moment we select. Even in considering examples of scientific observation methodically carried out, as in astronomy, meteorology, or geology, we never feel fully convinced of the causes of the phenomena observed until we can demonstrate the working of these same forces by actual experiment in the laboratory. Outside of experimental science we have not learned of a single new force. I believe that this fact is not without significance.

It is plain that, from the experience which we collect in the way I have been describing, we are able to learn as much of the meaning of sensible signs as can afterwards be verified by further experience—that is to say, all that is real and positive in our conceptions.

It has been hitherto supposed that the sense of *touch* confers the notion of space and movement. At first, of course, the only direct knowledge we acquire is that we can produce, by an act of volition, changes of which we are cognizant by means of touch and sight. Most of these voluntary changes are movements, or changes in the relations of space; but we can also produce changes in an object itself. Can we, however, recognize the movements of our hands and eyes as changes in the relations of space without knowing it beforehand? and can we distinguish them from other changes which affect the properties of external objects? I believe we can.

It is an essential and distinguishing characteristic of spatial relations that they are variable but do not depend on the quality or quantity of the substances related, while all other relations between objects depend upon their properties. The perceptions of sight prove this directly and easily. A movement of the eye which causes the retinal image to shift its place upon the retina always produces the same series of changes as often as it is repeated, whatever objects the field of vision may contain. The effect is that the impressions which had before the local signs a_0, a_1, a_2, a_3, receive the new local signs b_0, b_1, b_2, b_3; and this always occurs in the same way, whatever the quality of the impressions. By this means we learn to recognize such changes as belonging to the special phenomena which we call changes in space. This is enough for the purpose of empirical philosophy, and we need not enter further upon a discussion of the question of how much of our general conception of space is given a priori and how much a posteriori.

An objection to the empirical theory of vision might be found in the fact that illusions of the senses are possible, for if we have learned the meaning of our sensations from experience, they ought always to agree with experience. The explanation of the possibility of illusions lies in the fact that we apply our notions of external objects, which would be correct under normal conditions, to cases in which unusual circumstances have altered the retinal images. By "normal conditions" I mean to imply, not only that the rays of light pass in straight lines from each visible point to the cornea, but also that we use our eyes in the way they should be used in order to receive the clearest, most easily distinguishable images. We must successively bring the images of the separate points of the outline of the objects we are looking at upon the centers of both retinas (the yellow spots) and also move our eyes so as to obtain the surest comparison between their various positions.

Whenever we deviate from these conditions of normal vision, illusions result. Such are the long recognized effects of the refraction or reflection of rays of light before they enter the eye. But there are many other causes of mistake as to the position of the objects we see, including defective accommodation when looking through one or two small openings, improper convergence when looking with one eye only, and irregular position of the eyeball from external pressure or from paralysis of its muscles. Moreover, illusions may occur if certain elements of sensation are not accurately distinguished, such as the degree of convergence of the two eyes, which it is difficult to judge accurately when the muscles which produce it become fatigued.

The simple rule for all illusions of sight is that *we always believe that we see those objects which would, under conditions of normal vision, produce the retinal image of which we are actually conscious.* If these images are such as could not be produced by any normal kind of observation, we judge them according to their nearest resemblance; and in forming this judgment, we more easily neglect the parts of sensation which are imperfectly than those which are perfectly apprehended. When more than one interpretation is possible, we usually waver involuntarily between them; but we can end this uncertainty by choosing any one of the possible interpretations and bringing that idea as vividly as possible before the mind by a conscious effort of the will.

These illusions obviously depend upon mental processes which may be described as false inductions. But these are surely not inferences which depend upon our consciously thinking over previous observations of the same kind and determining whether they justify the conclusion we have drawn. I have therefore called them *unconscious inferences*—a term which, though accepted by other supporters of the empirical theory, has excited much opposition because, according to generally accepted psychological doctrines, an inference, or logical conclusion, is the highest of the conscious operations of the mind.

The inferences which play so great a part in the perceptions we infer from our senses cannot be expressed in the ordinary form of logically analytical statements, and it is necessary to deviate somewhat from the beaten paths of psychological analysis to convince ourselves that we really have here the same kind of mental operation as that involved in inferences usually recognized as such. There appears to me in reality only a superficial difference between the inferences of logicians and those inductive inferences whose results we recognize in the conceptions we gain of the outer world through our sensations. The chief difference is that the former inferences are capable of expression in words, while the latter are not, because instead of words they only deal with sensations and the memory of sensations. Indeed, it is precisely the impossibility of describing sensations, whether actual or remembered, in words which makes it so difficult to discuss this area of psychology at all.

Besides the knowledge which has to do with concepts and is therefore capable of expression in words, there is another result of our mental operations, which may be described as knowledge of the relations among those sensory impressions which are not capable of direct verbal expression. For instance, when we say that we "know"[17] a man, a road, a fruit, or a perfume, we mean that we have seen, or tasted, or smelled these objects. We keep the sensible impression fast in our memory and can recognize it when it is repeated, but we cannot describe the impression in words, even to ourselves. Yet it is certain that this kind of knowledge (*Kennen*) may attain the highest possible degree of precision and certainty. In this respect it is not inferior to any knowledge (*Wissen*) which can be expressed in words; but it is not directly communicable unless the object in question can actually be shown, or unless the impression it produces can be otherwise represented, as by drawing the portrait of a man instead of producing the man himself.

It is an important part of this kind of knowledge to be acquainted with the particular innervation of muscles which is necessary in order to produce any particular effect we intend by moving our limbs. As children we must learn to walk; afterward we must learn how to skate, or go on stilts, or ride, or swim, or sing, or pronounce a foreign language. Moreover, observation of infants shows that they must learn a number of things so well that later they forget there was ever a time when they were ignorant of them. For example, every one of us had to learn, when an infant, how to turn his eyes toward the light in order to see. This kind of knowledge (*Kennen*) we also call being able to do a thing (*können*)[18] and understanding how to do it (*verstehen*), as, "I know how to ride," "I am able to ride," and "I understand how to ride."

It is important to notice that this knowledge of the effort of the will to be exerted must attain the highest possible degree of certainty, accuracy, and precision if we are to maintain so artificial a balance as is necessary for walking on stilts or for skating, or if the singer is to strike a note with his voice, or the violinist with his finger, so exactly that its vibration shall not be out by a hundredth part.

Moreover, it is clearly possible, by using these sensible images of memory instead of words, to produce the same kind of combination which, when expressed in words, would be called a proposition or judgment. For example, I may know that a certain person with whose face I am familiar has a peculiar voice, of which I have an equally lively recollection. I could with the utmost

17. [In German this kind of knowledge is expressed by the verb *kennen* (cf. *cognoscere, connaître*), to be acquainted with, while *wissen* (cf. *scire, savoir*), means to be aware of. The former kind of knowledge is applicable only to objects directly cognizable by the senses, whereas the latter applies to notions or conceptions which can be formally stated as propositions.–P. H. P.-S.]

18. *Können* is said to be of the same etymology as *kennen,* in which case their likeness in form would be explained by their likeness in meaning.

certainty recognize his face and his voice among a thousand, and each would recall the other. But I cannot express this fact in words unless I am able to provide some conceptually defined characteristics of the person in question. Then I shall be able to make an inference and say, "This voice which I now hear belongs to the man whom I saw then and there."

Universal as well as particular propositions can be expressed in terms of sensible impressions instead of words. To prove this I need only refer to the effect of works of art. The statue of a god would not be capable of conveying a notion of a definite character and disposition if I did not know that the feature of its face and its expression have usually, or constantly, a definite signification. Or, to keep in the domain of sensory perceptions, if I know that a particular way of seeing, for which I have learned how to employ exactly the right kind of innervation, is necessary in order to bring into direct vision a point two feet away and so many feet to the right, this also is a universal proposition, which applies to every case in which I have fixed a given point at that distance before or may do so hereafter. It is a piece of knowledge which cannot be expressed in words but which sums up my previous successful experience. It may at any moment become the major premise of an inference—whenever, in fact, I fix on a point in this position and feel that I am doing so by looking as that major proposition states. This perception of what I am doing is then my minor premise, and the conclusion is that the object I am looking for will be found at the spot in question.

Suppose that I employ the same way of looking, but look into a stereoscope. I am now aware that there is no real object before me at the spot where I am looking, but I have the same sensible impression as if one were there; and yet I am unable to describe this impression to myself or others, or to characterize it otherwise than as "the same impression which would arise in the normal method of observation if an object were really there." It is important to notice this. No doubt the physiologist can describe the impression in other ways—by the direction of the eyes, the position of the retinal images, and so on—but there is no other way of defining and characterizing the sensation which we experience. We recognize it as an illusion, yet we cannot get rid of the sensation. We cannot erase our memory of its normal signification, even when we know that in this case it does not apply—just as little as we are able to drive out of our mind the meaning of a word in our mother tongue when it is employed as a sign for an entirely different purpose.

These conclusions in the domain of our sense perception appear as inevitable as one of the forces of nature, and their results seem to be directly perceived without any effort on our part; but this does not distinguish them from logical, conscious inferences, or at least from those which really deserve the name. All we can do by intentional, conscious effort to come to a conclusion is, after all, to supply complete materials for constructing the necessary premises. As soon as this is done, the conclusion forces itself upon us. Those conclusions which may be accepted or avoided at will are not worth much.

The reader will see that these investigations have led us to a field of mental operations which has been seldom entered by scientific explorers. The reason is that it is difficult to express these operations in words. They have been hitherto most discussed in writings on æsthetics, where they play an important part under such obscure designations as "intuition," "unconscious ratiocination," and "sensible intelligibility." Underlying all these phrases is the false assumption that the mental operations we are discussing take place in an undefined, obscure, half-conscious fashion—that they are, so to speak, mechanical operations and thus subordinate to conscious thought, which can be expressed in language. I do not believe that any difference in kind between the two operations can be proved.

The enormous superiority of knowledge which has become ripe for expression in language is sufficiently explained by two facts. In the first place, speech makes it possible to collect the experiences of millions of individuals and thousands of generations, to preserve them safely, and by continual verification gradually to prove them more and more certain and universal. In the second place, all deliberately combined actions of mankind, and so the greatest part of human power, depend on language. In neither of these respects can mere acquaintance with phenomena (*Kennen*) compete with the knowledge of them which can be communicated by speech (*Wissen*). Yet it does not necessarily follow that one kind of knowledge is of a different nature from the other or is any less clear.

The supporters of the nativistic theories of sensation often adduce the capabilities of newborn animals, many of which show themselves much more skillful than a human infant. It is quite clear that an infant, despite the greater size of its brain and its power of mental development, learns with extreme slowness to perform the simplest tasks, such as directing its eyes to an object or touching what it sees with its hands. Must we not conclude that a child has much more to learn than an animal, which is safely guided, but also restricted, by its instincts? It is said that the calf sees the udder and goes after it, though one might question whether it does not simply smell it and make those movements which bring it nearer the scent. At any rate, the child knows nothing of the meaning of the visual image presented by its mother's breast. It often turns obstinately away from it to the wrong side and tries to find it there.

The young chicken very soon pecks at grains of corn; but it pecked while it was still in the shell, and when it hears the hen peck, it pecks again, at first seemingly at random. Then, when it by chance hits upon a grain, it no doubt learns to notice the field of vision which is at the moment presented to it. The process is all the quicker because the whole of the mental furniture which it requires for its life is very small.

We need, however, further investigations on the subject in order to throw light upon this question. As far as the observations with which I am acquainted go, they seem to prove only that certain tendencies are inborn with animals. At all events, one distinction between animals and man lies precisely in this—that these innate or congenital tendencies, impulses, or instincts are in him reduced to the smallest possible number and strength.

There is a most striking analogy between the entire range of processes which we have been discussing and another system of signs, one which is not given by nature but is arbitrarily chosen and must undoubtedly be learned before it is understood. I mean the words of our mother tongue. Learning how to speak is obviously a much more difficult task than acquiring a foreign language in later life. First the child has to guess that the sounds it hears are intended to be signs, and then the meaning of each separate sound must be discovered by the same kind of induction as the meaning of the sensations of sight and touch. Yet we see children by the end of their first year, already understanding certain words and phrases, even if they are not yet able to repeat them. We may sometimes observe the same in dogs.

This connection between names and objects, which demonstrably must be learned, becomes just as firm and indestructible as that between sensations and the objects which produce them. We cannot help thinking of the usual signification of a word, even when it is used exceptionally in some other sense, and we cannot help feeling the emotions which a fictitious narrative calls forth, even when we know that it is not true—just as we cannot get rid of the normal signification of the sensations produced by an illusion of the senses, even when we know that the apparent cause is not real.

One other point of comparison is worth notice. The elementary signs of language are only twenty-six letters, yet what wonderfully varied meanings we can express and communicate by their combination! Consider, in comparison with this, the enormous number of elementary signs with which the machinery of sight is provided. We may take the number of fibers in the optic nerves as 250,000, each of which is capable of innumerable degrees of sensation of one, two, or three primary colors. It is obviously possible to construct an immeasurably greater number of combinations here than with the few letters which build up our words. Nor must we forget the extremely rapid changes of which the images of sight are capable. No wonder, then, if our senses speak to us in language which can express far more delicate distinctions and richer varieties than can be conveyed by words.

This is the solution of the riddle of how it is possible to see; as far as I can judge, it is the only one of which the known facts admit. Those striking and broad incongruities between sensations and objects, both as to quality and as to localization, on which we dwelt, are exactly the phenomena which are most instructive because they compel us to take the right road. Even those physiologists who try to save fragments of a pre-established harmony between sensations and their objects cannot but confess that the completion and refinement of sensory perceptions depend largely upon experience—so largely, indeed, that experience must be decisive whenever it conflicts with the supposed congenital arrangements of the organ. The utmost significance which may still be conceded to any such anatomical arrangements is that they may assist our first exercise of our senses.

The correspondence between the external world and the perceptions of sight rests, therefore, either in whole or in part, upon the same foundation as all our knowledge of the actual world—on experience, with constant verification of its accuracy by experiments which we perform with every movement of our body. It follows, of course, that we are warranted in accepting the reality of this correspondence only as far as these means of verification extend. For practical purposes that is really all we need.

Beyond these limits—for example, in the region of qualities—we can prove conclusively that in some instances there is no correspondence at all between sensations and their objects. Only the relations of time, space, and equality and those which are derived from them (number, size, and regularity of coexistence and of sequence)—mathematical relations, in short—are common to the outer and the inner worlds. Here we may indeed anticipate a complete correspondence between our conceptions and the objects which excite them.

But it seems to me that we should not quarrel with the bounty of Nature because the greatness, and also the emptiness, of these abstract relations have been covered by the gaily colored magnificence of a multifarious system of signs. For they can thus be the more rapidly surveyed and used for practical purposes, while enough traces remain evident to guide the theoretical spirit aright in its search after the meaning of sensible images and signs.

7.

THE AIM AND PROGRESS OF PHYSICAL SCIENCE [1869]

The opening address delivered at the Naturforscherversammlung in Innsbrück in 1869

IN accepting the honor you have done me in inviting me to deliver the first lecture at the opening session of this year's meeting of the association, it appears to me more in keeping with the import of the moment and with the dignity of this assembly that, instead of dealing with some particular line of research of my own, I should invite you to glance at the development of all of the branches of natural science represented on this occasion. These branches include vast areas of special study, involving material of a character almost too complex for comprehension. The scope and richness of these materials become greater with each year, and no limits can be set to their increase. During the first half of the present century we had an Alexander von Humboldt, who was able to survey even in its details the scientific knowledge of his time and bring all of it within one vast system. At the present time it is obviously very doubtful whether this undertaking could be carried out in a similar way, even by a man with gifts as peculiarly suited for this purpose as Humboldt's were and even if all of his time and work were devoted to the task.

Each of us, working as we do to advance a single branch of science, can devote but little time to the simultaneous study of other branches. As soon as we enter upon any investigation, all our powers have to be concentrated within a field of narrow scope. Not only must we, like the philologist or historian, seek out and search through books and gather from them what others have already discovered about the subject under inquiry; that is but a secondary part of our work. We must attack the subject itself, and each offers new and peculiar difficulties of quite a different kind from those the scholar encounters.

In addition, in the majority of cases most of our time and labor is taken up with side issues that are but remotely connected with the main purpose of an investigation. At one time we must study our instruments' errors with a view to eliminating them or, where that is impossible, to compensating for their harmful effects. At other times we must wait for the moment when an organism presents itself under circumstances most favorable for research. Again, during the course of an investigation we may learn for the first time of possible errors which will vitiate some result—or perhaps only suggest that it may be vitiated—and thus we may find ourselves compelled to do our work over again until every shadow of doubt is removed. For it is only when an observer takes such a grip on a subject, so fixes all his thought and all his interest upon it that he cannot separate himself from it for weeks, for months, even for years—in short, cannot force himself away from it until he has mastered every detail and feels secure in all the results which have been obtained—that a complete and valuable piece of work is done.

You all know how in every piece of good research the preparations, the subsidiary operations, the control of possible errors, and especially the separation of the results obtainable at each stage of work from those that cannot be obtained there, consume far more time than is required to make actual observations or experiments. You all know how much more thought and ingenuity are frequently spent in bringing a refractory piece of brass or glass under subjection than are spent in sketching out the plans for an entire investigation. Each of you has experienced the eager excitement which comes during work when all thought is focused on a narrow circle of questions, the import of which appears trifling and contemptible to an outsider because he does not know the goal of the work, the door that only it can open. I believe I am correct in thus describing the work and the mental conditions that preceded all those great achievements which have hastened so much the development of the sciences after their long quiescence and have given them so powerful an influence over every phase of human life.

The period of work, then, is not a time for a broad, comprehensive survey. When, however, a victory over difficulties happily has been gained and results are secured, a period of repose naturally follows, and then our attention turns toward an examination of the importance of the newly established facts and a wider survey of adjoining regions. This is necessary; indeed, only those who are capable of making such a survey can hope to find useful starting places for further investigations. Early work is thus followed by later, treating of other matters.

Even in a series of investigations, however, a scientist will not deviate far from a more or less narrow path. For it is not only important for an investigator to collect information from books concerning some region to be explored—and the human memory is, on the whole, rather powerful and can store up an almost incredibly large amount of learning. In addition to this knowledge which he gets from lectures and books, the student of science must have other knowledge which only broad, careful perception can give him, and he must have skills which come only after repeated experiments and long practice. His senses must be sharpened for certain kinds of observation, for detecting minute differences of form, color, solidity, smell, etc., in objects under examination. His hand must be equally trained in the work of the blacksmith, the locksmith, and the carpenter, or in that of the draughtsman and the violinist; and when adjusting a microscope, it must surpass the lacemaker's delicacy in handling the needle. Moreover, when he encounters powerful destructive forces or performs bloody operations upon animals or men, he must possess the courage and coolness of the soldier. Such qualities and capabilities, partly the result of natural aptitude, partly cultivated by long practice, are not so readily and so easily acquired as the mere amassing of facts in the memory. It is for these reasons that an investigator is compelled, during the labor of his entire lifetime, to limit his field carefully and to confine himself to those areas which suit him best.

We must not forget, however, that the more an individual is forced to narrow the scope of his activity, so much the more is it important that he be induced not to sever his connections with the rest of his field in its entirety. How is it possible for him to draw strength and pleasure from his arduous work, and how is it possible for him to know that what has occupied so much of his time will not decay uselessly away but will remain a thing of lasting value, unless he keeps alive within himself the conviction that he too has added a fragment to the great body of scientific knowledge which will make the forces of nature subservient to the moral goals of humanity?

One cannot count in advance on an immediate practical use for every individual investigation. Physical science, it is true, has transformed the entire life of modern man through the practical applications of its discoveries. As a rule, however, these applications have appeared under conditions where they were least expected. To search for practical uses usually leads to nothing, unless specific practical goals have definitely been fixed in advance and all that must be done is to remove specific obstacles in the way of their realization. If we investigate the history of the most important practical discoveries, we find that they were either—especially in earlier times—made by craftsmen who worked at one thing all their lives and who, by a lucky accident or by groping, tentative experiments, hit upon some new method advantageous to their particular handicraft; or they were—and this is especially the case with the most recent discoveries—the fruits of a mature scientific knowledge of some subject, a knowledge which in every case had originally been acquired without any direct view to any possible use.

Our association represents the whole of natural science. Assembled here today are mathematicians and physicists, chemists and zoologists, botanists and geologists, teachers of science and physicians, technologists and amateurs who find in scientific studies relaxation from other occupations. Here each of us hopes to find fresh stimulation for and encouragement in his special work. The man who lives in a small country place hopes here to meet with recognition, otherwise unobtainable, for having aided in the advance of science; he hopes by conversation with men pursuing more or less the same line of work to mark out some new researches. We are pleased to find among us also a large number of individuals from the cultivated classes of the nation; we see influential statesmen among us. They all have an interest in our labors; they look to us for further progress in civilization, further victories over natural forces. It is they who place at our disposal the actual means for carrying on our labors, and they are therefore entitled to inquire into the results of those labors. It appears to me, therefore, appropriate on this occasion to take account of the progress of science as a whole, the goals it aspires to reach, and the strides it has made toward achieving them.

Such a survey is desirable; but that it is beyond the capability of any one man to carry out, even incompletely, should be evident from what I have already said. If I stand here today with such a task entrusted to me, my excuse must be that no one else would risk it, and I believe that some attempt to accomplish it, even if it meets with but little success, is better than none at all. Besides, a physiologist more than others has, perhaps, a special motive for maintaining a clear, continuous view of the entire field of science, for at the present time it is peculiarly the lot of the physiologist to receive help from, and to stand in alliance with, all other branches of science. Indeed, the main force of the vast progress of which I shall speak has been felt chiefly in physiology, and some of the most valuable of the recent advances in science have developed out of the major controversies within this field.

If I leave large gaps in my survey, my excuse is partly the magnitude of the task and partly the fact that the pressing summons of the secretary of this association reached me but recently, during the course of my summer holiday in the mountains. The gaps which I leave will, in any case, be abundantly filled by the proceedings of the sections. Let us proceed with our task!

In discussing the progress of physical science as a whole, the first question which presents itself is: By what standard are we to judge this progress?

To the uninitiated, science is an immense, bewildering accumulation of facts, some of which are conspicuous for their usefulness in practice, while others are merely curiosities or objects of wonder. Even if this mass of unrelated facts were given some systematic order, however, as is done for example in the Linnaean system and in encyclopedias, so that each fact might readily be found when required, such knowledge would not deserve the name of science. It would satisfy neither the human mind's demand for scientific order nor our desire for progressive mastery over the forces of nature. The former requires an intellectual grasp of the connections of ideas; the latter requires an ability to anticipate results in untried cases and under conditions which are to be introduced in the course of experiments. Obviously, both of these requirements are satisfied only by a knowledge of the laws of phenomena.

Isolated facts and experiments have in themselves no value, however great their number may be. From a theoretical as well as a practical point of view, they become valuable only when they lead us to an understanding of a law governing a series of uniformly recurring phenomena—or, in some cases, when they give a negative result showing the incompleteness of our knowledge of such a law, which until then had been thought perfect. Given the strict, universal conformity of natural phenomena to laws, a single observation of a specific relationship that we may rightly assume to be lawlike suffices sometimes to establish a rule with the highest degree of probability, as in the case of the skeleton of a prehistoric animal, where we assume our knowledge of the species to be complete if we find only one complete skeleton of a single individual. But we must not lose sight of the fact that such an isolated observation is of value, not because it is isolated, but because with the assumption of lawful regularity, it leads to the knowledge of the bodily structure of an entire species of organisms. In like manner, the knowledge of the specific heat of one small piece of a new metal is important because we have no reason to doubt that all other pieces of the same metal, subjected to the same treatment, will yield the same result.

To find the law by which phenomena are related is to comprehend them. Basically, a law is a general concept under which a series of uniformly recurring natural processes are embraced. Just as we include under the concept *mammal* all that is common to the human, the ape, the dog, the lion, the hare, the horse, the whale, etc., so we embrace in the law of refraction all the phenomena which regularly recur when a ray of light of any color passes in either direction across the common boundary of any two transparent media.

A law of nature, however, is not merely a logical concept that we have adopted as a kind of *memoria technica* to enable us more readily to remember facts. We now have sufficient insight to know that the laws of nature cannot be discovered by any purely speculative method. On the contrary, we must discover them in the facts. We must test them by repeated observation or experiment in new cases and under changing conditions; and only insofar as they hold good under a constantly increasing change of conditions, in a constantly increasing number of cases, and with increasing delicacy in the means of observation, does our confidence in their reliability increase.

We experience the laws of nature as objective forces. They cannot be arbitrarily chosen by or defined in our minds, as one might devise various systems of classification of animals and plants, so long as the goal is only to remember their names. Before we can say that our knowledge of any law of nature is complete, we must know that it holds good without exception, and we must make this the test of its correctness. Whenever the conditions under which the law operates present themselves, the result must follow without arbitrariness, without choice, without our intervention; it must follow from the very necessity which governs the things of the external world as well as our perceptions.

When this occurs, we experience the law in the form of an objective power, and for that reason we think of it as a *force*. For instance, we regard the law of refraction objectively as a refractive force in transparent substances and the law of chemical affinity as a force of attraction exhibited by different substances toward one another. Similarly, we speak of an electric force of contact of metals, a force of adhesion, a capillary force, and so on. In this way, laws which for the most part embrace limited series of natural processes occurring under fairly complex conditions are objectified. Concept formation in science must proceed through the establishment of these specialized laws until it is possible to subsume a number of them under more general laws. Eventually we must try to eliminate all the accidents of form and of distribution in space which bodies under investigation may present by seeking to discover from the phenomena of large visible masses the laws of behavior of vanishingly small particles. Expressed objectively, we must resolve the forces of composite masses into the forces of their smallest elementary particles.

But precisely in the simplest manifestation of force, that is, in mechanical force acting on a mass point, it is especially clear that to refer to an objective force is only a way of expressing a law concerning some phenomena. The force arising from the presence of any given body is equal to the acceleration of the mass on which the body acts multiplied by that mass. The empirical meaning of such an equation is expressed in the following law: "If such and such masses are present and no others, then such and such acceleration of their individual points will occur." This empirical meaning can be compared with the facts and tested by them. Thus the abstract concept of force which we introduce only indicates that we did not establish the law arbitrarily—that, on the contrary, it is a law which the phenomena compel us to accept.

Our task of comprehending natural phenomena, that is, of finding their laws, may be expressed in another way: *we must search for the forces which are the causes of phenomena*. As soon as we recognize that nature is independent of our thought and will, its conformity to law may be thought of as the causal relatedness of natural phenomena. If we direct our attention to the question of the progress of physical science as a whole, we may judge it by the degree to which the recognition and knowledge of the causal connections embracing all natural phenomena have advanced.

If we look back over the history of our sciences, the first great example we find of the subjugation of an extensive area of facts to a comprehensive law occurred in theoretical mechanics, whose fundamental concepts were first clearly set forth by Galileo. The problem at that time was to find the general propositions which now seem to us so self-evident—that all matter is inert and that the magnitude of a force is to be measured, not by the velocity, but by the change in velocity it produces. At first the operation of a continuous force could be represented only as a series of small impulses. Not until Leibnitz and Newton, through the discovery of the differential calculus, had dispelled the ancient darkness which enveloped the concept of infinity, and had clearly established the concepts of continuity and continuous change, was any progress made toward a full, fruitful application of the newly found concepts of mechanics.

The most striking instance of such application was in connection with the motion of the planets, and I need scarcely remind you here how brilliant an example astronomy has been for the development of the other branches of science. In the case of the motion of the planets, a vast, complex set of facts was for the first time embraced under a single principle of great simplicity—the theory of gravitation—and a correspondence of theory and fact was established such as has never been achieved in any other department of science, either before or since. Almost all of our precise methods of measurement, as well as the principal advances made in modern mathematics, have originated in attempts to meet the demands of astronomy. And astronomy is, of course, especially suited to attract the attention of the general public, partly by the grandeur of the objects under investigation and partly by its practical utility in navigation and geodesy.

Galileo began with the study of terrestrial gravitation; Newton extended the application of the principle of gravitation, at first hesitantly and cautiously to the moon and then boldly to all of the planets. In more recent times, we have learned that the law of inertia and the law of gravitation of all ponderable masses hold good for the movements of the most distant double stars that we are able to see.

During the latter half of the last century and the first half of the present one came the great development of chemistry, which finally solved the ancient problem of determining the elementary substances—a task to which so much metaphysical speculation has been devoted. Reality has always far exceeded even the boldest and wildest speculation, and in place of the four primitive metaphysical elements (fire, water, air, and earth) we now have the sixty-five elements of modern chemistry. Science has shown that these elements are truly indestructible, inalterable both in their mass and in their other properties. From any state into which they may be changed they can invariably be isolated, and in every case they show those qualities which they possessed previously in the free state.

Through all the various changes of the phenomena of animate and inanimate nature, so far as we are acquainted with them—in all the amazing reactions of chemical composition and decomposition, whose number and diversity the chemists augment with unwearied industry every year—the law of the conservation of matter rules with a necessity that knows no exception. And chemistry has, through spectrum analysis, already pressed on into the depths of immeasurable space. It has detected in the most distant suns and nebulae indications of well-known terrestrial elements, so that doubts as to the homogeneity of the matter of the universe no longer exist, although certain elements may perhaps be restricted to certain groups of the heavenly bodies.

From this invariability of the elements follows another, more general consequence. Chemistry shows by empirical investigation that all matter is made up of the elements that have been discovered. These elements may exhibit many variations in their combinations and mixtures with one another, in their aggregate forms or molecular structures—that is, they may vary greatly in the way in which they are distributed in space—but in their properties they are altogether unchangeable. In the same compound, or isolated, or in the same state of aggregation, they invariably exhibit the same properties.

If all elementary substances are unchangeable with respect to their properties and change only in their combinations and states of aggregation—that is, only in their distribution in space—it follows that all changes in the world are changes in the local distribution of elementary matter and are, in the last analysis, brought about by motion. Further, if motion is the fundamental change which lies at the base of all other changes occurring in the world, then every elementary force is a motive force, and the ultimate aim of physical science must be to find the movements which are the real causes of all other phenomena and to determine the motive forces upon which these movements depend. In other words, its aim is to reduce all phenomena to mechanics.

Though this is clearly the ultimate consequence of the qualitative and quantitative immutability of matter, it is obviously an ideal goal whose realization is still very remote. Only in very limited areas has there been any success in tracing actually observed changes back to motion and to the forces of motion of the kind mentioned above. Besides astronomy, the purely mechanical part of physics may be cited; then acoustics, optics, and the theory of electricity. In thermodynamics and in chemistry, strenuous efforts are being made toward developing definite, specific ideas concerning the positions and motion of molecules. Physiology has taken scarcely a step in this direction.

This renders all the more important, therefore, an advance of great significance and great generality which was made during the last quarter of a century in the direction we are considering. If all elementary forces are forces of motion, all are therefore of the same nature, and they should all be measurable by the same standard—that is, by the standard of mechanical force. That this is actually the case is now regarded as established. The law stating this is known as the law of the conservation of force.

This law was propounded by Newton for a restricted group of natural phenomena. It was stated more clearly and in more general terms by Bernoulli, and from that time on it was recognized as valid for the majority of the purely mechanical processes known. Certain extensions of the law at times attracted attention, such as those of Rumford, Humphry Davy, and Montgolfier. The first man, however, to comprehend this law clearly and to venture to state its absolute universality was one whom we shall soon have the pleasure of hearing from this platform, Dr. Robert Mayer, of Heilbronn. While Dr. Mayer was led by physiological considerations to the discovery of the most general form of the law, technical questions in mechanical engineering simultaneously and independently led Mr. Joule, of Manchester, to the same reflections. It is to Mr. Joule that we are indebted for those important and laborious experimental researches in the area where the validity of the law of the conservation of force appeared most doubtful and where there were the greatest gaps in our knowledge, namely, concerning the production of work from heat and of heat from work.

To state the law clearly it was necessary to develop, in contradistinction to Galileo's concept of the *intensity* of force, a new mechanical concept which we may designate the *quantity* of force. (It has also been called the quantity of work, or energy.) The path to the concept of the quantity of force had already been prepared, partly in theoretical mechanics by the concept of the quantity of vis viva of a moving body and partly in practical mechanics by the concept of the motive power necessary to keep a machine in operation. Further, students of mechanics had already found, in the determination of the number of pounds that could be raised one foot in one second, a standard by which any motive power could be measured. As is known, a horsepower is defined as equivalent to the motive power required to lift seventy kilograms one meter in one second.

Machines, and the motive power required for their operation, furnish the most popular illustrations of the relatedness of all natural forces expressed by the law of the conservation of force. Any machine which is to be set in motion requires a mechanical motive power. Where this power comes from and what form it takes are of no consequence, provided that it is sufficiently great and acts continuously. At one time we use a steam engine, at another a water wheel or a turbine; here we use horses or oxen at a winch, there a windmill; or if only a little power is required, we use a human arm, a raised weight, or an electric motor. The choice of the machine depends entirely upon the amount of power we want or upon what power is available. In the water mill the weight of the water flowing downhill is the force; the water is raised by a meteorological process, and this process thus becomes the source of motive power for the mill. In the windmill it is the vis viva of the moving air which drives the sails; thus this motion also is due to a meteorological process. In the steam engine we have the pressure of the heated steam, which drives the piston back and forth; the steam itself is generated by the heat produced by the combustion of the coal in the firebox–in other words, by a chemical process, which is the source of the motive power. In the case of the work done by the horse or by the human arm, it is the muscles, stimulated through the nerves, that directly produce the mechanical force. In order for the living body to generate muscular power, however, it must be nourished and it must breathe. The food it consumes separates from the body again, after having combined with the oxygen inhaled from the atmosphere, in the form of carbonic acid and water. Here again, a chemical process is the essential factor in the production of muscular power.

Thus we obtain mechanical motive power from the most diverse natural processes and in the most diverse ways. We must notice, however, that it is always limited in quantity, and in every case we consume something that nature supplies to us. In the watermill we consume a quantity of water collected at an elevation. In the steam engine we consume coal; in the battery of the electric motor, zinc and sulphuric acid. The working horse consumes food, and in the windmill the motion of the wind, which is arrested by the sails, is consumed.

Conversely, if we have a motive force at our disposal, we can achieve the most diverse effects with it. I need not discuss here the vast number of different kinds of industrial machines and the many kinds of work which they can perform. Instead, let us note the many different possible effects of a motive force. With its help we can raise loads, pump water to an elevation, compress gases, set railroad trains in motion, and (through friction) generate heat. With its help we can turn generators and produce an electric current; and with the current we can decompose water and other chemical compounds having the most powerful affinites, make wires incandescent, magnetize iron, and so on.

Moreover, if we had at our disposal a sufficient mechanical motive force, we could restore all those states and conditions from which we obtained the mechanical motive power in the first place. However, the motive power obtainable from any given natural process is limited, so there is a limit to what we can do by the use of any given motive force.

These truths, first formulated in connection with machines and physical apparatus, have now been welded into a law of nature of the most general validity. Every change in nature is equivalent to a certain production, and to a certain consumption, of motive force. If motive power is produced, it may appear as such, or it may immediately be used up again producing other changes equivalent in magnitude. The main determination of this equivalence is based upon Joule's measurements of the mechanical equivalent of heat. If, by the application of heat, we set a steam engine in motion, an amount of heat proportional to the work done disappears. The heat which can warm a given weight of water 1° C. is able, if converted to work, to lift the same weight of water to a height of 425 meters. If we convert work into heat by friction, we again use, in heating a given weight of water 1° C., a motive force which the same quantity of water would have generated in flowing down from a height of

425 meters. Chemical processes generate specific amounts of heat, and we can determine the motive power equivalent to such chemical forces. Further, the energy of chemical binding forces is measurable relative to a mechanical standard. The same is true for all other kinds of natural forces—but we shall not discuss this subject any further here.

It has actually been established, in short, as a result of these investigations, that all the forces of nature are measurable according to the same mechanical standard and that all motive forces are, insofar as the performance of work is concerned, equivalent. Thus one very significant step has been taken toward the solution of the comprehensive theoretical problem of tracing all natural phenomena back to motion.

While I have tried in the preceding discussion mainly to clarify the logical status of the law of the conservation of force, its factual significance for our general conception of the processes of nature is to be found in the relationships which it reveals among all the transformations in the universe throughout all space and time. According to this law, the universe is endowed with a store of energy which all the various changes due to natural processes can neither increase nor diminish; a store of energy which is manifested in ever-changing phenomena but which, like matter, is from eternity to eternity unchanging in quantity; a store of energy which exists in space but which, unlike matter, is not divisible in space.

Every change in the world consists simply in a change in the outward form of this store of energy. Here we find one part of it as the vis viva of moving bodies, there as the regular vibrations of light and sound, and elsewhere as heat—that is, as the irregular motion of invisible particles. The energy appears as the weight of two masses gravitating toward each other, as the internal tension and pressure of elastic bodies, as chemical attraction, as electrical potential, and as magnetic distribution. If it disappears in one form, it reappears inevitably in another; and whenever it appears in some new form, we may be certain that it does so at another's expense.

Carnot's law in the mechanical theory of heat, as modified by Clausius, makes it clear that these changes on the whole tend continuously in a definite direction, so that a constantly increasing amount of the great store of energy in the universe is being transformed into heat. We can see, therefore, with the mind's eye, the original condition of things in which the matter constituting the celestial bodies was still cold and probably distributed chaotically throughout space in the form of vapor or dust. We can see that it must have developed heat when it condensed under the influence of gravity. By means of spectrum analysis (a method whose theoretical principles are derived from the mechanical theory of heat) we can still detect the remains of this loosely distributed matter in nebulae; it is evident too in meteor showers and in comets. The condensation and the resulting development of heat still continue, although in our part of the universe they have on the whole come to a stop.

The greatest part of the primordial energy of the matter belonging to our solar system now has the form of solar heat. This energy, however, will not remain locked up in our system forever; some of it is continually being radiated, in the form of light and heat, into infinite space. Our earth receives a share of this radiation. It is these solar heat rays which produce both the winds at the surface of the earth and the currents of the ocean; it is they which raise the water vapor from the tropical seas. This vapor, distributed over mountains and plains, returns to the sea again from springs and rivers. The rays from the sun give plants the power to form the combustible substances which serve as food for animals. Thus with the many processes of organic life, too, the motive force is derived from the infinitely vast store of the universe.

This sublime picture of the connections existing among all the processes of nature has often been presented to us in recent times; here I need only to direct your attention to its leading features. If the task of physical science is to discover laws, a step of the greatest significance has been taken here toward that goal.

The application of the law of the conservation of force to the vital processes of animals and plants, which has just been mentioned, leads us to another way in which progress has been made in our knowledge of nature's conformity to law. The law under discussion is of the most fundamental importance in physiology. Indeed, it was the basic problems in this area that led Dr. Mayer and myself to investigations concerning the conservation of force.

With regard to the phenomena of inorganic nature, all doubts concerning methodological principles have long since been laid to rest. It is apparent that rigorous laws govern these phenomena, and enough examples are already known to make it evident that such laws can be found. As regards the vital processes, however, their greater complexity, their connection with mental activities, and the unmistakable evidence of purposiveness which organic structures exhibit might easily make the existence of a strict conformity to law appear doubtful. Indeed, physiology has always had to struggle with the fundamental question: Are all vital processes absolutely conformable to law, or is there perhaps a certain range of phenomena within which freedom reigns?

More or less vaguely, the view of Paracelsus, Helmont, and Stahl was generally held—and is still held, particularly outside Germany—that there exists a life soul (*Lebensseele*), something like the conscious human soul, which regulates organic processes. The influence of the inorganic forces of nature on the organism was accepted by those who held this view, since it was assumed that the life soul exercised control over matter only by means of the physical and chemical forces of matter. Without the aid of these forces the life soul could accomplish nothing, but it had the power to permit or suspend their operation at will. After death, when the organism was no longer subject to the control of the life soul, or vital force, it was these very chemical forces in organic matter which brought about decomposition.

No matter what it was called—the *archeus*, the *anima inscia*, the vital force, the healing force of nature, or whatever—the power to develop the body according to plan and to accommodate it suitably to external conditions remained the most basic property of this hypothetical controlling principle postulated by vitalistic theory. (Because of its attributes, only a name using the word *soul* is fully suitable for this principle.)

It is apparent, however, that this notion runs directly counter to the law of the conservation of force. If a vital force were to suspend for a time the action of gravity on a body, it could be raised without work to any height desired, and if the action of gravity were subsequently restored, it could perform work of any amount desired. Thus work could be obtained without expense out of nothing. If a vital force could suspend for a time the chemical affinity of carbon for oxygen, carbonic acid could be decomposed without work being done, and the carbon and oxygen liberated could be used to perform new work.

In reality, however, there is no evidence that the living organism can perform the slightest trace of work without a corresponding consumption of energy. If we consider the work done by animals, we find it similar in every respect to that done by a steam engine. Animals, like machines, can move and do work only if they are continuously supplied with fuel (that is to say, food) and air containing oxygen. Both animals and machines give off this material again in a burned state, and both produce simultaneously heat and work. So far, no investigation concerning the amount of heat which an animal produces when at rest is in any way at variance with the assumption that this heat exactly corresponds to its equivalent in the chemical forces in operation.

As to the processes that occur in plants, a source of power exists in the solar rays which is in every way sufficient for their growth. It should be noted, of course, that exact quantitative determinations of the equivalence of the forces produced and consumed in the vegetable, as well as in the animal, kingdom must still be made in order to establish fully the exact agreement of these two values.

If, then, the law of the conservation of force holds good for living beings too, it follows that the physical and chemical forces of the material employed in building up the body are in continuous action without interruption and without choice and that their strict conformity to law never suffers a moment's interruption.

In their investigations of vital processes, physiologists should thus expect an unconditional conformity of the forces of nature to laws, and they must take seriously the investigation of the physical and chemical processes going on within the organism. Theirs is a task of vast scope and complexity, but a large number of enthusiastic men are at work, especially in Germany, and we can already state that their labors have not gone unrewarded and that our knowledge of vital phenomena has made greater progress during the past forty years than during the two preceding centuries.

Assistance of the utmost value toward the clarification of the fundamental principles of the theory of life has come from the area of descriptive natural history. I refer to Darwin's theory of the evolution of organic forms, which provides the possibility of an entirely new interpretation of organic purposiveness.

The wonderful–and, through the growth of science, the more and more evident–purposiveness in the structure and function of living beings was, indeed, the main reason behind the comparison of the vital processes with the behavior of something functioning under the control of an entity like a soul. In the whole world we know of but one series of phenomena possessing similar characteristics, namely, the acts and activities of intelligent human beings; and we must admit that in innumerable instances organic purposiveness appears to be so extraordinarily superior to the capacities of human intelligence that sometimes we are inclined to ascribe a higher rather than a lower level to it.

Before the time of Darwin, only two theories concerning organic purposiveness were in vogue, both of which pointed to the interference of free intelligence in the course of natural processes. On the one hand, it was held–in accordance with the vitalistic theory–that the vital processes were continuously directed by a life soul; on the other, recourse was had to belief in an act of supernatural intelligence in order to account for the origin of every living species. The latter view implied that the causal connectedness of natural phenomena had been broken less often than did the vitalistic theory, and it was compatible with a strictly scientific analysis of the processes observable in existing species. It was not able, however, to avoid exceptions to the law of causality entirely, and consequently it enjoyed no great favor in comparison to the vitalistic position, which was also apparently supported by powerful evidence and by our natural desire to find similar causes behind similar phenomena.

Darwin's theory contains an essentially new and fruitful line of thought. It shows how adaptation in the structure of organisms can result from the blind rule of a law of nature without any intervention of intelligence. I refer here to the law of transmission of individual characteristics from parents to offspring, a law which had long been known and correctly appreciated but which needed a more precise formulation. If both parents have characteristics in common, the majority of their offspring will also possess these characteristics; and even though some among the offspring may display them to a less marked degree, there will always be others (given a large enough number) in which they have been intensified. If the latter offspring are selected to propagate, a gradual intensification of the characteristics may be obtained and transmitted. This is, of course, the method employed in animal breeding and in gardening to obtain with greater certainty new breeds and varieties with well-marked characteristics. Indeed, the experience of animal breeders may be regarded, from a scientific point of view, as experimental confirmation of the law under discussion; in a vast number of instances these experiments have proved successful with species from every class of the animal kingdom and with respect to all the different organs of the body.

After the general applicability of the law of transmission had been established in this way, it only remained for Darwin to discuss the bearing it had on animals and plants in the wild state. The well-known conclusion he arrived at is that in the struggle for existence those individuals which are distinguished by some advantageous qualities are the most likely to produce offspring and thus to transmit to them their advantageous qualities. In this way, from generation to generation, a gradual adjustment is made in the adaptation of a species of living creature to the conditions under which it must live, until finally it reaches such a degree of perfection that any further major change would be disadvantageous. The species then remains unchanged as long as the external conditions under which it exists remain substantially the same. The plants and animals now living appear to have reached such an almost completely stable state, which accounts of course for the apparent general stability of the species, at least during historical times.

An animated controversy still continues, of course, concerning the truth or the probability of Darwin's theory, though the controversy now centers mainly on the scope that should be assigned to variations of species. The opponents of Darwin can hardly deny that hereditary differences in race can arise within a single species and, indeed, that many of the forms hitherto regarded as distinct species of the same genus have been derived from the same primitive form in the way stated in the theory. Whether we must restrict ourselves to this observation, however, or whether we may perhaps venture to derive all mammals from one original marsupial, or all vertebrates from a primitive lancelet, or all plants and animals together from the slimy protoplasm of the Protista, depends at the present time more on the leanings of individual students than on facts. Still, fresh links connecting classes of apparently incompatible types are always presenting themselves; and the actual transition of some forms into others that are quite different has already been traced in regularly deposited geological strata and has come to be beyond question. Since this line of research has been taken up, how numerous are the facts which fully accord with Darwin's theory and give specific effect to it!

Moreover, we must not forget the clear understanding that Darwin's grand conception has provided for the previously puzzling notion of natural kinship, for the natural system of classification, and for the homology of organs in various animals; nor how through its help the remarkable recurrence of the structural characteristics of lower animals in the embryos of others higher in scale, the natural relationships appearing in the series of paleontological forms, and the peculiar interrelationships among the fauna and flora of restricted areas have all received clarification. Formerly, natural kinship appeared to be an enigmatic, altogether groundless similarity of form; now it has become a matter of real consanguinity. The natural system of classification certainly forced itself upon the mind as natural, although theory strictly disavowed any real significance to it; now it represents an actual genealogy of organisms. The facts of paleontological and embryological evolution, and of geographical distribution, were mysteries as long as each species was regarded as the result of an independent act of creation, and they cast a scarcely favorable light on the Creator, to Whom a strange, hesitant way of acting had to be ascribed. Darwin has changed all of these areas from collections of puzzling mysteries into a great, unified system of development. He has established definite ideas in the place of artistic intuitions or insights, such as those which occurred to Goethe concerning comparative anatomy and the morphology of plants.

The theory also provides us with specific questions for further investigation. This in itself is a great gain, even if it should turn out that Darwin's theory does not represent the whole truth and that, in addition to the influences which he has indicated, others are found to operate in the modification of organic forms.

While the Darwinian theory treats exclusively of the gradual modification of a species after a succession of generations, we know that to a certain extent a single individual may adapt itself or become accustomed to the circumstances under which it must live. Thus even during the life of a single individual, a higher degree of organic adaptation may be attained. Moreover, it is especially in that area of organic phenomena where purposiveness of structure has reached its highest form and excited the greatest admiration—the area of sense perception—that, as the latest developments in physiology teach us, this individual adaptation has come to play a most important role.

Who has not marveled at the fidelity and accuracy of the information which our senses, especially the far-reaching eyes, convey to us about the external world? The information so gained furnishes the premises for all the inferences we make and for all the acts we preform. Only if our senses provide us with correct perceptions, therefore, can we expect to act in such a way that the consequences of our actions will correspond to our expectations. By the success or failure of our acts we test again and again the truth of the information with which our senses supply us; and experience, after millions of repetitions, shows that this fidelity is very great—indeed, almost free from exceptions. The exceptions, the so-called illusions of the senses, are rare and are produced only under very special or unusual circumstances.

Whenever we extend our hand in order to take hold of some object, or advance our foot in order to step upon something, we must first form accurate visual images of the position, form, and distance of the object; otherwise we shall fail. The certainty and accuracy of our sense perceptions must be at least equal to the certainty and accuracy which we can achieve in our actions after long practice. Our belief in the reliability of our senses, therefore, is by no means blind. It is a belief whose correctness, in a practical sense, has been tested and verified again and again in countless experiments.

If this correspondence between our sense perceptions and the objects causing them—this basis of all our knowledge—were a preordained product of an organic creative force, this force would have reached its highest degree of perfection in the formation of these perceptions. An examination of the facts, however, destroys at once and in the most merciless manner any belief in a pre-established harmony between the inner and the external world.

You are all familiar with the startling and unexpected results of ophthalmometrical and optical research, which prove that the eye is by no means a more perfect optical instrument than those constructed by human hands. On the contrary, they show that in addition to the faults inseparable from any refracting instrument, the eye possesses others which we would criticize severely if we found them in a manufactured instrument. Nor need I remind you that the ear does not convey sounds to us in the ratio of their actual intensities, but modifies and analyzes them and intensifies or weakens them in very different degrees depending upon their pitch.

These anomalies, however, are nothing compared with those we find when we examine the characteristics of the sensations through which we become acquainted with the various properties of the objects surrounding us. It can be shown very easily that there is no similarity whatsoever, either in kind or in degree, between the qualities of sensations and the qualities of external agents exciting and being portrayed by them.

In its leading features, this was established by Johannes Müller in his law of the specific energies of the senses. According to this law, each nerve of sense is excited by any one of a number of agents, and the same agent may affect different organs of sense. No matter how they are excited, however, we never have in nerves of sight any sensations but those of light; or in the nerves of the ear, any but those of sound. In short, in each individual nerve of sense we have only sensations of its specific energy. The most marked differences in the qualities of sensations, those among the sensations of the different senses, are thus in no way dependent upon the nature of the exciting agent. They depend only upon the nerve apparatus affected.

The range of Müller's law has been extended by subsequent research. It now appears highly probable that the sensations of different colors and of differences in pitch—that is, the qualitative differences among sensations of light and among sensations of tone—also depend upon the excitation of different systems of fibers with distinct specific energies among the nerves of sight and the nerves of hearing. The immensely great variety of objective combinations of colors is reducible in sensation to combinations involving only the three fundamental colors. Because of this reduction in the number of factors or elements, it follows that very different objective combinations of lights may appear the same. Thus it has been shown that there is no necessary physical similarity corresponding to the subjective similarity of combinations of light of similar colors.

From these and related facts we are led to the very important conclusion that our sensations are, insofar as their quality is concerned, only *signs* of external objects, not *images* with any sort of resemblance to them. An image must be similar in some respect to an object. A statue, for example, has the same bodily form as the human being after which it is modeled; a painting has the same color and perspective projection. For a sign, it is sufficient that it appear whenever that which it signifies makes an appearance, the correspondence between them being restricted to their appearing simultaneously. The correspondence existing between our sensations and the objects producing them is precisely of this kind. Sensations are signs which we have learned to decipher. They form a language given to us by our physical make-up, a language by which external objects speak to us. It is, however, a language, which, like our mother tongue, we can learn only by practice and experience.

Moreover, what has been said holds good, not only for the qualitative differences among sensations, but also for the greater and most important part, if not for all, of the *spatial* differences we perceive. In this connection the modern theory of binocular vision and the invention of the stereoscope have been important. All that the sensations of the two eyes can convey to us directly—that is, without mental activities—are two somewhat different plane images of a two-dimensional external world; yet we experience in perception three-dimensional images of the objects around us. We perceive the distance of objects which are not too far away from us, and we perceive their perspective juxtaposition. We compare the actual sizes of two objects of apparently unequal magnitude at different distances from us with greater certainty than the apparently equal sizes of a finger, say, and the moon.

In my judgment, there is only one possible explanation which is consistent with all the facts concerning our visual perceptions of space. With Lotze, we must assume that different local signs, whose spatial significance we must learn, are provided by the sensations of spatially different nerve fibers. A knowledge of their significance can be attained, with the help of course of the movements of our bodies; and we can therefore learn the movements necessary to bring about results we desire and can perceive when we have arrived at them—all this has been demonstrated in many ways.

That experience is of enormous importance in the interpretation of visual images and that in cases of doubt it is generally the final arbiter, is admitted even by those physiologists who wish to preserve as much as possible of an innate harmony between the senses and the external world. Controversy at the present time is confined almost entirely to the question as to whether or not there exists at birth some innate instinct which facilitates training in the understanding of sense impressions. The assumption that such an instinct exists is unnecessary, and it obscures rather than clarifies the interpretation of many well-authenticated phenomena in adults.

It follows, then, that with a few questionable exceptions the subtle, much admired harmony between our sense perceptions and their objects is in reality an adaptation individually acquired. It is the product of experience, of training, and of the recollection of previous cases of a similar kind.

This completes the circle of our observations and leads us back to the point from which we started. We found at the beginning that what our sciences strive after is the knowledge of laws, that is, the knowledge of how at different times but under the same conditions the same results occur. We saw that all laws ultimately must be reduced to laws of motion. We now find, in conclusion, that our sense impressions are only signs of changes taking place in the external world and that they can be regarded as images only in that they represent succession in time. However, it is just in this way—that is, with respect to succession in time—that they are in a position to show directly the conformity of natural phenomena to law. If, under the same natural conditions, the same effects recur, a person observing them under the same conditions will experience a recurrence of the same series of impressions. What our organs of sense provide is clearly sufficient to enable us to achieve the goal of science. It is also sufficient for the practical purposes of man, which depend upon a knowledge of natural laws, acquired in part involuntarily through daily experience and in part purposefully through the study of science.

Having now completed our survey, we may surely strike a not unsatisfactory balance. The natural sciences have made active progress, not only in this or in that direction, but as a vast whole; and what has already been accomplished guarantees the realization of further progress. Doubts concerning the conformity of all of nature to laws have gradually been dispelled, and increasingly general and comprehensive laws have been disclosed. That the direction which scientific study has taken is a healthy one is clearly demonstrated by its great practical results, and here I may be permitted to direct particular attention to the branch of science most especially my own. In physiology, scientific work had been especially crippled by doubts concerning necessary conformity to law, which means, as we have shown, doubts as to the intelligibility of vital phenomena. These doubts naturally extended to the practical science directly dependent upon physiology, that is, to medicine. Both have now seriously adopted the methods of physical science—exact observation of phenomena, and experimentation—and have consequently received an impetus such as had not been felt for a thousand years.

As a practicing physician in my earlier days, I can personally bear witness to this. I was educated at a period when medicine was in a transitional state, when the minds of the more thoughtful and exact were filled with despair. It was not difficult to see that the older, highly speculative methods of practicing medicine were altogether untenable; even the facts upon which this speculative medicine was based were so inextricably entangled with the theories that they too had, on the whole, to be discarded. How a new science could be built up was clear from the example of the other sciences; but the task was of colossal proportions, and only a few steps—themselves in part crude and clumsy—had been taken toward accomplishing it. We need not wonder that many serious, honest thinkers abandoned medicine at that time as unsatisfactory or resigned themselves on principle to an extreme empiricism.

Well-directed efforts, however, produced genuine results more quickly than many had dared to hope. The application of mechanical concepts to the study of circulation and respiration, a better understanding of thermal phenomena, and a more refined physiology of the nerves soon led to practical results of the greatest importance. Microscopic examination of parasitic organisms and a tremendous development of pathological anatomy led irresistibly from nebulous theories to reality. We found that we possessed much more definite ways of distinguishing among diseases and a much clearer insight into the mechanisms of pathological processes than the pulse beats, urinary deposits, and fever types of the older medical science had ever given us.

If I were to name one branch of medicine in which the influence of scientific methods has been perhaps most brilliantly displayed, it would be ophthalmic medicine. The eye is of such a nature that we are able to apply the methods of physics both to the study of the function of the living organ and to the correction of its anatomical flaws. Simple physical instruments (that is, spectacles), sometimes spherical, sometimes cylindrical or prismatic, suffice in many cases to cure disorders which once left the organ in a condition of chronic incapacity; and a great many changes which formerly did not attract attention until they had caused incurable blindness can now be detected and remedied when they first begin. Since it presented the most favorable place for the application of scientific methods, ophthalmology proved attractive to an unusually large number of excellent investigators, and it rapidly attained its present position. For the other branches of medicine it now offers an example, as brilliant as that which astronomy set for the other branches of physical science, of the possibilities opened up by the application of correct methods of investigation.

Although the several European nations showed nearly the same degree of advancement in the investigation of inorganic nature, the recent progress in physiology and medicine is primarily due to German scientists. I have already spoken of the obstacles which formerly delayed progress in these areas. Questions concerning the nature of life are closely bound up with psychological and ethical questions. They demand, moreover, that we work with unwearied diligence toward ideal goals without any narrow expectations of practical benefits. We may make it our boast that this exalted and self-denying diligence, this labor for inward satisfaction rather than outward success, has always distinguished the German scientist.

In my opinion, however, what really distinguishes us from others is something else. It is that among us there is a total lack of fear over the implications of the knowledge of the complete truth. In both England and France there are excellent investigators who are capable of working along scientific lines with complete dedication. Hitherto, however, they have almost always had to bow to social or ecclesiastical prejudice, and they have been able to express their convictions openly only at the expense of their social influence and usefulness.

Germany has advanced with bolder steps. She has had the confidence, which has never been shaken, that the truth, fully known, carries its own protection against the dangers and disadvantages which may occasionally attend some half-knowledge of the truth. A work-loving, frugal, moral nation may exercise such boldness; it may stand face to face with truth. It has nothing to fear from the presentation of some hasty and partial theories, even if they should appear to threaten the foundations of morality and society.

We are met here on the southern frontier of our country. In science, however, we recognize no political boundaries, for our country reaches as far as the German tongue is heard, to wherever German diligence and German intrepidity in striving after truth find favor. That they find favor here is shown by our hospitable reception and by the inspiring words with which we have been greeted. A new medical faculty is to be established here. In its career we wish it Godspeed, that it may grow strong in the cardinal virtues of German science, for then it will not only find remedies for bodily suffering but will become an active center for the strengthening of intellectual independence, of fidelity to conviction, and of love of truth. At the same time, it will be a means for deepening the sense of unity throughout our country.

8.

THE ORIGIN AND MEANING OF GEOMETRIC AXIOMS (I) [1870]

An article published in Mind, A Quarterly Review, *Volume I, Number 3 (July 1876), pp. 301-21**

MY purpose in this article is to discuss the philosophical significance of recent inquiries concerning geometric axioms and the possibility of working out analytically new systems of geometry with axioms other than Euclid's. The original works on the subject, addressed to experts only, are particularly abstruse, but I shall try to make it plain even to those who are not mathematicians. It is of course no part of my plan to prove the new doctrines correct as mathematical conclusions. Such proof must be sought in the original works themselves.

Among the first elementary propositions of geometry, from which the student is led by continuous chains of reasoning to the laws of more and more complex figures, are some which are held not to admit of proof, though sure to be granted by everyone who understands their meaning. These are the so-called axioms; for example, the proposition that if the shortest line drawn between two points is called straight, there can be only one such straight line. Again, it is an axiom that through any three points in space, not lying in a straight line, a plane may be drawn, that is, a surface which will wholly include every straight line joining any two of its points. Another axiom—about which there has been much discussion—affirms that through a point lying outside a straight line, only one straight line can be drawn parallel to the first, parallel lines being those straight lines which lie in the same plane and never meet, however far they may be produced. There are also axioms which determine the number of dimensions of space and its surfaces, lines, and points, showing how they are continuous. Among these are the propositions that a solid is bounded by a surface, a surface by a line, and a line by a point;

*A number of bibliographical footnotes are omitted.

that the point is indivisible; and that by the movement of a point a line is described, by that of a line a line or a surface, by that of a surface a surface or a solid, but by the movement of a solid a solid and nothing else is described.

What is the origin of such propositions, unquestionably true yet incapable of proof in a science where everything else is reasoned conclusion? Are they inherited from the divine source of our reason, as the idealistic philosophers think, or has the ingenuity of mathematicians simply not yet been penetrating enough to find the proof? Every new votary, coming with fresh zeal to geometry, naturally strives to succeed where all before him have failed. And it is quite right that each should make the trial afresh, for as the question has hitherto stood, it is only by the fruitlessness of one's own efforts that one can be convinced of the impossibility of finding a proof. Meanwhile, solitary inquirers are always appearing who become so deeply entangled in complicated trains of reasoning that they can no longer discover their mistakes and believe they have solved the problem. The axiom of parallels especially has called forth a great number of seeming demonstrations.

The main difficulty in these inquiries is and always has been the readiness with which results of everyday experience become entangled in the logical processes as apparent necessities of thought, so long as Euclid's method of constructive intuition is exclusively followed in geometry. In particular, it is extremely difficult by this method to be quite sure that in the steps prescribed for the demonstration we have not involuntarily and unconsciously drawn in some very general results of experience, which the power of executing certain parts of the operation has already taught us practically.

In drawing any subsidiary line for the sake of his demonstration, the well-trained geometrician asks always if it is possible to draw such a line. It is notorious that problems of construction play an essential part in the system of geometry. At first sight these appear to be practical operations, introduced for the training of students, but in reality they have the force of existential propositions. They declare that points, straight lines, or circles such as the problem requires to be constructed are possible under all conditions, or they determine any exceptions that may exist. The crux of the investigations which we are going to consider is essentially of this nature.

The foundation of all proof by Euclid's method consists in establishing the congruence of lines, angles, plane figures, solids, and so on. To make the congruence evident, the geometric figures are supposed to be superimposed on one another, of course without changing their form and dimensions. That this is in fact possible we have all experienced from our earliest youth. But when we would build necessities of thought upon this assumption of the free translation of fixed figures with unchanged form to every part of space, we must see whether the assumption does not involve some presupposition for which no logical proof is given. We shall see later that it does contain one of very serious import. Every proof by congruence thus rests upon a fact which is obtained from experience only.

I offer these introductory remarks only to show what difficulties attend the complete analysis of the presuppositions we make in employing the common constructive method. We avoid them when we apply to the investigation of principles the analytical method of modern algebraic geometry. The whole process of algebraic calculation is a purely logical operation; it can yield no relation between the quantities submitted to it that is not already contained in the equations to which it is applied. The recent investigations have accordingly been conducted almost exclusively by means of the purely abstract methods of analytic geometry.

However, after identifying by the abstract method the points in question, we shall best get a distinct view of them by taking a region of narrower limits than our own world of space. Let us, as we logically may, suppose reasoning beings of only two dimensions to live and move on the surface of some solid body. We shall assume that they have not the power of perceiving anything outside this surface, but that upon it they have perceptions similar to ours. If such beings worked out a geometry, they would of course assign only two dimensions to their space. They would ascertain that a point in moving describes a line and that a line in moving describes a surface. But they could as little represent to themselves what further spatial construction would be generated by a surface moving, as we can represent what would be generated by a solid moving out of the space we know. By the much abused expression *to represent* or *to be able to think how something happens* I understand—and I do not see how anything else can be understood by it without loss of all meaning—the power of imagining the whole series of sensible impressions that would be had in such a case. As no sensible impression is known relating to such an unheard-of event as the movement to a fourth dimension would be to us, or as a movement to our third dimension would be to the inhabitants of a surface, such a representation is as impossible as the representation of colors would be to one born blind, though a description of them in general terms might be given to him.

Our surface-beings would also be able to draw shortest lines in their superficial space. These would not necessarily be straight lines in our sense, but what are technically called geodesic lines of the surface on which they live, lines such as are described by a stretched thread which is laid along the surface and can slide upon it freely. I shall henceforth speak of such lines as the *straightest* lines of any particular surface or given space, so as to bring out their analogy with the straight line in a plane.

If beings of this kind lived on an infinite plane, their geometry would be exactly the same as our planimetry. They would affirm that only one straight line is possible between two points, that through a third point lying outside this line only one line can be drawn parallel to it, that the ends of a straight line never meet though it is produced to infinity, and so on. Their space might actually be infinitely extended; but even if there were limits to their movement and perception, they would be able to represent to themselves a continuation beyond these limits, and thus their space would still appear to them infinitely extended, just as ours does to us, although our bodies cannot leave the earth and our sight only reaches as far as the visible fixed stars.

But intelligent beings of this kind might also live on the surface of a sphere. Their shortest or straightest line between two points would then be an arc of the great circle passing through them. Every great circle passing through two points is divided by them into two parts. If the parts are unequal, the shorter is certainly the shortest line on the sphere between the two points; but the other, or larger, arc of the same great circle is also a geodesic, or straightest, line; that is, every smaller part of it is the shortest line between its ends. Thus the notion of the geodesic, or straightest, line is not quite identical with that of the shortest line.

If the two given points are the ends of a diameter of the sphere, every plane passing through this diameter cuts semicircles on the surface of the sphere, all of which are shortest lines between the ends; in this case there is an infinite number of equal shortest lines between the given points. Accordingly, the axiom that there is only one shortest line between two points would not hold without a certain exception for the dwellers on a sphere.

Of parallel lines the sphere-dwellers would know nothing. They would declare that any two straightest lines, if sufficiently extended, must finally intersect not only in one but in two points. The sum of the angles of a triangle would be always greater than two right angles, increasing as the surface of the triangle grew greater. They could thus have no conception of geometric similarity between greater and smaller figures of the same kind, for with them a greater triangle must have different angles from a smaller one. Their space would be unlimited, but would be found to be finite or at least represented as such.

It is clear, then, that such beings must set up a very different system of geometric axioms from that of the inhabitants of a plane or from ours, with our space of three dimensions, though the logical powers of all were the same; nor are more examples necessary to show that geometric axioms must vary according to the kind of space inhabited. But let us proceed still further.

Let us think of reasoning beings existing on the surface of an egg-shaped body. Shortest lines could be drawn between three points of such a surface and a triangle constructed. But if the attempt were made to construct congruent triangles at different parts of the surface, it would be found that two triangles with three pairs of equal sides would not have equal angles. The sum of the angles of a triangle drawn at the sharper pole of the body would depart further from two right angles than if the triangle were drawn at the blunter pole or at the equator. Hence it appears that not even such a simple figure as a triangle could be moved on such a surface without change of form. It would also be found that if circles of equal radii were constructed at different parts of such a surface (the length of the radii being always measured by shortest lines along the surface), the periphery would be greater at the blunter than at the sharper end.

We see accordingly that if a surface admits of the figures lying on it being freely moved without change of any of their lines and angles as measured along it, the property is a special one and does not belong to every kind of surface. The condition under which a surface possesses this important property was pointed out by Gauss in his celebrated treatise on the curvature of surfaces. The measure of curvature, as he called it—that is, the reciprocal of the product of the greatest and least radii of curvature—must be everywhere equal over the whole extent of the surface.

Gauss showed at the same time that this measure of curvature is not changed if the surface is bent without distention or contraction of any part of it. Thus we can roll up a flat sheet of paper into the form of a cylinder or a cone without any change in the dimensions of the figures taken along the surface of the sheet. Or the hemispherical fundus of a bladder may be rolled into a spindle shape without altering the dimensions on the surface. Geometry on a plane will therefore be the same as on a cylindrical surface; though in the latter case we must imagine that any number of layers of this surface, like the layers of a rolled sheet of paper, lie one upon another and that after each entire revolution around the cylinder a new layer is reached.

These observations are meant to give the reader a notion of a kind of surface whose geometry is, on the whole, similar to that of the plane but in which the axiom of parallels does not hold good. This is a kind of curved surface which geometrically is, as it were, the counterpart of a sphere and which has therefore been called the *pseudospherical surface* by the distinguished Italian mathematician E. Beltrami, who has investigated its properties. It is a saddle-shaped surface of which only limited pieces or strips can be connectedly represented in our space, but which may still be thought of as infinitely continued in all directions, since each piece lying at the limit of the part constructed can be conceived as drawn back to the middle of it and then continued. The piece displaced must in the process change its flexure but not its dimensions, just as happens with a sheet of paper moved about a cone formed out of a plane rolled up. Such a sheet fits the conical surface in every part, but it must be more bent near the vertex and cannot be displaced so as to be adapted at the same time to the existing cone and to its imaginary continuation beyond.

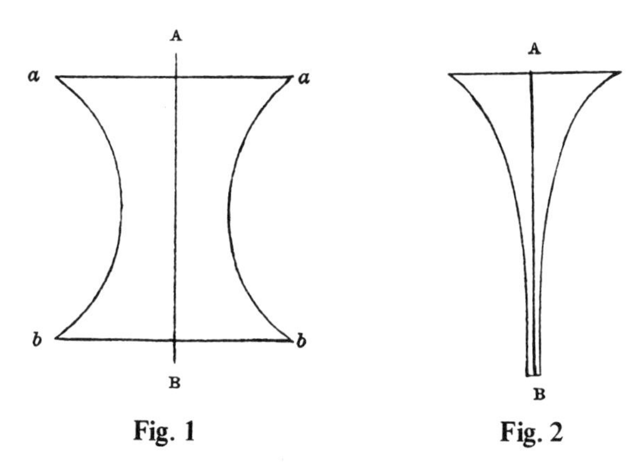

Fig. 1 **Fig. 2**

Like the plane and the sphere, the pseudospherical surface has a constant measure of curvature, so that every piece of one can be exactly superimposed on every other piece. All figures constructed at one place on the surface can therefore be transferred to any other place with perfect congruity of form and perfect equality of all dimensions lying in the surface itself. The measure of curvature as laid down by Gauss, which is positive for the sphere and zero for the plane, would have a constant negative value for the pseudospherical surface because the two principal curvatures of a saddle-shaped surface have their concavities turned opposite ways.

A strip of a pseudospherical surface may, for example, be represented by the inner surface (turned toward the axis) of a solid anchor ring. If the plane figure *aabb* (Fig. 1) is made to revolve on its axis of symmetry *AB*, the two arcs *ab* will describe a pseudospherical concave-convex surface like that of the ring. Above and below, toward *aa* and *bb,* the surface will turn outward with ever-increasing curvature until it becomes perpendicular to the axis and ends at the edge with one curvature infinite. Again, half of a pseudospherical surface may be rolled up into the shape of a champagne glass (Fig. 2) with a tapering stem infinitely prolonged. But the surface is always necessarily bounded by a sharp edge beyond which it cannot be directly continued. Only by supposing each single piece of the edge cut loose and drawn along the surface of the ring or glass can it be brought to places of different curvature at which farther continuation of the piece is possible.

In this way, too, the straightest lines of the pseudospherical surface may be infinitely produced. They do not, like those on a sphere, return upon themselves; rather, as on a plane, only one shortest line is possible between two given points. The axiom of parallels, however, does not hold good. If a straightest line is marked on the surface and a point outside it, a whole pencil of straightest lines may pass through the point, no one of which, though infinitely produced, intersects the first line. The pencil itself is limited by two straightest lines, one of which intersects one of the ends of the given line at an infinite distance, the other the other end.

As it happened, a system of geometry excluding the axiom of parallels was devised using Euclid's synthetic method as far back as 1829 by N. I. Lobachevsky, professor of mathematics at Kazan, and it was proved that this system could be carried out as consistently as Euclid's. It agrees exactly with the geometry of the pseudospherical surfaces worked out recently by Beltrami.

Thus we see that in the geometry of two dimensions, a surface is marked out as a plane, or a sphere, or a pseudospherical surface by the assumption that any figure may be moved about in all directions without change of dimensions. The axiom that there is only one shortest line between any two points distinguishes the plane and the pseudospherical surface from the sphere, and the axiom of parallels marks off the plane from the pseudosphere. These three axioms are in fact necessary and sufficient to define as a plane the surface to which Euclid's planimetry has reference, as distinguished from all other kinds of space in two dimensions.

The difference between plane and spherical geometry has been long evident, but the meaning of the axiom of parallels could not be understood until Gauss had developed the notion of surfaces flexible without extension and consequently the notion of the possibly infinite continuation of pseudospherical surfaces. Inhabiting a space of three dimensions and endowed with organs of sense for their perception, we can represent to ourselves the various cases in which surface-beings might have developed their perception of space, for we have only to limit our own perceptions to a narrower field. It is easy to think away the perceptions we have, but it is very difficult to imagine perceptions to which there is nothing analogous in our experience. When, therefore, we pass to space of three dimensions we are stopped in our power of representation by the structure of our organs and the experiences obtained through them, which correspond only to the space in which we live.

There is, however, another way of treating geometry scientifically. All known space relations are measurable; that is, they can be brought to determination of magnitudes (lines, angles, surfaces, volumes). Problems in geometry can therefore be solved by finding methods of calculation for arriving at unknown magnitudes from known ones. This is done in analytic geometry, where all forms of space are treated only as quantities and are determined by means of other quantities. Even the axioms themselves make reference to magnitudes. The straight line is defined as the *shortest* between two points, which is a determination of quantity. The axiom of parallels states that if two straight lines in a plane do not intersect (are parallel), the alternate angles, or the corresponding angles, made by a third line intersecting them are equal; or it may be laid down instead that the sum of the angles of any triangle is equal to two right angles. These are determinations of quantity.

We may start with this view of space–according to which the position of a point is determined by measurements in relation to some given figure (system of coordinates) taken as fixed–and then inquire, what are the special characteristics of our space as manifested in the measurements that must be made, and how does it differ from similar extended quantities? This path was first entered by one too early lost to science. B. Riemann, of Göttingen. It has the peculiar advantage that all its operations consist of pure calculation of quantities, which obviates entirely the danger of habitual perceptions being taken for necessities of thought.

The number of measurements necessary to give the position of a point is equal to the number of dimensions of the space in question. In a line the distance from one fixed point is sufficient, that is to say, one quantity; in a surface the distances from two fixed points must be given; and in space we require the distances from three–as, on the earth, longitude, latitude, and height above the sea; or as is usual in analytic geometry, the distances from three coordinate planes. Riemann calls a system in which one individual can be determined by n measurements an $n$fold extended aggregate, or an aggregate of *n* dimensions. Thus the space in which we live is a threefold, a surface is a twofold, and a line is a simple extended aggregate of points. Time also is an aggregate of one dimension. The system of colors is an aggregate of three dimensions, inasmuch as each color, according to the investigations of Thomas Young and Clerk Maxwell, may be represented as a mixture of three primary colors in definite quantities. The particular mixtures can be actually made with the color wheel.

In the same way we may consider the system of simple tones as an aggregate of two dimensions, if we distinguish only pitch and intensity and leave out of account differences of timbre. This generalization of the idea is well suited to bring out the distinction between space of three dimensions and other aggregates. We can, as we know from daily experience, compare the vertical distance of two points with the horizontal distance of two others because we can apply a measure first to the one pair and then to the other. But we cannot compare the difference between two tones of equal pitch and different intensity with that between two tones of equal intensity and different pitch.

Riemann showed by considerations of this kind that the essential foundation of any system of geometry is the expression it gives for the distance between two points lying in any direction from one another, beginning with one which is infinitesimal. He took from analytic geometry the most general form for this expression, namely, that which leaves altogether open the kind of measurements by which the position of any point is given.[1] Then he showed that the kind of free mobility without change of form which belongs to bodies in our space can only exist when certain quantities yielded by the calculation[2]—quantities which coincide with Gauss's measure of surface curvature when they are expressed for surfaces—have everywhere an equal value. For this reason Riemann calls these quantities, when they have the same value in all directions for a particular point, the measure of curvature of the space at this point. (To prevent misunderstanding I shall once more observe that this so-called measure of space curvature is a quantity obtained by purely analytical calculation and that its introduction involves no suggestion of relations which would have a meaning only for sense perception. The name

1. For the square of the distance of two infinitely near points, the expression is a homogeneous quadric function of the differentials of their coordinates.

2. They are algebraic expressions compounded from the coefficients of the various terms in the expression for the square of the distance of two contiguous points and from their differential quotients.

is merely taken, as a short expression for a complex relation, from the one case in which the quantity designated admits of sensible representation.)

Whenever the value of this measure of curvature in any space is everywhere zero, that space everywhere conforms to the axioms of Euclid; it may be called a *flat (homaloid)* space, in contradistinction to other spaces, analytically constructible, which may be called *curved* because their measure of curvature has a value other than zero. Analytic geometry may be as completely and consistently worked out for such spaces as ordinary geometry is for our actually existing homaloid space.

If the measure of curvature is positive, we have *spherical* space, in which straightest lines return upon themselves and there are no parallels. Such a space would, like the surface of a sphere, be unlimited but not infinitely great. A constant negative measure of curvature, on the other hand, gives *pseudospherical* space, in which straightest lines run out to infinity and a pencil of straightest lines may be drawn in any flattest surface through any point without intersecting another given straightest line in that surface.

Beltrami has rendered these last relations imaginable by showing that the points, lines, and surfaces of a pseudospherical space of three dimensions can be so portrayed in the interior of a sphere in Euclid's homaloid space that every straightest line or flattest surface of the pseudospherical space is represented by a straight line or a plane, respectively, in the sphere. The surface itself of the sphere corresponds to the infinitely distant points of the pseudospherical space; and the different parts of this space, as represented in the sphere, become smaller the nearer they lie to the spherical surface, diminishing more rapidly in the direction of the radii than in that perpendicular to them. Straight lines in the sphere which intersect only beyond its surface correspond to straightest lines of the pseudospherical space which never intersect.

Thus it appears that space, considered as a region of measurable quantities, does not correspond at all with the most general conception of an aggregate of three dimensions. It involves special conditions, depending, not only on the perfectly free mobility of solid bodies without change of form to all parts of it and with all possible changes of direction, but also on the special value of the measure of curvature, which for our actual space equals, or at least is not distinguishable from, zero. This latter definition is given in the axioms of straight lines and parallels.

While Riemann entered upon this new field from the side of the most general and fundamental questions of analytic geometry, I myself arrived at similar conclusions, partly through seeking to represent in space the system of colors (involving the comparison of one threefold extended aggregate with another) and partly through inquiries into the origin of our ocular measure for distances in the field of vision. Riemann starts by assuming the above-mentioned algebraic expression, which represents in the most general form the distance between two infinitely close points, and deduces therefrom the conditions of mobility of rigid figures. I, on the other hand, starting from the observed fact that the movement of rigid figures is possible in our space, with the degree of freedom that we know, deduce the necessity of the algebraic expression taken by Riemann as an axiom. The assumptions I had to make as the basis of the analysis were the following:

First, to make algebraic treatment possible, it must be assumed that the position of any point *A* can be determined, in relation to certain given figures taken as fixed bases, by measurement of some kind of magnitudes, such as lines, angles between lines, angles between surfaces, and so forth. The measurements necessary for determining the position of *A* are known as its coordinates. In general the number of coordinates necessary for the complete determination of the position of a point marks the number of the dimensions of the space in question. It is further assumed that with the movement of the point *A*, the magnitudes used as coordinates vary continuously.

Secondly, the definition of a solid body, or rigid system of points, must be made in such a way as to admit of magnitudes being compared by congruence. As we must not at this stage assume any special methods for the measurement of magnitudes, our definition can, in the first instance, run only as follows: Between the coordinates of any two points belonging to a rigid body, there must be an equation which, however the body is moved, expresses a constant spatial relation (proving at last to be the distance) between the two points and which is the same for congruent pairs of points, that is, such pairs as can be made to coincide successively in space with the same fixed pair of points.

In spite of its indeterminate appearance, this definition has most important consequences, for with an increase in the number of points, the number of equations grows much more rapidly than the number of point coordinates determined by them. Five points *(A, B, C, D, E)* result in ten different pairs of points *(AB, AC, AD, AE, BC, BD, BE, CD, CE, DE)* and thus ten equations, which in a space of three dimensions contain fifteen variable coordinates. Of these fifteen, however, six must remain unfixed if the system of five points is to admit of free movement and rotation; we thus have ten equations for determining the coordinates of nine points as functions of six variables. With six points we have fifteen equations for twelve quantities; with seven points, twenty-one equations for fifteen, and so on. If we have more than n equations, the number in excess of n must be derivable from the first n. Hence the equations which exist between the coordinates of each pair of points of a rigid body must have a special character, since when (in space of three dimensions) they are satisfied for nine pairs of points as formed out of any five points, the equation for the tenth pair follows as a logical consequence. Thus our assumption for the definition of rigidity is quite sufficient to determine the kind of equations holding between the coordinates of two points rigidly connected.

Thirdly, the analysis must further be based on the fact of a peculiar characteristic of the movement of rigid bodies—a fact so familiar to us that but for this inquiry it might never have been thought of as something that need not be. When in our space of three dimensions two points of a rigid body are kept fixed, its movements are limited to rotations around the straight line connecting them. If we turn it completely around once, it again occupies exactly the position it had at first. This fact—that rotation in one direction always brings a solid body back into its original position—needs special mention, for a system of geometry is possible without it. This is most easily seen in the geometry of a plane. Suppose that with every rotation of a plane figure, its linear dimensions increased in proportion to the angle of rotation. The figure after one whole rotation through 360 degrees would no longer coincide with itself as it was originally; but any second figure that was congruent with the first in its original position might be made to coincide with it in its second position by being also turned through 360 degrees. A consistent system of geometry would be possible upon this supposition, which does not come under Riemann's formula.

On the other hand, I have shown that the three assumptions taken together form a sufficient basis for the starting point of Riemann's investigation and thence for all his further results relating to the distinction of different spaces according to their measure of curvature.

It still remained to be seen whether the laws of motion as dependent on moving forces could also be consistently transferred to spherical or pseudospherical space. This investigation has been carried out by Professor Lipschitz, of Bonn. It is found that the comprehensive expression for all the laws of dynamics, Hamilton's principle, may be transferred directly to spaces whose measure of curvature is other than zero. In this respect also, the disparate systems of geometry lead to no contradiction.

We have now to seek an explanation of the special characteristics of our own flat space, since it appears that they are not implied in the general notion of an extended quantity of three dimensions and of the free mobility of specific figures therein. Necessities of thought involved in such a conception they are not. Let us then examine the opposite assumption—that their origin is empirical—and see whether they can be inferred from facts of experience and so established or whether, when tested by experience, they are perhaps to be rejected. If they are of empirical origin, we must be able to represent to ourselves a connected series of facts indicating a different value for the measure of curvature from that of Euclid's flat space. But if we can imagine such spaces of other sorts, it cannot be maintained that the axioms of geometry are necessary consequences of an a priori transcendental form of intuition, as Kant thought.

The distinction among spherical, pseudospherical, and Euclidean geometry depends, as was observed above, on the value of a certain constant which Riemann called the measure of curvature of the space in question. The value must be zero for Euclid's axioms to hold good. If it were not zero, the sum of the angles of a large triangle would differ from that of the angles of a small one, being larger in spherical, smaller in pseudospherical space. Again, the geometrical similarity of large and small solids or figures is possible only in Euclid's space. All systems of practical mensuration that have been used for the angles of large rectilinear triangles, especially all systems of astronomical measurement which make the parallax of the immeasurably distant fixed stars equal to zero (in pseudospherical space the parallax even of infinitely distant points would be positive), confirm empirically the axiom of parallels and show the measure of curvature of our space thus far to be indistinguishable from zero. It remains, however, a question, as Riemann observed, whether the result might not be different if we could use other than our limited base lines, the greatest of which is the major axis of the earth's orbit.

Meanwhile, we must not forget that all geometric measurements rest ultimately upon the principle of congruence. We measure the distance between points by applying to them the compass, rule, or chain. We measure angles by bringing the divided circle or theodolite to the vertex of the angle. We also determine straight lines by the path of rays of light, which in our experience is rectilinear; but that light travels in shortest lines as long as it continues in a medium of constant refraction would be equally true in space of a different measure of curvature. Thus all our geometric measurements depend on our instruments being really, as we consider them, invariable in form, or at least on their undergoing no other than the small changes we know of as arising from variation of temperature or from gravity acting differently at different places.

In measuring, we are simply employing the best and surest means we know to determine what we otherwise are in the habit of making out by sight and touch or by pacing. Here our own body with its organs is the instrument we carry about in space. Now it is the hand, now the leg that serves for a compass, while the eye turning in all directions is our theodolite for measuring arcs and angles in the visual field.

Every comparative estimate of magnitudes or measurement of their spatial ralations proceeds, therefore, upon a supposition as to the behavior of certain physical things, either the human body or other instruments employed. The supposition may be in the highest degree probable and in closest harmony with all other physical relations known to us, yet it passes beyond the scope of pure space intuition.

It is in fact possible to imagine conditions for bodies apparently solid such that the measurements in Euclid's space become what they would be in spherical or pseudospherical space. Let me first remind the reader that if all the linear dimensions of other bodies and our own were diminished or increased at the same time in like proportion, as for instance to half or double their size, we should with our means of space perception be utterly unaware of the change. This would also be the case if the distention or contraction were different in different directions, provided that our own body changed in the same manner and provided further that a body in rotating assumed at every moment, without suffering or exerting mechanical resistance, the amount of expansion in its different dimensions corresponding to its position at the time.

Think of the image of the world in a convex mirror. The common silvered globes set up in gardens give the essential features, distorted only by some optical irregularities. A well-made convex mirror of moderate aperture represents the objects in front of it as apparently solid and in fixed positions behind its surface. But the images of the distant horizon and of the sun in the sky lie behind the mirror at a limited distance, equal to its focal length. Between these and the surface of the mirror are found the images of all the other objects before it, but the images are diminished and flattened in proportion to the distance of their objects from the mirror. The flattening, or decrease in the third dimension, is relatively greater than the decrease of the surface dimensions. Yet every straight line or plane in the outer world is represented by a straight line or plane in the image. The image of a man measuring with a rule a straight line from the mirror would contract more and more the farther he went, but with his shrunken rule the man in the image would count out exactly the same number of centimeters as the real man. And, in general, all geometric measurements of lines or angles made with regularly varying images of real instruments would yield exactly the same results as in the outer world; all congruent bodies would coincide on being superimposed upon one another in the mirror as in the outer world; and all lines of sight in the outer world would be represented by straight lines of sight in the mirror. In short I do not see how men in the mirror could discover that their bodies are not rigid solids and their experiences good examples of the correctness of Euclid's axioms. But if they could look out upon our world, as we can look into theirs, without overstepping the boundary, they would surely declare ours a picture in a spherical mirror and would speak of us just as we speak of them; and if two inhabitants of the different worlds could communicate with each other, neither, so far as I can see, would be able to convince the other that he had the true, the other the distorted relations. Indeed, I cannot see that such a question would have any meaning at all, so long as mechanical considerations are not mixed up with it.

Beltrami's representation of pseudospherical space in a sphere of Euclid's space is quite similar, except that the background is not a plane, as in the convex mirror, but the surface of a sphere and that the proportion in which the images contract as they approach the spherical surface has a different mathematical expression. Let us imagine, then, conversely, that in the sphere (for the interior of which Euclid's axioms hold good) moving bodies contract as they depart from the center like the images in a convex mirror and in such a way that their representatives in pseudospherical space retain their dimensions unchanged. Observers whose bodies were regularly subjected to the same change would obtain the same results from the geometric measurements they could make as if they lived in pseudospherical space.

We can even go a step further and infer how the objects in a pseudospherical world, were it possible to enter one, would appear to an observer whose habits of visual measurement and experiences of space had been gained, like ours, in Euclid's space. Such an observer would continue to look upon rays of light or the lines of vision as straight lines, such as are met with in flat space and as they really are in the spherical representation of pseudospherical space. The visual image of the objects in pseudospherical space would thus make the same impression upon him as if he were at the center of Beltrami's sphere. He would think he saw the most remote objects round about him at a finite distance,[3] let us suppose a hundred feet off. But as he approached these distant objects, they would dilate before him, though more in the third dimension than superficially, while behind him they would contract. He would know that his eye judged wrongly. If he saw two straight lines which in his estimate ran parallel for the hundred feet to his world's end, he would find on following them that the farther he advanced, the more they diverged, because of the expansion of all the objects which he approached. On the other hand, behind him their distance would seem to diminish, so that as he advanced they would appear always to converge more and more. But two straight lines which from his first position seemed to converge at the same point of the background a hundred feet distant would continue to do this, however far he went; he would never reach their point of intersection.

We can obtain exactly similar images of our real world if we look through a large convex lens of corresponding negative focal length, or even through a pair of convex spectacles if ground somewhat prismatically to resemble pieces of one continuous larger lens. With these, as with the convex mirror, we see remote objects as if they were near us, the most remote appearing no farther distant than the focus of the lens. In going about with this lens before the eyes, we find that the objects we approach dilate exactly in the manner I have described for pseudospherical space. Anyone using a lens, were it even so strong as to have a focal length of only sixty inches, to say nothing of a hundred feet, would perhaps observe for the first moment that he saw objects brought nearer. But after he went about a little, the illusion would vanish, and in spite of the false images he would judge distances rightly. We have every reason to suppose that what happens in a few hours to anyone beginning to wear spectacles would soon be experienced in pseudospherical space. In short, pseudospherical space would not seem to us very strange, comparatively speaking; we should only at first be subject to illusions in measuring by eye the size and distance of the more remote objects.

There would be illusions of an opposite description if, with eyes practised to measure in Euclid's space we entered a spherical space of three dimensions. We should suppose the more distant objects to be more remote and larger than they are, and we should find on approaching them that we reached them more quickly than we expected from their appearance. But we should also see before us objects that we can fixate only with diverging lines of

---

3. The reciprocal of the square of this distance, expressed in negative quantity, would be the measure of curvature of the pseudospherical space.

sight, namely, all those at a greater distance from us than the quadrant of a great circle. Such an aspect of things would hardly strike us as very extraordinary, for we can have it even as things are if we place before the eye a slightly prismatic glass with the thicker side toward the nose: the eyes must then diverge to take in distant objects. This excites a certain feeling of unwonted strain in the eyes but does not perceptibly change the appearance of the objects thus seen. The strangest sight, however, in the spherical world would be the back of our own head, in which all visual lines not stopped by other objects would meet again and which must fill the extreme background of the whole perspective picture.

At the same time it must be noted that as a small, elastic, flat disc, say of India rubber, can be fitted to a slightly curved, spherical surface only with relative contraction of its border and distention of its center, so our bodies, developed in Euclid's flat space, could not pass into curved space without undergoing similar distentions and contractions of their parts—their coherence being maintained, of course, only insofar as their elasticity permitted their bending without breaking. The kind of distention must be the same in passing from a small body imagined at the center of Beltrami's sphere to its pseudospherical or spherical representation. For such passage to appear possible, it must always be assumed that the body is sufficiently elastic and small in comparison with the real or imaginary radius of curvature of the curved space into which it is to pass.

These remarks will suffice to show how we can infer from the known laws of our sensible perceptions the series of sensible impressions which a spherical or pseudospherical world would give us, if it existed. In doing so, we nowhere meet with inconsistency or impossibility, any more than in the calculation of its metrical proportions. We can represent to ourselves the look of a pseudospherical world in all directions, just as we can develop the conception of it. Therefore it cannot be allowed that the axioms of our geometry depend on the native form of our perceptive faculty or are in any way connected with it.

It is different with the three dimensions of space. As all our means of sense perception extend only to space of three dimensions, and a fourth is not merely a modification of what we have but something perfectly new, we find ourselves by reason of our bodily organization quite unable to represent a fourth dimension.

In conclusion, I would again urge that the axioms of geometry are not propositions pertaining only to the pure doctrine of space. As I said before, they are concerned with quantity. We can speak of quantities only when we know some way by which we can compare, divide, and measure them. All space measurements, and therefore all ideas of quantities applied to space, assume the possibility of figures moving without change of form or size. It is true that we are accustomed in geometry to call such figures purely geometric solids, surfaces, angles, and lines because we abstract these from all the other characteristics, physical and chemical, of natural bodies. Only one physical quality, rigidity, is retained. We have no mark of rigidity of bodies or figures other than congruence whenever they are superimposed on one another, at any time or place and after any revolution. We cannot, however, decide by pure geometry and without mechanical considerations whether the coinciding bodies may not both have varied in the same way.

If it were useful for any purpose, we might with perfect consistency look upon the space in which we live as the apparent space behind a convex mirror with its shortened and contracted background. We might also consider a bounded sphere of our space, beyond the limits of which we perceive nothing, as infinite pseudospherical space. We should then, however, have to ascribe to the bodies which appear as solid–and to our own bodies, at the same time–corresponding distentions and contractions. We would also have to change our system of mechanical principles entirely, for even the proposition that every point in motion, if acted upon by no force, continues to move with unchanged velocity in a straight line is not adapted to the image of the world in the convex mirror. The path would indeed be straight, but the velocity would depend upon the place.

Thus the axioms of geometry are concerned, not only with space relations, but also with the mechanical behavior of solid bodies in motion. The concept of a rigid geometric figure might indeed be conceived as transcendental in Kant's sense, that is, as formed independently of actual experience, which need not exactly correspond to it, any more than natural bodies ever in fact correspond exactly to the abstract conception we have obtained of them by induction. Taking the concept of rigidity thus as a mere ideal, a strict Kantian might look upon the geometric axioms as propositions given a priori by transcendental intuition, which no experience could either confirm or refute, because it must first be decided by them whether any natural bodies can be considered rigid. But then we should have to maintain that the axioms of geometry are not synthetic propositions, as Kant held them: they would merely define what qualities and behavior a body must have to be recognized as rigid.

But if to the geometric axioms we add propositions relating to the mechanical properties of natural bodies—if only the axiom of inertia or the single proposition that the mechanical and physical properties of bodies and their mutual reactions are, other circumstances remaining the same, independent of place—such a system of propositions has a real import which can be confirmed or refuted by experience, but for the same reason can also be got by experience. The mechanical axiom just cited is, in fact, of the utmost importance for our whole system of mechanical and physical conceptions. That rigid solids, as we call them (they are really elastic solids of great resistance), retain the same form in every part of space if no external force affects them is a single case falling under the general principle.

For the rest, I do not, of course, suppose that mankind first arrived at space intuitions in agreement with the axioms of Euclid by any carefully executed system of exact measurement. It was rather a succession of everyday experiences—especially the perception of the geometric similarity of great and small bodies, possible only in flat space—that led to the rejection as impossible of every geometric representation at variance with this fact. For this no knowledge of the necessary logical connection between the observed fact of geometric similarity and the axioms was needed, but only an intuitive apprehension of the typical relations among lines, planes, angles, etc., obtained by numerous, attentive observations—an intuition of the kind the artist possesses of the objects he is to represent and by means of which he decides surely and accurately whether a new combination which he tries corresponds to their nature. It is true that we have no word but *intuition* to mark this, but it is knowledge empirically gained by the aggregation and reinforcement of similar recurrent impressions in memory, not a transcendental form given before experience. That other such empirical intuitions of fixed typical relations, when not clearly comprehended, have frequently been taken by metaphysicians for a priori principles is a point on which I need not insist.

To sum up, the final outcome of the whole inquiry may be thus expressed:

1. The axioms of geometry, taken by themselves out of all connection with mechanical propositions, represent no relations of real things. When thus isolated, if we regard them with Kant as forms of intuition transcendentally given, they constitute a form into which any empirical content whatever will fit and which therefore does not in any way limit or determine beforehand the nature of the content. This is true, however, not only of Euclid's axioms, but also of the axioms of spherical and pseudospherical geometry.

2. As soon as certain principles of mechanics are conjoined with the axioms of geometry, we obtain a system of propositions which has real import and which can be verified or overturned by empirical observations, as it can be inferred from experience. If such a system were to be taken as a transcendental form of intuition and thought, there must be assumed a pre-established harmony between form and reality.

9.

THE ORIGIN OF THE PLANETARY SYSTEM [1871]

A lecture delivered at Heidelberg and Cologne in 1871

IT is my intention today to present to you a subject which has already been frequently discussed. I refer to the hypothesis of Kant and Laplace concerning the formation of celestial bodies, more especially concerning the formation of our planetary system.

The choice of this subject needs, perhaps, a small amount of justification or apology. In popular lectures, like the present one, the audience has the right to expect that the lecturer will present well-established facts or the final results of some investigation, not half-developed assumptions, hypotheses, or dreams. Of all of the subjects to which the thought and imagination of man could be directed, the question of the origin of the world has, since remote antiquity, been the favorite area for the wildest speculations. Beneficent and malignant deities; giants; Kronos, who devoured his children; the ice-giant Ymir, who is killed by the gods so that out of him the world might be constructed–these are all figures that fill the cosmogonic systems of the more cultivated of peoples. The universality of the fact that all nations develop their own cosmogonies, sometimes in great detail, is an expression of the interest we all feel in knowing our own origin and in knowing the ultimate origin of the things around us. With the question of the beginning of things, of course, the question of the end is closely connected, for that which comes into being may also pass away. Indeed, the question of the end of things is perhaps of even greater practical interest than that of the beginning.

I should also like to note here that the theory which I plan to discuss today was first presented by Immanuel Kant, who has become known as the most abstract of philosophical thinkers and as the originator of transcendental idealism and of the categorical imperative. The work in which he developed this theory, the *General Natural History and Theory of the Heavens* (1755), was one of his first publications, appearing in his thirty-first year. Looking at the writings of this first period of his intellectual activity, a period which lasted until approximately his fortieth year, we find that they belong mostly to natural philosophy and that in a number of the happiest ideas they are far in advance of their time. His philosophical writings during this period were few and, like his habilitation lecture, were done in connection with some special event. Comparatively unoriginal in content, they are notable only for their destructive, or indeed scathing, criticism.

It cannot be denied that early in his career Kant was a natural philosopher by instinct and inclination; it was probably only due to the force of circumstances, due to the lack of the means necessary for independent scientific research and to the tone of thought prevalent at the time, that he kept to philosophy. It was only much later, as you know, that he produced anything original and important, his *Critique of Pure Reason,* which appeared in his fifty-seventh year. Even in the later periods of his life, between his great philosophical works, he wrote occasional memoirs on natural philosophy and regularly delivered a course of lectures on physical geography. He was handicapped in this latter area because of the meager knowledge and limited research techniques available at his time and, of course, because of the out-of-the-way place in which he lived. With his large, intelligent mind, however, he strove for more general points of view such as Alexander von Humboldt later worked out. It is a perversion of history when Kant's name is sometimes used to support the recommendation that natural philosophy leave the inductive method, by which it has become great, and return to the airy speculations of the so-called deductive method. No one would have objected more energetically and more incisively than Kant himself to such a misuse, if he were still among us.

The same hypothesis concerning the origin of our planetary system was advanced a second time, apparently quite independently of Kant, by the Marquis de Laplace, the most celebrated of French astronomers. It formed the ultimate conclusion of his work, executed with tremendous industry and great mathematical ingenuity, on the mechanics of our solar system. You can see already from the names of these two men, whom we meet as the experienced, tested guides along the path we are about to take, that in approaching a subject concerning which they both agree, we are dealing, not with some fanciful dream, but with a careful, well-considered attempt to deduce conclusions as to the unknown past from the known conditions of the present time.

It is in the nature of things that a hypothesis concerning the origin of the world which we inhabit, a hypothesis which deals with things in the most distant past, cannot be verified by direct observation. It may, however, receive indirect confirmation if, in the progressive development of scientific knowledge, new facts accrue to those already known which like them are explained by the hypothesis. This is particularly the case if survivals of the processes assumed to have taken place in the formation of the heavenly bodies can be proved to exist at the present time. And, indeed, various kinds of indirect confirmation have been found for the view we are about to discuss and have materially increased its probability.

It is partly this fact–and partly the fact that the hypothesis in question has recently been mentioned in popular and scientific books in connection with philosophical, ethical, and theological questions–that has emboldened me to speak of it here. I intend, not so much to tell you anything substantially new concerning the hypothesis, as to endeavor to give, as completely as possible, the reasoning which has led to and has confirmed it.

The apologies which I feel compelled to make apply only to the fact that I treat a theme of this kind in a popular lecture. Science is not only justified in making such an investigation; it is obligated to do so. For science, there is a very specific and very important question at issue here–the question whether there are limits to the range of validity of the laws of nature which rule all that now surrounds us. Have these laws always held in the past? Will they always hold in the future? Or do our inferences from present conditions to the past and to the future, on the supposition of an eternal uniformity of natural laws, lead inevitably to impossible conclusions, to the necessity of infractions of natural laws, or to a beginning which could not have been due to processes now known to us? An investigation of the possible or probable primeval history of our present world is, from the point of view of science, no idle speculation; it is an investigation concerning the limits of scientific methods and the extent to which existing laws are valid.

It may perhaps seem presumptuous that we–restricted in our observations, as we are in space, by our position on this little earth, which is like a speck of dust in our Milky Way, and in time by the short span of human history–should attempt to take laws which we have established on the basis of the restricted range of phenomena open to us and apply them to the whole of immeasurable space and to time from eternity to eternity. All our thoughts and all our actions, however, in the greatest of things as well as in the least, are based upon our confidence in the inalterable, lawful order of nature–a confidence which has always been increasingly justified, the deeper we have penetrated into the interrelationships of natural phenomena. As to the validity throughout the most distant reaches of space of the general laws that have been established, there have been important factual confirmations of this during the past half-century.

In the front rank of all is the law of gravitation. The celestial bodies, as you all know, float and move in infinite space. Compared with the enormous distances between them, each one, even the largest, is but a speck of dust. The nearest fixed stars, even under the most powerful magnification, have no visible diameter; and since the masses of these stars, in the cases where this has been determined, have not been found significantly different from that of the sun, we may be certain that even our sun, looked at from the nearest fixed stars, would appear only as a single, luminous point. In spite of these enormous distances, however, there exists an invisible bond among these bodies, a bond which links them together and brings them into mutual interdependence. This bond is the force of gravitation, with which all ponderable masses attract one another. We know this force as weight when it is operative between a material body and the mass of our earth. The force which causes a body to fall to the ground, however, is the same as that which continually forces the moon to accompany the earth in its path around the sun and which keeps the earth itself from flying off into space away from the sun.

It is possible to illustrate the path of planetary motion by means of a simple mechanical model. At an appropriate height fasten a silk thread to the branch of a tree or to a rigid bar fixed horizontally in the wall, and at the end of the thread fasten a small, heavy body, such as a lead ball. If you allow the body to hang at rest, it stretches the thread. This is the position of equilibrium of the ball. To mark this position of equilibrium and to keep track of it, put some other object, such as a terrestrial globe mounted on a stand, in the place of the ball. In order to do this, you must push the ball aside; it will press against the globe, however, and if it is pulled away, it will tend to return to it. The reason is that gravity impels the ball toward its position of equilibrium, which is at the center of the sphere. In whatever direction it may be pulled aside, the same thing will always happen. The force which pulls the ball toward the globe represents in our model the attraction which the earth exerts on the moon or the sun on the planets.

After you have convinced yourself of the accuracy of these facts, give the ball, when it is a short distance from the globe, a slight push in a lateral direction. If you have correctly judged the strength of the push, the small ball will move around the large one in a circular path and will retain this motion for some time, just as the moon persists in its course around the earth and the planets in theirs around the sun. In the model, of course, the circles described by the lead ball will become narrower and narrower, due to the fact that the opposing forces (the resistance of the air, the rigidity of the thread, and friction) cannot be eliminated so as to match the conditions in the planetary system completely.

If the path around the center of attraction is exactly circular, the attracting force will always act on the planet or on the lead ball with the same intensity. In that case, the law according to which the intensity of the force increases or decreases at varying distances from the center is of no consequence. If the original impulse was not of the right intensity, however, the path of the planet or of the lead ball will not be circular. It will be elliptical, that is, of the form of the curved line in Fig. 1.

Fig. 1

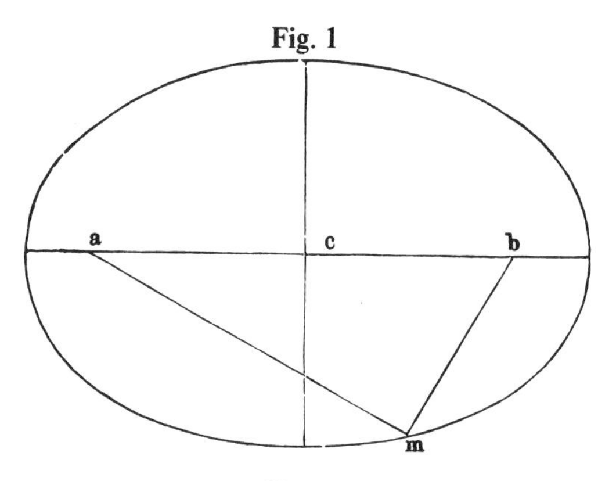

There is now one important difference, however, between the two cases. In the model, the farther the lead sphere is removed from its position of equilibrium, the stronger the force of attraction will be. The path of the ellipse will be such that the center of attraction coincides with the geometrical center, *c,* of the ellipse. In the case of the planet, on the other hand, the attracting force will be weaker, the farther the planet is removed from the attracting body. This is why an ellipse is described, one of whose foci lies at the center of attraction. The foci, *a* and *b,* are two points which are symmetrically placed with respect to the ends of the ellipse; they are further characterized by the property that the sum of the distances $am + bm$ is a constant for all points on the ellipse.

Kepler had found that the paths of the planets are ellipses of this kind; and since, as the above example shows, the form and position of the orbital path relative to the center depends upon the law according to which the magnitude of the attracting force changes, Newton was able to deduce from the form of the planetary orbits the well-known law of gravitational attraction, according to which the force that attracts the planets to the sun decreases with increase of distance as the square of that distance. Terrestrial bodies obeyed this same law, but Newton had the wonderful patience to refrain from publishing his important discovery until it had received direct confirmation; this followed from the observation that the force which attracts the moon toward the earth bears toward the weight of a terrestrial body the ratio required by the law.

In the course of the eighteenth century, techniques of mathematical analysis and methods of astronomical observation improved so much that all the complicated interactions between and among the planets and their satellites due to their influence upon one another (astronomers call these *disturbances,* that is, deviations from the simpler elliptical paths around the sun which each would follow if the others were absent) could be predicted theoretically from Newton's laws and could be compared precisely with what actually took place in the heavens. The development of the theory of planetary motion in its greatest detail was, as has already been noted, the achievement of Laplace.

The agreement between the theory, which was developed from the simple law of gravitational attraction, and the most varied, complicated phenomena was more complete and more accurate than any which had ever been attained previously in any other branch of human knowledge. Emboldened by this agreement between theory and fact, astronomers concluded that wherever slight but constant disturbances were still found, there must exist unknown causes. Thus, because of the discrepancies between the calculations of Bessel and the observed motion of Uranus, it was inferred that there must be still another planet. The position of this planet was calculated by Leverrier and Adams, and there Neptune, the most distant planet known at that time, was discovered.

But it was not only with respect to the force of attraction of our sun that the law of gravitation was found to hold. In the region of the fixed stars, it was found that double stars move around each other in elliptical orbits and that therefore the same law of gravitation must hold for them as for our planetary system. The distance of some of these stars can be calculated. The nearest of them, Alpha in the constellation of the Centaur, is 226,000 times farther from the sun than is the earth. Light, which has the enormous velocity of 186,000 miles per second and which travels the distance between the sun and the earth in eight minutes, would require three years to travel from Alpha Centauri to us. The increasingly delicate methods of modern astronomy have made it possible to determine distances which light would take thirty-five years to travel, as, for example, the distance to the North Star. Thus the law of gravitation is seen to hold, ruling the motion of double stars at distances in the heavens which we have been unable to measure with all the means that we possess.

The knowledge of the law of gravitation has here in the region of the fixed stars, as in the case of Neptune, led to the discovery of new bodies. Peters, in Altona, confirming a conjecture of Bessel, found that Sirius, the most brilliant of the fixed stars, moves in an elliptical path around an invisible center. This was thought to be a dark companion; and when the excellent, powerful telescope at Cambridge, Massachusetts, was finished, such a body was indeed found to exist. It is not quite dark, but its light is so faint that it can be seen only with the most powerful instruments. The mass of Sirius is found to be 13.76, and that of its companion 6.71, times the mass of the sun. The distance between them is equal to thirty-seven times the radius of the earth's orbit, or somewhat greater than the distance of Neptune from the sun. Another fixed star, Procyon, is similar to Sirius in having a companion, but it has not been seen as yet.

You can see that in gravitation we have discovered a property common to all matter, a property which is not confined to bodies in our solar system but which extends as far into celestial space as our means of observation have been able to penetrate.

The most distant heavenly bodies and our terrestrial bodies have more properties in common, however, than just this property of all ponderable masses. Spectrum analysis has taught us that a number of well-known terrestrial elements are found in the atmosphere of the fixed stars and even in distant nebulae.

You all know that a narrow beam of light, seen through a glass prism, looks like a colored band, red and yellow at one end, blue and violet at the other, and green in the middle. Such a colored image is called a color spectrum. The rainbow is also an example of a color spectrum; while it is not produced by a prism, it is produced by the refraction of light, and it exhibits the series of colors into which white sunlight can be analyzed. The formation of the prismatic spectrum depends upon the fact that the light of the sun and of most luminous bodies is made up of various kinds of light, the rays of which are separated from one another when refracted by a prism. These different rays appear to our eyes as different colors.

If a solid or a liquid is heated to the point where it becomes incandescent, the spectrum of its light is (like the rainbow) a broad, unbroken, colored band exhibiting the well-known series of colors, red, yellow, green, blue, and violet. The spectrum is in no way determined by the nature of the body which emits the light.

The case is different, however, if the light is emitted by an incandescent gas or an incandescent vapor, that is, by a substance which has been vaporized by heat. The spectrum of such a substance consists of one or more, sometimes even a large number, of quite distinct bright lines, the position and arrangement of which in the spectrum is characteristic of the substance constituting the gas or vapor. It is therefore possible to determine the chemical constitution of an incandescent gaseous material by means of spectrum analysis. Gas spectra of this kind are displayed by many nebulae in the universe; for example, they display spectra which have the bright lines of incandescent hydrogen and oxygen, generally along with a line which has never yet been found in the spectrum of any terrestrial element. In addition to proving the existence of two well-known terrestrial elements in the nebulae, these investigations are of the utmost importance, since they furnish the first unmistakable proof that the cosmic nebulae are not just multitudes of fine stars and that the greater part of the light which they emit is really due to materials in a gaseous state.

The gas spectrum is different in appearance if the gas is in front of an incandescent solid whose temperature is far higher than that of the gas. In this case one sees the continuous spectrum of a solid, but it is cut by fine dark lines in just those places where the spectrum of the gas alone, seen in front of a dark background, would display bright lines. Kirchhoff has proved that these two kinds of spectra are necessarily related. It is therefore possible to tell by the dark lines in a spectrum what gases are to be found in front of some body emitting light. The solar spectrum is one with dark lines, as is that of a large number of fixed stars. The dark lines of the solar spectrum, originally discovered by Wollaston, were first investigated and measured by Fraunhofer, and are now known as Fraunhofer lines.

Far more powerful apparatus has since been developed by Kirchhoff, and more recently by Angstrom, in order to extend the analysis of light even further. Fig. 2 shows an instrument with four prisms, constructed for Kirchhoff by Steinheil. At the far end of telescope *A* is a screen with a fine slit

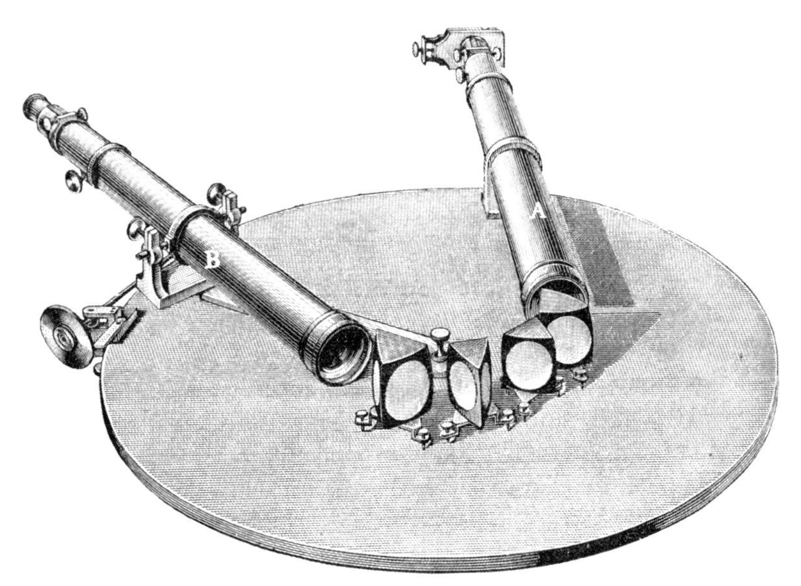

Fig. 2

which can be narrowed or widened by a small screw. The light to be investigated enters the instrument through this slit in the form of a narrow beam. It passes through telescope *A*, through the four prisms, and finally through telescope *B*, from which it reaches the eye of the observer.

Figs. 3, 4, and 5 are small sections of the solar spectrum as mapped by Kirchhoff, taken from the region of green, yellow, and golden yellow. The chemical symbols–Fe (iron), Ca (calcium), Na (sodium), and Pb (lead)–and the lines below the spectrum indicate the positions in which the vapors of these metals, when made incandescent either by heat or by an electrical discharge, show bright lines. The scale of numbers above the spectrum indicates where the lines are located in the total spectrum as mapped by Kirchhoff. Note the large number of iron lines; in the whole spectrum Kirchhoff has found no fewer than 450. It follows from this that the solar atmosphere contains large amounts of iron vapor, which among other data justifies us in concluding that an enormously high temperature must prevail there. The spectrum also shows the presence of calcium and sodium (these are indicated in Figs. 3, 4, and 5), as well as hydrogen, zinc, copper, magnesium, aluminum, barium, and other terrestrial elements. Lead, on the other hand, is absent, as are gold, silver, mercury, antimony, tin, arsenic, and certain other elements.

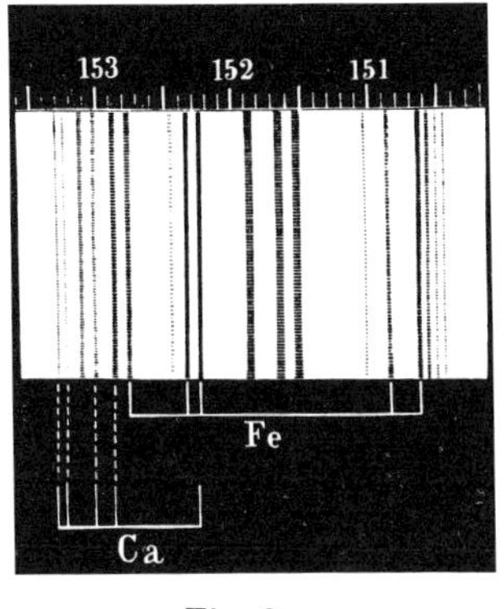

Fig. 3

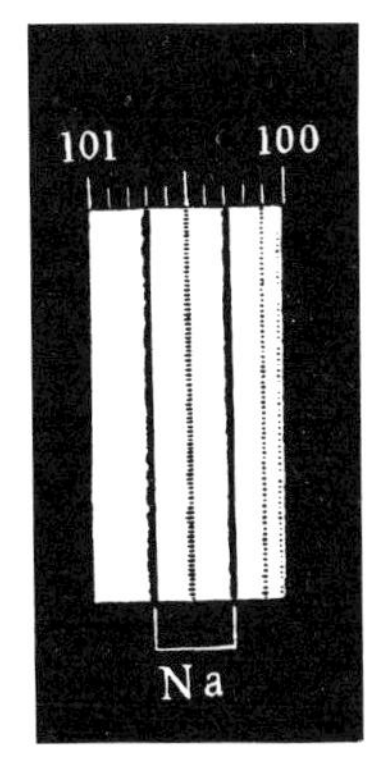

Fig. 4

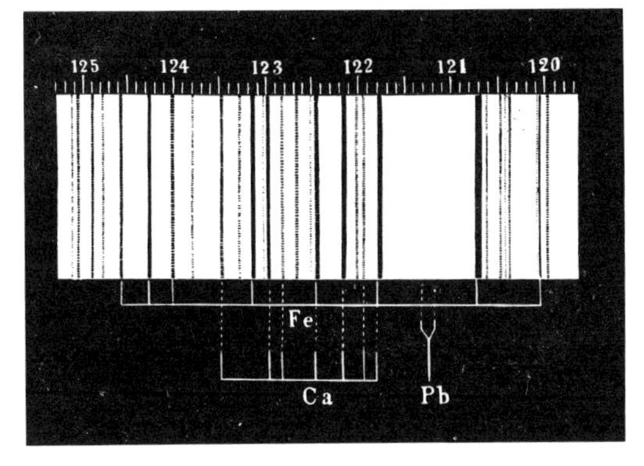

Fig. 5

The spectra of many fixed stars are similar; they show systems of fine lines which can be identified with those of terrestrial elements. In the atmosphere of Aldebaran in Taurus, for example, there are hydrogen, iron, magnesium, calcium, and sodium, as well as mercury, antimony, and bismuth; and according to H. C. Vogel, in Alpha Orionis there is the rare metal thallium.

We cannot, to be sure, say that we have explained all spectra. Many fixed stars exhibit peculiar banded spectra, which apparently belong to gases whose molecules have not been completely resolved into their atoms by the high temperature. In the spectrum of the sun, also, there are many lines which we cannot identify with those of terrestrial elements. It is possible that they are produced by the extremely high temperature of the sun, far transcending anything we can produce on earth. It is certain, however, that the known terrestrial substances are widely diffused in space—especially nitrogen, which constitutes the greater part of our atmosphere, and hydrogen, one of the elements in water. Both have been found in the irresolvable nebulae, which, judging from the inalterability of their shape, must be masses of enormous dimensions and at enormous distances. It was the very fact that they cannot be resolved with a telescope, of course, that led Sir W. Herschel to conclude that they did not belong to our system of fixed stars, the Milky Way, but were indeed other galactic systems.

Further, spectrum analysis has provided us with new information about the sun by which it has been, as it were, brought nearer to us than formerly seemed possible. You know, of course, that the sun is an enormous sphere, whose diameter is 112 times as great as that of the earth. We may consider what we see as its surface to be a layer of incandescent vapor; to judge from the appearance of sunspots, it has a depth of about five hundred miles. This layer of vapor, which is continually radiating heat into space, is certainly cooler than the inner masses of the sun. It is hotter, however, than all our terrestrial flames, hotter even than the incandescent carbon points of an electric arc, which is the highest temperature attainable by terrestrial means. This conclusion can be inferred with certainty, using Kirchhoff's law concerning radiation from opaque bodies, from the greater luminous intensity of the sun. The older assumption—that the sun is a dark, cool body, surrounded by a photosphere which radiates heat and light only externally—is a physical impossibility.

Outside of the opaque photosphere, the sun appears to be surrounded by layers of transparent gases which are hot enough to show brightly colored lines in the spectrum. These layers are therefore called the chromosphere; their spectrum shows the bright lines of hydrogen, sodium, magnesium, and iron. Enormous storms occur in these layers of gas and vapor surrounding the sun—storms which are as much greater, both in extent and in velocity, than those on our earth as the sun is greater in size than the earth. Currents of incandescent hydrogen, like gigantic jets or tongues of flame with clouds of smoke above them, burst out for several thousands of miles.[1] Formerly these currents, forming what is called the corona, could be seen only at the time of a total eclipse of the sun. We now possess a method, developed by Janssen and Lockyer, by which they may be seen at any time with the aid of the spectroscope.

1. According to H. C. Vogel's observations, to a height of 70,000 miles. The spectroscopic displacement of the lines indicate velocities of 18 to 23 miles per second; according to Lockyer, even of 37 to 42 miles per second.

In addition, there are usually separate and distinct darker places on the sun's surface, the so-called sunspots. These were first noticed by Galileo. They are funnel-shaped, the sides of the funnel not being as dark as the core or deepest part. Fig. 6 represents such a spot according to Father Secchi, as seen under powerful magnification. The diameter of a sunspot is often several thousand miles, so that two or three earths could be placed in one of them.

Fig. 6

These spots may last for weeks or months, slowly changing, before they are finally resolved, the sun in the meantime making several revolutions. Sometimes, however, there are very rapid changes in them. That the core is deeper than the edges of the surrounding penumbra is proved by their relative displacements as they come near the edge of the sun and are thus seen from an increasingly oblique angle. Fig. 7 shows the changing appearance of such a spot as it approaches the edge of the sun.

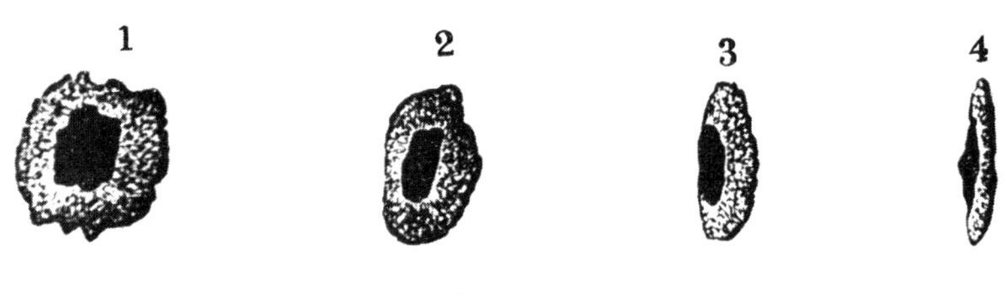

Fig. 7

There are spectroscopic indications of very violent motion just at the edge of these spots, and near them there are often large outbursts which frequently display a circular motion. These are thought to be places where the cooler gases from the outer layers of the sun's atmosphere sink down and perhaps produce temporary local cooling of the sun's mass. To understand the origin of these phenomena it must be remembered that the gases, as they rise from the hot body of the sun, are laden with vapors of volatile metals. As these vapors ascend, they expand and are cooled, partly by their expansion, partly by radiation of their heat into space. As a result of this cooling, the least volatile constituents condense and form fog or clouds. (This cooling, of course, can only be regarded as comparative; the temperature is probably still higher than any temperature attainable on earth.) If the uppermost layers, freed from the heavier vapors, sink down, there will be a place on the sun's surface that is free from vapor. This place then appears as a depression, since layers of incandescent vapor as much as five hundred miles high surround it.

Violent storms cannot fail to occur in the sun's atmosphere, for it is cooled from the outside, and thus the cooler, denser, and heavier layer of its atmosphere comes to rest on the hotter and lighter. This is the same reason, of course, why we have incessant, at times sudden and violent, movements in the earth's atmosphere: it is cooled above and warmed below by the sun's heat reflected from the ground. The sun's size and temperature being far more colossal, its meteorological processes are obviously on a far larger scale and far more violent.

We shall now turn to the question of the permanence of the present structure of our planetary system. For a long time the view was generally held that in its main features, at least, it was absolutely unchangeable. This opinion was based mainly upon the conclusions which Laplace stated as the final results of his long, laborious investigations of the effects of planetary disturbances. (As I have already mentioned, by disturbances of planetary motion, astronomers understand those deviations from purely elliptical motion caused by the attraction of the various planets and satellites upon one another.) The attraction of the sun, which is by far the largest body in our solar system, is by far the main force producing the motion of the planets. If this force alone were operative, each planet would move continuously in a non-varying ellipse, whose axes would retain the same direction and magnitude, and would make its revolution always in the same period of time. In point of fact, however, in addition to the attraction of the sun there are the forces of attraction of all of the other planets. While these are small, they nevertheless effect slow changes over long periods of time in the plane, direction, and magnitude of the axes of the elliptical orbits.

The question has been raised as to whether these changes in the planetary orbits might cause two neighboring planets to collide or cause one to fall into the sun. Laplace was able to show that this could not happen, since all the orbital changes produced by this kind of disturbance vary periodically, the orbits always returning eventually to a given state. It must not be forgotten, however, that this result of Laplace's investigations applies only to disturbances due to the reciprocal attraction of the planets and is based upon the assumption that no other forces of any kind have any influence on the motion of the planets.

Here on the earth we cannot produce an everlasting motion like that which seems to be characteristic of the planets, for resisting forces are continually opposing all movements of terrestrial bodies. The best known of these forces are friction, resistance of the air, and inelastic impact. Hence the fundamental law of mechanics, according to which a body upon which no force acts will move forever in a straight line with uniform velocity, holds only under conditions which are never fully realized on earth.

Ignoring, for example, the influence of gravity on a ball which is rolling along a plane surface, we see the ball roll for a time, the distance depending upon the smoothness of the surface. Simultaneously, however, we hear the ball make a clattering sound; that is, it produces vibrations in neighboring bodies. And there is friction on even the smoothest surface, which sets the surrounding air in vibration while at the same time detracting from the motion of the ball. Thus it happens that the velocity of the ball decreases more and more until finally it ceases to roll. In like manner, even the most carefully constructed wheel, rotating upon the finest bearings, may turn for a quarter of an hour, or even more, but then stops. There is always some friction on the axles, and there is the resistance of the air, due mainly to the friction of the air particles in the vicinity of the wheel.

If we could once set a body in rotation and keep it from falling without having it supported by another body, and if we could transfer the whole system to an absolute vacuum, it would continue to move forever with undiminished velocity. These conditions, which cannot be realized with terrestrial bodies, are apparently realized in the case of the planets and their satellites. They appear to move in the perfect vacuum of cosmic space, without contact with any body which could produce friction. Hence their motion seems to be one which never diminishes.

You can see, however, that whether this conclusion is justified or not depends upon the answers to two questions: Is cosmic space really a vacuum? And is it really true that there is no friction in the motion of the planets? As a result of the progress which has been made since the time of Laplace in the understanding of nature, we must now answer both questions in the negative.

Celestial space is not absolutely empty. In the first place, it is filled with that continuous medium whose vibrations constitute light and radiant heat; physicists call it the luminiferous ether. In the second place, fragments of tangible matter, ranging in size from huge stones to dust particles, are scattered throughout space, or at least in those parts of space through which our earth moves.

As to the luminiferous ether, its existence cannot be doubted. That light and radiant heat are wavelike movements spreading in all directions has been sufficiently established, and in order for these movements to pass through space, there must be something in space through which they are transmitted. Indeed, from the magnitude of the effects of this motion—from what is called its vis viva in the science of mechanics—we can deduce certain limits of the density of this medium. Such calculations were made by Sir W. Thomson, the celebrated Glasgow physicist. He found that though its density may be far less than that of the air in the most perfect partial vacuum obtainable by a good air pump, the mass of the ether cannot be absolutely equal to zero. A volume of ether equal to the volume of the earth cannot contain less than 2775 pounds of luminiferous ether.[2]

The phenomena in celestial space are in agreement with this. Just as a heavy stone thrown through the air shows scarcely any effect of the resistance of the air, while a light feather is appreciably retarded, so the medium which fills space is far too attenuated for any diminution to have been noticed in

2. The foundation upon which these calculations were based would collapse if Maxwell's hypothesis, according to which light depends upon electric and magnetic oscillations is confirmed.

the motion of the planets during the time in which we have been making astronomical observations of their paths. It is different, however, with the smaller bodies in our system. For example, Encke has shown that the small comet named after him travels around the sun in ever-diminishing orbits and in ever shorter periods of time. Its motion is similar to that of the circular pendulum which we have already mentioned: as the velocity of the pendulum is gradually diminished by the resistance of the air, it describes smaller and smaller circles about its center of attraction.

The reason for this phenomenon is as follows: The force which offers a resistance to the sun's attraction on all comets and planets and which prevents them from getting continually nearer to the sun is called their centrifugal force, that is, their tendency to continue in motion in a straight line along the tangent to their orbit. As this force diminishes due to resistance, the planets and comets yield by a corresponding amount to the sun's attraction and are drawn nearer to it. If the resistance continues, they will continue to get nearer to the sun until they fall into it. This is obviously the case with Encke's comet. The same resistance, of course, whose existence is indicated by this comet has long been acting and must still be acting, although much more gradually, on the far larger masses of the planets.

The presence of both large and small ponderable masses throughout all of cosmic space is indicated still more definitely by the existence of asteroids and meteorites. We now know that these are bodies which travel about in cosmic space before they come within the region of the earth's atmosphere. They are retarded in their motion in the more strongly resistant medium which this atmosphere offers, and at the same time they are heated by the resulting friction. Many of them may escape from the earth's atmosphere and continue their path through space with an altered and smaller velocity. Others fall to the earth; the larger ones come to rest as meteorites, while the smaller are probably transformed by the heat into dust and as such fall without being seen.

According to Alexander Herschel's estimate, we may think of falling stars as being on the average the same size as paving stones. Their incandescence occurs mostly in the higher, more attenuated regions of the atmosphere, eighteen miles and more above the earth's surface. Since they move in space under the influence of the same laws as the planets and comets, they possess a planetary velocity of from eighteen to forty miles per second. They are, indeed, *stelle cadenti,* falling stars, as they have long been called by poets.

The enormous velocity with which they enter our atmosphere is undoubtedly the reason for their becoming heated. As is well known, friction heats bodies that are rubbed. Every match we ignite, every badly greased coach wheel, every auger we turn in hard wood, teaches us this. The air, like solid bodies, is heated not only by friction but also by the work expended to compress it. One of the most important discoveries of modern physics, the empirical proof of which is due mainly to the Englishman Joule, is that the heat generated because of friction is directly proportional to the mechanical work expended. If, like the students of mechanics, we define work as the product of the weight necessary to produce it multiplied by the height from which the weight must fall, Joule showed that the work produced by a given weight of water falling through a height of 425 meters is just sufficient to raise the temperature of the same weight of water 1° C.

The equivalent in work of a velocity of eighteen to twenty-four miles per second may easily be calculated given well-known mechanical laws. This work, transformed into heat, would be sufficient to raise the temperature of a piece of meteoric iron from 900,000° to 2,500,000° C., provided that all the heat were retained by the iron and did not (as it undoubtedly does) mainly dissipate into the atmosphere. The temperatures attainable by terrestrial means barely exceed 2,000°. These calculations show, at any rate, that the velocity of falling stars is sufficient to raise them to the most brilliant incandescence. In fact, the outer crusts of meteoric stones generally show traces of incipient fusion; and when observers have examined the stones quickly enough, they have found them hot on the surface, while the interior of those that were split seemed to show the intense cold of cosmic space.

To the observer who looks only casually at the starry skies, meteorites appear to be a rare and exceptional phenomenon. If one observes more systematically, however, they are seen with fair regularity, especially toward morning. A single observer can look at only a small part of the sky, but calculations made for the entire surface of the earth indicate that about 7,500,000 meteorites fall every day. In our regions of space they are somewhat scarce and distant from one another: according to Alexander Herschel's estimate, each stone is on the average a distance of 450 miles from its neighbors. The earth, however, travels eighteen miles every second and has a diameter of 7820 miles; it therefore sweeps through 876,000,000 cubic miles of space every second and carries with it whatever stones are contained therein.

Many groups of meteorites are irregularly distributed in space and are probably those which have already been disturbed by the moving planets.

There are also denser swarms of meteorites, however, which travel in regular elliptical paths, cutting the earth's orbit in definite places and therefore on specific days of the year. Thus August 10 is noteworthy each year, and every thirty-three years the splendid fireworks of November 12–14 are repeated for a few years. It is remarkable that certain comets accompany the paths of these meteors, giving rise to the conjecture that the comets gradually split up into large numbers of smaller bodies.

This is an important process. What occurs with respect to the earth undoubtedly occurs also with the other planets and, to a far higher degree, with the sun, toward which all the smaller bodies enmeshed in our system must move, the smaller faster than the larger. The earth and the planets have been sweeping together the loose masses in space for millions of years, and they hold fast what they have once attracted. It follows that the earth and the planets were once smaller than they are now and that more bodies were once diffused in space. If we pursue this line of thought, we are led back to a time when perhaps all the material now accumulated in the sun and the planets wandered loosely diffused in space. If we consider, further, that the small masses of meteorites which now fall have perhaps been formed by the gradual aggregation of fine dust, we see ourselves led to a primeval state of fine nebulous masses.

From this point of view the fall of shooting stars and of meteorites assumes a far greater significance, for these may be small survivals of the process by which our world was once formed. This would be a hypothesis to which we might grant some possibility but for which we could not claim any great degree of probability, if we did not find that our predecessors, starting from quite different considerations, had arrived at the same hypothesis.

You know that a fairly large number of planets revolve around the sun. Besides the eight major planets (Mercury, Venus, Earth, Mars, Jupiter, Saturn, Uranus, and Neptune) there are, as far as we now know, 156 small planets or planetoids in the region between Mars and Jupiter. Moons also revolve around the larger planets—that is, around the earth and the four most distant planets, Jupiter, Saturn, Uranus, and Neptune. And the sun and at least the larger planets rotate about their own axes.

It is remarkable that all the planes of rotation of the planets and their satellites, as well as the equatorial planes of these planets, vary little from one another and that in these planes all motion is in the same direction. The only major exceptions to this rule that we know of are the moons of Uranus, whose orbital plane is almost at right angles to the planes of the larger planets. It must also be stressed that the similarity in the direction of these planes is on the whole greater, the larger the bodies and the longer the orbital path, while for the smaller bodies and the shorter paths, especially for the rotations of the planets about their own axes, there are more marked variations. Thus the orbital planes of all the planets, with the exception of Mercury and the small ones between Mars and Jupiter, differ at most by 3° (Venus) from the path of the earth. The equatorial plane of the sun deviates by only 7.5°, that of Jupiter by only half as much. The equatorial plane of the earth deviates, it is true, to the extent of 23.5° from its orbital plane, that of Mars by 28.5°; and the separate paths of the small planets and satellites differ still more. In these paths, however, they all move directly, that is, in the same direction as the earth around the sun; and as far as we can determine, they also move around their own axes from west to east, like the earth.

If the planets had originated independently of one another and had just come together accidentally, any orientation of their individual planes would have been equally probable. The reverse direction of revolution would have been just as probable as the direct, and very elliptical paths would have been as probable as the almost circular ones we meet with in the case of all of the bodies mentioned. There is, to be sure, a complete randomness in the movements of the comets and the swarms of meteors, but we have many reasons for believing that these have entered only accidentally into the sphere of attraction of the sun.

The number of coincidences in the behavior of the planets and their satellites is too great to be ascribed to chance. We must search for the causes of these coincidences, and these causes can only be found in the original state or condition of the entire system. We are acquainted with forces and processes which can condense an originally diffuse mass, but we know of none that can force such large masses as the planets as far as into space as the orbits in which we now find them. Furthermore, the planets should have more markedly elliptical orbits if they had been closer to the sun when they were first detached from the common mass. We must assume, therefore, that in its primeval state this mass extended at least to the orbit of the outermost planet.

These are the basic points the consideration of which led Kant and Laplace to their hypothesis. In their view, our solar system was originally a ball of nebulous matter existing in a chaotic state. Originally, when this ball extended to the orbit of the most distant planet, many billions of cubic miles of space contained scarcely a gram of matter. When the ball became detached from the nebulous masses of the adjacent fixed stars, it had a slow rotational motion. It condensed under the influence of the mutual attraction of its parts, and as it condensed, the rotational motion increased, causing the ball to change into a flat disc. From time to time, masses at the circumference of this disc were detached due to the influence of the increasing centrifugal force. These took the form of rotating nebulous masses, which either simply condensed to form a planet or, during condensation, spun off other masses from their periphery, which in turn became satellites or in one case (that of Saturn) continued as a stable ring. In still another case, the mass which separated from the outside of the main mass split into many detached parts, which eventually came to form the swarms of small planets between Mars and Jupiter.

Our more recent observations of showers of falling stars indicate that this process of condensation of loosely diffused masses to form larger bodies is by no means complete. It is still going on, although on a much smaller scale. The form which this condensation takes has changed, too, due to the fact that the gaseous or dustlike masses diffused in space have, under the influence of the forces of attraction and crystallization, come to consist of larger pieces than those which originally existed.

The showers of falling stars, as examples now occurring of the processes which formed the heavenly bodies, are important from another point of view. They produce light and heat—a fact that directs our attention to a third series of considerations which lead again toward the same general conclusions.

All life and all motion on our earth are, with a few exceptions, maintained by a single force, that of the sun's rays, which bring us light and heat. These rays warm the air of the more temperate zones, which becomes lighter and rises, being replaced by the colder air flowing from the poles. In this way the great circuit of air known as the trade winds is formed. Local differences of temperatures over land and sea, over plains and mountains, produce various disturbances in these great movements of air, giving rise to capricious changes in the winds. Warm water vapors rise with the warm air, are condensed into clouds, and fall as rain and snow in the cooler zones and upon the snow-covered tops of mountains. The water collects in streams and rivers, irrigates the plains and makes life possible; it also breaks up rocks and carries their fragments along, thus causing geological transformations of the earth's surface. Only under the influence of the sun's rays do the various plants which cover the earth grow; and as they grow, they store up organic materials which serve the whole animal kingdom for food and man for fuel. Even coal and lignite, the sources of power of our steam engines, are the remains of primeval plants, the ancient product of the sun's rays.

Need we wonder if, to our forefathers of the Aryan race in India and Persia, the sun appeared the most appropriate symbol for the Deity? They were right in regarding it as the giver of all life, as the ultimate source of all that has happened on earth.

But where does the sun acquire this force? It radiates a light far more intense than any which can be produced by any terrestrial means. It generates as much heat as if 1500 pounds of coal were being burned every hour upon every square foot of its surface. Of the heat which issues from it, the small fraction which enters our atmosphere does a tremendous amount of mechanical work. The fact that heat can do work is known from the operation of the steam engine, and the sun drives a kind of steam engine here on the earth, the output of which is far greater than that of any man-made machine. As has already been said, the water circulation in the atmosphere raises the water evaporated from the warm tropical seas to the tops of mountains; this is in effect a water-pumping operation of the most powerful kind, one with which the work done by an artificial machine cannot be even remotely compared. I have previously explained the mechanical equivalence of heat. Computed by that standard, the work which the sun does by its radiation is equal to the continuous expenditure of seven thousand horsepower for every square foot of the sun's surface.

For a long time, experience had impressed upon scientists and technicians the fact that a motive force cannot be produced out of nothing. Whether it be obtained from rushing water, or from the wind, or from the beds of coal, or from men and animals (which cannot work without consuming food), it can be obtained only from the stores which nature possesses—stores which are strictly limited and cannot be increased at will. Modern physics has attempted to prove the universality of these observations, to show that they apply to the totality of natural processes and are independent of the special interests of man. They have been generalized and given expression in the all-encompassing law of the conservation of force.

According to this law, there is no single natural process and no series of natural processes, however sweeping the changes which take place in them, by which a motive force can be continuously produced without some corresponding consumption of force. Just as the human race finds on earth only a limited supply of motive force capable of doing work, a supply which can be utilized but cannot be increased, so must this be the case too in the totality of nature. The universe has its definite store of force, which works within it under ever-varying forms; it is indestructible, not to be increased, everlasting, and unchangeable, like matter itself. It is almost as if Goethe had an intuition of this when he made the Earth Spirit speak of himself as the agent of the forces of nature:

> In the tides of Life, in Action's storm,
> A fluctuant wave,
> A shuttle free,
> Birth and the Grave,
> An eternal sea,
> A weaving, flowing
> Life, all-glowing,
> Thus at Time's humming loom 't is my hand prepares
> The garment of Life which the Deity wears![3]

> *In Lebensfluten, im Tatensturm*
> *Wall ich auf und ab,*
> *Webe hin und her,*
> *Geburt und Grab,*
> *Ein ewiges Meer,*
> *Ein wechselnd Weben,*
> *Ein glühend Leben:*
> *So schaff ich am sausenden Webstuhl der Zeit*
> *Und wirke der Gottheit lebendiges Kleid.*

Let us return to the special question which concerns us here: Whence does the sun derive this enormous store of force which it pours forth?

On earth the processes of combustion are the richest and main source of heat. Does the sun's heat, perhaps, originate in a process of this kind? This question can be answered completely and decidedly in the negative. We now know that the sun contains the terrestrial elements with which we are familiar. Let us select from among these elements the two which, for the smallest mass, produce the greatest amount of heat when they combine: let us assume that the sun consists of hydrogen and oxygen, mixed in the proportion in which they would unite to form water. The mass of the sun is known and also the quantity of heat produced by the union of known weights of oxygen and hydrogen. Calculations show that under the assumptions just mentioned, the heat resulting from the combustion of hydrogen and oxygen would be sufficient to sustain the radiation of heat from the sun for 3,021 years. That is, to be sure, a long period of time, but even the history of man teaches us that the sun has lighted and warmed us for much longer than 3,000 years, and geology puts it beyond doubt that this period must be extended to millions of years.

---

3. *Faust, Part I.* Translated by Bayard Taylor.

Known chemical forces are so completely inadequate, even on the most favorable assumption, to explain the production of heat which takes place in the sun that we must surely drop this hypothesis.

We must search for forces of far greater magnitude—and these can be found only in cosmic forces of attraction. We have already seen that the comparatively small masses of falling stars and meteorites can produce extraordinarily large amounts of heat when their cosmic velocities are arrested by our atmosphere. The force which produces these great velocities is, of course, gravitation. We are familiar with this force as one acting on the surface of our planet when it appears as weight. We know that a weight raised above the earth can drive our clocks, and in a similar fashion the weight of the water rushing down from the mountains drives our mills.

If a body falls from a height and strikes the ground, it loses the visible motion which it had as a whole. In fact, however, this motion is not lost; it is transferred to the smaller elementary particles of the body, and this invisible vibration of the molecules which make up the body is the motion of heat. Visible motion is changed by impact into the motion of heat.

That which holds true in this respect for weight, holds also for gravitation. A heavy mass of any kind which is separated in space from another heavy mass represents a force capable of doing work. Both masses attract each other, and if unrestrained by centrifugal force, they will move toward one another under the influence of this attraction with ever-increasing velocity. If the velocity is finally destroyed, whether suddenly by collision or gradually by the friction of moving parts, the corresponding quantity of the motion of heat will be produced. The precise amount of heat produced can be calculated from the equivalence (discussed above) between heat and mechanical work.

We may assume that in all probability a great many more meteors fall upon the sun than upon the earth and that they travel with greater velocity and therefore produce more heat. Yet the hypothesis that all the heat which the sun is continually emitting in radiation is due to the fall of meteors—a hypothesis which was propounded by Robert Mayer and which has been favorably received by several other physicists—runs into serious difficulties. According to the investigations of Sir William Thomson, if this hypothesis were true, the sun's mass would increase so rapidly that the effects of this increase would be evident in the acceleration of the planets. The entire emission of heat from the sun cannot be explained in this way; at the most the hypothesis can account for only part of it, although perhaps a not inconsiderable part.

If, then, there is no known existing process of generating energy sufficient to account for the production of the sun's heat, the sun must originally have had a store of heat which it gradually emits. But whence this store? We know that only cosmic forces could have produced it, and here the hypothesis concerning the sun's origin which we have already discussed comes to our aid. If the mass of the sun was once diffused in cosmic space and then condensed—that is, came together under the influence of celestial gravity—and if the gravitational motion was destroyed by friction and impact, with the production of equivalent heat, then the new body produced by such condensation must have acquired a store of heat, not only of considerable, but even of colossal magnitude.

Calculations show that, taking the thermal capacity of the sun to be the same as that of water, the sun's temperature could have been raised by condensation to 28,000,000°, provided of course that none of the heat was lost or dissipated away. This surely did not occur, however, for without some heat loss an increase in temperature would have offered the greatest resistance to condensation. It is much more probable that a significant part of the heat produced by the condensation began, after a time, to radiate into space. If the present line of thought is correct, further calculations show that the heat which the sun produced by its condensation is sufficient to account for an emission of heat at the present rate for a period of not less than 22,000,000 years.

The sun is by no means as dense as it can still become. Spectrum analysis shows the presence of large masses of iron and other known constituents of minerals in the sun. The pressure compressing the material in the interior of the sun is about eight hundred times as great as the pressure at the center of the earth, yet the density of the sun, owing probably to its enormous temperature, is less than a quarter of the mean density of the earth. We may therefore assume with a high degree of confidence that the sun will continue its condensation. Even if it only attains the density of the earth—in all probability it will become far denser in the interior, due to the enormous pressure—this will mean the development of additional heat sufficient to maintain the same intensity of sunshine as that which is now the source of all terrestrial life for an additional 17,000,000 years.

The smaller bodies of our solar system are less able to produce heat than the sun, for their force of attraction for new materials is weaker. A body like the earth could, if we assume its thermal capacity to be as high as that of water, reach a temperature of only 9000° in its interior, which is, to be sure, a higher temperature than that of a fire. The smaller bodies would also cool more rapidly, at least as long as they are still in a fluid state. The increase in temperature at greater depths in bores and mines, like the existence of hot wells and volcanic eruptions, show that in the interior of the earth the temperature is very high. This heat can hardly be anything but a residual effect of the high temperature which prevailed at the time the earth was formed. At least, attempts to account for the earth's internal heat as being of more recent origin and due to chemical processes have so far rested on very arbitrary assumptions and are insufficient to account for the general, uniform distribution of the internal heat.

Like the sun, the larger planets–Jupiter, Saturn, Uranus, and Neptune–have relatively small densities, while the smaller planets and our moon have approximately the same density as the earth. We are reminded again of the higher initial temperature and the slower rate of cooling which characterize larger masses. The moon, on the one hand, exhibits formations on its surface which are strikingly suggestive of volcanic craters, formations which point to a former condition of incandescent heat. Its mode of rotation, however, by which the same side is always kept toward the earth, is a peculiarity which might have been caused by the friction within a fluid. There is no longer, of course, any trace of a fluid on its surface.

You can see by what different paths we are continually being led to the same primeval conditions. The hypothesis of Kant and Laplace proves to be one of the happiest ideas in science, one which at first astonishes us by its boldness but which is then seen to be interrelated on every side with other discoveries which tend to confirm it. To add to this general confirmation, there is still another point to be mentioned–the observation that the process of transformation which the theory presupposes is still going on, although on a smaller scale. All the states of the process, in fact, can still be observed.

As we saw earlier, the larger bodies which are already formed continue to increase in size by attracting the meteoric masses diffused in space, developing heat in the process. Even now small bodies are slowly being drawn toward the sun against the various resistances to be met with in space. According to Sir John Herschel's newest catalogue, there are over five thousand nebulae in the region of the fixed stars. Those whose light is sufficiently strong display a colored spectrum of fine, bright lines, similar to the spectrum of incandescent gases. The nebulae are sometimes round, the so-called planetary nebulae (Fig. 8); sometimes they are completely irregular in form, as for example the large nebula in Orion (Fig. 9); and sometimes they are partly annular, as the

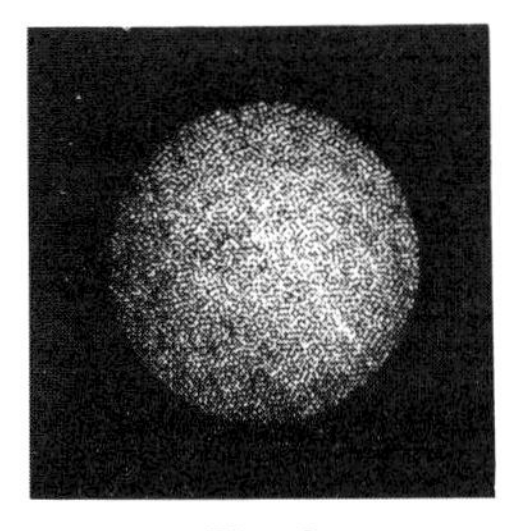

Fig. 8

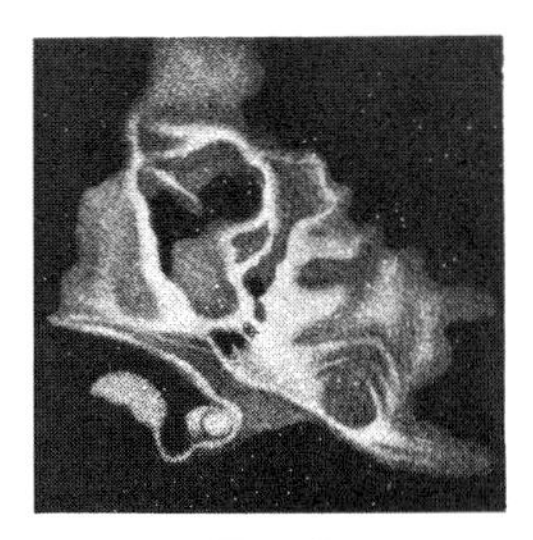

Fig. 9

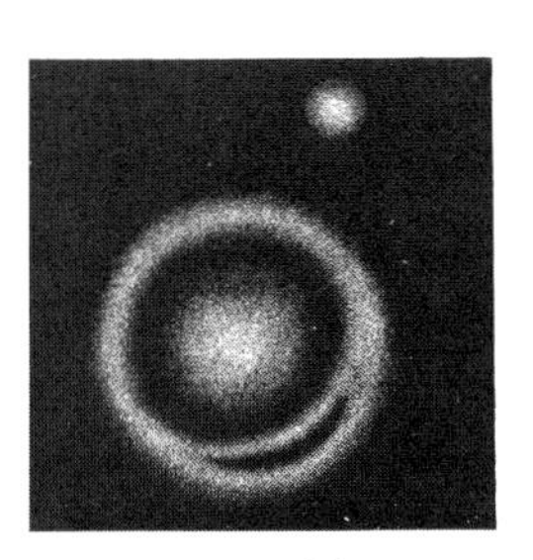

Fig. 10

one in Canes Venatici (Fig. 10). For the most part they are feebly luminous over their entire surface, in contrast to the fixed stars, which appear only as luminous points.

In many nebulae small stars can be distinguished, such as those in Sagittarius and Aurigo (Figs. 11 and 12). As better telescopes come to be used in analyzing the nebulae, more stars are continually being discovered in them. Thus, before the discovery of spectrum analysis, Sir William Herschel's early suggestion that the nebulae we see are made up only of multitudes of

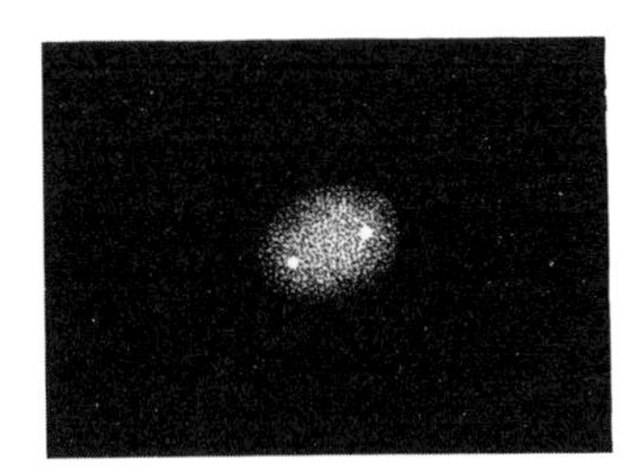

Fig. 11

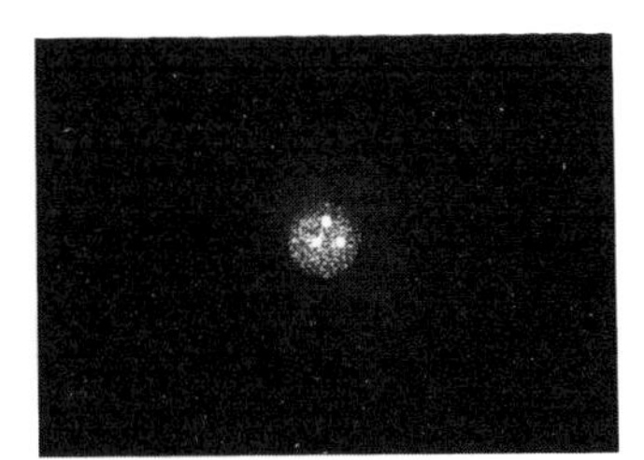

Fig. 12

very fine stars—that they are in reality other Milky Ways—might have been regarded as the most probable. Spectrum analysis, however, has shown gas spectra for many nebulae containing distinct stars, whereas groups of stars alone would show only the continuous spectra of incandescent solids.

Nebulae in general have three distinct, recognizable lines: one (in the blue) belongs to hydrogen, the second (in bluish green) to nitrogen,[4] while the

4. Perhaps also to oxygen. The line occurs in the spectrum of atmospheric air, and according to H. C. Vogel's observations, it is absent in the spectrum of pure oxygen.

third (between the other two) is of unknown origin. Fig. 13 shows such a spectrum of a small but bright nebula in the Dragon. Traces of other bright lines can also be seen along with these three, and sometimes (as in Fig. 13) traces of a continuous spectrum too. These latter are always too weak, however, to admit of precise investigation. It should also be noted that in the very feeble light of some celestial objects which give off a continuous spectrum, when this light is spread by a spectroscope over a large surface and thus greatly weakened or even extinguished, the bright gas lines remain unaltered and hence can still be seen. In any case, analysis of the light of the nebulae shows that by far the greater part of the light from their surface is due to incandescent gases, of which hydrogen is the main one.

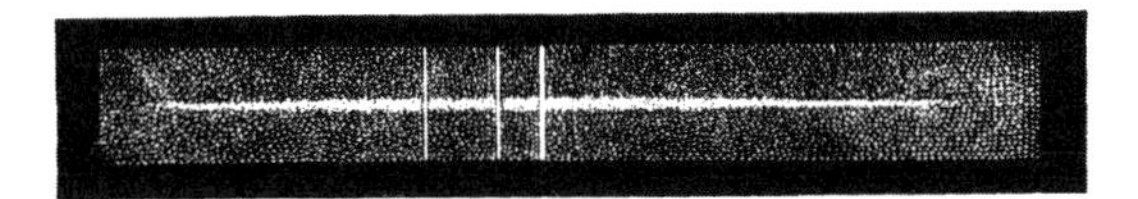

Fig. 13

In the case of the planetary nebulae, both the spherical and the disc-shaped, it may be assumed that the gaseous mass has reached a condition of equilibrium. Most other nebulae, however, exhibit highly irregular forms, which by no means resemble states of equilibrium. Since their shape has changed not at all, or at least not appreciably, since they have been known and studied, they must either have very little mass, or they must be of colossal size and at great distances. The first alternative does not seem very probable, because small masses very quickly give off their store of heat. We are left with the second alternative—that they are of huge dimensions and at great distances. The same conclusion has already been drawn by Sir William Herschel, on the assumption that the nebulae are groups of stars.

Those nebulae which, besides the lines of gases, show the continuous spectra of denser incandescent bodies, also show spots which are sometimes irresolvable and sometimes resolvable into groups of stars which display only continuous spectra.

The countless bright stars of the heavenly firmament, whose number increases with each new and more perfect telescope, have all passed through these primeval stages in the formation of worlds. These stars are like our sun in magnitude, in luminosity, and on the whole in the chemical composition of their surface, although there may be differences in the quantities of the individual elements.

We also find in the universe another stage, that of extinct suns—a phenomenon for which there is factual evidence. First of all, in the course of history there have been fairly frequent examples of the appearance of new stars. In 1572, Tycho Brahe observed one that stood still like a fixed star; although it gradually got paler and finally reverted to the darkness from which it had so suddenly emerged, it was visible for two years. The largest seems to have been that observed by Kepler in 1604; it was brighter than a star of the first magnitude and was observed from September 27, 1604, to March 1606. Perhaps the cause of its brilliance was a collision with a smaller body. In a more recent case, in which on May 12, 1866, a small star of the tenth magnitude in the Corona quickly increased in brightness to the second magnitude, spectrum analysis showed that it was a burst of incandescent hydrogen which produced the light. This star was brighter than normal for only twelve days.

In other cases the presence of dark heavenly bodies has become evident due to the attraction they exert on neighboring bright stars, an attraction which produces a characteristic motion in the latter. Such a motion has been observed in the case of both Sirius and Procyon. By means of a new refracting telescope Alvan Clark and Pond, of Cambridgeport, Massachusetts, discovered the companion of Sirius. It is scarcely visible, having very little luminosity, yet it is almost seven times as heavy as the sun, having about half of the mass of Sirius. Its distance from Sirius is about equal to that of Neptune from the sun. The companion of Procyon has not yet been seen; it is apparently quite dark.

Thus there are extinct suns. This fact lends new weight to the reasoning which leads us to conclude that our sun, too, is slowly giving out its store of heat and will someday become extinct. The period of 17,000,000 years, which I have already mentioned, might be considerably extended, due to a gradual reduction in radiation, to the further accretion of falling meteors, and to the possibility of still greater condensation than that which I assumed in my calculations. We know of no natural process, however, which can spare our sun the fate which has manifestly fallen upon other suns.

This is a thought which we have difficulty in admitting to ourselves. It seems to us a flaw in the beneficent creative force which otherwise we find at work everywhere, especially in living beings. Still, we must learn to reconcile ourselves to the thought that, however we may feel ourselves to be the center and the final purpose of creation, we are like dust on the earth, which is itself only a speck of dust in the immensity of space. The whole past history of our race, even if we trace it back far beyond written history and into the era of the lake dwellings or of the mammoth, is still an instant compared with the primeval times of our planet, when the living beings whose strange, unearthly remains still gaze at us from their ancient tombs existed upon it. Far more does the duration of our race sink into insignificance when we compare it with the enormous periods during which whole worlds were formed in the past—and during which they will continue to be formed when our sun is extinguished and our earth is either completely solidified with cold or united with the fiery central body of our planetary system.

But who can say whether the first living inhabitants of the warm sea on the young earth, whom we ought perhaps to honor as our ancestors, would not have regarded the present cooler state of the earth with as much horror as we feel in thinking about a world without a sun? Considering the wonderful adaptability to the conditions of life which all organisms possess, who can say to what degree of perfection our posterity will have developed in 17,000,000 years; and whether our fossilized bones will not perhaps seem to them as monstrous as those of the ichthyosaurus seem to us; and whether they, adjusted to a more sensitive stage of equilibrium, will not consider the extremes of temperature within which we now live to be just as violent and destructive as those of the older geological periods seem to us?

Indeed, even if sun and earth should solidify and cease to move, who can say what new worlds will not be ready to develop life? Meteoric stones sometimes contain hydrocarbons, and the light from the heads of comets exhibits a spectrum which is very much like that of electric discharges through gases containing hydrogen and carbon. Carbon, of course, is the element which is characteristic of the organic compounds which go to constitute living bodies. Who can say whether the comets and meteors, which swarm everywhere throughout space, may not scatter germs of life wherever a new world has reached the stage in which it is a suitable dwelling place for organic beings? We might, perhaps, consider such life to be allied to ours, at least in germ, however different the form it might assume in adapting itself to its new dwelling place.

However this may be, that which most arouses our moral feelings at the thought of a future (though perhaps very remote) cessation in the generation of living things upon the earth is surely the question whether life itself is a purposeless game, one which will ultimately fall prey to destruction by natural forces. Yet we are beginning to see, given the light of Darwin's great theory, that not only pleasure and joy but also pain, struggle, and death are the powerful means by which nature has built up her finer, more perfect forms of life. And we, as men, know that in our knowledge, in our civic order and morality, we are living on the heritage which our forefathers gained for us through work, struggle, and sacrifice. We know that what we acquire in the same way will, in the same manner, ennoble the lives of our posterity. The individual who works for the ideal goals of humanity, even if in a modest position and in a limited sphere, may bear without fear the thought that the thread of his own consciousness will one day break. This is true, even though men of such free and great minds as Lessing and David Strauss found it impossible to reconcile themselves to the thought of a final destruction of the living race and, with it, of all the fruits of all past generations.

Hitherto we knew of no fact, established by scientific observation, which showed that the delicate, complex ingredients of living beings could exist except as the materials of organic bodies—that they could propagate themselves like the sound wave of a violin string, which leaves its original confined and stationary home and spreads out in the sea of air, always retaining its pitch and the most delicate shading of its overtones, and, when it meets another string attuned to it, is absorbed in it or excites a flame ready to sing to the same tone. A flame, too, which of all of the processes in inanimate nature is the most similar to life, may become extinct, but the heat which it produces continues to exist—indestructible, imperishable—as an invisible motion, sometimes moving the molecules of ponderable matter and sometimes radiating into boundless space as vibrations in the ether. Even there it will retain the characteristics peculiar to its origin and reveal its history to the investigator who questions it with the aid of the spectroscope. United again, these rays may ignite a new flame and thus, as it were, acquire a new bodily existence.

Just as the flame remains the same in appearance and continues to exist with the same form and structure, although it continues to draw fresh combustible gases and fresh oxygen into the vortex of its ascending currents, and just as the wave goes on in unaltered form and yet is reconstructed every moment from fresh particles of water, so also in the living being it is not the definite mass of substance making up the body at any one moment that constitutes the individual. For the material of the body, like that of the flame, is subject to continuous, comparatively rapid changes—changes which are the more rapid, the livelier the activity of the organs in question. Some constituents of the body are renewed from day to day, some from month to month, and others only after years. That which continues to exist as a particular individual is, like the flame and the wave, only the form of motion which continually attracts fresh matter into its vortex and expels the old. The observer with a deaf ear recognizes the vibrations of sound only as long as they are sensible and can be felt, bound up with tangible matter. Is our understanding like the deaf ear in this respect, with reference to life?

10.

THE RELATION OF OPTICS TO PAINTING [1871]

The substance of a series of lectures delivered in Cologne, Düsseldorf, Berlin, and Bonn, 1871

I am afraid that the announcement of my intention to speak on the subject of the art of painting may have caused no little surprise among some of my audience. Indeed, I am sure that many of you have had much more frequent opportunity to look at works of art than I have had, and have made a more thorough study of its historical aspects than I can rightly claim to have done. Some of you, too, have had personal experience in the actual practice of art, experience which I am completely lacking. I have come to my studies of art by a path which is rarely taken, that is, by way of the physiology of the senses. Compared with those who have had a long acquaintance with, and are quite at home in, the beautiful fields of art, I think of myself as a traveler who has entered upon them by a steep, stony mountain path, but one who, in doing so, has passed many a point from which a good view may be obtained. As I discuss with you now what I believe I have learned, however, I do so only with the understanding that I regard myself always as open to instruction from those more experienced than myself.

The physiological study of the ways in which our sense perceptions originate—of how external impressions are transmitted along our nerves and how the state, or condition, of these nerves is thereby altered—presents many points of contact with the study of the fine arts. On an earlier occasion I endeavored to establish such relations between the physiology of the sense of hearing and the theory of music. In that art the relations between physiology and theory are particularly clear and quite specific, for the basic elements of musical form depend more completely upon the nature and peculiarities of our sensations than is the case in any of the other arts, where the nature of the materials used and of the objects represented also has considerable importance. Even in the other branches of art, however, the peculiar mode of operation of the organs of sense by which impressions are received is not without significance. A theoretical investigation into an art and into the principles by which it achieves its effects would be incomplete if the physiological aspect were not taken into account. Next to music, this seems to be particularly true of painting, and that is why I have chosen painting as the subject of my present lecture.

The painter's main aim is to produce, by the use of his palette of colors, a vivid visual impression of the objects he wishes to represent. His aim, in a certain sense, is to produce a kind of optical illusion—not that, like the birds who pecked at the painted grapes of Apelles, we are to believe that we have the real objects themselves before us and not a painting, but that the artistic representations produce a conception of their subjects as vivid and powerful as the subjects themselves would if they were actually before us. As you know, the investigation of what are called illusions of the senses is a very large and important part of the study of the physiology of the senses. Cases in which external impressions evoke ideas in us which do not correspond to reality are particularly instructive when one is trying to determine the laws of the processes and means by which normal perceptions occur.

We should look upon painters as individuals whose awareness of sense impressions is unusually vivid and exact and whose memory of such impressions is unusually accurate and true. What a long tradition has handed down to the men most gifted in this art and what they have discovered themselves by incessant experimentation concerning various means and techniques of representation—all this forms a body of very important information which the physiologist, who here has much to learn from the artist, cannot afford to neglect. The study of works of art throws great light upon the question: which elements and relations among our visual impressions are most important in determining our conceptions of what we see, and which are less important? In so far as it lies within his power an artist tries, of course, to emphasize the former at the expense of the latter.

In this sense, then, a careful examination of the works of the great masters is as useful to physiological optics as, conversely, the investigation of the laws of sense impressions and perceptions is to the theory of art—that is, to the study of its techniques and effects. We are concerned here, not with the ultimate aims and purposes of art, but only with analyzing the elementary materials and techniques with which the artist works. It is self-evident, however, that a knowledge of these materials and techniques provides an indispensable foundation for understanding the more fundamental issues, if one happens to be interested in comprehending fully the problems which an artist must solve and the ways in which he attempts to attain his goals.

I need not emphasize the fact, following as it does from what I have already said, that it is not my intention to furnish instructions for artists to follow in their work. I consider it a mistake to believe that any aesthetic analysis can ever do this—a mistake which is very frequently made, however, by those who have only practical interests and goals.

I. FORM

An artist endeavors, in a painting, to present a representation of external objects. The first goal of our analysis will be to try to determine just what kind and degree of similarity it is possible for him to achieve between a painting and these objects, and what limits are imposed upon him by the very techniques or methods he uses.

An untrained observer usually desires only a very close reproduction of nature in a painting; the more this is achieved, the more he delights in it. An observer whose taste in works of art has been more finely educated expects, either consciously or unconsciously, something more and something different. He regards a faithful copy of nature as at most an artistic tour de force. In order to satisfy him there must be an artistic selection, grouping, and even idealization of the subjects represented. The human figures must not be common or ordinary, such as we see in photographs; they must have a characteristic, expressive development and, if possible, beauty of form, so as to produce a vivid, undistracting impression of some particular aspect of human existence. Even though a painting may present idealized types, however, must it not still convey at least the true forms of the objects represented?

Since paintings are done on plane surfaces, it is obvious that the "true forms" can at best be only faithful perspective images of the objects. Our eyes—which in their optical properties are similar to the well-known apparatus of the photographer, the camera obscura—produce on their retinas, which are their sensitive plates, only perspective images of the external world. These retinal images, like the general design of a painting, remain stationary as long as the position of the eyes is not changed. In fact, if we disregard color for a moment and restrict ourselves to the *forms* of the objects perceived, a well-executed perspective drawing presents to the eyes of an observer the same visual image that the objects themselves would present if they were viewed from the same position.

Disregarding the fact that every movement which changes the position of an observer's eyes produces changes in his retinal images (whether of real objects or of objects in a painting), it is only in the case of vision with one eye that the impressions of objects and the impressions of a painting of those objects may be the same. We observe the world, however, with two eyes, which occupy slightly different positions in space and thus provide us with somewhat different perspective images of the objects around us. The difference between the retinal images of the two eyes is one of our most important means of judging the distances of objects and their extension in depth. But this, of course, is precisely what is lacking when we look at a painting. Indeed, the absence of two different retinal images produces the opposite effect, for in binocular vision we unquestionably perceive a painting as a flat surface.

You are all familiar with the striking sharpness and clarity of the forms of objects when two good stereoscopic pictures of them are placed in a stereoscope. It is a kind of clarity which neither picture displays when studied alone without the instrument. The illusion is most striking and instructive in the case of simple linear figures, such as models of crystals, where no other kind of illusion is possible. The reason for the illusion we experience with a stereoscope is simply that when we look with two eyes, we see the world simultaneously from two somewhat different positions and consequently receive different perspective images of it. With the right eye we see a little more of the right side of an object and a little more of anything behind it on that side than we do with the left eye; and conversely, with the left eye we see more of the left side of an object and of anything behind and partially concealed by its left side. A flat painting, however, provides the right eye with exactly the same images of all of the objects represented in it as it does the left eye.

If we prepare a separate picture for each eye just like the one each sees when looking at an object, and if the pictures are placed in a stereoscope so that each eye sees only its appropriate picture, then insofar as form is concerned, the same impression will be produced as would be produced by the object itself. If, on the other hand, we look with both eyes at a drawing or a picture, we are quickly aware that we are seeing a representation on a plane surface, which is quite a different experience from seeing an actual object with both eyes. This is also the reason for the familiar increase in the vividness of a picture when it is viewed—while remaining absolutely stationary—with only one eye through a dark tube, for by this means we make it impossible to compare the distance of the picture with the distance of other objects in the room.

It should also be noticed that just as we use the different perspective images of the two eyes to perceive depth, so the successive retinal images of each eye as we move from place to place serve the same purpose. In moving, whether on foot or riding, objects closer to us are clearly displaced relative to more distant ones: the former appear to move backwards, the latter to move along with us. In this way there arises a far sharper distinction between what is near and what is distant than could be obtained by looking with only one eye from an unchanging position. If, on the other hand, we move relative to a picture, the fact that it is a flat surface hanging against a wall forces itself more strongly upon us than it does when we look at it while remaining stationary.

In the case of large pictures at a considerable distance, all the effects which depend upon binocular vision and upon the movement of the body are less marked, because with distant objects the differences between the retinal images of the two eyes and between the retinal images from neighboring points of view are less marked. Hence large pictures provide a less distorted impression of their subjects than small ones, even though the retinal image produced on a stationary eye by a small picture close at hand may be just the same as that produced by a large one at a distance. The difference is that, with respect to the painting close at hand, the fact that it is a flat surface is forced upon us clearly, powerfully, and continuously.

The fact that a perspective drawing which is made from a point of view too close to an object may easily produce a distorted impression of it is, I believe, related to what has just been said. In this case, the lack of two quite markedly different retinal images is too disturbing. On the other hand, so-called geometrical projections–that is, perspective drawings which represent objects as they would appear from an infinite distance–in many cases provide an unusually good impression of objects, even though these projections are made from a point of view which in reality is impossible. (In a geometrical projection the images of an object are still, of course, the same for both eyes.)

There is, then, a fundamental difference, one that cannot simply be ignored, between the appearance of a picture and the appearance of real objects. This difference may be reduced, but it can never be entirely eliminated. Due to the special characteristics of binocular vision, our most important natural means of judging the depth of objects is lacking when we look at a painting. The painter has only a number of subordinate ways, some of which are of limited applicability or of slight effect, to convey the impression of depth. In studying the art of painting from a theoretical point of view, it is important to become acquainted with these techniques, for in practice they are obviously related in very important ways to the composition of a painting, that is, to the selection, arrangement, and mode of illumination of the objects represented.

Concern for the clarity or comprehensibility of what is represented is, of course, of secondary importance when compared with the ideal goals of art. Yet clarity must not be underrated, for it is the first condition which must be satisfied in order to provide an observer with an impression at once effortless and forceful of the objects represented. This immediate clarity or comprehensibility is the basic requirement which must be met for a painting to affect the feelings and mood of an observer.

The subordinate means which have been referred to for giving the impression of depth depend upon perspective relationships. First of all, nearby objects partially conceal more distant ones but are never concealed by them. If a painter skillfully groups his subjects so that this relationship is stressed, it immediately provides a very clear distinction between objects that are close and those that are farther away. This kind of masking was found to be more important than other depth relationships, according to experiments on binocular perception using stereoscopic pictures which contradict one another in depth.

Secondly, the perspective projections of objects of regular or well-known shape indicate quite clearly their extension in depth. If we look at houses or other products of man's craftsmanship, we know of course that typically the shapes consist of plane surfaces at right angles to one another, with occasional surfaces that are circular or even spherical. A correct perspective drawing is usually sufficient to enable an observer to visualize the complete shape of such objects. This is also true with figures of men and animals which are familiar to us and which have two symmetrical halves. On the other hand, even the best perspective drawing is of little use with irregular shapes—rough blocks of rock or ice, masses of foliage, and the like. This is most apparent in photographs, where the perspective and shading may be absolutely correct and yet the total impression indistinct and confused.

If buildings are visible in a picture, they indicate the direction of the horizontal plane and also the slope of the ground, which otherwise would often be difficult to determine.

The apparent sizes in different parts of a picture of objects whose actual sizes are known is also important. Men and animals, as well as trees of familiar species, are useful to the painter in this connection. In the middle distance of a landscape they appear smaller than in the foreground, and their apparent size provides a measure of the distance at which they are placed.

Shading, especially shadows, is also of great importance. You all know how much more distinct an impression is conveyed by a well-shaded drawing than by an outline drawing. Shading is one of the most difficult, but at the same time one of the most effective, techniques in the repertory of a draftsman or painter. He tries to imitate the fine gradations and transitions of light and shade on curved surfaces, for this is his chief means of showing the modeling of these surfaces in all its fine changes of curvature. To achieve this, he must also take into account the range of light sources, the boundaries of these sources, and the mutual reflection of surfaces upon one another.

The use of shadows is also a most effective technique. While variations in the brightness of the surface of bodies are often difficult to determine—for example, an intaglio of a medal may in certain cases give an incorrect impression of the direction from which it is illuminated—shadows indicate unambiguously that the body which throws the shadow is nearer the source of light than that which is in the shadow. This rule is so completely without exception that an incorrectly placed shadow in a stereoscopic picture will confuse or destroy the illusion completely.

Not all kinds of illumination are equally good in producing the full effect of shading. If an observer looks at objects from the same direction as that in which the light falls upon them, he sees only their bright sides and nothing of their shadows; the relief which shadows could produce is entirely absent. If the objects are between the observer and the source of light, he sees only the shadows. We need lateral illumination for pictorially effective shading. For surfaces with figures which vary only slightly, or for flat or only gently hilly land, we need lighting that is almost parallel to the surface itself, for only in this way will shadows be produced. This is one reason why illumination by the rising or setting sun is so effective: the forms in the landscape then stand out more distinctly. To this must be added the influence of color and of aerial light, which we shall discuss shortly.

Direct illumination from the sun or from a flame produces shadows that are sharply defined and hard. Illumination from very broad, luminous surfaces, such as a cloudy sky, blur them or destroy them almost entirely. Between these extremes are various intermediate conditions. Illumination by a portion of the sky, or lighting restricted by a window or by trees, for example, produces shadows which stand out more or less prominently depending upon the nature of the subject. You all know how important this is to photographers, who must adjust and control their lighting by all sorts of screens and curtains in order to obtain well-modeled portraits.

Of much greater importance for the representation of extension in depth than the techniques discussed so far, all of which are more or less of local, accidental significance, is what is called *aerial perspective.* This is the optical effect produced by the passage of light through the air between an observer and objects at a distance. The effect itself is due to a very slight haziness in the atmosphere, which never entirely disappears. If fine, transparent particles with various densities and various refractive powers are suspended in a transparent medium, they deflect the light passing through it, partly by reflection and partly by refraction; to use an expression from optics, they *scatter* the light in all directions. If the particles are sparsely distributed in the medium, so that most of the light passes through without being deflected, distant objects are seen in sharp, well-defined outline, even though a small part of the light—that which is scattered—is distributed in the medium as a faint haziness. Water rendered turbid by a few drops of milk shows this scattering of light and this cloudiness very distinctly.

The turbidity in the air of an ordinary room is very easily seen whenever the room is darkened and a ray of sunlight is admitted through a narrow opening. We perceive what are called motes, many of which are large enough to be seen by the naked eye, while others form a fine, homogeneous cloudiness. These motes are apparently, for the most part, suspended particles of organic substances, for according to an observation by Tyndall, they can be burned. If the flame of a spirit lamp is placed directly beneath the path of the rays of the sun, the air rising from the flame stands out quite darkly in the adjoining bright rays; this means, of course, that the air rising from the flame has been freed of particles.

In the open air, besides dust and occasionally smoke, there is frequently haziness caused by the formation of small drops of water. This occurs whenever the temperature of moist air falls to such an extent that the water retained in it can no longer exist as an invisible vapor. Part of the water settles in the form of fine drops, like a very fine aqueous dust, creating a more or less dense fog or even clouds. The haziness which appears when the sun is hot and the air is dry may be due partly to dust which the ascending currents of warm air whirl about, partly to the irregular mixture of cold and warm layers of air of different densities, as is seen in the shimmering motion of the lower layers of air over surfaces warmed by the sun. Science cannot yet explain precisely the turbidity of the higher regions of the atmosphere which causes the blue of the sky. We do not know whether it is due solely to suspended particles of foreign substances or whether the air molecules themselves act to produce turbidity in the luminiferous ether.

The color of the light reflected by the particles depends mainly upon their size. If we produce small circular waves near a block of wood floating in water, the waves will be thrown back by the wood as if by a solid wall. In the long waves of the sea, however, the block of wood will be tossed about without disturbing the waves materially in their progress. It is well known that light is an undulatory motion in the ether which fills all of space and that the red and yellow rays of light have the longest wave lengths, the blue and violet the shortest. Very fine particles which disturb the uniformity of the ether will therefore reflect the latter more strongly than the red and yellow rays. Indeed, the finer the particles, the bluer the medium will appear, while larger particles will reflect light of all colors and thus produce a whitish cloudiness.

The presence of particles in the atmosphere is the reason for the blue of the sky, that is, for the color of the atmosphere as seen against dark cosmic space. The purer and more transparent the air is, the bluer the sky appears. The sky is bluer and darker, too, at the tops of mountains, partly because the air at great heights is less hazy and partly because there is less air above them.

The very same blue which is seen against dark celestial space is seen against the background of dark terrestrial objects—for instance, when a thick layer of illuminated air lies between us and deeply shaded or wooded hills. The same aerial light makes the mountains appear blue. The only difference is that the sky's blue is pure, whereas the mountains' is mixed with the light from the distant objects; it is also mixed with the coarser turbidity of the lower regions of the atmosphere, so that it is whiter. In those hot countries where the air is dry and where there is very little haziness in the lower regions, the blue in front of distant objects is more like that of the sky; the clarity and the very pure colors of Italian landscapes are attributable to this fact. On high mountains, particularly in the morning, the haziness of the air is often so slight that the colors of the most distant objects can scarcely be distinguished from those of the nearest; the sky may then appear almost bluish black.

A denser cloudiness, on the other hand, is due mainly to larger particles and is therefore whitish. This is the usual condition in the lower regions of the atmosphere, especially where the water vapor in the air is near the point of condensation.

Light from distant objects, which reaches our eyes after having traveled an extended path through the air, has been robbed of part of its violet and blue by scattered reflections. It therefore appears yellowish to reddish yellow or red, the former when the turbidity is fine, the latter when it is coarse. The color of the sun and moon at their rising and setting—and of distant, brightly lighted mountaintops, especially those covered with snow—are explained in this way.

These various effects are not peculiar to the air; they occur whenever one transparent substance is rendered turbid by being mixed with another transparent substance. We see them in diluted milk and in water to which a few drops of eau de Cologne have been added. In the latter case, the ethereal oils and resins dissolved in the eau de Cologne separate out and produce the cloudiness. Also, extremely fine blue clouds, bluer even than the air, may be produced, as Tyndall has observed, when the sun's light is allowed to exert its decomposing action on the vapors of certain carbon compounds. Many years ago Goethe emphasized the universality of the phenomena of turbidity and sought to base his theory of color upon them.

Aerial perspective is the artistic representation of the haziness of the atmosphere to indicate depth, for variations in the strength of the color of the atmosphere, imposed on the colors of objects, show relative distances very clearly. Landscapes in particular take on various appearances of depth. Depending upon the weather, the haziness of the air may be greater or less, more white or more blue. Very clear air, such as that encountered after continued rain, makes distant mountains appear smaller and nearer; when the air contains more vapor, they appear larger and more distant.

This is important for a painter to know. The high, clear landscapes of mountainous regions, which so often lead Alpine climbers to underestimate the distance and size of mountain peaks, are difficult to realize in a painting. Views looking upward from valleys, seas, or plains, in which the aerial light is faintly but clearly shown, are far better. Not only do such perspectives allow the various distances and sizes of objects to stand out; they also help to produce an artistic unity in the general coloration of a painting.

Aerial color is most evident in the greater depths of landscapes, but given the proper lighting it is not entirely absent in front of objects in a room. Objects which appear to be sharp and well-defined when sunlight passes into a dark room through an opening in the shutters are less sharp and well-defined when the whole room is lighted. If the aerial light in an interior scene is strong enough, it will be set off against the background and will deaden the colors there somewhat, compared to those of objects in the foreground. These differences, although far slighter than in a landscape, are important for the historical, genre, or portrait painter. When carefully observed and imitated, they greatly heighten the character of the objects represented.

II. VARIATIONS IN BRIGHTNESS

The facts which we have just discussed point to thoroughgoing differences—differences which are exceedingly important in the perception of the forms of objects—between the visual images we have when we stand before objects, on the one hand, and before paintings, on the other. Because of these differences, the choice of subjects that might be represented in a painting is immediately greatly restricted. Artists are well aware, of course, that many things cannot be represented using the techniques at their disposal. Part of their skill as artists consists in overcoming the unfavorable conditions imposed upon them by suitably arranging objects, choosing a judicious point of view, and carefully controlling the light in a painting.

Turning again to the question of the fidelity of a painting to reality, it might seem that a painting should at least provide us with the same distribution of light, color, and shading—and thus, in these respects, the same retinal images—as the objects represented in it would provide if they were at the same places in our field of vision. It might appear to be the aim of artistic work to produce (within the limits already discussed, of course) the same physiological effect by a painting as is produced by the objects themselves. But when we proceed to examine, as we shall now do, whether—and if so, how far—a painting can produce such an effect, we come upon difficulties from which we should perhaps shrink, if we did not know that they have already been overcome.

Let us begin with the simplest issue, that of the quantitative relations between and among different intensities of light. If an artist were to imitate *exactly* the impressions which objects produce upon our eyes, he would have to use the same absolute degrees of brightness and darkness as are encountered in the subjects he paints. This, however, is absolutely impossible. Consider the following example:

Imagine, in a picture gallery, a desert scene with a procession of white-shrouded Bedouins and dark Negroes marching under a burning sun. Imagine close to this another painting: a bluish moonlight scene, with the moon reflected in some water and groups of trees and human forms seen faintly in the distance. We all know from experience that both paintings, if they are well done, can produce with surprising vivacity the conception of their subjects; yet in both paintings the brightest parts are produced with the same white lead, only slightly altered by other tones, while the darkest parts are produced with the same black pigment. Since both paintings are hung on the same wall, they share the same lighting, and thus the brightest as well as the darkest parts of them barely differ from one another insofar as their brightness is concerned.

What, however, about the actual, objective brightness of the subjects represented in each of these paintings? The relation between the intensity of the light of the sun and that of the full moon was measured by Wollaston, who compared them with that of the light of candles of a standard material. He found that the light of the sun is 800,000 times that of the brightest light of a full moon.

An opaque body illuminated by some source of light can, even under the most favorable conditions, reflect only as much light as falls upon it. Actually, according to Lambert's observations, even the whitest bodies reflect only about two fifths of the incident light. The sun's diameter is 864,000 miles; its rays, at the point when they reach us, are distributed uniformly over a sphere 195,000,000 miles in diameter. The density and illuminating power of the sunlight at the surface of the earth, therefore, are only 1/40,000th of what they were when it left the sun's surface. Lambert's figures lead to the conclusion that the sun's disc is 100,000 times more intense than the brightest white surface, a surface upon which the sun's rays are falling perpendicularly.

The moon, however, is a gray body; its mean intensity is only about one fifth of that of the brightest white. When the moon shines on a body of the purest white, the brightness of that body is only 1/100,000th of that of the moon itself. Hence the sun's disc is eighty billion times brighter than a white which is lighted by the full moon.

Paintings which are hung in a gallery are illuminated, not by direct sunlight, but by light reflected from the sky and clouds. I do not know of any direct measurements of the normal intensity of the light in a gallery, but estimates may be made from known data. With strong overhead light, or strong light from a cloudy sky, the brightest white on a picture probably has one twentieth of the intensity of white directly lighted by the sun; generally it has only one fortieth, or even less.

Thus the painter of the desert scene, even if he does not try to represent the sun's disc (which in any case can be done only very poorly), will have to represent the glaringly bright garments of the Bedouins with a white which, even under the most favorable conditions, is only one twentieth as bright as the real garments would be. If he were to take his painting, with its illumination unchanged, into the desert and place it near something white, the white paint would appear dark gray. In fact, I have found by experiment that lampblack lighted by the sun is no less than half as bright as something white in a bright part of a room.

In the painting of the moonlight scene, the same white which was used to depict the Bedouins' garments must be used, with a slight tinting, to represent the moon's disc and its reflection in the water, although the real moon has only one fifth of the brightness of the white pigment and its reflection in the water even less. White garments or marble surfaces under moonlight in a painting, even when the artist tones them strongly with gray, are always ten to twenty times brighter than they are in reality.

Indeed, the deepest black that an artist could possibly use would scarcely be dull enough to represent the actual brightness of a white object upon which the moon is shining. Even the deadest coating of lampblack, or black velvet, appears gray when powerfully lighted, as we often enough discover to our sorrow in optical experiments when we wish to shut out superfluous light. I investigated a coating of lampblack and found its brightness to be about one hundredth of that of white paper. The painter's brightest colors are only about one hundred times brighter than his darkest shades.

The statements I have made may seem exaggerated, but they depend upon measurements, and they can be checked by well-known observations. According to Wollaston, the light of the full moon is equal to that of a candle burning at a distance of twelve feet. You know that we cannot read by the light of the full moon, but we can read at a distance of three or four feet from a candle. Suppose that you suddenly passed from a brightly lighted room into a vault which was perfectly dark except for the light of a single candle. At first you would think you were in absolute darkness, or at most you would see only the flame of the candle. Surely you would not notice the slightest trace of any objects at a distance of twelve feet from the candle, though their brightness is the same as that of objects illuminated by the full moon. You would, of course, become accustomed to the darkness after a while, and then you could find your way about without difficulty. If you then returned to the daylight, which before was perfectly comfortable, it would be so dazzling that you might be forced to close your eyes; at best you would be able to look around only with your eyes almost closed.

It is obvious that we are dealing here, not with small, but with colossal differences in light intensities. How, then, given these conditions, can anyone possibly imagine that the impressions produced by a painting and those produced by real objects are the same?

Our discussion of our experience in the vault, where at first we saw little or nothing and only gradually came to perceive things, illustrates a most important point, namely, the various degrees to which our senses may be dulled by light. This is a condition to which we can attach the same name—*fatigue*—that we apply to the corresponding condition involving our muscles. Any activity of any part of our body diminishes its power for a period of time. Muscles grow tired with work, the brain is tired by thinking and by emotion; analogously, the eyes are tired by light, and the more so, the more powerful the light. Fatigue makes them dull and insensitive to new impressions, so that they are only moderately sensitive to strong ones and not at all to those which are weak.

You can appreciate, when all this is taken into consideration, how difficult the task of an artist is. The eyes of the traveler looking at the real caravan have been dulled to the last degree by the dazzling sunlight, while those of the moonlight wanderer have been raised to the extreme of sensitiveness. The eyes of a person looking at a painting, however, differ from both: they possess a certain mean degree of sensitivity. The painter must try by the proper use of his colors to produce on the moderately sensitive eyes of a spectator both the impression which the desert produces on deadened eyes and the impression which the moonlight produces on eyes that are unusually sensitive.

Thus, not only the objective relations among light intensities in the external world, but also the different physiological conditions of the eye are extremely important, and the artist must take them into consideration in his work. He must produce, not just a copy of some object, but a translation of impressions into another *scale* or *degree* of sensitivity possessed by the eyes. With paintings the eyes respond to an entirely different language from that to which they respond when the impressions are those of the external world.

In order to make what follows comprehensible, I must first explain the law which Fechner discovered concerning the degrees of sensitivity of the eyes. It is a special case of the more general psycho-physical law, which this ingenious scientist also discovered, relating various sense impressions to the stimuli which produce them. The law in question may be stated as follows: *Within broad limits, all changes in the intensity of light are equally distinct—that is, they appear equal in experience—if they are equal fractions of some given total quantity of light.* Variations in intensity of one hundredth of the total light, for example, can be recognized without any great difficulty in cases involving very different light intensities. Whether we are dealing with the brightest daylight or with illumination by a good candle makes no difference in the ease or accuracy of the judgment.

The easiest way to produce precise, measurable differences in the brightness of white surfaces is by using rapidly rotating discs. If a disc like that in Fig. 1 is allowed to rotate very rapidly (twenty to thirty times per second), it appears to the eye to be covered with three gray rings, as in Fig. 2. (The reader must imagine the gray of these rings to be only a scarcely perceptible shading of the background.) When the rotation is sufficiently rapid, each ring of the disc appears to be illuminated in such a way that the light which falls upon it is uniformly distributed over its entire surface.

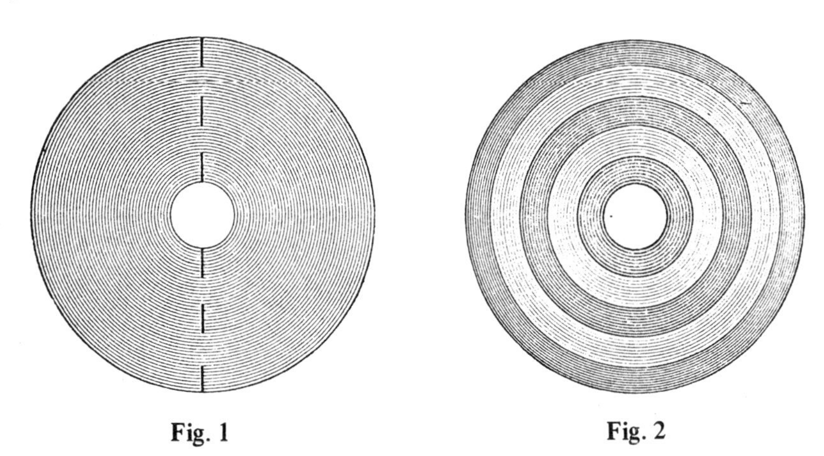

Fig. 1 **Fig. 2**

The rings which contain the black lines reflect slightly less light, of course, than those which are completely white. If the breadth of the lines is compared with half the circumference of each ring, we can calculate the proportion by which the intensity of the light of the white background of the disc will be diminished in that shaded ring. If the lines are all equal in breadth, as in Fig. 1, the inner rings will appear darker than the outer ones, since the same loss of light is distributed over a smaller area. With such a disc, extremely delicate variations in brightness can be obtained, and no matter what the strength of the illumination, the brightness will always be diminished in the same proportion of the maximum value. It is found, in accordance with Fechner's law, that for very different strengths of illumination, the distinctness of the rings is virtually the same. We exclude, of course, cases where the light is too dazzling or too dim, for then the finer distinctions can no longer be detected by the eye.

The situation is quite different if, starting with different intensities of illumination, we produce equal changes in intensity. For example, if during the day we close the shutters of a room so that it is quite dark, and then light it by a candle, we can see without difficulty the shadows thrown by the candlelight, for instance, the shadow of our hand on a sheet of white paper. If the shutters are then opened again, so that the daylight enters the room, we are no longer able to recognize the shadow cast by our hand in the candlelight, although on the part of the paper not in the shadow there is still the same excess of candlelight as there was before. This small amount of light disappears in comparison with the newly added daylight, provided that the latter strikes all parts of the white sheet uniformly. In short, while the difference between candlelight and darkness can easily be perceived, the equally great difference between daylight, on the one hand, and daylight plus candlelight, on the other, cannot.

These relations are of great importance in distinguishing among the various degrees of brightness of natural objects. A white object appears white because it reflects a large fraction, and a gray one appears gray because it reflects a smaller fraction, of incident light. With changing intensities of illumination, the differences in brightness between white and gray will always be related to the difference in their maximum brightness, and these differences will always be perceived so by us, provided we do not approach too closely to the upper or lower limits of intensities, beyond which Fechner's law does not hold. Because of these relationships, a painter can in general produce what appear to be equal differences in brightness, notwithstanding the changing intensities of light in a gallery, provided he gives his colors the same *ratios* of brightness as those which actually exist.

In our normal observation of objects in nature, their apparent brightness varies very widely, according to the actual intensity of the light and the sensitivity of our eyes. All that remain constant are the ratios among the brightnesses of different colored surfaces when lighted to the same degree. These ratios of brightness, therefore, are the only sensible signs we use in judging the darkness or brightness of the colors of the objects we see.

A painter can imitate these ratios without distortion and in conformity with nature, and he is thus able to produce the same impressions as those of the actual objects he represents. Within the range of light intensities where Fechner's law holds, an imitation which is correct in this respect can be achieved if an artist reproduces the fully lighted parts of the objects he wishes to represent with pigments that, given the same light, are the same as the colors to be represented. On the whole, this is what is done. The painter chooses colored pigments which almost exactly reproduce the colors of the bodies represented, especially for objects of no great depth, such as portraits, and for objects which are darker only in shaded areas. Children begin to paint on this principle: they imitate one color by another. This is also the way painting is done among nations in which the art of painting has remained at a primitive stage.

Artistic painting of the highest type, however, is achieved only when an artist succeeds, not just in reproducing colors, but in imitating *the action of light upon the eyes.* Only when we consider the goal of pictorial representation from this standpoint is it possible to understand the deviations from nature that artists must introduce in their choice of colors and light intensities.

In the first place, as we have noted several times already, these deviations are necessary due to the fact that Fechner's law is true only for moderate degrees of brightness; appreciable deviations from the law are experienced with intensities that are either very high or very low. At both extremes of intensity the eyes are less sensitive to differences in intensity than is required by the law. With a very strong light they are dazzled—that is, their internal activity cannot keep pace with the external excitations, and the nerves are soon exhausted. Very bright objects almost always appear equal in brightness, even when there are in fact appreciable differences among them. The light at the edge of the sun, for example, is only about half as bright as that at the center, yet none of you would have been aware of this fact had you not looked at the sun through colored glasses, which reduce the brightness so that the difference can be detected. With a weak light the eyes are also less sensitive to differences in intensity, but for the opposite reason. If a body is so feebly illuminated that we can scarcely perceive it, we surely will not be able to perceive that a shadow lessens its intensity by a factor of one part in a hundred or even one in ten.

It follows from these observations that under weak illumination dark objects are more like the darkest ones, while under strong illumination bright objects are more like the brightest, than would be the case according to Fechner's law, which holds only for mean degrees of intensity. And from this follows what, in the art of painting, is a very characteristic difference between the impressions of very strong and very weak illumination.

If painters wish to represent glowing sunshine, they make all objects almost equally bright, and thus with only moderately bright colors they produce the same impression as that which the sun's glow would have upon the dazzled eyes of an observer. If, on the other hand, they wish to represent moonlight, they show only the very brightest objects, especially the reflection of moonlight on shiny surfaces. They make everything else so dark as to be almost unrecognizable; that is, they make all dark objects as dark as possible with the colors available to them. In both cases, by making alterations in the intensity of light, painters give expression to the eye's insensitivity to differences in the intensity of light when that light is either too strong or too weak.

If they could use color of the dazzling intensity of full sunlight, or of the real paleness of moonlight, they would not have to represent the gradations of light in a painting other than as they are in nature. A painting would then make exactly the same impressions upon the eyes as would real objects in light of the same intensity. The alterations in brightness which I have just described are necessary because the colors in paintings are to be seen in the moderate light of an average room, for which Fechner's law holds, while the paintings represent objects whose actual brightness lies outside the scope of the law.

The older masters, especially Rembrandt, employed the alterations in light intensity which are needed to paint moonlight landscapes, even when they had no desire to produce the idea of moonlight or any similar weak light. The brightest parts of the objects in these paintings are produced in bright, luminous yellowish colors. The gradations in color toward the black, however, are very marked, so that the darker objects are almost lost in an impenetrable darkness. Yet this darkness itself is covered with the yellowish haze of powerfully lighted air masses. As a result, in spite of their darkness, these paintings give the impression of sunlight, and due to the very marked gradation of the shadows, the contours of the faces and figures in them appear unusually clear. The deviation from strict fidelity to nature in this shading is very great, yet these paintings lead to unusually clear and vivid ideas of the objects represented. They are thus of special interest to anyone wishing to understand the principles of lighting in art.

In order to explain the phenomena I have just been discussing, I must point out that while Fechner's law is approximately correct for those intensities of light which are usually encountered, the deviations from the law which are so marked in the case of high and low intensities are not completely absent in the case of normal illumination. It is just that we must observe more carefully in order to notice such deviations. The fact of the matter is that when the very slightest variations in shading are produced on a rotating disc, they are perceptible only under a certain intensity of illumination; namely, illumination like that of a piece of white paper which is lighted by the sky on a bright day but which is not struck directly by the sun's rays. With such illumination, variations of 1/150th or 1/180th of the total intensity can be recognized.

The light in which paintings are hung, however, is much more feeble. Consequently, if we are to retain the same sharpness of the finest shading and the resultant modeling, the gradations of shading in a painting must be somewhat greater than the gradations in actual light intensities. The darkest objects in a painting must be made unnaturally dark. This is not, of course, in conflict with the artist's aim, if that is to direct the observer's attention to the brighter parts of a painting. The great artistic effectiveness of this technique shows us that the main emphasis is to be placed on imitating differences of brightness, not on imitating absolute brightness. It also shows us that large differences in absolute brightness can be represented without great distortion, provided the gradations are carefully done.

III. COLOR

Related to these alterations in intensity are certain alterations in color which, considered physiologically, are necessary because we have different sensitivities to different colors. The strength of the sensation produced by light of a particular color and at a given intensity of light depends upon the specific reaction of a complex of nerves which are set in operation by the action of the light. All our sensations of color are combinations of three simple sensations—red, green, and violet—which, according to a not improbable hypothesis of Thomas Young, we experience quite independently of one another through three different systems of nerve fibers.

Corresponding to these independent color sensations are correlative differences in our sensitivity to variations in intensities of light. Recent measurements have shown, for example, that our eyes' sensitivity to gentle shading is greatest in the blue tones and least in the red. A difference of 1/205th to 1/268th of a given intensity can be detected in the blue tones. With untired eyes a difference of only 1/16th can be observed in the red tones, and even when the color impression grows dim after having been observed for a longer time, a difference of only 1/50th to 1/70th can be detected.

Our eyes, therefore, are relatively less sensitive to changes in the intensity of red than to changes in the intensity of blue. Moreover, as the intensity of light increases, the glare of red is greater than that of blue. Given a piece of blue and a piece of red paper which appear equally bright under an average amount of ordinary light, Dove observed that if the light is made dimmer, the blue will appear brighter, whereas if the light is strengthened, the red will seem to grow brighter. I have found in my own studies that these differences are even more evident with red and violet; indeed, they are clearly evident with only slight changes in the intensity of light.

The impression of white is made up of the impressions which the individual spectral colors constituting white light have upon our eyes. If we increase the brightness of white, the strength of the sensations of red, yellow, and green will increase more than those of blue and violet. With a bright white, therefore, the former will produce relatively stronger impressions than the latter; with a dull white, on the other hand, the blue and bluish colors will be stronger. Thus a very bright white appears yellowish, and a dull white appears bluish. In our ordinary observations of the objects around us we are not readily conscious of this, for direct comparison of colors under very different lighting conditions is difficult. We are accustomed to seeing familiar white objects under constantly changing conditions of illumination, and we have learned to discount the influence of the intensity of the lighting in judging the colors of these objects.

If a painter has the problem in a painting of representing with weak colors a white irradiated by the sun, he can easily obtain a striking resemblance to reality. By mixing some yellow with his white, he can make this color predominate to the extent it appears to do in actual light because of the way in which our sensory apparatus functions. It will be the same impression as that produced when we look at a landscape under a cloudy sky through a yellow glass, which gives the landscape the appearance of sunniness. An artist will, on the other hand, give a bluish tint to moonlight or to any other faint white, for (as we have seen) faint lights in a painting are far brighter than the actual colors to be represented. Indeed, in moonlight scarcely any color other than blue can be recognized; the sky or blue flowers may still appear quite distinctly colored, while yellow and red can be seen only as variations of a general bluish white or gray.

Let me repeat that an artist would not have to make these color changes if he had at his disposal colors of the same brightness, or the same faintness, as the actual colors of objects illuminated by the sun or by the moon. The changes in color, like the changes in brightness which we discussed earlier, are subjective effects which an artist must represent objectively on his canvas, since he cannot produce them directly with only moderately bright colors.

We observe something quite similar with *contrast phenomena.* By this term we understand cases where the color or brightness of a surface appears changed by the proximity of some other light or color. A color appears darker when it is near a brighter one, for example, and brighter next to a darker one. Again, a color tends to give emphasis to its complement when they are near each other in a painting.

There are a number of contrast phenomena, each resulting from a different cause. One kind, Chevreul's simultaneous contrast, is independent of the movements of the eyes and occurs with surfaces upon which there are very slight differences in color and shading. This kind of contrast appears both with paintings and with actual objects and is well known to artists. Colors, spread out and mixed on a palette, often appear quite different from the way they appear later in a painting. The color changes that occur are often quite striking; I shall not discuss them here, however, since they do not produce any differences between a painting and reality.

A second kind of contrast phenomenon, one which is more important for our purposes, occurs when we change the direction of our gaze. The effect is particularly striking when there are large differences in shade and color. As the eye glides over bright and dark colored objects and surfaces, the impression of each color changes, due to the fact that the light strikes portions of the retinas which immediately before had been excited by other colors and lights, by which their sensitivity was changed. This kind of contrast, clearly dependent upon the movements of the eyes, has been called successive contrast by Chevreul.

We have already mentioned that in the dark the retinas become more sensitive to feeble light. In strong light, on the contrary, they are dulled and become less sensitive. This phenomenon is called retinal fatigue. The retinas are exhausted by their own activity, just as muscles grow tired by theirs. It should be noted, however, that the retinal fatigue due to stimulation by light does not necessarily cover the entire surface of the retinas. It may be restricted to a small area, if that is the only one excessively stimulated.

You must all have observed the dark spot which moves about in the field of vision after you have looked toward the setting sun for even a short time. Physiologists call this the negative afterimage of the sun. It appears because those parts of the retinas which have been struck by the sunlight are temporarily insensitive to any further action of light. If, with eyes which are thus locally tired, we look toward a uniformly bright surface, such as the sky, the insensitive parts of the retinas are more feebly affected than the other parts, so that we think we see a dark spot in the sky which moves about with our gaze. The dark spot in the sky corresponds to the tired parts of the retinas, the bright areas to the parts that are still sensitive.

Bright objects like the sun produce negative afterimages that are quite striking; but with a little attention, afterimages may be seen even after much more moderate excitation. A longer time is required to produce one that can be easily recognized, of course, and it is necessary to concentrate upon a definite point on an object without moving the eyes, so that its image is distinctly formed on the retinas and so that only limited portions of the retinas are excited and thus fatigued. This is similar to what must be done to obtain sharp portraits in photography: the subject must remain stationary during the entire time of exposure so that the image on the sensitive plate is not blurred. The afterimages in the eyes are, as it were, photographs on the retinas, which become visible owing to the retinas' altered sensitivity to fresh light; they last, however, for only a short time. The exact length of time depends upon the intensity and duration of the original excitation.

If the object one stares at is colored—for example, a piece of red paper—the afterimage on a gray background will be of the complementary color, in this case bluish green.[1] Rose-red paper produces a pure green afterimage, green a rose-red, blue a yellow, and yellow a blue. These phenomena indicate the possibility of partial fatigue of the retinas for each of the several colors. According to Thomas Young's hypothesis concerning the existence of three systems of fibers among the optic nerves (one of which is sensitive to red, no matter what the excitation; the second to green, and the third to violet), when one looks at something green only those fibers of the retina which are sensitive to green are powerfully excited and fatigued. If this same part of the retina is afterward excited by ordinary light, the sensation of green will be weakened, while those of red and violet remain strong and vivid. The combination of these two colors gives the sensation of purple, which, combined with the white background, produces rose-red.

Ordinarily, in looking at brightly lighted and colored objects, we do not concentrate for long on any one point. On the contrary, following the play of our attention we are always shifting our gaze to new parts of the objects as they come to interest us. This manner of observation, in which the eyes are continually moving, thus making the images shift about on the retinas, prevents the disturbances of vision that cause strong, continuous afterimages.

1. In order to see this kind of afterimage as distinctly as possible, it is important to avoid any movement of the eyes. On a large sheet of dark gray paper, draw a small black cross, the center of which can be fixed in vision. From the side insert a square of paper of the color whose afterimage is to be observed, so that one of its corners touches the cross. Allow the sheet to remain in this position for a minute or two (keeping your eyes steadily upon the cross), and then draw it quickly away (without shifting your gaze). In the place where the sheet was a moment before, the afterimage will appear against the gray background.

Yet even in ordinary vision afterimages are not completely absent; they are merely shadowy in their contours and of very short duration.

If a red surface is placed upon a gray background, and if we shift our gaze from the red across the boundary to the gray, the edges of the gray will seem tinted by the afterimage of red; that is, they will appear faintly bluish green. Since the afterimage rapidly disappears, however, only those parts of the gray which are nearest the red will show the afterimage to any marked degree.

This phenomenon is produced more strongly by bright light and brilliantly saturated colors than by faint light and duller colors. On the whole, however, an artist works with the latter. Most color tones are produced by the mixture of pigments, and each mixed color is grayer and duller than the pure colors from which it is made. Even the few pigments of highly saturated color which can be used in oil painting, such as cinnabar and ultramarine blue, are comparatively pale; and the colors employed in water colors and the colors of chalks are, of course, even paler. Thus one cannot generally expect to find the same strong contrast effects in a painting that one finds with strongly colored and lighted objects in nature.

If an artist wishes to reproduce as strikingly as possible, with the pigments at his command, the impression which real objects produce, he must indicate with paint the contrasts which the real objects naturally display. If the colors in a painting were as strong and brilliant as those of actual objects, the contrasts which appear in reality would appear automatically in a painting. Here, once more, subjective visual phenomena must be introduced objectively into a painting, since the colors and light intensities in it are different from reality.

With a little attention you will notice that painters and draftsmen generally make a plain, uniformly lighted surface brighter where it meets a dark object and darker where it meets a light one. You will find that uniformly gray surfaces are given a yellowish tint at the edge where there is a background of blue and a rose-red tint where there is green, provided of course that none of the light reflected from the blue or green falls upon the gray. Where the sun's rays, passing through the green, leafy shade of trees, strike against the ground, they appear to the eye—fatigued by looking at the predominant green—of a rose-red tint. Reddish-yellow candlelight, on the other hand, appears blue in the daylight entering a room through a small slit. All these phenomena are represented objectively by the painter, since the colors in his paintings are not bright enough by themselves to produce contrasts as a result of fatigue.

To the number of subjective phenomena which artists must represent objectively in their paintings should be added that of *irradiation.* By this we understand cases in which a bright object in the field of vision spreads its light or color over the surrounding region. The phenomenon is the more marked, the brighter the object giving off the light; and the halo of light, brightest in the immediate neighborhood of the object, is diminished at greater distances. The phenomenon is most striking around a very bright light against a dark background.

If the sight of a flame is blocked off by a narrow dark object, such as a finger or a ruler, the bright misty halo which covers the whole region near the flame will also disappear, and at the same time the objects in the dark part of the field of vision will be seen more distinctly. If the flame is only partly screened by the ruler, these objects will appear jagged or irregular where the flame projects into the darkness. The intensity of the light in the neighborhood of the flame will be so great that it can scarcely be distinguished from that of the flame itself. Moreover, as is true of any very bright object, the flame will appear to be magnified and at the same time to spread over adjacent darker objects.

The cause of this phenomenon is quite similar to that of aerial perspective. It is due to the scattering that results from the passage of the light through a hazy medium. The only difference is that with the phenomenon of aerial perspective, the haziness is in the air in front of the eyes, while with irradiation it is in the transparent media of the eyes. If even the healthiest human eye is examined under a powerful light (the best being a pencil of sunlight concentrated on the side of the eye by a condensing lens), it is evident that the cornea and the crystalline lens are not perfectly clear. If strongly illuminated, they both appear whitish and as if rendered turbid by a fine mist. Both are, in fact, tissues which are fibrous in structure and thus not as homogeneous as a pure liquid or a pure crystal. Every irregularity, however small, in the structure of a transparent body can reflect some of the incident light—that is, can scatter it in all directions.[2]

Irradiation also occurs with only moderate amounts of light. A dark hole in a sheet of paper illuminated by the sun, for example, or a small dark object which is held against the clear sky on a colored glass plate, appear to be the color of the adjacent surface.

The phenomenon of irradiation is unquestionably very similar to that due

---

2. I disregard here the view that irradiation in the eye is due to a diffusion of the excitation in the substance of the nerves, as this view appears to me too hypothetical. In any case, we are concerned here with the phenomenon, not with its cause.

to the haziness of the air. The only essential difference is that the effect of haziness is stronger in viewing distant objects (that is, objects which have a greater mass of air in front of them) than in viewing those that are close by, whereas the irradiation in the eye sheds its halo uniformly over both near and distant objects.

Irradiation must be included among the subjective visual phenomena which the artist must represent objectively, because the lights—even the sunlight—in a painting are not bright enough to produce any noticeable irradiation in the eyes of an observer.

I have already used the word *translation* to characterize the representation a painter must give of the lighting and colors of his subjects, and I have urged that it is generally impossible to achieve a copy which is true to nature in all respects. The altered scale of light intensities which an artist is forced to use in so many cases stands in the way. It is not the colors of objects, however, but the visual impressions they produce which must be imitated in order to produce as distinct and vivid an idea as possible of the objects represented. A painter is forced to alter light and color relationships when he executes a painting, but he only alters what in any case would be subject to various changes depending upon the lighting and the sensitivity of the eyes. He preserves the more essential element, *the ratios or gradations of brightness and color.*

A number of other effects also occur as a result of the ways in which the eyes respond to external stimulation. Since these effects depend upon the intensity of the stimulation, they are not automatically produced by the altered intensities of light and color which the artist must use in a painting. These subjective phenomena, which occur naturally when one looks at real objects, would be missing when one looks at a painting if the painter did not represent them on the canvas. The fact that they can be so represented is particularly significant in showing the kinds of problems that must be solved in artistic representation.

In all translations the personality of the translator plays a part. In artistic translations many important points are left to the artist's choice, and he can make his decision according to his individual tastes or according to the requirements of his subject. Within certain limits he is free to select the relative brightness of his colors, as well as the gradation or degrees of shading. Like Rembrandt he may exaggerate the shading in order to achieve strong relief; or he may, with Fra Angelico and his modern imitators, lessen the shading in order to soften material shadows in the representation of sacred subjects. Like the Dutch school he may, by his representation of the varying light of the atmosphere–now bright and sunny, and now pale; now warm and now cold–evoke in the observer moods which depend upon the illumination and upon the state of the weather; or he may, by representing undisturbed air, make his figures stand out with objective clarity, uninfluenced by subjective impressions. By these means it is possible to achieve great variety, both in general and in the constituent elements, in what artists call *treatment* or *style.*

IV. COLOR HARMONY

At this point, the following question quite naturally arises: If the small amount of light in and the weak saturation of his colors force an artist to struggle (by using various indirect techniques and by imitating subjective effects) toward a close but necessarily always incomplete correspondence with reality, would it not be more sensible for him to look for ways of avoiding these drawbacks? And, indeed, such ways exist. Frescoes are sometimes viewed in direct sunlight; transparencies and paintings on glass can utilize far higher intensities of light and far more saturated colors; and in dioramas and for theatrical presentations we can use powerful artificial light–if need be, even an electric arc.

As I mention these branches of art, however, it should immediately be obvious that the works which we admire as the greatest masterpieces of painting are of a different kind. Most great paintings were executed with comparatively dull water or oil colors and were intended for rooms with only a moderate amount of light. If it were possible to achieve the higher artistic effects with colors lighted by the sun, we should undoubtedly have paintings which took advantage of this illumination. Fresco painting would have led to them; and surely the experiments by Munich's celebrated optician Steinheil to produce oil paintings intended to be seen in bright sunlight–experiments which he made as a scientific investigation–would not have remained isolated curiosities.

Experience seems to indicate that there are advantages to the moderate intensities of the lights and colors of paintings, and we need only look at frescoes (such as those in the new museum in Munich) in direct sunlight to see what these advantages are. Specifically, the brightness of these frescoes is so great that we cannot look at them steadily for any length of time. What is so painful and tiring to the eye with frescoes, however, would have the same effect, though to a smaller degree, if brilliant colors were used in a painting even locally and moderately to represent the average intensity of bright sunlight and a normal amount of atmospheric light. (It is much easier, of course, in dioramas and stage settings, to produce with artificial light an accurate imitation of the more feeble light of the moon.)

What we think of as the *resemblance* of a beautiful painting to nature can be more accurately considered *an ennobled fidelity to nature.* A painting can provide all the essential impressions–indeed, a full visual sense–of reality without injuring our eyes or tiring them by the harsh lights of reality. The differences between art and nature are, as we have already seen, restricted mainly to relations, such as the absolute intensity of light, which we can judge at best only in an arbitrary, uncertain fashion.

It has already been mentioned that we are best able to make delicate discriminations in shading, and in the modeling which such shading produces, under light of moderate intensity. In sense perception, the most pleasant yet not exhausting excitations of our nerves and the highest feeling of well-being in them arise under precisely the conditions which permit the most delicate discriminations and observations.

I should also like to direct your attention here to another point of great importance in the analysis of the art of painting. I refer to our natural tendency to delight in colors, which undoubtedly plays a large role in the pleasure we receive from the works of painters. In its simplest forms, as with gaudy flowers, feathers, stones, fireworks, or Bengal lights, this delight in colors has very little to do with man's artistic endeavors. It is only the natural enjoyment on the part of a sensitive organism of the varied and changing excitations of its nerves. This delight in colors is also necessary, of course, for the health and efficiency of these nerves.

The thoroughgoing purposiveness of the structure of a living organism, whatever its origin, excludes the possibility that any tendency, including this one, would be developed or maintained in the majority of healthy members of a species if it did not serve some specific purpose. We need not seek far to find the reason for our delight in light and colors–and for our dread of darkness: it coincides with our drive to see and understand the objects around us. Darkness owes the greatest part of the terror it arouses to our fear of the unknown and the unknowable.

A painting with its colors provides a richer, more accurate, and more easily obtained perceptual awareness of the objects represented than does a drawing, which shows only contrasts of light and shade. A painting shows these too, but in addition it provides means for making distinctions which only colors afford. Through differences in colors, surfaces which appear equally bright in a drawing are seen in a painting to be made up of different objects; and through similarities in color, different areas are seen to be similar objects or parts of the same object. Taking advantage of these natural relations, an artist may use color emphases to direct and hold the observer's attention upon the main elements in a painting, and by variations in dress he may distinguish various figures from one another and unify each in itself.

Our natural pleasure in pure, strongly saturated colors has its rationale along these lines. These colors are analogous to a full, pure, beautiful voice in music. Such a voice is very expressive: even the smallest change in pitch or quality, any slight interruption, any vibrato, any swelling or falling, is more clearly and immediately noticed with such a voice than with sounds that are less refined. It also seems to be true that the powerful excitations which such a voice produce in a listener's ears arouse ideas and passions better than do more feeble excitations. A pure primary color with even the smallest mixture of some other color is like a background upon which a slight amount of light is visible. All of the ladies present know how sensitive clothes of the same saturated color are to dirt, in comparison to gray or grayish brown materials.

This is also in agreement, of course, with Young's theory of colors. According to this theory, the sensation of each of the three primary colors is due to the excitation of only one kind of color-sensitive nerve fiber, while the other two kinds remain at rest or are at best only feebly excited. Thus a brilliant, pure color produces a powerful stimulus and, at the same time, a high degree of sensitivity to the other colors, the nerve fibers of which are at rest.

The modeling of a colored surface depends mainly upon the reflection of light of other colors which falls upon it. When a piece of cloth glistens, it is the color of the incident light which predominates. In the depths of the folds of the material, on the other hand, the colored surfaces reflect upon themselves, and their own color is thus accentuated. A white surface of great brightness, of course, produces a dazzling effect and masks all small differences in shading. In general, strong colors, by the powerful excitation which they produce, can hold the observer's attention and at the same time express the slightest changes in modeling or illumination. They are, in other words, expressive in the artistic sense of the term.

If we cover too large an area with the same pigment, we find that the eyes are fatigued by the color and grow less sensitive toward it. The color itself then tends to seem grayish, and the color complementary to it appears on all differently colored surfaces, especially on surfaces which are gray or black. Thus clothes and especially carpets which are of a single, over-bright color have a disturbing, tiring effect. Such clothing has a further disadvantage in that it casts a color complementary to its own over the face and hands of the person wearing it. Blue casts yellow, violet casts greenish yellow, bright purple casts green, scarlet casts blue-green, yellow casts blue, and so on.

Another thing an artist must keep in mind is that color is an important means of attracting the attention of spectators. He must therefore be sparing in the use of saturated colors, for they distract the attention and at the same time make a painting garish. It is also necessary to avoid the partial retinal fatigue caused by colors which are too prominent. This fatigue may be avoided either by using the main color only moderately and against a dull, weakly colored background or by juxtaposing several saturated colors, thus producing a certain equilibrium in the excitation of the eyes.

In the latter case, the contrasts of the afterimages will also strengthen the various colors. A green surface, for example, upon which the green afterimage of a purple surface falls will appear a far purer green than it does without such an afterimage. Due to the eyes' fatigue toward purple, that is, toward red and violet, the effect of any admixture of these colors in the green is weakened, while the green itself produces its full effect. In this way the sensation of green is purified from any foreign elements. Even the purest and most saturated green, the green that appears in the prismatic spectrum, can be made still more highly saturated in this way. The other pairs of complementary colors which we have mentioned also make each other more brilliant by contrast. Colors which are close together in the spectrum, on the other hand, are detrimental to each other and produce a grayish effect.

These relations of various colors to one another are obviously of great importance in determining how pleasing different combinations of colors will be. It is possible, without bad effect, to juxtapose two colors which are so similar that they look like varieties of the same color produced by different amounts of light and shade. Thus one can paint the shaded parts of a scarlet object carmine and the shaded parts of a straw-colored object golden yellow. If we pass just beyond this point, however, we get unpleasant combinations, such as carmine and orange, or orange and straw-yellow. Colors at a greater distance from each other in the spectrum must then be chosen in order to obtain pleasing combinations again.

Pairs of colors which stand at the greatest distance from each other on the color wheel are called complementary colors. When they are used together–

for example, straw-yellow and ultramarine blue, or verdigris green and purple—the result is somewhat insipid and common. Perhaps this is because we already expect to see the second color as the afterimage of the first, and so it does not appear to be a significantly new, independent element in the combination. On the whole, combinations in which the second color is close to, but clearly different from, the complement of the first are more pleasing. Thus scarlet and greenish blue are complementary, but a more satisfactory combination is formed when the greenish blue is allowed to shade over either into ultramarine blue or into yellowish green (leaf green). In the latter case the combination has a tendency toward yellow; in the former, toward rose-red.

Still more satisfying than such pairs of colors are combinations of three colors whose impressions are in equilibrium and which, in spite of the greater concentration of colors, avoid both a one-sided tiring of the eyes and the insipidness of complementary colors. To this group belongs the combination of red, green, and violet which the Venetian masters used so much, as well as Paolo Veronese's purple, greenish blue, and yellow. The colors in the first of these triads, insofar as they can be produced by pigments, correspond closely to the three physiologically fundamental colors; the second triad may be obtained by mixing only two of these fundamental colors.

It should be emphasized, of course, that it has not yet been possible to establish rules for the harmony of colors with the same precision and certainty as those for the consonance of tones. Indeed, a consideration of the facts shows that a number of other factors besides color become important as soon as colored surfaces are used to represent natural objects and material forms or even to suggest the differences between shaded and unshaded surfaces. Moreover, it is often difficult to establish as a matter of fact precisely what colors produce harmonious impressions. This is especially true with paintings in which the aerial color and the colored reflections and shadings so alter the colors of irregular surfaces that it is hardly possible to say clearly what the local colors of the surfaces are. In such paintings, moreover, the direct action of the colors upon the eyes is only of secondary importance, for the main colors and lights must also serve to direct attention to the more important parts of the painting. Compared with these more substantive factors, considerations of the pleasing effect of the colors is of secondary importance. Only in the ornamentation of carpets, draperies, or ribbons and of some buildings is there free scope for pure pleasure in colors. Only there can it be developed according to its own laws.

As a rule, in paintings there is no complete equilibrium among the various colors; rather, one of them, because it is the color of the dominant light, stands out to some extent. This results directly from the attempt to represent actual conditions. For example, if the illumination is rich in yellow light, yellow colors will appear brighter and more brilliant than blue ones, since yellow objects will reflect the light, while in general those painted blue will absorb it. Further, the yellow light will tend to make the shaded parts of blue objects yellow, and it will change the blue parts more or less to gray. The yellow light will have the same effect, although to a lesser extent, on red and green, so that in shaded areas these colors tend to become more yellowish.

This, again, is clearly in accordance with the aesthetic requirements of unity in the composition of colors. In general, the colors which change, especially in the shaded areas, show a relation to the dominant color and tend, in changing, toward it. When this does not occur, the various colors fall harshly upon one another, and since each calls attention to itself, they produce either a colorful but disordered impression or one that is cold, since there is no unifying light thrown over all the objects in the painting.

The light of the setting sun, which casts a flood of light and color over regions that otherwise would be dimly seen and brightens them harmoniously, is a natural example of the general aesthetic harmony which the well-executed illumination of air masses can produce. With the setting sun, the increase in the aerial illumination is due to the fact that the lower, more opaque layers of air lie in the direction of the sun and consequently the light is reflected more fully. At the same time, the yellowish-red color of the light which has passed through the atmosphere grows more pronounced as the length of the path it has traveled increases. This color stands out even more strongly as the background falls into shadow.

To summarize briefly once more the preceding analysis and observations: We saw, first of all, that certain limitations are imposed upon any attempt on the part of an artist to represent nature as it really is. We saw how binocular vision, the chief means nature has chosen to enable us to perceive depth in our field of vision, fails the painter. Indeed, we saw how it is turned against him, for with binocular vision we always see the flatness of a canvas. We saw how the painter must therefore plan carefully both the perspective arrangement of his subjects (their location and position) and their light and shading in order to give us an immediately intelligible image of their size, shape, and distance. And we saw how a faithful representation of aerial light is one of the most important means for attaining these ends.

We then saw that the scale of light intensities encountered in looking at real objects must be altered in a painting, sometimes by a hundredfold. We saw how, consequently, the colors of objects cannot be simply matched by the colors of pigments–indeed, how it is necessary to introduce important changes in the distribution of light and darkness and in the distribution of yellowish and bluish tones.

An artist cannot copy nature; he must translate it. Nevertheless, this translation can give us an impression which is in the highest degree clear and forceful, not only of objects themselves, but also of the different lights under which we see them. Indeed, the altered scale of light intensities is in many cases an advantage, for it eliminates everything which in actual objects is too dazzling and too tiring to the eyes. Thus the imitation of nature in a painting is more than a copy: it is an ennobling of the impressions of the senses.

Furthermore, and because of this, we can indulge ourselves more quietly and for longer periods of time in the contemplation of a work of art than in the contemplation of reality. A work of art can provide those gradations of light and those color tones which make the modeling of forms most distinct and therefore most expressive. It can present an abundance of vivid colors and yet, by contrasting them skillfully, maintain the eyes' sensitivity in a state of equilibrium. It can employ the entire force of powerful sense impressions and the accompanying feeling of delight to capture and direct the attention. It can heighten the immediate understanding of the subjects represented and still keep the eyes in the condition of excitation most advantageous and agreeable for delicate perception.

When, in the preceding analysis and observations, I continued to give great emphasis to effortless, delicate, and precise perceptual clarity or comprehensibility in artistic representations, this may have seemed to many of you to be focusing on a very minor point–a point which, if mentioned at all by writers on aesthetics, is treated only as a side issue. I think, of course, that these writers are mistaken in dismissing this point. Perceptual clarity is by no means a trivial or subordinate element in the effect of a work of art. Its importance has forced itself more and more strongly upon me as I have sought to investigate the physiological aspects of the effects of works of art.

What effect should be produced by a work of art, using this phrase in its highest sense? It should capture and excite our attention. It should arouse in us a host of slumbering conceptions and the feelings related to them, and it should direct them easily and freely toward a common point. Thus it should present us with a vivid, unified perception of all the features of an ideal type, the separate fragments of which normally lie scattered chaotically in our minds.

The only way it seems possible to explain the power, so frequently discussed, of art over the human mind is to note that reality always presents some element that is disturbing, distracting, or injurious, while art can unify every relevant factor for a desired effect and can allow each to act without hindrance. The power of these effects will no doubt be the greater, in proportion as the sense impressions aroused by the artistic images—and the emotions related to them—are more penetrating, finer, and truer to nature. A work of art must act surely, quickly, unequivocally, and precisely if it is to produce a vivid, powerful effect. Essentially, this is what I mean by the *comprehensibility of a work of art.*

The specific elements of artistic technique to which investigations in physiological optics have led us are closely related to the highest goals of art. Indeed, we may perhaps entertain the thought that even the ultimate mystery of artistic beauty—that is, the wonderful pleasure we feel in its presence—is really based upon a sense of the smooth, harmonious, and vivid current of our ideas which, in spite of many changes, flows toward a common point and brings to light laws hitherto concealed, allowing us to gaze into the deepest recesses of our own nature.

11.

THE ENDEAVOR TO POPULARIZE SCIENCE [1874]

Preface to the German translation of John Tyndall's Fragments of Science *(Braunschweig, 1874)*

ALTHOUGH my name appears on the title page simply as the editor of this volume of translations of Tyndall's essays, I have nevertheless taken the trouble to provide the same assistance here as I have with earlier volumes; that is, I have gone through the translation to make sure that the factual scientific materials were rendered correctly, and where it seemed necessary, I have tried to improve the translation. Though overwhelmed by other official and scientific work, I have not withdrawn from this enterprise, for I believe that the wide dissemination of excellent popular expositions of the most significant and most instructive parts of the natural sciences is a very important undertaking.

I consider the awakening desire for instruction in the natural sciences, which is now finding expression among the educated classes of Germany, not to be just a striving after new forms of amusement or the result of an empty, barren curiosity. It is, rather, a fully justified intellectual necessity, one which is closely related to the most important source of intellectual development at the present time. For the natural sciences have come to be a powerful influence on the social, industrial, and political life of civilized nations. This is not simply because the great forces of nature have been subordinated to the goals of man and now supply him with a host of new means for attaining them, although this fact is, to be sure, so important that even the statesman, the historian, and the philosopher, not to mention the manufacturer and the merchant, can no longer avoid participating in at least the practical results of scientific investigations. It is rather because the natural sciences are influential in a different way—a way which goes much deeper and much further, even though it may be slower in manifesting itself. I refer, of course, to their influence on the direction of the intellectual progress of man.

It has often been said—and even brought as a charge—that through the natural sciences a schism, formerly unknown, has been introduced into modern education. And, indeed, there is truth in this claim. A schism is perceptible. Still, such a break must mark every new step in intellectual development, every point where something new has become a force and demands a settlement of its just claims against the just claims of an older tradition.

Until now, programs of education in civilized nations have had their center in the study of language. Language is the great instrument which most clearly separates man from the lower animals. Through its use each man is able to share the experience and knowledge of other individuals of his own time and even of earlier generations. Without it each man would be limited, like the lower animals, to his instincts and his own particular experience. That the improvement of language was the primary and most necessary work of a developing race, and that the most refined training in its comprehension and use is and must ever be the main goal in the education of each individual, is self-evident. Further, the culture of modern European nations has a peculiarly intimate connection with the study of the remains of antiquity and thereby with the study of language. With the latter, moreover, has been associated the study of the forms of thought, as these are to be found in speech. Thus logic and grammar—that is, the art of speech and the art of writing (using these words in their original meaning)—taken in their highest sense, have hitherto quite naturally received the main emphasis in education.

Nevertheless, while language is the means of preserving and handing down truth once it is accepted, we must not forget that its study teaches us nothing about how new knowledge is to be found. Similarly, logic shows how, starting from a proposition which forms the major premise of a syllogism, conclusions are to be drawn, but it can tell us nothing about the source of the proposition. On the contrary, whoever wishes to assure himself of its truth by independent study must gain a knowledge of the individual cases which fall under it and which afterward, when it has been established, may of course be accepted as instances of it. Only in cases where such knowledge has been communicated by others does a universal proposition actually take precedence over knowledge of these separate instances, and in such cases the old treatises of formal logic do have undeniable practical importance.

Thus these studies themselves do not lead us to the real source of knowledge; they do not bring us face to face with the reality we wish to know. Moreover, there is a danger in transmitting knowledge to someone who has had no personal experience of its source. Comparative mythology and the critique of metaphysical systems tell us a great deal about how figurative linguistic expressions come in time to be treated as having genuine significance and how they even come to be prized as ultimate wisdom.

Thus, with respect to the intellectual development of mankind, we recognize fully the great importance of the beautifully developed art of exchanging and transmitting knowledge which has already been acquired. We also recognize the importance of the contents of classical writings for the cultivation of moral and aesthetic sentiments and for the development of a clear appreciation of human feelings, ways of thinking, and cultural patterns. Nevertheless, we must emphasize that there is an important element missing in an exclusively logical-literary education. I refer to the methodical discipline of the activity by which we reduce the confused materials we find in the real world, at first sight apparently ruled by chance rather than reason, to ordered concepts and thus make them suitable for expression in language. Such an art of observation and experiment, systematically developed, has as yet been found only in the natural sciences. The hope that the psychology of individuals and of nations, as well as the practical sciences of education and of social and political government which are based upon them, will attain the same goal, can only be realized in the distant future.

This new enterprise, prosecuted by the natural sciences along new paths, has very quickly produced fresh and unprecedented results, indicating what achievements the human intellect is capable of when it can proceed under favorable conditions from facts to a full knowledge of laws, testing and establishing everything itself. The simpler phenomena, especially those of inorganic nature, permit of our possessing such a penetrating, precise knowledge of their laws and such far-reaching deductions of consequences from them—consequences which can be tested and verified by careful reference to reality—that, given the systematic conceptual development of such knowledge (for example, the deduction of astronomical phenomena from the law of gravitation), there is hardly any other edifice of human thought which, for strict logic, certainty, precision, and fruitfulness, can in any way be compared with them.

I point out these matters only to show in what sense the natural sciences are a new and essential element in education, an element that will remain of unquestionable importance in all future developments in education. The complete training of individual men, as well as of nations, is no longer possible without a combination of the traditional logical-literary studies with instruction in the new natural sciences. The majority of educated and cultured people, however, have been trained only in the older way and have hardly come in contact with the kind of thought characteristic of the natural sciences. At the most, perhaps, they have studied a little mathematics. It is men educated in this way who are usually appointed by our governments to educate our children, to maintain reverence for moral order, and to preserve the treasures of knowledge and wisdom of our forefathers. It is they, too, who must organize the changes in the education of the rising generation, wherever such changes are required. They must be encouraged or directed to do this by the public opinion of the informed classes of the whole community, both men and women.

Apart from the natural impulse of every civilized man to lead others to that which he has found to be true and right, there is a strong desire in every friend of the natural sciences to share in such work because he realizes that the future development of these sciences, the increase in their influence on education, and, insofar as they are a necessary element in this education, even the health of the future intellectual development of the race, require that as much insight as possible into the nature and results of scientific investigations —short of a personal occupation with these subjects—be afforded the cultured classes. The necessity of such insight is felt even by those who have grown up under the predominantly linguistic and literary course of instruction. One may cite as proof both the large number of popular books of natural science which appear annually and the eagerness with which popular lectures on subjects in the natural sciences are attended.

It lies in the nature of things, however, that the essential part of this necessity, corresponding to the depth of its roots, is not easily satisfied. It is true that what science has established and worked up in the form of solid results can be brought together by intelligent compilers and put into a form in which a reader without previous knowledge of the subject may, with some patience and perseverance, understand it. Such knowledge, however, limited as it is to factual results, is not really what we have in mind. Indeed, these books, compiled with the best intentions, often lead into false paths. To avoid boredom they must seek to hold the attention of the reader by an accumulation of curiosities, and in this way the image of science is quite incorrectly presented. One often feels this when a reader begins, on his own impulse, to tell what he thought was important in one of these books. There is the further objection that books can give only verbal descriptions or, at best, drawings, representing more or less imperfectly the things and processes of which they treat. The reader's powers of imagination are thereby subjected to a much greater strain, with much less satisfactory results, than those of a scientist or student who, in museum collections and laboratories, sees things before him in their living reality.

It seems to me, however, that what the most intelligent and well-educated of those not trained in the sciences wish is, not so much a knowledge of the results of scientific investigations, as an understanding of the intellectual activities of the investigators, the characteristics of their scientific procedures, the goals toward which they strive, and the fresh points of view which their work affords concerning the great problems of human existence. There is hardly anything about all this in purely scientific treatises on subjects within the natural sciences. On the contrary, the strict discipline of the exact methods requires that in standard works, only that which is completely established—or, at the most, hypotheses or queries for further investigation—be discussed. Whether a scientist says "I know" or "I suppose" is immaterial to the majority of even educated readers; they are concerned only about results and the authority by which these results are supported, not about evidence or doubts. The natural prudence of a serious investigator dictates therefore the greatest care in this connection.

It should not be overlooked that the special discipline of scientific thought which is necessary for the most complete understanding of newly found abstract concepts and laws and for separating out all the accidental occurrences found within the sensuous order of phenomena, as well as the habitual residence of the scientific mind among a circle of ideas far removed from general interest, are not a favorable preparation for a generally intelligible presentation of the insights obtained in the natural sciences to an audience which does not have the same discipline. For this task an artistic talent for exposition, a certain kind of eloquence, is required. The lecturer or writer must find generally familiar images through which he can call forth new ideas with the most vivid clarity, and he must then allow the abstract principles which he wishes to make intelligible to derive concrete life from them. This is almost the opposite mode of treatment from that which obtains in scientific treatises, and it can readily be understood that men who are equally fitted for both kinds of intellectual labor are rare indeed.

Due to all these circumstances, a kind of barrier has been raised between men of science and others who might wish to obtain instruction and guidance from them. Given such a state of affairs, it seems to me all the more gratifying when among those who have the highest qualifications for independent scientific work there is found at times a man like Tyndall, full of enthusiasm to make the newly acquired insights and findings of his science available to a wider circle of people and, at the same time, endowed with eloquence and the gift of lucid exposition, the other qualities necessary for success in achieving this goal.

In England the custom of popular scientific lectures has been in existence much longer than it has in Germany. Because the constitution of the English universities is very different from ours, fewer individuals there are in a position to carry on scientific research as their profession or to give scientific instruction to regularly prepared students. This generally makes it much more difficult for someone to go deeply into a special department of study, although genius can of course overcome these and other obstacles. The same circumstance, on the other hand, has fostered a closer connection between the workers in science and all the other classes of the nation, and it has given rise to a more generous concern for the instruction of students who are not regularly trained. This has hitherto been quite rare in Germany, but in England there have long been solid, well-furnished institutions for this purpose.

The one that stands out most prominently is the Royal Institution in London. It has been called "Royal" only since King George III assumed the patronage of the institution; prior to that, it was established and supported by private funds. This institution has its own buildings, with a large scientific library, a lecture hall, a collection of instruments for physics and chemistry, a laboratory, and so on. A professor of physics and a professor of chemistry (at the present time Tyndall and Frankland) are regularly installed. Sometimes there are single lectures, which are held (on Friday evenings) only for the members or invited guests and which are devoted mainly to the communication of new scientific findings. Sometimes there are courses of from six to twelve lectures on a single subject, chiefly but not exclusively from the natural sciences. Anyone who has the entrance fee may attend these lectures. The lecturer is sometimes a member of the institution who is required to present such a course each year, sometimes an English or foreign scholar who is invited for this purpose. In England, then, courses of a moderate number of connected lectures can be delivered in buildings well-suited for demonstrations and experiments of every kind—a great advantage over the general custom in Germany, where each lecturer delivers only a single lecture.

It is understandable that in the seventy years during which these conditions have prevailed, and because of many other favorable circumstances, the English public has educated its lecturers—and the lecturers their public—much better than has hitherto been the case in Germany. The Royal Institution has had among its professors two men of the first rank, Sir Humphry Davy and Faraday, who have cooperated to that end. Their successor, Professor Tyndall, is held in especially high esteem both in England and in the United States because of his talent for popular presentation of scientific subjects. Anyone who is conscious of having within himself the gift and the power to work in a particular direction for the intellectual development of mankind usually finds pleasure in such activity and its consequences and is ready to devote to it a good share of his time and energies. This is especially the case with Professor Tyndall. He has therefore remained loyal to the Royal Institution, although other honorable posts have been offered to him. It would be quite erroneous, however, to think of him as merely an able, popular lecturer. The greater part of his energies have always been devoted to scientific researches, and we owe to him a series of rather highly original and remarkable investigations and discoveries in physics and physical chemistry.

These are the reasons which led me to believe that the publication of Tyndall's popular lectures in Germany would contribute toward satisfying a real intellectual need of the present period in history. The result so far—that is, Tyndall's book on heat—seems to me clearly to have helped to do so. I have heard men from very different walks of life speak enthusiastically of how useful they have found it. The present volume contains a variety of lectures, delivered upon a variety of occasions. In part they present new discoveries by the author himself; in part they are a discussion of his ideas on the methods of scientific investigation or are illustrations of these methods; and in part they deal with the relationships between natural scientific knowledge and other areas of intellectual activity.

The essay on the scientific use of the imagination is especially important in indicating the characteristics of the author's thought. According to this essay, there are two ways of searching for lawful interrelations in nature: one by the use of abstract concepts, the other by carrying out thoroughgoing experimental research. The first method, involving the use of mathematical analysis, leads ultimately to a precise quantitative knowledge of phenomena. It can only advance, however, where the second method has already at least partially opened up a region—that is, where experimentation has already resulted in an inductive knowledge of the laws of at least some groups of phenomena belonging to it—and where the problem is to test and refine the laws already found, to pass from them to the most general and ultimate principles of the region in question, and to develop fully the implications of these principles. The second method leads to a rich knowledge of the behavior of natural substances and forces; but by this method the laws or regularities are understood at first only in the way artists grasp them, that is, by means of vivid sensuous intuitions of types of action or behavior. Only later are these intuitions worked out in the form of concepts.

These two sides of the physicist's work are never quite separate, although sometimes the diversity of individual talent will make one man tend toward mathematical deduction and another toward the inductive activity of experimentation. Should the first method become wholly divorced from actual observations, it runs the risk of laboriously building air castles on unstable foundations and of not recognizing the points at which it can check to see whether its deductions agree with the facts. The second method, on the other hand, may lose sight of the proper aim of science if it does not work toward bringing its observations ultimately into the precise form of concepts.

The first discovery of a law of nature, that is, the discovery of a new regularity or uniformity among apparently unconnected phenomena, is a matter of wit (taking this word in its broadest meaning) and is nearly always accomplished only through the analysis of rich sensory experience. The completion and refinement of what has been discovered devolves afterward on the deductive work of concept formation—preferably on mathematical analysis, for the final problem is always one concerning the equality of quantities.

Tyndall is predominantly an experimenter: he forms his generalizations on the basis of extensive observation of the play of natural forces and then, as he indicates in the lecture just referred to, extends what he has seen to both the greatest and the smallest spatial dimensions. It would be a serious mistake, however, to consider what he calls imagination to be mere fancy. Precisely the opposite meaning is intended—that is, intuition based upon rich empirical experience. The clarity of his lectures on physical phenomena and his success as a popular lecturer are undoubtedly to be attributed to this kind of imagination.

Tyndall is close to us in Germany in that he completed part of his studies here, chiefly in Marburg. His love for German literature and science manifests itself at many places in his writings. He has also shown his gratitude toward us in that he has sought many times to get the work of Continental investigators, such as Robert Mayer and Kirchhoff, the recognition it deserves in his own country. He came immediately to the defense of the investigations of glaciers carried out by the Swiss scientists Rendu, Agassiz, and Desor. The same gratitude to Germany is documented in the bequest which he made from the profits of his earnings in his tremendously successful course of lectures in America. He directed, at the conclusion of that tour, that from these funds "two American students who show definite talent for physics and who declare their intention to make this their life's work be assisted or supported at such European universities as, in the opinion of the trustees of the bequest, appear most suitable for that purpose. My wish would be that each of these students spend four years at a German university, three of which should be used for their formal education and one for independent research."

Thus I find it all the more to be deplored that in Germany, Tyndall has been made the target of an attack which, in the name of German patriotism, has been directed against the penetration of foreign scientific tendencies. This attack is to be found in J. C. F. Zöllner's book on the nature of the comets. The reason for the attack, insofar as it depends upon scientific differences, is the author's philosophical opposition to the inductive method of science, which was first carefully formulated by Bacon and which was followed earliest and has been followed most seriously by his fellow Englishmen. This is, of course, an old point of contention, over which many rivers of bitter polemic have already flowed.

The natural sciences have made extensive and rapid progress precisely to the degree in which they have freed themselves from the influence of so-called a priori deductions. We in this country have been the last to break away but at the same time the most resolute, and German physiology in particular bears witness to the range and significance of this freedom. The break occurred, however, in opposition to the last great systems of metaphysical speculation, which have harnessed and fettered the expectations and interest of the educated part of the nation. It occurred in a fight against the belief that only pure thought involves important, intelligent work, while the collection of empirical facts is, on the contrary, low and common.

I restrict the name *metaphysics* here to the so-called science whose goal is to establish, by pure thought, conclusions concerning the ultimate principles of the universe. I wish in particular to guard against the belief that what I say against metaphysics is applicable to philosophy in general. It seems to me that nothing has been so unfortunate for philosophy as its repeated confusion with metaphysics. The latter has played to the former somewhat the same role that astrology once played to astronomy. It is chiefly metaphysics which has drawn the attention of the majority of scientific dilettantes to philosophy, and this mass of devotees and hangers-on have actually brought more harm to it than the bitterest opponent could have done. It has been a delusive hope—that of being able to obtain, in a comparatively quick and effortless way, insight into the most fundamental order of the universe and the ways of the human spirit, both in the past and in the future—that has aroused the interest which has led so many to the study of philosophy, just as the hope of gaining insight into the future once provided authority and support for astronomy.

What philosophy can teach us—or what, by continued study of the facts, it will someday be able to teach us—is of the highest importance for the scientific thinker, for he must learn to know fully the instrument with which he works, that is, the human intellect. This abstract and difficult study, however, will continue in the future to offer only scant and hard-won reward for the satisfaction of a dilettantish curiosity—or, more to the point, human egotism—just as the mathematical mechanics of the planetary system and the calculation of perturbations were once much less popular, in spite of their admirable systematic completeness, than astrological sophistry.

To be sure, the new metaphysics has given up the bold and, in its very boldness, astonishing plan to develop a system of everything worth knowing out of pure thought. There is now a readiness to absorb great masses of material from the empirical sciences and to formulate hypotheses which are explicitly recognized as hypotheses. On the other hand, a number of a priori propositions still remain, among which Zöllner, for example, includes the law of gravitation and the biological principle of the generation of like out of like.

There are perhaps many readers, to whom the natural sciences are foreign and in whose hearts there remains a residue of hope for the future fulfillment of the bold scheme of a great speculative system, who are inclined to grant some credence to Zöllner's exposition of the principle of natural scientific method and the history of scientific discoveries. To do so, however, would only set back the hope of a final solution of the problems in our present program of education. There is no room here to go into details; I must restrict myself to the plea that no exposition be accepted without criticism and to the hope that men who are accustomed to scientific rigor will be able to tell, even when they are not familiar with the subject matter, where such rigor is present and where it is not.

12.

THOUGHT IN MEDICINE [1877]

An address delivered on August 2, 1877, on the thirty-fifth anniversary of his doctorate at the Royal Friedrich-Wilhelm Institute of Medicine and Surgery in Berlin

---

IT is now thirty-five years since, on the second of August, I stood on the rostrum in the hall of this institute, before another audience such as this one, and read a paper on the operation for tumors of the blood vessels. I was then still a student at the institute, just at the end of my studies. I had never seen a tumor operation performed, and the substance of my lecture was taken entirely from books. Knowledge gained from books played at that time, of course, a far greater and far more influential role in medicine than we are now inclined to assign to it. That earlier period was one of fermentation; it was the period of the struggle between the learned tradition and the spirit of natural science, which would have no more of tradition and wished to rely upon personal experience. The authorities at that time judged more favorably of my lecture than I did myself, and I still possess the books which were awarded to me as the prize.

The memories which have crowded in upon me on this occasion have brought vividly before my mind a picture of the state of our science at that time and of our endeavors and hopes, and they have led me to compare the state of things then with that which has since developed. Much indeed has been accomplished. Although not all that we hoped for has been realized and many things have turned out differently from what we then wished, we have gained much for which we could not have dared to hope. Just as the history of the world has taken one of its rare giant steps before our very eyes, so also has our science. An old student like myself scarcely recognizes the somewhat matronly aspect of Dame Medicine when he accidentally comes in contact with her again, so vigorous and so capable of growth has she become in the fountain of youth of the natural sciences.

I may, perhaps, retain the impression of those earlier struggles more freshly than do those among my contemporaries (some of whom I have the honor to see assembled before me) who, having remained permanently connected with the science and practice of medicine, have been less struck and less surprised by the great changes, which have naturally taken place by slow steps. In any case, this will be my excuse for speaking to you about the metamorphosis which has taken place in medicine during the past thirty-five years (and with whose specific results you are better acquainted than I am). In particular, I should not like the impression of this development and its causes to be completely lost on the younger members of my audience. If you were to glance sometime at the literature of that earlier period, you would meet with principles which appear to be written in a lost language—so much so that it is by no means easy to adapt oneself to the mode of thought of that period, which in reality lies so closely behind us.

The course of development of medicine is an instructive lesson in the true principles of scientific inquiry, and the positive part of this lesson has, perhaps, never been so impressively taught as during the past generation. Since it is my duty to teach that branch of the natural sciences which makes the most inclusive generalizations and deals with the most fundamental concepts (for which reason it is still quite appropriately called natural philosophy by English-speaking people), it does not fall too far out of the range of my official duties and of my own studies if I attempt to discuss here the most general methodological principles of the empirical sciences.

With regard to my acquaintance with the tone of thought of medicine thirty-five years ago, I had at that time a special incentive over and above the general obligation incumbent upon every educated physician to understand the literature of his science and to recognize the direction of, as well as the conditions necessary for, its progress. In my first professorship at Königsberg, from 1849 to 1855, I had to lecture each winter on general pathology, that is, on that part of medicine dealing with the most general theoretical concepts concerning the nature of disease and the principles of its treatment. General pathology was regarded at that time as the fairest blossom of medical science. Now, of course, what once formed the essence of medicine is only of historical interest to the disciples of modern natural science.

Many of my predecessors, in particular Henle and Lotze, had already broken a lance in questioning the scientific qualifications of general pathology as it was then taught. Indeed, Lotze, who also left medicine, presented in his work on general pathology and therapeutics (published in 1842) a methodical and thoroughly devastating critique of general pathology.

My original and basic inclination was toward physics; external circumstances forced me to take up the study of medicine, which the liberal provisions of the Friedrich-Wilhelm Institute made possible. It was the custom at that time, however, to combine the study of medicine with that of the natural sciences, and I now consider those compelling circumstances to have been a stroke of good fortune. This is so, not only because I entered medicine at a time when anyone who was even moderately at home in physical investigations found a fruitful virgin soil for cultivation, but because the study of medicine, it seems to me, taught me more vividly and more convincingly than any other training could have done the everlasting principles of all scientific work, the principles which are so simple and yet are always being forgotten, so clear and yet are always hidden by a deceptive veil.

Perhaps one must personally face the fading eyes of one who is approaching death and witness the grief of a distracted family; perhaps one must ask himself the difficult question whether he has done everything possible to ward off the dread event—and whether science has done everything it could have done—in order to realize that epistemological problems concerning the methodology of science are of great importance and have serious practical consequences. The purely theoretical scientist, provided he remains undisturbed in his study, may smile with aristocratic calm when, for a time, imagination and fantasy run wild within a science. Or he may look upon the effects of ancient prejudices as interesting and pardonable, the products of poetic romanticizing or youthful enthusiasm. One who must deal with the hostile forces of reality, however, must thrust aside both indifference and romanticism. What one knows and what one can do are exposed to severe tests. One must rely only on the hard, clear light of facts and never be lulled by pleasant illusions.

I am pleased, therefore, to be able to address once more an assembly consisting almost exclusively of physicians who have gone through the same training I once underwent. Medicine was once my intellectual home, the one in which I grew up, and the wanderer best understands and is best understood in his native land.

If I were called upon to state in a few words the fundamental error of that earlier period, I should be inclined to say that it lay in the pursuit of a false ideal of science. It had a one-sided, mistaken reverence for the deductive method. Medicine, to be sure, was not the only science in which this error was found, but in no other science were its consequences so glaring or so detrimental to progress. The history of medicine occupies, therefore, a special place in the history of the development of the human mind. No other science is, perhaps, in a better position to show that a proper critique of the sources of knowledge is in actual practice an exceedingly important task of true philosophy.

"Godlike is the physician who is also a philosopher"—these proud words of Hippocrates served as a banner, as it were, of the old deductive medicine. We must surely assent to this maxim, once we understand what the term *philosopher* meant at the time of Hippocrates. In ancient times philosophy embraced all theoretical knowledge; philosophers then pursued, in close connection with actual philosophical and metaphysical investigations, mathematics, physics, astronomy, and natural history. The physician-philosopher of Hippocrates was one who has a complete insight into the causal connections of natural processes, and we may indeed agree that such a man can give godlike aid. Understood in this sense, the aphorism states in a few words the ideal toward which our science must strive. Who can say whether that goal will ever be attained?

Those disciples of medicine, however, who considered themselves divine in their own lifetime and wished to impose themselves as such upon others were not content to delay their apotheosis for a long period of study. The requirements for status of philosopher were therefore considerably relaxed. Every adherent of any cosmological system into which the facts of reality could be fitted in some fashion or other held himself to be a philosopher. The philosophers of that time knew little more about the laws of nature than the untaught laymen; the emphasis in their work was placed upon thinking, upon the logical rigor and completeness of their systems.

It is not difficult to understand how, in periods of youthful development, such a one-sided overestimation of the importance of pure thought could come to be accepted. The superiority of man over other animals and of the educated man over the barbarian depends upon thinking. Sensation, feeling, and perception, on the other hand, are shared with the lower animals, and in acuteness of the senses many of them are even superior to man. The development of thought to the utmost became man's goal, upon whose achievement the feeling of his own dignity, as well as his own practical power, depended. It was a natural error to consider unimportant the dowry of those mental capacities which nature had also given to animals. It was natural to believe that man could liberate thought from its natural basis, observation and perception, in order to begin an Icarian flight of metaphysical speculation.

It is, in fact, no easy task to ascertain completely the origins of our knowledge. An enormous amount of it is transmitted by speech and writing. Man's power of gathering together the stores of knowledge of generations is the chief reason for his superiority over other animals, who are restricted to inherited blind instincts and to individual experience. All the knowledge which is transmitted, however, is handed on already formed; where the transmitter obtained it and how much criticism he gave it can seldom be determined, especially if it has been handed down through several generations. We must accept such knowledge upon good faith, for we cannot get at the source of it; and after many generations have been content with it, have brought no criticism to bear upon it, and have, indeed, gradually added all sorts of small alterations which finally have accreted into large ones—after all this, strange things have often been reported and accepted upon the authority of primeval wisdom. A curious case of this kind is the history of the attempts to explain the circulation of the blood, of which we shall speak later.

Those who reflect on the origins of knowledge are also aware of another kind of linguistic tradition, one which is more complicated and which remained unnoticed for a long time. Names for classes of objects and of processes would not be formed readily within a language if we did not have to refer frequently and under varying circumstances to the separate members of these classes and to indicate what they have in common. These individuals must, therefore, possess many traits in common. If, reflecting scientifically upon them, we select a few of these characteristics upon which to base a definition of the class, common possession of these selected characteristics implies that a large number of other characteristics will also be regularly met with and that there is a natural relation between the former and the latter characteristics. If, for example, we assign the name *mammal* to those animals which, when young, are suckled by their mothers, we can further assert that they are all warm-blooded, that they are born alive, that they have a spinal column but no quadrate bone, that they breathe by means of lungs, that they have separate chambers of the heart, and so on. In short, the fact that a certain class of things is included under a common name indicates that they fall under a general natural law. In this way a host of experiences are transmitted from generation to generation without our being aware that such a transmission has actually taken place.

The adult, moreover, when he begins to reflect upon the origins of his knowledge, finds himself in possession of an enormous number of everyday experiences which extend, on the whole, back into the obscurity of early childhood. Particular events and individual cases have long since been forgotten, while the traces which the daily repetition of such cases have left in the memory have been deeply engraved. Since only that which is in conformity with laws is repeated with regularity, these deeply impressed traces of previous perceptions provide us with a conception of what aspects of things and processes are conformable to law. Thus a man, when he begins to think, finds that he is in possession of an extensive body of knowledge whose origin he does not know but which he has possessed as long as he can remember. To explain this does not require any reference to the possibility of hereditary transmission.

The concepts which he has formed—the concepts which his native language has transmitted to him—act like regulative principles even in the objective world of fact; and since he does not realize that he or his forefathers formed these concepts on the basis of the things themselves, the world of facts seems to him to be governed, like his concepts, by intellectual forces. We can recognize this psychological anthropomorphism all the way from the "ideas" of Plato to the "immanent dialectic of the world processes" of Hegel and the "unconscious will" of Schopenhauer.

Natural science, which in earlier times was identical with medicine, followed the path of philosophy: the deductive method seemed to be capable of achieving everything. Socrates, to be sure, developed the inductive procedure of concept formation in a most instructive manner. The best of what he succeeded in accomplishing was, however, as so frequently happens, never fully understood.

I shall not lead you through the confusing maze of theories of pathology which, depending upon the particular inclinations of their authors, sprang up in consequence of one or another development within the sciences. These theories on the whole were put forward by physicians who, independently of their theories, attained fame as great observers and empirics. Following these men came the less gifted students who, undisturbed by any discordance with nature, imitated and exaggerated their masters' theories, making them more logical but also more one-sided. The more rigorous the system, the fewer and more general were the methods to which the healing art was restricted. The more the schools were driven into a corner by the increase in real knowledge, the more they depended upon ancient authorities and the more intolerant they became of innovations. The great reformer of anatomy, Vesalius, was cited before the theological faculty of Salamanca; Servetus was burned at Geneva along with the book in which he described the pulmonary circulation of the blood; and the faculty of the University of Paris prohibited the teaching of Harvey's analysis of the circulation of the blood in its lecture halls.

The foundations of the systems from which these schools started were, on the whole, scientific concepts whose use within restricted areas would have been quite legitimate. What was not correct was the belief that it was more scientific to explain all diseases by using only one explanatory principle rather than several. What was called solidar pathology wanted to reduce everything to alterations in the operation of the solid parts of the body; specifically, to alterations in their tension, to *strictum* and *laxum*, to tone and lack of tone, and finally to strained or unstrained nerves and to obstructions in the vascular systems. Humoral pathology was acquainted only with changes in various mixtures. The four cardinal fluids (blood, phlegm, black bile, and yellow bile) which were representatives of the four classical elements; the acrimonies or dyscrasies, which had to be driven out by sweating and purging; at the beginning of our modern period, the acids and alkalis, or the alchemical spirits and the occult qualities of assimilated substances—all were elements in this system.

Among such ideas, all sorts of physiological insights were to be found, some of which, such as the innate vital force of Hippocrates (the source of all motion, maintained by nutritive substances digesting in the stomach), contained remarkable foreshadowings of subsequent concepts. Here, for example, is to be found the path which led later to the discovery of the equivalence between mechanical work and animal heat, as well as to the law of the conservation of force. On the other hand, the *pneuma*, which was thought of as being half spirit and half air and as being driven from the lungs and filling the arteries, produced a great deal of confusion. The fact that air is generally found in the arteries of dead bodies (it only enters them, as a matter of fact, the moment the vessels are cut) led the ancients to believe that air is also present in the arteries during life. This left only the veins in which the blood could circulate. It was believed that the blood was formed in the liver, moved from there to the heart, and then flowed through the veins to the various organs of the body. Careful observation of the operation of bloodletting should have shown that the blood in the veins comes from the periphery and flows toward the heart. This false theory of circulation, however, had become so closely interwoven with the explanation of fever and inflammation that it had acquired the status of a dogma and was dangerous to attack.

Still, the really fundamental error of these systems remained the erroneous kind of logical inference to which they supposedly were forced. I refer to the belief, common to all these schools, that it was possible to build a comprehensive system, embracing all forms of disease and their cures, upon one simple explanatory principle. The complete knowledge of the causal connections of one class of phenomena does, indeed, result in a logically coherent system. There is no prouder edifice of rigorous thought than modern astronomy, within which even the smallest disturbances are deduced from Newton's law of gravitation. Newton, however, was preceded by Kepler, who related the facts to one another by induction; and astronomers have never believed that Newton's force excludes the simultaneous action of other forces. They have been continually on the watch to see whether or not friction, or a resisting medium, or perhaps a swarm of meteors might also have some influence.

The ancient philosophers and physicians believed they could make deductions before they had established their general principles by induction. They forgot that the conclusion of a deductive argument can have no more certainty than the principles from which it is deduced and that each new deductive conclusion is basically a new experiential test of its own premises. The fact that a conclusion follows by the strictest logic from a questionable premise does not give it the slightest validity or value.

One characteristic of the schools which built their systems upon hypotheses which were accepted as dogmas was their intolerance, which I have already mentioned briefly. Anyone who starts from a well-established foundation can readily admit an error; he need give up nothing more than that concerning which he is mistaken. If, however, the starting point is a hypothesis which either seems to be guaranteed by authority or is held only because it agrees with what one wishes to believe is true, any flaw may hopelessly unravel all the whole structure of belief. The strict adherent of such a school must therefore claim the same degree of infallibility for each individual part of a system, e.g., for the anatomy of Hippocrates just as much as for the concept of fever crises. Every opponent appears to be either stupid or evil; and the polemic, according to an old rule, will be so much the more passionate and personal, the more questionable the theory which is being defended.

We have had frequent opportunities to confirm these general reflections while studying the schools of dogmatic, deductive medicine. They turned their intolerance partly against one another and partly against the eclectics, who found different explanations for different kinds of disease. Eclecticism, which in reality is completely justified, had in the eyes of systematists the defect of being illogical. Yet the greatest physicians and observers—Aretaeus, Galen, Sydenham, and Boerhaave, with Hippocrates at their head—were eclectics, or at any rate, very lax systematists.

At the time when the oldest among us today began the study of medicine, it was still under the influence of the important discoveries which Albrecht von Haller had made concerning the excitability of the nerves, discoveries which he had related to the vitalistic theory of the nature of life. Haller had observed the excitability in the nerves and muscles of amputated arms and legs. What most surprised him was that the most varied external agents—mechanical, chemical, and thermal, to which electrical were subsequently added—always had the same result; that is, they all produced muscular contractions. These actions differed only quantitatively, that is, only in the intensity of the effect. Haller therefore used the term *stimulus* to refer to all these agents; he called the altered condition of the nerves their *excitation* and their capacity (which was lost at death) of responding to a stimulus their *excitability.*

This excitability, which, considered physically, really indicates only that the nerves have a state of equilibrium which is easily disturbed by excitation, was taken to be the fundamental property of animal life and was unhesitatingly extended to the other organs and tissues of the body, an extension for which there was no justification in fact. It was believed that none of the organs and tissues of the body were active in themselves and that they had to receive impulses from external sources; air and nourishment were considered the normal stimuli. The *kind* of activity displayed by an excited organ seemed, on the other hand, to be determined by its specific energy, which was under the influence of the vital force. Increase and diminution of excitability were the categories under which all the acute diseases were placed, these categories also indicating the type of treatment to be recommended. The complete one-sidedness and the unrelenting logic with which Robert Brown once worked out this system had been discredited, but the system still furnished the most important general point of view.

The vital force was once thought of as an ethereal spirit, a *pneuma*, residing in the arteries. Later, with Paracelsus, it took the form of an *archeus*, a kind of useful kobold or indwelling alchemist. Still later it received its clearest scientific formulation as the life soul, or *anima inscia*, in the work of Georg Ernst Stahl, who was professor of chemistry and pathology at Halle in the first half of the last century. Stahl had a clear, acute mind, and he is informative and stimulating in the way he asks the right questions, even in those cases in which he decided against our present views. It was he who established the phlogiston theory, the first comprehensive system of chemistry. If we translate his *phlogiston* into *latent heat*, the theoretical basis of his system may on the whole be translated into the system of Lavoisier. (Stahl was not familiar with oxygen, which occasioned some false hypotheses, such as the one about the negative weight of phlogiston.)

Stahl's life soul was constructed in part on the model which the pietist communities of his time used to represent the sinful human soul. It was subject to errors and passions, to sloth, fear, impatience, sorrow, indiscretion, despair. The physician had either to soothe it, or goad it, or punish it and compel it to repent. The way Stahl established the necessity of the action of physical and chemical forces in the body was well thought out: the life soul governed the body and acted only through the physico-chemical forces of assimilated substances. It had the power to bind and release these forces, to restrain them or allow them full play. After death the restrained forces were freed, and putrefaction and decomposition began. In order to refute this hypothesis of binding and releasing, it was first necessary to formulate clearly the law of the conservation of force.

The second half of the eighteenth century was too much under the influence of the principles of the Enlightenment to acknowledge openly Stahl's life soul. The concept, retaining in the main the same functions, was presented more scientifically as *vis vitalis*, the vital force. Under the name of "nature's healing power," it played a prominent role in the treatment of diseases.

The doctrine of vital force entered into theories of pathology in connection with changes in excitability. An attempt was made to distinguish between those pathological conditions which were the immediate effects of blind natural forces (*symptomata morbi*) and those which were due to the reaction of the vital force (*symptomata reactionis*). The chief among the latter were inflammation and fever. The physician's principal function was to observe the strength of these reactions and to stimulate or moderate them according to circumstances.

The treatment of fever seemed to be the main thing, the field of medicine which had a real scientific foundation and relative to which all local treatments were considered unimportant. The treatment of all diseases which produced a fever tended to be very much the same, although bloodletting, which since that time has been almost completely abandoned, was still extensively used. Therapeutics became still more impoverished as the younger, more critical generation grew up and came to question the presuppositions of what had earlier been considered scientific. Among the younger generation were many who, in despair concerning the state of their science, either gave up therapeutics almost entirely or accepted on principle an empiricism such as the one Rademacher then taught, which considered any expectation of a scientific explanation a vain hope.

What we learned thirty-five years ago were only the ruins of the older dogmatism, whose questionable features soon manifested themselves.

The vitalistic physicians believed that the most fundamental part of the vital processes was not subject to natural forces—that is, to forces which, acting with blind necessity and according to fixed laws, determined what occurred. What such forces could do appeared to them of secondary importance, scarcely worthy of careful study. They thought they had to deal with a soul-like entity, an entity which fell within the jurisdiction of a thinker, a philosopher, or a man of spiritual insight. May I illustrate this by a few examples?

At that time auscultation and percussion of the organs of the chest were regularly being practiced in the clinical wards. Often, however, I heard it maintained that these were coarse, mechanical methods of investigation, which a physician with a clear mental vision did not need, and that these methods debased the patient, who was after all a human being, by treating him as a machine. To feel the pulse seemed the most direct method of learning the mode of action of the vital force, so this was practiced as by far the most important means of investigation. To measure the pulse rate with some kind of chronometer was quite a common practice, although this method seemed to the older gentlemen not quite in good taste. At that time there was no thought of measuring temperature in cases of disease. In reference to the ophthalmoscope, a celebrated surgeon and colleague told me that he would never use the instrument, as it was too dangerous to admit crude light into diseased eyes. Another declared that the instrument might be useful for physicians with bad eyes, but his were good, and he did not need it.

A professor of physiology at that time, celebrated for his literary activities and noted as an orator and man of intelligence, had an argument over the images in the eye with a colleague who was a physicist. The latter challenged the physiologist to visit him and witness an experiment. The physiologist, however, refused his request with indignation, announcing that a physiologist need have nothing to do with experiments, which were of value only to the physicist. Another aged and learned professor of therapeutics, who occupied himself a great deal with the reorganization of the universities in order to restore the good old days, urged me to divide physiology into two parts, to lecture on the really intellectual part myself, and to hand over the lower, experimental part to a colleague whom he regarded as good enough for that purpose. He quite gave up on me when I said that I considered experiments to be the true basis of science.

I mention these things, which I experienced myself, in order to indicate the feelings of the older schools–indeed, of the most illustrious representatives of medical science–toward the developing ideas of the natural sciences. In the literature these feelings naturally found fainter expression, for the old gentlemen were cautious and worldly-wise.

You can appreciate how great a hindrance to progress such feelings on the part of influential, respected men must have been. Medical education at that time was based mainly on the study of books. There were lectures, of course, but they were restricted mainly to dictation; and as to experiments and demonstrations in connection with the lectures, the provisions made were sometimes good and sometimes poor. There were no laboratories of physiology or physics in which the student himself might go to work. Liebig's great achievement, the foundation of the chemistry laboratory at Giessen, had been completed, but his example had not been followed elsewhere. Yet in anatomical dissections medicine possessed an important method of training in independent observation–an advantage which was not available to the other academic faculties and to which I am inclined to attach great weight. Microscopic demonstrations in the lectures were unsystematic and infrequent. Microscopes were costly and scarce. I came to own one by having spent my autumn vacation in 1841 in the Charité, prostrated by typhoid fever; as a student I was nursed without expense, and upon my recovery I found myself still in possession of my small savings. The instrument I bought with them was not beautiful, but I was able to use it to study the prolongations of the ganglionic cells in the invertebrates, which I described in my dissertation, and to carry out investigations of the vibrios in my research on putrefaction and fermentation.

Any of my fellow students who wished to make experiments had likewise to do so at their own expense. One thing we learned thereby–the younger generation perhaps does not learn this as well in the laboratories–was to consider every possible way of achieving a desired end and to exhaust all possibilities in searching for a useful path. We had, to be sure, an almost totally uncultivated field before us, a field in which almost every stroke of the spade could produce fruitful results.

There was one man in particular who aroused our enthusiasm for work in the right direction–the physiologist Johannes Müller. On theoretical issues he still favored the vitalistic hypothesis, but on the most essential points he was a natural philosopher, firm and immovable; to him, all theories were only hypotheses which had to be tested by facts and concerning which only facts could decide. Even opinions about those points which tended most easily to crystallize into dogmas, that is, opinions about the mode of operation of the vital force and the activity of the conscious soul, he tried continually to define more precisely and to prove or refute by referring to facts. Although he relied most heavily upon techniques of anatomical investigation, which were most familiar to him, he familiarized himself also with the alien methods of chemistry and physics. He furnished the proof that fibrin is dissolved in blood, experimented on the propagation of sound in such mechanisms as are found in the inner ear, and treated the action of the eye according to the principles of optics.

His most important achievement for the physiology of the nervous system, as well as for the theory of knowledge, was the firm empirical foundation he gave to the principle of the specific energies of the nerves. With respect to the distinction between motor nerves and sensory nerves, he showed how to make the experiments necessary to prove Bell's law concerning the roots of the spinal cord in such a way that they were free from error; and with respect to the specific energies of the sensory nerves, he not only established the general law but carried out a large number of specific investigations in order to eliminate objections, close loopholes, and refute erroneous interpretations. What earlier had been assumed to be true on the basis of everyday experience and had been expressed in a vague, misleading manner—or what had been established only within a narrow region, such as Young's work in the theory of colors or that of Sir Charles Bell concerning the motor nerves—emerged from Müller's hands in a state of classical perfection, a scientific achievement whose value I am inclined to consider equal to that of the discovery of the law of gravitation.

His scientific spirit and especially his example had a strong influence upon his students. My fellow students and I were preceded by Schwann, Henle, Reichert, Peters, and Remak; as fellow students I met E. du Bois-Reymond, Virchow, Brücke, Ludwig, Traube, J. Meyer, Lieberkühn, and Hallmann; we were succeeded by A. von Graefe, W. Busch, Max Schultze, and A. Schneider.

Microscopic and pathological anatomy, experimental pathology and therapeutics, the study of organic types, physiology, and ophthalmology all developed in Germany under the influence of this powerful impulse far beyond where they were in other countries. This advance was helped along, of course, by the labor of a number of Müller's contemporaries who held similar views.

Among these the three Weber brothers of Leipzig, who achieved solid results in investigations of the mechanism of circulation and of the muscles, the joints, and the ear, must be mentioned first of all.

An attack was made wherever a way could be seen to make one of the vital processes comprehensible; it was presupposed that they could be understood, and success justified this presupposition. Since that time a large number of delicate techniques have been developed in microscopy, in physiological chemistry, and for vivisection, the latter being greatly facilitated by the use of ether as an anesthetic and of curare as a paralyzing agent. By the use of these drugs a number of profound problems, which to my generation seemed impossible to solve, have been opened to attack. The thermometer, the ophthalmoscope, the aural speculum, the laryngoscope, and nervous stimulation of the living body—all give the physician means of delicate, reliable diagnosis where earlier there was nothing. The ever-increasing number of established parasitical organisms means the substitution of tangible objects for mystical entities, the knowledge of which helps the surgeon to forestall the fearful, insidious diseases of decomposition.

But please do not think, gentlemen, that the struggle is at an end. As long as there are people of such astonishing conceit as to imagine that they can establish by some flash of genius what really can be achieved only by hard labor, there will be hypotheses formulated which, propounded as dogmas, promise to solve all problems at once. And as long as there are people who believe implicitly in whatever they wish were true, these hypotheses will find credence. Both classes of people will certainly not die out, and the majority of men will always belong to the latter class.

There are two motives in particular, either of which leads to metaphysical systems. In the first place, men always wish to think of themselves as beings of a higher order, above the level of the rest of nature; this wish is satisfied by some form of spiritualism. Secondly, they like to believe that by the use of their conceptual systems—those conceptual systems, of course, which have been developed at their particular time in history—they are unquestionably lords of the world; this belief leads to materialism.

But anyone who, like the physician, must actively face up to the forces of nature which produce good or ill is under an obligation to search for a knowledge of the truth and of the truth only, regardless whether what he finds is agreeable to the spiritualists or to the materialists. His aim is firmly fixed; for him factual results are the final test. He must try to determine, before undertaking any course of action, what the consequences of that action will be. To gain a knowledge of what has not yet been observed but will surely occur, there is no other method than that of learning by observation the laws of phenomena, and these laws can be learned only by induction—only by the careful production, observation, and investigation of the cases which fall under them. When, finally, we believe that we have arrived at a law, the business of deduction may properly begin. Then it is our duty to develop the consequences of our law as completely as possible, but only so that we can apply the test of experience to it and determine whether it can be tested, whether it holds or not, and if so, under what conditions. This testing never really ceases. The true natural scientist, when some new kind of phenomena appears, wonders whether even the most firmly established laws of the most familiar forces might not have to be changed, though no change may be contemplated, of course, which contradicts the complete body of our previous experience.

Our knowledge never reaches the status of unconditional truth; it reaches only a degree of probability so high as to be practically equal to certainty. The metaphysicians may amuse themselves at this; we shall take their mocking to heart, however, only when they are in a position to do better, or even as well as can be done by the inductive method. The saying about Socrates, that great master of inductive concept formation, is as fresh today as it was two thousand years ago: "Others imagined that they knew what they did not know; he had the advantage in not pretending to know what he did not know." Socrates was surprised that others could not see clearly that it is impossible for men to discover ultimate truth. Those who most prided themselves on their theories, he noted, did not agree among themselves and behaved toward one another like madmen. Schopenhauer, on the other hand, called himself a Mont Blanc by the side of a molehill when he compared himself to a natural philosopher. Students admired these big words and tried to imitate the master.

When I speak against the empty manufacture of hypotheses, please do not imagine that I wish to deny the value of genuine, original thought. The discovery of a new law is the discovery of a uniformity which hitherto was concealed in the course of natural processes. Such a discovery is a manifestation of what was once, in a serious sense, described as wit; it is of the same nature as the highest achievements of artistic intuition manifested in the discovery of new forms of expression. It is something which cannot be forced and which cannot be deliberately accomplished by any known method. For this reason, all those who wish to pass as the favored children of genius strive after it. It seems so easy, so effortless, to gain an advantage over one's contemporaries by sudden mental flashes.

Real artists and true investigators know, however, that great achievements can be accomplished only by hard work. The proof that while some new ideas merely scrape together superficial resemblances, others are the result of a deep insight into the relations underlying phenomena, is only given when the ideas are completely developed. In the case of newly discovered natural laws, this involves showing their agreement with facts. Ultimately the test is not a matter of public success; it is a matter of how deep and complete the insight is.

To find superficial resemblances is easy; they are amusing to society, and clever ideas soon gain for their author a reputation as an ingenious man. Among a large number of such ideas there are always some which will ultimately be found to be partially or wholly correct; it would show remarkable skill *always* to be wrong. If he happens on a good idea, a man can loudly claim his priority in making a discovery; in the opposite case, a fortunate oblivion conceals the erroneous conclusions. Fellow adherents of this procedure are always glad to certify the value of an "initial insight." Conscientious workers, meanwhile, since they are hesitant to present their thoughts to the public before they have been thoroughly tested and their truth established beyond question, are at a distinct disadvantage. The present practice of settling questions of priority only by the date of first publication, without any consideration of the maturity of the research, has seriously favored this sort of abuse.

In the printer's type-case is to be found all the wisdom of the world, all that has been or ever will be discovered; it is only necessary to know how the letters are to be arranged. Analogously, in the hundreds of books and pamphlets published each year about the ether, the structure of atoms, the theory of perception, the nature of asthenic fever and carcinoma, and so on, are to be found the most subtle variations of every possible hypothesis. Among these there must necessarily be many fragments of the correct theories. Who knows, however, how to find them?

I insist upon this in order to make clear to you that all this literature—this mass of untried, unconfirmed hypotheses—has no value in the progress of science. On the contrary, the few sound ideas which this literature may contain are concealed by the rubbish of the rest. Anyone who wants to publish some new and well-confirmed facts sees himself open to the danger of countless claims of priority, unless he is prepared to waste both time and energy in reading a pile of absolutely useless books and to destroy his readers' patience by a multitude of useless citations.

Our generation has had to suffer under the tyranny of spiritualistic metaphysics; the younger generation will probably have to guard against materialistic metaphysics. Kant's rejection of the claims of pure reason has gradually made an impression, but Kant allowed one escape. It was as clear to him as it had been earlier to Socrates that all the metaphysical systems propounded before his time were networks of fallacious thought. His *Critique of Pure Reason* is a continual sermon against the use of the categories of reason beyond the limits of possible experience. Geometry, however, seemed to him to provide something which metaphysics was striving after. He believed that the axioms of geometry, which he looked upon as a priori principles antecedent to all experience, were given by transcendental intuition and that they provided the innate form of all external intuition.

Since Kant's time, pure a priori intuition has been the anchoring ground of metaphysicians. As such, it is even more convenient than pure thought, since anything and everything can be built upon a priori intuitions without bothering with chains of reasoning, which of course are always open to proof or to refutation. The nativistic theory of sense perception is an example of an appeal to a priori intuition in physiology. All metaphysicians unite to fight against any attempt to resolve these intuitions—whether pure or empirical; whether the axioms of geometry, the principles of mechanics, or the perceptions of vision—into their elements.

It is for this reason, of course, that I consider the investigations of Lobachevsky, Gauss, Riemann, and others concerning the changes which are logically possible in the axioms of geometry—and especially their proof that the axioms are principles which must be confirmed, or perhaps even refuted, by experience and thus are *principles acquired from experience*—to be such very important steps forward. The fact that all metaphysical sects fly into a rage on this issue must not lead you astray. These investigations lay the axe at the base of what is apparently the firmest support which their claims still possess.

Please do not forget that materialism is also a metaphysical hypothesis. It has proved itself very fruitful in the area of the natural sciences, but it is nevertheless a hypothesis. If one forgets this, materialism becomes a dogma which hinders the progress of science and, like all dogmas, leads to violent intolerance. The danger of a hypothesis becoming a dogma arises as soon as facts are denied or glossed over, either in the interest of the epistemological principles of some system or in the interest of some special theory which is scientifically vacuous but which someone wishes to apply to some area. Against those investigators, for example, who tried to determine what part or aspect of sense perception depended on memory or on the repetition of similar impressions—in short, what part was the result of experience—arose the partisan cry that they were spiritualists. As if memory, experience, and training were not themselves facts whose laws must be sought! They are facts, however, which cannot be simply and smoothly reduced to established laws of nervous excitation and transmission. Problems raised by the structure and function of the brain present a most favorable area for the spinning of fantasies.

No matter how self-evident and important the principle that *natural science must search for the laws of phenomena* may seem, it is nevertheless often forgotten. When we finally acknowledge some law that has been discovered to be a power ruling the processes of nature, we treat it objectively as a force; and when we explain individual cases by referring to a force which produces the same effect under the same conditions, we call this a causal explanation of the phenomena. We cannot always trace these forces back to those which are atomic; we also speak of refractive, electromotive, and electrodynamic forces. But unless the specific conditions and the specific effects can be given, the attempted explanation is no more than a modest confession of ignorance, and it is certainly better to confess this openly.

If, for example, some botanical process is explained by reference to forces in the cells, without a closer specification of the conditions under which, and of the direction in which they work, the most that is asserted is that the more remote parts of the plant have no influence on the process. It would be difficult to confirm claims like this with certainty in more than a few cases. Again, the original, specific meaning which Johannes Müller gave to the idea of reflex action has gradually evaporated; and now, when an impression is made on one part of the nervous system and an effect occurs in any other part, this is supposedly explained by saying that it is a reflex action. Much may be attributed to the irresolvable complexity of the nerve fibers of the brain, but the resemblance here to the *qualitates occultae* of ancient medicine is very suspicious.

From the general tenor of my remarks it can be seen that my strictures against metaphysics are not meant to be directed against philosophy. Metaphysicians, however, have always tried to give the impression that they are philosophers; and dilettantes in philosophy have for the most part been interested in the soaring speculations of the metaphysicians, by which they hope in a short time and with no great trouble to learn the whole of what is worth knowing. On another occasion[1] I compared the relationship of metaphysics

1. Preface to the German translation of Tyndall's *Fragments of Science*.

to philosophy with that of astrology to astronomy. Astrology has excited the greater interest among the public at large and especially in the fashionable world, where its practitioners have gained positions of influence. Astronomy, on the other hand, although it has become the ideal of scientific research, has had to be content with a small number of modest but diligent disciples.

Philosophy, if it gives up metaphysics, still has a wide, important field—the study of cognitive and psychic processes and their laws. Just as the anatomist, when he has reached the limits of microscopic vision, must try to gain an insight into the operation of his optical instruments, so every scientific investigator must study carefully the chief instrument of his research, the human mind, in order to determine its capacities. The groping of the medical schools for the past two thousand years is one of many good examples of the harm caused by erroneous views in this matter. Moreover, the physician, the statesman, the jurist, the clergyman, and the teacher must build upon a knowledge of psychic processes if they wish to have a truly scientific basis for their practical activities. The real science of philosophy, however, has had to suffer even more perhaps than medicine itself from the evil ways and false ideals of metaphysics.

One word of caution. I should not like you to think that my statements are influenced by personal irritation. I need not remind you that anyone who holds opinions such as I have laid before you—and who impresses upon his students, whenever he can, the principle that a metaphysical proposition is either false or a concealed experimental one—is not exactly beloved by the votaries of metaphysics and of a priori intuitions. Metaphysicians, like all the others who cannot give any cogent reasons to their opponents, are usually not very polite in their polemics, and one's own success can be gauged with some accuracy by their increasing lack of politeness. My researches have led me, more than other disciples of the school of natural science, into controversial regions. Expressions of metaphysical discontent have therefore, as many of you well know, been addressed more toward me than toward my friends.

In order, therefore, to keep my own personal opinions out of the picture, I have already allowed two trustworthy authorities to speak for me. They are Socrates and Kant, both of whom, convinced that all metaphysical systems presented before their times were empty networks of fallacies, guarded themselves against adding any new system. To show that matters have still not changed, neither in the last two thousand nor in the past hundred years, let me conclude with a quotation from Friedrich Albert Lange, the author of *The History of Materialism*, who unfortunately has been taken from us too soon.

In his posthumous *Logical Studies*, which he wrote knowing of his approaching death, Lange presents the following description, which struck me because it could refer equally well to solidar or humoral pathology or to any of the other old dogmatic schools of medicine.

> The Hegelian sees the Herbartian as being less perfect in knowledge than himself, and vice versa. Neither hesitates, however, to judge the other's knowledge as higher than that of the empiricist and to see in it at least an approximation to the only true knowledge. From this it can be seen that metaphysicians pay no attention to the weight of evidence; the mere statement that something is a deduction within a system makes it apodictic knowledge for them.

Let us, therefore, cast no stones at our early forebears in the science of medicine who, in dark ages and with very little antecedent knowledge to work from, fell into precisely the same errors as the great intellects of the supposedly enlightened nineteenth century. They did no worse than their contemporaries; the nonsense of their method was simply more evident in comparison with the natural sciences.

Let us work on. In the work of true enlightenment, physicians must play a prominent role, for among those who are continually called upon to apply their knowledge actively to nature, it is the physicians who begin with the best mental preparation and are familiar with the most varied regions of natural phenomena.

Finally, let me conclude our consultation concerning the condition of Dame Medicine in the proper fashion with the epicrisis: I think we have every reason to be content with the success of the treatment which the school of natural science has applied, and we can only recommend that the younger generation continue the same therapy.

13.

THE ORIGIN AND MEANING OF GEOMETRIC AXIOMS (II) (INTRODUCTION AND SECTION I) [1878]

In 1878, in response to some criticisms of "The Origin and Meaning of Geometric Axioms (I)," Helmholtz published in Mind *(Volume III, No. 10 [April 1878], pp. 212-25) another article on the foundations of geometry. This article, "The Origin and Meaning of Geometric Axioms (II)," consists of some introductory paragraphs and three sections. The introduction and the first section follow; the second and third sections were also published as Appendix 3 of his essay on "The Facts of Perception" (Chapter 14)*

My article on "The Origin and Meaning of Geometric Axioms" in *Mind*, No. 3, was critically examined by Professor Land in No. 5, and I shall now try to answer his objections. We differ substantially on two points. I am of opinion that the recent mathematical investigations—or, as they have been called, metamathematical investigations[1]—as to further kinds of geometry have established the following propositions:

1. Kant's proof of the a priori origin of geometric axioms, based on the assumption that no other space relations are conceivable, is insufficient, the assumption being at variance with fact.

2. If, in spite of the defective proof, it is still assumed hypothetically that the axioms are really given a priori as laws of our space intuitions, two kinds of equivalence of space magnitudes must be distinguished: *subjective equality* given by the hypothetical transcendental intuition, and *objective equivalence*

1. The name has been given by opponents in irony, as suggesting *metaphysical*; but as the founders of "non-Euclidean geometry" have never maintained its objective truth, they can very well accept the name.

of the real substrata of such space relations, proved by the equality of physical states or actions existing or occurring in what appear to us as congruent parts of space. The coincidence of the second with the first could be proved only by experience; and as the second alone concerns us in our scientific or practical dealings with the objective world, the first, in case of discrepancy, must be discounted as a deceptive appearance.

For the rest, it is a misunderstanding on Professor Land's part if he thinks I wished to raise any objection to the notion that space is for us an a priori and necessary, or (in Kant's sense) transcendental, form of intuition. I had no such intention. True, my view of the relations between this transcendental form and reality, as I shall set it forth in the third section of this paper, does not quite coincide with that of many followers of Kant and Schopenhauer. But space may very well be a form of intuition in the Kantian sense and yet not necessarily involve the axioms. To cite a parallel instance, it undoubtedly lies in the organization of our optical apparatus that everything we see can be seen only as a spatial distribution of colors. This is the innate form of our visual perceptions. But it is not in the least thereby predetermined how the colors we see shall coexist in space and follow one another in time. And just so, in my view, the representation of all external objects in space relations may be the only possible form in which we can represent the simultaneous existence of a number of discrete objects, though there is no necessity that a particular space perception should coexist with or follow upon certain others; e.g., that every rectilineal equilateral triangle should have angles of 60°, whatever the length of the sides.

According to Kant, indeed, the proof that space is an a priori form is based essentially on the position that the axioms are synthetic propositions a priori. But even if this assertion with the dependent inference is dropped, the space representation might still be the necessary a priori form in which every coextended manifold is perceived. This is not surrendering any essential feature of the Kantian system. On the contrary, the system becomes more consistent and intelligible if the proof of the possibility of metaphysics derived from the evidence of geometric axioms is seen to break down. Kant himself, as is well known, limited the scope of metaphysical science to the geometric and physical axioms. But the physical axioms are either of doubtful validity, or they are mere consequences of the principle of causality, that is to say, of our intellectual impulse to view everything that happens as conforming to law and thus as conceivable. And as Kant's *Critique* is otherwise hostile to all metaphysical reasoning, his system seems to be freed from inconsistency, and a clearer notion of the nature of intuition is obtained, if the a priori origin of the axioms is abandoned and geometry is regarded as the first and most perfect of the natural sciences.

I pass accordingly to the proof of the two theses enunciated above.

I

Kant's proof of the a priori origin of the geometric axioms is based on the assertion that it is impossible to form a mental representation of space relations at variance with Euclid's geometry. But the metamathematical investigations reviewed in my former paper have shown that it is quite possible to devise and consistently work out systems of geometry which differ from Euclid's both in the number of space dimensions and in their axioms, with their related systems of mechanics. I myself have tried to show what would be the sensible appearance of objects in spherical or in pseudospherical space. The mathematical correctness of those geometric deductions (carried out for the most part analytically) is, as far as I can see, beyond question, and the same may be said as to the perfect validity of the corresponding systems of mechanics, which afford the same degree of free mobility for solid bodies and the same independence of mechanical and physical processes from mere position that are presupposed in the Euclidean geometry.

Nor is there the least difficulty or uncertainty as to the nature of the space perceptions that human beings would have in these other circumstances. In particular, Beltrami's discovery of the way of representing pseudospherical space in a sphere of Euclidean space shows directly what would be the appearance of visual images in pseudospherical or spherical space. Every visual image of objects at rest as seen by a spectator at rest would, in fact, be exactly the same as that of the corresponding representation in Beltrami's sphere as seen from the center (supposing always that the distance of the two eyes may be neglected in comparison with the imaginary radius of curvature of the space). There would be a difference only in the order of succession of the images, according as the observer or the solid objects moved. Nothing would be changed but the rule for inferring what images would succeed others in case of movement. And, as I have maintained, such differences are not necessarily considerable, nor need they excite attention.

Men lived for a long time on what they thought was the flat earth before they discovered its spherical form, and they struggled long enough against this truth, just as our Kantians at the present day will not listen to the possibility of representing pseudospherical space. The discrepancies in pseudospherical space would be of a somewhat similar kind and not necessarily more striking (if the measure of curvature tallied) than are those betrayed by the spherical surface of the earth to an observer whose movements are limited to a few miles.

In discussing the question whether space relations can be imagined in metamathematical spaces, the first thing to settle is the rule by which we shall judge the imaginability of an object that we have never actually seen.

I advanced a definition to this effect—that we need the power of fully representing the sense impressions which the object would excite in us according to the known laws of our sense organs under all conceivable conditions of observation and by which it would be distinguished from other similar objects. I am of opinion that this definition contains stricter and more definite requirements for the possibility of imagination than any previous one; and as far as I can see, Professor Land does not contend that these requirements cannot be satisfied for objects in spherical or pseudospherical spaces. At the same time, the representation of objects that we have often perceived, or that resemble such in whole or in parts, will necessarily be superior in one respect to the representation of objects of which this cannot be said, namely, in the swiftness and ease with which we can imagine beforehand the various aspects of the objects under different conditions of observation or run them over in memory. This ease and swiftness in the imagination of an object never actually seen will be wanting, just in proportion as the observer has more rarely perceived and less carefully apprehended anything like it.

We have absolutely never had before us constructions of three dimensions in spherical or pseudospherical space. The geometrician, however, who has trained himself in the power of representing surfaces that can be bent without stretching and without change of their measure of curvature, as well as the figures that can be drawn upon them, finds relations in these that are closely analogous to the relations in those other spaces. The physiologist, too, who has studied the combinations of sense impressions under every possible variety of conditions, including those which never occur in daily experience, is more practiced in representing unusual (yet strictly determinate) series of sense impressions than one who has never had the same training. I may perhaps be pardoned, then, if I do not see why the fact that I come "fresh from the physiology of the senses" to epistemological inquiries should be a positive bar to my dealing with such questions as the one before us.

Since, then, the metamathematical space relations have never been actually perceived by us, we must not expect to have that power of swift, easy representation of the varying aspects of objects in them that can come only from daily experience and practice. The utmost we can expect is to arrive by slow steps and careful reflection at a full, consistent representation of the corresponding series of sense impressions. But in point of fact, we strike upon similar and equally great difficulties of representation when we seek to picture to ourselves the course of an intricately knotted thread, or a many-sided crystal model, or a complex building that we have never seen, although the possibility of imagining all these is proved by the fact of actual perception.

Unfortunately, Professor Land does not say whether he has any objection to my definition of imaginative representation, nor does he himself offer any other, though he several times hints that he means something different by "imaginability." Thus he says: "We do not find that they [the non-Euclideans] succeed in this [making metamathematical spaces imaginable], unless the notion of imaginability be stretched far beyond what Kantians and others understand by the word." At the same place, he asserts that only that which can be connectedly constructed in our space can be regarded as "imagined." He adds: "Non-Euclideans try to make imaginable that which is not so in the sense required for argumentation in this case." If by "argumentation" is here meant the discussion of the question whether our conviction of the actual validity of Euclid's axioms in our objective world justifies a conclusion as to their a priori origin, I am of opinion that my definition of imaginability is the only one that can decide the question. If we should define thus: "Nothing is to be held as imaginable in space, of which we cannot actually construct a model with existing bodies," all discussion of the question in dispute is, no doubt, cut short. But this imaginability, ascribed by the definition to Euclid's space alone, affords not the least ground for deciding whether its origin is to be sought in a law of the objective world or in the constitution of our minds. Accordingly, I do not believe that Professor Land means to postulate this, though his words bear the interpretation.

I can only suppose him to object to my definition of imaginability that it does not include a reference to the apparently spontaneous readiness with which the various aspects of any common object are represented when we have sensible experience of some one of them. But we know that such an association of different impressions can be acquired and strengthened by frequent repetition–notably, for instance, between the sound of a word and its meaning. I therefore do not see that we have the right to consider this readiness of suggestion as essential to imaginability. The fact, moreover, that

Lobachevsky by pure synthesis–that is to say, by actual geometriç constructions–worked out a complete system of pseudospherical geometry, agreeing exactly with the results of analytical inquiry, shows that such a geometry can be grasped in all its details by the imagination.

As regards the use of analytical methods in metamathematical inquiries, this is justified by the circumstance that we have here to do with the representation of an object which has never been perceived–an object whose notion, or (so to speak) architectural plan, has first to be developed, to be shown inherently consistent, and to be elaborated in sufficient detail that for every particular case it is made clear what the corresponding sensible impression would be in the circumstances. This ideal development of the ground plan is best attained by the methods of analytic geometry, securing as these do most effectively universality and completeness of demonstration. No doubt a manipulation of notions by means of the calculus does not suffice to prove the existence of the object so treated, but the process is sufficient to the extent of proving the possibility of a consistent series of sensible pictures, whence it follows that the space relations actually perceived in a real world by organs analogous to our own might correspond with a geometry different from Euclid's.

Since the relations obtaining in metamathematical spaces of three dimensions satisfy the conditions of imaginability required by my definition–and more cannot be demanded in the case of objects never actually perceived–Kant's proof of the transcendental character of the axioms and their a priori origin must be pronounced insufficient.

14.

THE FACTS OF PERCEPTION [1878]

An address given on the anniversary of the founding of the University of Berlin in 1878

TODAY, on the birthday of its founder, the sorely tried King Friedrich Wilhelm III, we are celebrating the anniversary of the founding of our university. The year of its founding, 1810, was the time of our country's greatest oppression–a time when considerable territory had been lost and the land exhausted by war and foreign occupation; a time when the king's martial pride, reminiscent of the days of the electors and of the great kings, had been deeply humbled. Yet when we now look back upon that period, it seems so rich in intellectual gifts, inspiration, energy, ideal aspirations, and creative thought that we, despite the comparatively prosperous condition of our state and nation today, must think of it almost with envy. That the king, under the distressing pressure of other material claims, gave first priority to the founding of the university–and that he offered, at that time, both his throne and his life in order to evince his trust in the nation's resolute determination against the enemy–shows how deeply his trust in the intellectual forces of his people affected this simple, yet brilliant and steadfast man.

During his reign Germany could boast of a distinguished roster of great names in both art and science, some of which are to be numbered among the foremost in the whole history of man's intellectual endeavors. Goethe and Beethoven were still living, while Schiller, Kant, Herder, and Haydn had lived to see the first years of the new century. Wilhelm von Humboldt was laying the foundations for the new science of comparative linguistics; Niebuhr, Friedrich August Wolf, and Savigny were teaching ancient history, poetry, and law with a lively, penetrating understanding; Schleiermacher was seeking to comprehend the profound intellectual contents of religion; and the powerful, intrepid orator Johann Gottlieb Fichte, the second rector of our University, was carrying his audience away on the currents of his moral enthusiasm and the bold flights of his idealism.

Even in the aberrations of this system of thought, aberrations which found expression in the easily recognized weaknesses of the Romantics, there is something attractive in comparison to a dry, calculating egotism. One marvels at the beautiful emotions in which man delights; one looks for the power to have or to cultivate such feelings; one believes the imagination to be so much more wonderful a creative force, the more one has freed himself from the rules of the understanding. There was great vanity in this way of thinking, but it was always a vanity which aspired after high ideals.

The oldest among us knew the men of that period—the men who were the first volunteers to enter the army, who were always ready to delve into the investigation of metaphysical problems, who were well read in Germany's great poets, and who glared with anger when conversation turned to the first Napoleon or smiled with enthusiasm and pride when it turned to the war for freedom.

How things have changed! We may surely proclaim this at a time when an astonishing cynical contempt for all the ideal values of mankind is spread in the streets and in the press, a time that has been climaxed by two horrible crimes which obviously have occurred in the capital of our king[1] only because in him is united all that mankind previously considered worthy of respect and gratitude.

It is therefore almost with pain that we recall that only eight years have passed since the great hour when our whole population, unhesitant and full of self-sacrificing, enthusiastic patriotism, answered the call of this same monarch in a hazardous war against an enemy whose strength and bravery were not unknown to us. It is almost with pain that we think of the many efforts, both political and humane, to prepare a secure existence—one worthy of a human being—for the poorer classes of our people, of the many efforts to include them in the activities and ideas of the cultivated classes, and of how very much their lot has been bettered in both a material and a legal sense.

It seems to be the nature of mankind that next to great light, great darkness should be found. Political liberty allows the baser motives more freedom to display themselves and, provided they are not too energetic, to generate forces opposed to prevailing public opinion. Even in the years of the war of independence, when Fichte preached penetential sermons to his own age, these forces were not absent. Fichte paints a picture of conditions and con-

1. Two attempts were made on the life of Wilhelm I in 1878. He was seriously wounded by the second.

victions which remind us sadly of our own times: "The present period has as its main principle an arrogant despising of ideals, as if by merely dreaming of virtue it can cure itself of pleasure-seeking. It prides itself that it has gotten away from ideals and is no longer bound by them." Of the pleasure which surpassed the purely sensual, the representatives of that period could recognize only one—that which Fichte called "the comfort of their own craftiness." Yet that same period prepared a mighty movement which belongs among the most glorious events of our history.

We need not consider our own period hopelessly lost, but we must not console ourselves with the thought that in other times things were no better than they are now. Meanwhile, considering such sobering events, it is well that each of us examines the field with which he is familiar and in which he works, in order to see whether through its endeavors it is approaching the eternal goals of mankind, which lie always before our eyes. During the youth of our university, science too was boldly young and hopeful, and its eyes were turned toward the highest goals. These have not been as easy to reach as that generation hoped; the immense amount of work which has been done has only broadened the way and, by changing the very nature of the problems facing science, has forced us to turn toward other kinds of work less immediate to the goal. Still, it would undoubtedly be most unfortunate if our generation were to lose sight of the eternal ideals of mankind among subordinate occupations which have only practical value.

The problems which that earlier period considered fundamental to all science were those of the theory of knowledge: *What is true in our sense perceptions and thought?* and *In what way do our ideas correspond to reality?* Philosophy and the natural sciences attack these questions from opposite directions, but they are the common problems of both. Philosophy, which is concerned with the mental aspect, endeavors to separate out whatever in our knowledge and ideas is due to the effects of the material world, in order to determine the nature of pure mental activity. The natural sciences, on the other hand, seek to separate out definitions, systems of symbols, patterns of representation, and hypotheses, in order to study the remainder, which pertains to the world of reality whose laws they seek, in a pure form. Both try to achieve the same separation, though each is interested in a different part of the divided field.

The natural scientist no more than the philosopher can ignore epistemological questions when he is dealing with sense perception or when he is concerned with the fundamental principles of geometry, mechanics, or physics. Since my work has entered many times into both the region of science and the region of philosophy, I should like to attempt to survey what has been done from the side of the natural sciences to answer the questions which have just been stated. The laws of thought, after all, are the same for the scientist as for the philosopher.

In all cases where the facts of daily experience, which are already very copious, afford a clear-sighted thinker with a disinterested sense of the truth sufficient information for making correct judgments, the scientist must be satisfied to recognize that a methodologically complete collection of the facts of experience will simply confirm those judgments, though there are occasionally, of course, some conflicting cases. This is my excuse (if it must be excused) for the fact that in general, in the following paper, no completely new answers—on the contrary, only rather old answers, long since given to the questions to be dealt with—will be presented to you. Often enough, of course, even old concepts gain new illumination and new meaning from newly ascertained facts.

Shortly before the beginning of the present century, Kant expounded a theory of that which, in cognition, is prior or antecedent to all experience; that is, he developed a theory of what he called the *transcendental* forms of intuition and thought. These are forms into which the content of our sensory experience must necessarily be fitted if it is to be transformed into ideas. As to the qualities of sensations themselves, Locke had earlier pointed out the role which our bodily and mental structure or organization plays in determining the way things appear to us. Along this latter line, investigations of the physiology of the senses, in particular those which Johannes Müller carried out and formulated in the law of the specific energies of the senses, have brought (one can almost say, to a completely unanticipated degree) the fullest confirmation. Further, these investigations have established the nature of—and in a very decisive manner have clarified the significance of—the antecedently given subjective forms of intuition. This subject has already been discussed rather frequently, so I can begin with it at once today.

Among the various kinds of sensations, two quite different distinctions must be noted. The most fundamental is that among sensations which belong to different senses, such as the differences among blue, warm, sweet, and high-pitched. In an earlier work I referred to these as differences in the *modality* of the sensations. They are so fundamental as to exclude any possible transition from one to another and any relationship of greater or less similarity. For example, one cannot ask whether sweet is more like red or more like blue.

The second distinction, which is less fundamental, is that among the various sensations of the same sense. I have referred to these as differences in *quality.* Fichte thought of all the qualities of a single sense as constituting a *circle of quality*; what I have called differences of modality, he designated differences between circles of quality. Transitions and comparisons are possible only within each circle; we can cross over from blue through violet and carmine to scarlet, for example, and we can say that yellow is more like orange than like blue.

Physiological studies now teach that the more fundamental differences are completely independent of the kind of external agent by which the sensations are excited. They are determined solely and exclusively by the nerves of sense which receive the excitations. Excitations of the optic nerves produce only sensations of light, whether the nerves are excited by objective light (that is, by the vibrations in the ether), by electric currents conducted through the eye, by a blow on the eyeball, or by a strain in the nerve trunk during the eyes' rapid movements in vision. The sensations which result from the latter processes are so similar to those caused by objective light that for a long time men believed it was possible to produce light in the eye itself. It was Johannes Müller who showed that internal production of light does not take place and that the sensation of light exists only when the optic nerve is excited.

Every sensory nerve, then, when excited by even the most varied stimuli, produces a sensation only within its own specific circle of quality. The same external stimulus, therefore, if it strikes different nerves, produces diverse sensations, which are always within the circles of quality of the nerves excited. The same vibrations of the ether which the eye experiences as light, the skin feels as heat. The same vibrations of the air which the skin feels as a flutter, the ear hears as sound. In the former case the differences between the sensations are so great that physicists once felt justified in postulating two agents, analogous and in part equivalent to each other, one of which appears to us as light and the other as radiant heat. Only later, after careful, exhaustive experimental investigations, was the complete similarity of the physical characteristics of these two agents established.

Within the circle of quality of each individual sense, where the nature of the stimulating object determines at least in part the quality of the resulting sensation, the most unexpected incongruities have also been found. In this connection a comparison of sight and hearing is instructive, for the objects of both—light and sound—are vibrational movements which, depending upon the frequency of the vibrations, produce sensations of different colors in vision and differences of pitch in hearing. If, for greater clarity, we refer to the relationships among the vibrations of *light* in terms of the musical intervals formed by *sound* vibrations, the following points are evident: The ear is sensitive to about ten octaves of different tones, while the eye is sensitive to only a musical sixth. With both sound and light, however, vibrations exist outside of these ranges, and their physical existence can be demonstrated.

In its short scale the eye has only three independent, fundamental sensations—red, green, and blue-violet—out of which all of the other colors are formed by various combinations. These three sensations are combined in vision without being altered or disturbed. The ear, on the other hand, distinguishes an enormous number of tones of different pitch, and no one chord sounds exactly like another made up of different tones. In vision, the same sensation of white can be produced by combining the red and the green-blue of the spectrum; or green, red, and violet; or yellow and ultramarine blue; or green-yellow and violet; or any two, or three, or indeed all of these combinations together. If the same thing occurred in hearing, the simultaneous striking of *c* and *f* with *d* and *g*, or with *e* and *a*, or with *c, d, e, f, g, a,* and so on, would all produce the same sound. Thus it should be emphasized, with reference to the objective significance of colors, that except for the effect on the eye there is no single objective combination of colors which can be related invariantly to any one sensation of color.

Finally, consonance and dissonance in music are due entirely to the phenomenon of beats. These in turn are due to the rapid variations in the intensity of sound which result when two tones of almost equal pitch are alternatively in and out of phase, thus causing first strong and then weak vibrations in any body oscillating harmonically with them. As a physical phenomenon, beats can be produced just as readily by the interaction of two trains of light waves as by the interaction of two trains of sound waves. In order to be aware of them, however, the nerves would have to be affected by both wave trains, and the alternations between strong and weak intensities would have to follow each other at just the right intervals. In this respect the auditory nerves are greatly superior to the optic nerves.

Each fiber among the auditory nerves is sensitive to only a single tone from a narrow interval of the scale, so that in general only tones lying close together can interact with one another, while those at a distance cannot. If the latter do interact, they produce not beats but an overtone or some combination tone. It is in connection with these, as you know, that the difference between harmonic and non-harmonic intervals, that is, between consonance and dissonance, makes its appearance. In contrast again, every optic nerve fiber is sensitive to the entire spectrum, although, to be sure, they are sensitive in different degrees to different parts of the spectrum. If it were possible to detect by means of the optic nerves the enormously rapid beats resulting from the interaction of different vibrations of light, every mixed color would appear as a dissonance.

It is apparent that all these differences among the effects of light and sound are determined by the way in which the nerves of sense react. Our sensations are simply effects which are produced in our organs by objective causes; precisely how these effects manifest themselves depends principally and in essence upon the type of apparatus that reacts to the objective causes. What information, then, can the qualities of such sensations give us about the characteristics of the external causes and influences which produce them? Only this: our sensations are signs, not images, of such characteristics. One expects an image to be similar in some respect to the object of which it is an image; in a statue one expects similarity of form, in a drawing similarity of perspective, in a painting similarity of color. A sign, however, need not be similar in any way to that of which it is a sign. The sole relationship between them is that the same object, appearing under the same conditions, must evoke the same sign; thus different signs always signify different causes or influences.

To popular opinion, which accepts on faith and trust the complete veridicality of the images which our senses apparently furnish of external objects, this relationship may seem very insignificant. In truth it is not, for with it something of the greatest importance can be accomplished: we can discover the lawful regularities in the processes of the external world. All natural laws assert that from initial conditions which are the same in some specific way, there always follow consequences which are the same in some other specific way. If the same kinds of things in the world of experience are indicated by the same signs, then the lawful succession of equal effects from equal causes will be related to a similar regular succession in the realm of our sensations. If, for example, some kind of berry in ripening forms a red pigment and sugar at the same time, we shall always find a red color and a sweet taste together in our sensations of berries of this kind.

Thus, even if in their qualities our sensations are only signs whose specific nature depends completely upon our makeup or organization, they are not to be discarded as empty appearances. They are still signs of something—something existing or something taking place—and given them we can determine the laws of these objects or these events. And that is something of the greatest importance!

Thus, our physiological makeup incorporates a pure form of intuition, insofar as the qualities of sensation are concerned. Kant, however, went further. He claimed that, not only the qualities of sense experience, but also space and time are determined by the nature of our faculty of intuition, since we cannot perceive anything in the external world which does not occur at some time and in some place and since temporal location is also a characteristic of all subjective experience. Kant therefore called time the a priori and necessary transcendental form of the inner, and space the corresponding form of the outer, intuition. Further, Kant considered that spatial characteristics belong no more to the world of reality (the *dinge an sich*) than the colors we see belong to external objects. On the contrary, according to him, space is carried to objects by our eyes.

Even in this claim scientific opinion can go along with Kant up to a certain point. Let us consider whether any sensible marks are present in ordinary, immediate experience to which all perception of objects in space can be related. Indeed, we find such marks in connection with the fact that our body's movement sets us in varying spatial relations to the objects we perceive, so that the impressions which these objects make upon us change as we move. The impulse to move, which we initiate through the innervation of our motor nerves, is immediately perceptible. We *feel* that we are doing something when we initiate such an impulse. We do not know directly, of course, all that occurs; it is only through the science of physiology that we learn how we set the motor nerves in an excited condition, how these excitations are conducted to the muscles, and how the muscles in turn contract and move the limbs. We are aware, however, without any scientific study, of the perceptible effects which follow each of the various innervations we initiate.

The fact that we become aware of these effects through frequently repeated trials and observations can be demonstrated in many, many ways. Even as adults we can still learn the innervations necessary to pronounce the words of a foreign language, or in singing to produce some special kind of voice formation. We can learn the innervations necessary to move our ears, to turn our eyes inward or outward, to focus them upward or downward, and so on. The only difficulty in learning to do these things is that we must try to do them by using innervations which are unknown, innervations which have not been necessary in movement previously executed. We know these innervations in no form and by no definable characteristics other than the fact that they produce the observable effects intended. This alone distinguishes the various innervations from one another.

If we initiate an impulse to move—if we shift our gaze, say, or move our hands, or walk back and forth—the sensations belonging to some circles of quality (namely, those sensations due to objects in space) may be altered. Other psychical states and conditions that we are aware of in ourselves, however, such as recollections, intentions, desires, and moods, remain unchanged. In this way a thoroughgoing distinction may be established in our immediate experience between the former and the latter. If we use the term *spatial* to designate those relations which we can alter directly by our volition but whose nature may still remain conceptually unknown to us, an awareness of mental states or conditions does not enter into spatial relations at all.

All sensations of external senses, however, must be preceded by some kind of innervation, that is, they must be spatially determined. Thus space, charged with the qualities of our sensations of movement, will appear to us as that through which we move or that about which we gaze. In this sense spatial intuition is a subjective form of intuition, just as the qualities of sensation (red, sweet, cold) are. Naturally, this does not mean that the determination of the position of a specific object is only an illusion, any more than the qualities of sensation are.

From this point of view, space is the necessary form of outer intuition, since we consider only what we perceive as spatially determined to constitute the external world. Those things which are not perceived in any spatial relation we think of as belonging to the world of inner intuition, the world of self-consciousness.

Space is an a priori form of intuition, necessarily prior to all experience, insofar as the perception of it is related to the possibility of motor volitions, the mental and physical capacity for which must be provided by our physiological make-up before we can have intuitions of space.

There can be no doubt about the relationship between the sensible signs or marks mentioned above and the changes in our perception of objects in space which result from our movements. (See Appendix 1.) We still must consider the question, however, whether it is *only* from this source that all the specific characteristics of our intuition of space originate. To this end we must reflect further upon some of the conclusions concerning perception at which we have just arrived.

Let us try to set ourselves back to the state or condition of a man without any experience at all. In order to begin without any intuition of space, we must assume that such an individual no longer recognizes the effects of his own innervations, except to the extent that he has now learned how, by means of his memory of a first innervation or by the execution of a second one contrary to the first, to return to the state out of which he originally moved. Since this mutual self-annulment of different innervations is completely independent of what is actually perceived, the individual can discover how to initiate innervations without any prior knowledge of the external world.

Let us assume that the man at first finds himself to be just one object in a region of stationary objects. As long as he initiates no motor impulses, his sensations will remain unchanged. However, if he makes some movement (if he moves his eyes or his hands, for example, or moves forward), his sensations will change. And if he returns (in memory or by another movement) to his initial state, all his sensations will again be the same as they were earlier.

If we call the entire group of sensation aggregates which can potentially be brought to consciousness during a certain period of time by a specific, limited group of volitions the temporary *presentabilia*—in contrast to the *present,* that is, the sensation aggregate within this group which is the object of immediate awareness—then our hypothetical individual is limited at any one time to a specific circle of *presentabilia,* out of which, however, he can make any aggregate present at any given moment by executing the proper movement. Every individual member of this group of *presentabilia,* therefore, appears to him to exist at every moment of the period of time, regardless of his immediate present, for he has been able to observe any of them at any moment he wished to do so. This conclusion—that he could have observed them at any other moment of the period if he had wished—should be regarded as a kind of inductive inference, since from any moment a successful inference can easily be made to any other moment of the given period of time.

In this way the idea of the simultaneous and continuous existence of a group of different but adjacent objects may be attained. *Adjacent* is a term with spatial connotations, but it is legitimate to use it here, since we have used *spatial* to define those relations which can be changed by volition. Moreover, we need not restrict the term *adjacent* so that it refers only to material objects. For example, it can legitimately be said that "to the right it is bright, to the left dark," and "forward there is opposition, behind there is nothing," in the case where "right" and "left" are only names for specific movements of the eyes and "forward" and "behind" for specific movements of the hands.

At other times the circles of *presentabilia* related to this same group of volitions are different. In this way circles of *presentabilia,* along with their individual members, come to be something given to us, that is, they come to be *objects.* Those changes which we are able to bring about or put an end to by familiar acts of volition come to be separated from those which do not result from and cannot be set aside by such acts. This last statement is negative: in Fichte's quite appropriate terminology, the Non-Ego forces the recognition that it is distinct from the Ego.

When we inquire into the empirical conditions under which our intuition of space is formed, we must concentrate in particular upon the sense of touch, for the blind can form complete intuitions of space without the aid of vision. Even if space turns out to be less rich in objects for them than for people with vision, it seems highly improbable that the foundation of the intuition of space is completely different for the two classes of people. If, in the dark or with our eyes closed, we try to perceive only by touch, we are definitely able to feel the shapes of the objects lying around us, and we can determine them with accuracy and certainty. Moreover, we are able to do this with just one finger or even with a pencil held in the hand the way a surgeon holds a probe. Ordinarily, of course, if we want to find our way about in the dark we touch large objects with five or ten fingertips simultaneously. In this way we get from five to ten times as much information in a given period of time as we do with one finger. We also use the fingers to measure the sizes of objects, just as we measure with the tips of an open pair of compasses.

It should be emphasized that with the sense of touch, the fact that we have an extended skin surface with many sensitive points on it is of secondary importance. What we are able to find out, for example, about the impression on a medal by the sensations in the skin when our hand is stationary is very slight and crude in comparison with what we can discover even with the tip of a pencil when we move our hand. With the sense of sight, perception is more complicated due to the fact that besides the most sensitive spot on the retina, the *fovea centralis,* or pit, which in vision rushes as it were about the visual field, there are also a great many other sensitive points acting at the same time and in a much richer way than is the case with the sense of touch.

It is easy to see that by moving our fingers over an object, we can learn the sequences in which impressions of it present themselves and that these sequences are unchanging, regardless which finger we use. Further, these are not single-valued or fixed sequences, whose elements must always be covered, either forward or backward, in the same order. They are not linear sequences;

on the contrary, they form a plane coextension or, using Riemann's terminology, a manifold of the second order. The fingers are moved over a surface by means of motor impulses which differ from those necessary to carry them from one point on the surface to another, and different surfaces require different movements for the fingers to glide over them. Consequently, the space in which the fingers move requires a manifold of a higher order than that of a surface; the third dimension must be introduced.

Three dimensions are sufficient, however, for all our experience, since a closed surface completely divides space as we know it. Moreover, substances in a gaseous or fluid state, which are not dependent at all on the nature of man's mental faculties, cannot escape from a completely closed surface. And, just as a continuous line can enclose only a surface and not a space—that is, a spatial form of two and not of three dimensions—so a surface can enclose only a space of three and not of four dimensions.

It is thus that our knowledge of the spatial arrangement of objects is attained. Judgments concerning their size result from observations of the congruence of our hand with parts or points of an object's surface, or from the congruence of the retina with parts or points of the retinal image.

A strange consequence—a characteristic of the ideas in the minds of individuals with at least some experience—follows from the fact that the perceived spatial ordering of things originates in the sequences in which the qualities of sensations are presented by our moving sense organs: the objects in the space around us appear to possess the qualities of our sensations. They appear to be red or green, cold or warm, to have an odor or a taste, and so on. Yet these qualities of sensations belong only to our nervous system and do not extend at all into the space around us. Even when we know this, however, the illusion does not cease, for *it is the primary and fundamental truth.* The illusion is quite simply the sensations which are given to us in spatial order to begin with.

You can see how the most fundamental properties of our spatial intuition can be obtained in this way. Commonly, however, an intuition is taken to be something which is simply given, something which occurs without reflection or effort, something which above all cannot be reduced to other mental processes. This popular interpretation, at least insofar as the intuition of space is concerned, is due in part to certain theorists in physiological optics and in part to a strict adherence to the philosophy of Kant. As is well known, Kant taught, not only that the general form of the intuition of space is given transcendentally, but also that this form possesses, originally and prior to all possible experience, certain more specific characteristics which are commonly given expression in the axioms of geometry. These axioms may be reduced to the following propositions:

1. Between two points there is only one possible shortest line. We call such a line *straight.*

2. A plane is determined by three points. A plane is a surface which contains completely any straight line between any two of its points.

3. Through any point there is only one possible line parallel to a given straight line. Two straight lines are parallel if they lie in the same plane and do not intersect upon any finite extension.

Kant used the alleged fact that these propositions of geometry appear to us necessarily true, along with the fact that we cannot imagine or represent to ourselves any irregularities in spatial relations, as direct proof that the axioms must be given prior to all experience. It follows that the conception of space contained in them or implied by them must also constitute a transcendental form of intuition independent of all experience.

I would like to emphasize here, in connection with the controversies which have sprung up during the past few years as to whether the axioms of geometry are transcendental or empirical propositions, that this question is absolutely different from the one mentioned earlier, namely, whether space in general is a transcendental form of intuition or not. (See Appendix 2.)

Our eyes see everything in the field of vision as a number of colored plane surfaces. That is their form of intuition. However, the particular colors that appear at any one time, the relationships among them, and the order in which they appear are the effects of external causes and are not determined by any law of our organization. Equally, the fact that space is a form of intuition implies just as little concerning the facts which are expressed by the axioms. If these axioms are not empirical propositions but rather pertain to a necessary form of intuition, this is a further and quite specific characteristic of the general form, and the same reasoning which was used to establish that the general form of intuition of space is transcendental is not necessarily sufficient to establish that the axioms also have a transcendental origin.

In his assertion that it is impossible to conceive of spatial relations which contradict the axioms of geometry, as well as in his general interpretation of intuition as a simple, irreducible mental process, Kant was influenced by the mathematics and the physiology of the senses of his time.

In order to try to conceive of something which has never been seen before, it is necessary to know how to imagine in detail the series of sense impressions which, in accordance with well-known laws, would be experienced if the thing in question—and any changes in it—were actually perceived by any of the sense organs from all possible positions. Further, these impressions must be such that all possible interpretations of them except one can be eliminated. If these series of sense impressions can be specified completely and uniquely in this way, then in my opinion one must admit that the object clearly is conceivable.

Since by hypothesis the object has never been observed before, no previous experience can come to our aid and guide our imagination to the required series of impressions. Such guidance can be provided only by the concepts of the objects and relationships to be represented. Such concepts are first developed analytically as much as is necessary for the investigation at hand. Indeed, the concepts of spatial forms to which nothing in ordinary experience corresponds can be developed with certainty only by the use of analytic geometry. It was Gauss who, in 1828 in his treatise on the curvature of surfaces, first presented the analytical tools necessary for the solution of the present problem, the tools which Riemann later used to establish the logical possibility of his system of geometry. These investigations have been called, not improperly, metamathematical.

Furthermore, in 1829 and in 1840 Lobachevsky, using the ordinary, intuitive, synthetic method, developed a geometry without the axiom of parallels which is in complete agreement with the corresponding parts of the new analytical investigations. Beltrami has given us a method for representing metamathematical spaces in parts of Euclidean space, a method by which it is possible to imagine the appearance of such spaces in perspective vision with relative ease. Finally, Lipschitz has pointed out how the general principles of mechanics can be transferred to such spaces, so that the series of sense impressions which would occur in them can be specified completely. Thus, in my opinion, the conceivability of such spaces in the sense just indicated has been established.

There is considerable disagreement, however, on this issue. For a demonstration of conceivability I require only that, for every means of observation, the corresponding sense impressions be sketched out clearly and unambiguously, if necessary with the aid of scientific knowledge of the laws of these methods of observation. To anyone who knows these laws, the objects or relationships to be represented seem almost real. Indeed, the task of representing the various spatial relationships of metamathematical spaces requires training in the understanding of analytical methods, perspective constructions, and optical phenomena.

This, however, goes counter to the older conception of intuition, according to which only those things whose ideas come instantly–that is, without reflection and effort–to consciousness along with the sense impressions are to be regarded as given through intuition. It is true that our attempts to represent metamathematical spaces do not have the effortlessness, speed, or immediate clarity of our perceptions of, say, the shape of a room which we enter for the first time or of the arrangement and shape of the objects in it, the materials out of which they are made, and many other things. If this kind of immediate evidence is really a fundamental, necessary characteristic of all intuition, we cannot rightly claim the conceivability of metamathematical spaces.

But upon further consideration we find that there are a large number of experiences which show that we can develop speed and certainty in forming specific ideas after receiving specific sense impressions, even in cases where there are no natural connections between the ideas and the impressions. One of the most striking examples of this is learning a native language. Words are arbitrarily or accidentally selected signs, and in every language they are different. Knowledge of these signs is not inherited; to a German child who has been raised among French-speaking people and who has never heard German spoken, it is a foreign language. A child learns the meanings of words and sentences only by examples of their use; and before he understands the language, it is impossible to make intelligible to him the fact that the sounds he hears are signs which have meaning. Finally, however, after he has grown up, he understands these words and sentences without reflection, without effort, and without knowing when, where, or through what examples he learned them. He understands the most subtle shifts in their meaning, shifts which are often so subtle that any attempt to define them logically could be carried out only with difficulty.

It is not necessary for me to add further examples; our daily life is more than rich enough in them. Art, most clearly poetry and the plastic arts, is based directly upon such experiences. The highest kind of perception, that which we find in the artist's vision, is an example of this same basic kind of understanding, in this case the understanding of new aspects of man and nature. Among the traces which frequently repeated perceptions leave behind in the memory, the ones conforming to law and repeated with the greatest regularity are strengthened, while those which vary accidentally are obliterated. In a receptive, attentive observer, intuitive images of the characteristic aspects of the things that interest him come to exist; afterward he knows no more about how these images arose than a child knows about the examples from which he learned the meanings of words. That an artist has beheld the truth follows from the fact that we too are seized with the conviction of truth when he leads us away from currents of accidentally related qualities. An artist is superior to us in that he knows how to find the truth amid all the confusion and chance events of daily experience.

So much to remind ourselves how effective these mental processes are, from the lowest to the highest reaches of our intellectual life. In some of my earlier works I called the connections of ideas which take place in these processes *unconscious inferences.* These inferences are unconscious insofar as their major premise is not necessarily expressed in the form of a proposition; it is formed from a series of experiences whose individual members have entered consciousness only in the form of sense impressions which have long since disappeared from memory. Some fresh sense impression forms the minor premise, to which the rule impressed upon us by previous observations is applied. Recently I have refrained from using the phrase *unconscious inference* in order to avoid confusion with what seems to me a completely obscure and unjustified idea which Schopenhauer and his followers have designated by the same name. Obviously we are concerned here with the elementary processes which are the real basis of all thought, even though they lack the critical certainty and refinement to be found in the scientific formation of concepts and in the individual steps of scientific inferences.

Returning now to the question of the origin of the axioms of geometry, our lack of facility in developing ideas of metamathematical spatial relations because of insufficient experience cannot be used validly as an argument against their conceivability. On the contrary, these spatial relations are completely conceivable. Kant's proof of the transcendental nature of the geometrical axioms is therefore untenable. Indeed, investigation of the facts of experience shows that the axioms of geometry, taken in the only sense in which they can be applied to the external world, are subject to proof or disproof by experience.

The memory traces of previous experience play an even more extensive and influential role in our visual observations. An observer who is not completely inexperienced receives without moving his eyes (this condition can be realized experimentally by using the momentary illumination of an electric discharge or by carefully and deliberately staring) images of the objects in front of him which are quite rich in content. We can easily confirm with our own eyes, however, that these images are much richer and especially much more precise if the gaze is allowed to move about the field of vision, in this way making use of the kind of spatial observations which I have previously described as the most fundamental. Indeed, we are so used to letting our eyes wander over the objects we are looking at that considerable practice is required before we succeed in making them—for purposes of research in physiological optics—fix on a point without wandering.

In my work on physiological optics I have tried to explain how our knowledge of the field open to vision is gained from visual images experienced as we move our eyes, given that there are some perceptible differences of location on the retina among otherwise qualitatively similar sensations. Following Lotze's terminology, these spatially different retinal sensations were called *local signs.* It is not necessary to know prior to visual experience that these signs are local signs, that is, that they are related to various objective differences in place. The fact that people blind from birth who afterward gain their sight by an operation cannot, before they have touched them, distinguish between such simple forms as a circle and a square by the use of their eyes has been confirmed even more fully by recent studies.

Investigations in physiology show that with the eyes alone we can achieve rather precise and reliable comparisons of various lines and angles in the field of vision, provided that through the eyes' normal movements the images of these figures can be formed quickly one after another on the retina. We can even estimate the actual size and distance of objects which are not too far away from us with considerable accuracy by means of changing perspectives in our visual field, although making such judgments in the three dimensions of space is much more complicated than it is in the case of a plane image. As is well known, one of the greatest difficulties in drawing is being able to free oneself from the influence which the idea of the true size of a perceived object involuntarily has upon us. These are all facts which we would expect if we obtain our knowledge of local signs through experience. We can learn the changing sensory signs of something which remains objectively constant much more easily and reliably than we can the signs of something which changes with every movement of the body, as perspective images do.

To a great many physiologists, however, whose point of view we shall call nativistic, in contrast to the empirical position which I have sought to defend, the idea that knowledge of the field of vision is acquired is unacceptable. It is unacceptable to them because they have not made clear to themselves what even the example of learning a language shows so clearly, namely, how much can be explained in terms of the accumulation of memory impressions. Because of this lack of appreciation of the power of memory, a number of different attempts have been made to account for at least part of visual perception through innate mechanisms by means of which specific sensory impressions supposedly induce specific innate spatial ideas. In an earlier work I tried to show that all hypotheses of this kind which had been formulated were insufficient, since cases were always being discovered in which our visual perceptions are more precisely in agreement with reality than is stated in these hypotheses. With each of them we are forced to the additional assumption that ultimately experience acquired during movement may very well prevail over the hypothetical inborn intuition and thus accomplish in opposition to it what, according to the empirical hypothesis, it would have accomplished without such a hindrance.

Thus nativistic hypotheses concerning knowledge of the field of vision explain nothing. In the first place, they only acknowledge the existence of the facts to be explained, while refusing to refer these facts to well-confirmed mental processes which even they must rely on in certain cases. In the second place, the assumption common to all nativistic theories—that ready-made ideas of objects can be produced by means of organic mechanisms—appears much more rash and questionable than the assumption of the empirical theory that the noncognitive materials of experience exist as a result of external influences and that all ideas are formed out of these materials according to the laws of thought.

In the third place, the nativistic assumptions are unnecessary. The single objection that can be raised against the empirical theory concerns the sureness of the movements of many newborn or newly hatched animals. The smaller the mental endowment of these animals, the sooner they learn how to do all that they are capable of doing. The narrower the path on which their thoughts must travel, the easier they find their way. The newborn human child, on the other hand, is at first awkward in vision; it requires several days to learn to judge by its visual images the direction in which to turn its head in order to reach its mother's breast.

The behavior of young animals is, in general, quite independent of individual experience. Whatever these instincts are which guide them—whether they are the direct hereditary transmission of their parents' ideas, whether they have to do only with pleasure and pain, or whether they are motor impulses related to certain aggregates of experience—we do not know. In the case of human beings the last phenomenon is becoming increasingly well understood. Careful and critically employed investigations are most urgently needed on this whole subject.

Arrangements such as those which the nativistic hypotheses assume can at best have only a certain pedagogical value; that is, they may facilitate the initial understanding of uniform, lawful relations. And the empirical position is, to be sure, in agreement with the nativistic on a number of points–for example, that local signs of adjacent places on the retina are more similar than those farther apart and that the corresponding points on the two retina are more similar than those that do not correspond. For our present purposes, however, it is sufficient to know that complete spatial intuition can be achieved by the blind and that for people with vision, even if the nativistic hypotheses should prove partially correct, the final and most exact determinations of spatial relations are obtained through observations made while moving in various ways.

I should like, now, to return to the discussion of the most fundamental facts of perception. As we have seen, we not only have changing sense impressions which come to us without our doing anything; we also perceive while we are being active or moving about. In this way we acquire knowledge of the uniform relations between our innervations and the various aggregates of impressions included in the circles of *presentabilia.* Each movement we make by which we alter the appearance of objects should be thought of as an experiment designed to test whether we have understood correctly the invariant relations of the phenomena before us, that is, their existence in definite spatial relations.

The persuasive force of these experiments is much greater than the conviction we feel when observations are carried out without any action on our part, for with these experiments the chains of causes run through our consciousness. One factor in these causes is our volitions, which are known to us by an inner intuition; we know, moreover, from what motives they arise. In these volitions originates the chain of physical causes which results in the final effect of the experiment, so we are dealing with a process passing from a known beginning to a known result. The two essential conditions necessary for the highest degree of conviction are (1) that our volitions not be determined by the physical causes which simultaneously determine the physical processes and (2) that our volitions not influence psychically the resulting perceptions.

These last points should be considered more fully. The volition for a specific movement is a psychic act, and the perceptible change in sensation which results from it is also a psychic event. Is it possible for the first to bring about the second by some purely mental process? It is certainly not absolutely impossible. Whenever we dream, something similar to this takes place.

While dreaming we believe that we are executing some movement, and then we dream further that the natural results of this movement occur. We dream that we climb into a boat, shove it off from shore, guide it over the water, watch the surrounding objects shift position, and so on. In cases like this it seems to the dreamer that he sees the consequences of his actions and that the perceptions in the dream are brought about by means of purely psychical processes. Who can say how long and how finely spun, how richly elaborated, such dreams may be! If everything in dreams were to occur in ultimate accordance with the laws of nature, there would be no distinction between dreaming and waking, except that the person who is awake may break off the series of impressions he is experiencing.

I do not see how a system of even the most extreme subjective idealism, even one which treats life as a dream, can be refuted. One can show it to be as improbable, as unsatisfactory as possible (in this connection I concur with the severest expressions of condemnation), but it can be developed in a logically consistent manner, and it seems to me important to keep this in mind. How ingeniously Calderón carried out this theme in *Life Is a Dream* is well known.

Fichte also believed and taught that the Ego constructs the Non-Ego, that is, the world of phenomena, which it requires for the development of its psychical activities. His idealism is to be distinguished from the one mentioned above, however, by the fact that he considered other individuals not to be dream images but, on the basis of moral laws, to be other Egos with equal reality. Since the images by which all these Egos represent the Non-Ego must be in agreement, he considered all the individual Egos to be part of or emanations from an Absolute Ego. The world in which they find themselves is the conceptual world which the World Spirit constructs. From this a conception of reality results similar to that of Hegel.

The realistic hypothesis, on the other hand, accepts the evidence of ordinary personal experience, according to which the changes in perception which result from an act have more than a mere psychical connection with the antecedent volition. It accepts what seems to be established by our daily perception, that is, that the material world about us exists independently of our ideas. Undoubtedly the realistic hypothesis is the simplest that can be formulated. It is based upon and confirmed by an extraordinarily large number of cases. It is sharply defined in all specific instances and is therefore unusually useful and fruitful as a foundation for behavior.

Even if we take the idealistic position, we can hardly talk about the lawful regularity of our sensations other than by saying: "Perceptions occur as if the things of the material world referred to in the realistic hypothesis actually did exist." We cannot eliminate the "as if" construction completely, however, for we cannot consider the realistic interpretation to be more than an exceedingly useful and practical hypothesis. We cannot assert that it is necessarily true, for opposed to it there is always the possibility of other irrefutable idealistic hypotheses.

It is always well to keep this in mind in order not to infer from the facts more than can rightly be inferred from them. The various idealistic and realistic interpretations are metaphysical hypotheses which, as long as they are recognized as such, are scientifically completely justified. They may become dangerous, however, if they are presented as dogmas or as alleged necessities of thought. Science must consider thoroughly all admissible hypotheses in order to obtain a complete picture of all possible modes of explanation. Furthermore, hypotheses are necessary to someone doing research, for one cannot always wait until a reliable scientific conclusion has been reached; one must sometimes make judgments according to either probability or aesthetic or moral feelings. Metaphysical hypotheses are not to be objected to here either. A thinker is unworthy of science, however, if he forgets the hypothetical origin of his assertions. The arrogance and vehemence with which such hidden hypotheses are sometimes defended are usually the result of a lack of confidence which their advocates feel in the hidden depths of their minds about the qualifications of their claims.

What we unquestionably can find as a fact, without any hypothetical element whatsoever, is the lawful regularity of phenomena. From the very first, in the case where we perceive stationary objects distributed before us in space, this perception involves the recognition of a uniform or lawlike connection between our movements and the sensations which result from them. Thus even the most elementary ideas contain a mental element and occur in accordance with the laws of thought. Everything that is added in intuition to the raw materials of sensation may be considered mental, provided of course that we accept the extended meaning of *mental* discussed earlier.

If "to conceive" means "to form concepts," and if it is true that in a concept we gather together a class of objects which possess some common characteristic, then it follows by analogy that the concept of some phenomenon which changes in time must encompass that which remains the same during that period of time. As Schiller said, the wise man

Seeks for the familiar law amidst the awesome multiplicity
of accidental occurrences,
Seeks for the eternal Pole Star amidst the constant flight
of appearances.

Sucht das vertraute Gesetz in des Zufalls grausenden Wundern,
Sucht den ruhenden Pol in der Erscheinungen Flucht.[2]

That which, independently of any and everything else, remains the same during all temporal changes, we call a *substance*; the invariant relation between variable but related quantities we call a *law*. We perceive only the latter directly. Knowledge of substances can be attained only through extensive investigation, and as further investigation is always possible, such knowledge remains open to question. At an earlier time both light and heat were thought to be substances; later it turned out that both were only transitory forms of motion. We must therefore always be prepared for some new analysis of what are now known as the chemical elements.

The first product of the rational conception of phenomena is its lawfulness or regularity. If we have fully investigated some regularity, have established its conditions completely and with certainty and, at the same time, with complete generality, so that for all possible subsequent cases the effect is unequivocally determined—and if we have therefore arrived at the conviction that the law is true and will continue to hold true at all times and in all cases—then we recognize it as something existing independently of our ideas, and we label it a *cause,* or that which underlies or lies behind the changes taking place. (Note that the meaning I give to the word *cause* and its application are both exactly specified, although in ordinary language the word is also variously used to mean antecedent or motive.)

Insofar as we recognize a law as a power analogous to our will, that is, as something giving rise to our perceptions as well as determining the course of natural processes, we call it a *force.* The idea of a force acting in opposition to us arises directly out of the nature of our simplest perceptions and the way in which they occur. From the beginning of our lives, the changes which we cause ourselves by the acts of our will are distinguished from those which are neither made nor can be set aside by our will. Pain, in particular, gives us the most compelling awareness of the power or force of reality. The emphasis falls here on the observable fact that the perceived circle of *presentabilia* is

2. Friedrich von Schiller, *Der Spaziergang* ("The Walk").

not created by a conscious act of our mind or will. Fichte's *Non-Ego* is an apt and precise expression for this. In dreaming, too, that which a person believes he sees and feels does not appear to be called forth by his will or by the known relations of his ideas, for these also may often be unconscious. They constitute a Non-Ego for the dreamer too. It is the same for the idealists who see the Non-Ego as the world of ideas of the World Spirit.

We have in the German language a most appropriate word for that which stands behind the changes of phenomena and acts, namely, "the real" (*das Wirkliche*). This word implies only action; it lacks the collateral meaning of existing as substance, which the concept of "the actual" (*das Reelle*) or "the essential" (*das Sachliche*) includes. In the concept of "the objective" (*das Objective*), on the other hand, the notion of the complete form of objects is introduced, something that does not correspond to anything in our most basic perceptions. In the case of the logically consistent dreamer, it should be noted, we must use the words "effective" and "real" (*wirksam* and *wirklich*) to characterize those psychical conditions or motives whose sensations correspond uniformly to, and which are experienced as the momentary states of, his dreamed world.

In general, it is clear that a distinction between thought and reality is possible only when we know how to make the distinction between that which the ego can and that which it cannot change. This, however, is possible only when we know the uniform consequences which volitions have in time. From this fact it can be seen that conformity to law is the essential condition which something must satisfy in order to be considered real.

I need not go into the fact that it is a *contradictio in abjecto* to try to present the actual (*das Reelle*) or Kant's *ding an sich* in positive statements without comprehending it within our forms of representation. This fact has been pointed out often enough already. What we can attain, however, is knowledge of the lawful order in the realm of reality, since this can actually be presented in the sign system of our sense impressions.

All things transitory
But as symbols are sent.[3]

Alles Vergängliche
Ist nur ein Gleichnis.

I take it to be a propitious sign that we find Goethe with us here, as well as

3. *Faust, Part II.* Translated by Bayard Taylor.

further along on this same path. Whenever we are dealing with a question requiring a broad outlook, we can trust completely his clear, impartial view as to where the truth lies. He demanded of science that it be only an artistic arrangement of facts and that it form no abstract concepts concerning them, for he considered abstract concepts to be empty names which only hide the facts. In somewhat the same sense, Gustav Kirchhoff has recently stated that the task of the most abstract of the natural sciences, mechanics, is to describe completely and in the simplest possible way the kinds of motion appearing in nature.

As to the question whether abstract concepts hide the facts or not, this indeed happens if we remain in the realm of abstract concepts and do not examine their factual content, that is, if we do not try to make clear what new and observable invariant relations follow from them. A correctly formulated hypothesis, as we observed a moment ago, has its empirical content expressed in the form of a general law of nature. The hypothesis itself is an attempt to rise to more general and more comprehensive uniformities or regularities. Anything new, however, that an hypothesis asserts about facts must be established or confirmed by observation and experiment. Hypotheses which do not have such factual reference or which do not lead to trustworthy, unequivocal statements concerning the facts falling under them should be considered only worthless phrases.

Every reduction of some phenomenon to underlying substances and forces indicates that something unchangeable and final has been found. We are never justified, of course, in making an unconditional assertion of such a reduction. Such a claim is not permissible because of the incompleteness of our knowledge and because of the nature of the inductive inferences upon which our perception of reality depends.

Every inductive inference is based upon the belief that some given relation, previously observed to be regular or uniform, will continue to hold in all cases which may be observed. In effect, every inductive inference is based upon a belief in the lawful regularity of everything that happens. This uniformity or lawful regularity, however, is also the condition of conceptual understanding. Thus belief in uniformity or lawful regularity is at the same time belief in the possibility of understanding natural phenomena conceptually. If we assume that this comprehension or understanding of natural phenomena can be achieved—that is, if we believe that we shall be able to discern something fundamental and unchanging which is the cause of the changes we observe—then we accept a regulative principle in our thinking. It is called the law of causality, and it expresses our belief in the complete comprehensibility of the world.

Conceptual understanding, in the sense in which I have just described it, is the method by which the world is submitted to our thoughts, facts are ordered, and the future predicted. It is our right and duty to extend the application of this method to all occurrences, and significant results have already been achieved in this way. We have no justification other than its results, however, for the application of the law of causality. We might have lived in a world in which every atom was different from every other one and where nothing was stable. In such a world there would be no regularity whatsoever, and our conscious activities would cease.

The law of causality is in reality a transcendental law, a law which is given a priori. It is impossible to prove it by experience, for, as we have seen, even the most elementary levels of experience are impossible without inductive inferences, that is, without the law of causality. And even if the most complete experience should teach us that everything previously observed has occurred uniformly—a point concerning which we are not yet certain—we could conclude only by inductive inferences, that is, by presupposing the law of causality, that the law of causality will also be valid in the future. We can do no more than accept the proverb, "Have faith and keep on!"

> The earth's inadequacies
> Will then prove fruitful.
>
> *Das Unzulängliche*
> *Dann wird's Ereignis.*[4]

That is the answer we must give to the question: what is true in our ideas? In giving this answer we find ourselves at the foundation of Kant's system and in agreement with what has always seemed to me the most fundamental advance in his philosophy.

I have frequently noted in my previous works the agreement between the more recent physiology of the senses and Kant's teachings. I have not meant, of course, that I would swear *in verba magistri* to all his more minor points. I believe that the most fundamental advance of recent times must be judged to be the analysis of the concept of intuition into the elementary processes of thought. Kant failed to carry out this analysis or resolution; this is one reason why he considered the axioms of geometry to be transcendental propositions. It has been the physiological investigations of sense perception which have

4. *Faust, Part II.*

led us to recognize the most basic or elementary kinds of judgment, to inferences which are not expressible in words. These judgments or inferences will, of course, remain unknown and inaccessible to philosophers as long as they inquire only into knowledge expressed in language.

Some philosophers who retain an inclination toward metaphysical speculation consider what we have treated as a defect in Kant's system, resulting from the lack of progress of the special sciences in his time, to be the most fundamental part of his philosophy. Indeed, Kant's proof of the possibility of metaphysics, the alleged science he did nothing further to develop, rests completely upon the belief that the axioms of geometry and the related principles of mechanics are transcendental propositions, given a priori. As a matter of fact, however, Kant's entire system really conflicts with the possibility of metaphysics, and the more obscure points in his theory of knowledge, over which so much has been argued, stem from this conflict.

Be that as it may, the natural sciences have a secure, well-established foundation from which they can search for the laws of reality, a wonderfully rich and fertile field of endeavor. As long as they restrict themselves to this search, they need not be troubled with any idealistic doubts. Such work will, of course, always seem modest to some people when compared to the high-flown designs of the metaphysicians.

For with Gods must
Never a mortal
Measure himself.
If he mounts upwards,
Till his head
Touch the star-spangled heavens,
His unstable feet
Feel no ground beneath them;
Winds and wild storm-clouds
Make him their plaything;–

Or if, with sturdy,
Firm-jointed bones, he
Treads the solid, unwavering
Floor of the earth; yet
Reaches he not
Commonest oaks, nor
E'en with the vine may
Measure his greatness.[5]

5. The second and third verses of Goethe's "The Limits of Man." This translation, by John S. Dwight, appears in *The Permanent Goethe,* edited, selected, and with an Introduction by Thomas Mann (New York: The Dial Press, 1948), p. 17.

Doch mit Göttern
Soll sich nicht messen
Irgend ein Mensch.
Hebt er sich aufwärts
Und berührt
Mit dem Scheitel die Sterne,
Nirgends haften dann
Die unsicheren Sohlen,
Und mit ihm spielen
Wolken und Winde.

Steht er mit festen
Markigen Knochen
Auf der wohlgegründeten
Dauernden Erde:
Reicht er nicht auf,
Nur mit der Eiche
Oder der Rebe
Sich zu vergleichen.

The author of this poem has provided us with a model of a man who still retains clear eyes for the truth and for reality, even when he touches the stars with the crown of his head. The true scientist must always have something of the vision of an artist, something of the vision which led Goethe and Leonardo da Vinci to great scientific thoughts. Both artists and scientists strive, even if in different ways, toward the goal of discovering new uniformities or lawful regularities. But one must never produce idle swarms and mad fantasies in place of artistic vision. The true artist and the true scientist both know how to work steadily and how to give their work a convincing, truthful form.

Moreover, reality has always unveiled the truth of its laws to the sciences in a much richer, more sublime fashion than she has painted it for even the most consummate efforts of mystical fantasy and metaphysical speculation. What have all the monstrous offspring of indiscreet fancy, heapings of gigantic dimensions and numbers, to say of the reality of the universe, of the period of time during which the sun and earth were formed, or of the geological ages during which life evolved, adapting itself always in the most thoroughgoing way to the increasingly more moderate physical conditions of our planet?

What metaphysics has concepts in readiness to explain the effects of magnetic and induced electrical forces upon each other—effects which physics is now struggling to reduce to well-established elementary forces, without having reached any clear solution? Already, however, in physics light appears to be nothing more than another form of movement of these two agents, and the ether (the electrical and magnetic medium which pervades all space) has come to have completely new characteristics or properties.

And in what schema of scholastic concepts shall we put the store of energy capable of doing work, whose constancy is stated in the law of the conservation of energy and which, indestructible and incapable of increase like a substance, is acting as the motive power in every movement of inanimate as well as animate material—a store of energy which is neither mind nor matter, yet is like a Proteus, clothing itself always in new forms; capable of acting throughout infinite space, yet not infinitely divisible like space; the effective cause of every effect, the mover in every movement? Did the poet have a notion of it?

In the tides of Life, in Action's storm,
A fluctuant wave,
A shuttle free,
Birth and the Grave,
An eternal sea,
A weaving, flowing
Life, all-glowing,
Thus at Time's humming loom 't is my hand prepares
The garment of Life which the Deity wears![6]

In Lebensfluten, im Tatensturm
Wall ich auf und ab,
Webe hin und her.
Geburt und Grab,
Ein ewiges Meer,
Ein wechselnd Weben,
Ein glühend Leben;
So schaff ich am sausenden Webstuhl der Zeit
Und wirke der Gottheit lebendiges Kleid.

We are particles of dust on the surface of our planet, which is itself scarcely a grain of sand in the infinite space of the universe. We are the youngest species among the living things of the earth, hardly out of the cradle according to the time reckoning of geology, still in the learning stage, hardly half-grown, said to be mature only through mutual agreement. Nevertheless, because of the mighty stimulus of the law of causality, we have already grown beyond our fellow creatures and are overcoming them in the struggle for existence. We truly have reason to be proud that it has been given to us to un-

6. *Faust, Part I*; the Earth Spirit is speaking. Translated by Bayard Taylor.

derstand, slowly and through hard work, the incomprehensibly great scheme of things. Surely we need not feel in the least ashamed if we have not achieved this understanding upon the first flight of an Icarus.

APPENDICES

Appendix 1: The Localization of the Sensations of the Internal Organs

The question arises whether the physiological and pathological sensations of the internal organs of the body must fall into the same category with mental states or conditions, for many of these sensations are not—at least, not to any appreciable degree—altered by movement. There are, of course, sensations which are rather unspecific in character, such as depression, melancholy, and anxiety, which can arise just as easily from bodily as from psychic causes and of which we do not have any sense of specific location. At the most, as in the case of anxiety, the region of the heart may be vaguely suggested as the seat of the sensation, since in ancient thought it was generally believed that the heart was the seat and source of many emotions and since the motion of this organ is frequently changed by those emotions, the change being felt sometimes directly and sometimes indirectly by the hand held on the breast. In this way there arose a kind of false bodily localization for what were really mental conditions. In a diseased state this sort of false localization occurs fairly frequently. (I remember, as a young doctor, seeing a melancholy shoemaker who believed that his conscience was located between his heart and his stomach.)

On the other hand, there is a group of bodily sensations—such as hunger, thirst, satiety, and neuralgic and inflammatory pains—which we cannot localize precisely, yet which we consider to be located in the body and not to be purely mental, although they can rarely be changed by bodily movements. Most inflammatory and rheumatic pains are considerably intensified by pressure on or movement of the general part of the body where they have their seat. In other cases (this is true even of neuralgic pains), these sensations are looked upon as normal but highly intensified feelings of pressure or tension in the affected part. The general location of these sensations frequently gives us a hint as to their causes, and this enables us to learn something about the precise seat of the sensation.

Almost all sensations of the abdominal viscera, for example, are displaced to specific places on the anterior abdominal wall, even in the case of such organs as the duodenum, the pancreas, and the spleen, which lie nearer the posterior wall of the body. Pressure from the outside, however, can usually affect these organs only through the flexible posterior wall of the body and not through the thick layers of muscle between the ribs, the spine, and the hip bone. It is also worth noticing that with a toothache of the periosteum of a tooth, the patient is at first completely uncertain as to which of a pair of opposing teeth is causing the trouble. It is necessary to press firmly on each separately in order to determine which one is causing the pain. Is this not a consequence of the fact that pressure on the periosteum of the root of a tooth in normal condition occurs only when we are chewing, that is, when both teeth of each pair simultaneously experience the same pressure?

The feeling of satiety is the sensation of fullness in the stomach, a sensation which is clearly intensified by pressure on the pit of the stomach. The same pressure is lessened somewhat with the feeling of hunger. We may therefore come to perceive this sensation as localized in the pit of the stomach. And if we tend to believe that the same kinds of local signs always come from the same places in the body, a clearly established localization of one sensation, such as satiety, will lead us to assume the same location for all future sensations of the same kind.

This holds also for thirst, which is in part a sensation of dryness in the throat. The corresponding general feeling of a lack of water in the body, on the other hand—a feeling which is not removed by moistening the mouth and throat—is not definitely localized.

The peculiar feeling of respiratory deficiency, or what is called shortness of breath, is lessened by breathing and thus is localized. The sensation of an obstruction in breathing, however, is only partly distinguishable from the sensation of a restriction in circulation, where the latter is not related to sensible alterations in heartbeat. Perhaps this distinction is incomplete for the very reason that, as a rule, disturbances of breathing cause intensified heart action, while abnormal heart action increases the difficulty in satisfying respiratory deficiencies.

Furthermore, since we cannot easily feel them and cannot see them without optical apparatus, we have no conception—unless we carry out anatomical and physiological studies—of the form and function of such extraordinarily sensitive and well-adapted parts of our bodies as the soft palate, the epiglottis, and the larynx. Indeed, despite all our scientific investigations, we still cannot describe all the movements of these organs with certainty. We cannot describe, for example, the operation of the larynx during the production of a falsetto voice. If we had an innate knowledge of the location of our tactile organs or of all the organs which operate by contact, we should expect such knowledge for the larynx as well as for the hands. Our knowledge of the form, size, and operation of our own organs, however, extends only as far as we are able to see or to feel them.

The extraordinarily diverse and finely detailed movements of the larynx also teach us something about the relationship between an act of the will and its effects. What we intend to effect is not usually an innervation of a particular nerve or muscle, nor is it always an innervation at a particular place on some movable part of our body. On the contrary, we intend to produce an observable external effect. Insofar as we can ascertain the positions of the parts of our body by the use of our eyes and hands, the movement of these parts is the immediate observable effect produced by an intention of the will. When we cannot determine the position of a part of the body, as in the case of the larynx and the organs at the rear of the mouth, various modifications of the voice, the breath, swallowing, etc., are the immediate effects.

Thus movements of the larynx, although they are called forth by innervations which are completely similar to those used to produce movements of the limbs, do not come to our attention as a result of our observation of spatial changes. (The question whether the very clear and varied impressions of movement which music produces are perhaps to be explained by the fact that alterations in pitch are produced in singing by the innervation of muscles, and thus by the same kind of inner activity as that which results in the movements of the limbs, is not under consideration here.)

The situation is similar for the movements of the eyes. We all know quite well how to direct our gaze upon a particular place in the field of vision, that is, how to cause the image of that place to fall on the yellow spot of the retina. Untrained people, however, are not aware how they move their eyes in doing this, and they never respond to the ophthalmic surgeon's request to roll their eyes to the left, if the request is phrased exactly that way. More educated individuals know that to see an object held near their nose, they must focus inward; even they, however, do not know how to comply with the summons to squint inward if there is no object there.

Appendix 2: Space May Be Transcendental Without the Axioms of Geometry Being So Too

Almost all philosophical opponents of metamathematical investigations treat these two propositions as identical, which they by no means are. Benno Erdmann[7] has already quite clearly established this fact, using the current

7. *The Axioms of Geometry* (Leipzig, 1877), Chapter III.

mode of expression of philosophers. I myself have called attention to it in a reply addressed to the objections of Mr. Land, of Leyden.[8]

The author of the most recent critical comments, Albrecht Krause,[9] has quoted both these essays, yet he devotes the first five of his book's seven sections to a defense of the transcendental nature of space as a form of intuition; in only two sections is he concerned with the axioms of geometry. To be sure, he is not a pure Kantian but a spokesman for the most extreme nativistic theories in physiological optics. He considers the entire content of these theories to be included in Kant's system of epistemology. Such theories, however, do not have the most genuine title to that position, even though Kant's own opinion, reflecting the undeveloped state of physiological optics of his time, is approximately the same. The question whether or not intuition can be reduced to conceptual elements had not even been raised in Kant's time.

Further, Mr. Krause attributes to me ideas of local signs, of memory (*Sinnengedächtnis*), of the influence of retinal size, etc., which I never had or never used—or which I have, indeed, expressly troubled myself to reject. By *Sinnengedächtnis* I have consistently meant only the memory of immediate sense impressions, impressions which are not conceptualized, and I have consistently protested against the belief that this kind of memory has its seat in the peripheral sense organs. I have conducted and described experiments in order to show that we can quickly learn to overcome and correct illusions caused by false retinal images—the kind of images we get, for example, in looking through lenses or through convergent, divergent, or laterally displaced prisms. And then (on page 41 of his work) Mr. Krause interprets me as saying that a child, while he is growing, must see everything smaller because his eyes are smaller! Perhaps he will be convinced by the present lecture that he has completely misunderstood the meaning of my empirical theory of perception.

What Mr. Krause objects to in the sections of his book on the axioms of geometry has in part been settled in this lecture—for example, the reason why clear ideas of an object never seen before can be so difficult to obtain. Mr. Krause then proceeds, in connection with my supposition of two-dimensional beings living on a plane or on a sphere (a supposition which I made in my paper on the axioms of geometry in order to clarify the relations among different geometries), to a discussion of how two or more "straightest"[10] lines between two points can exist on a sphere, the axioms of Euclid on the contrary admitting only one "straight" line. For two-dimensional beings living

8. [See Chapter 13.]
9. *Kant and Helmholtz* (Lahr, 1878).
10. I used this term for "shortest," or geodesic, lines.

on the surface of a sphere, however, given the assumptions made for such beings, a *straight* line connecting two points on the spherical surface would have no real existence in their world. The *straightest* lines of their world would be for them just what *straight* lines are for us. Mr. Krause even makes an attempt to define a straight line as a line with only one direction. But how are we to define "direction"? Only by means of "straight line"! We are moving here in a vicious circle. "Direction" is, in fact, the derivative concept, and every straight line has two opposing directions.

Then there follows a discussion leading to the point that, if the axioms were really empirical propositions, we could not be as absolutely convinced of their truth as we in fact are. Here, then, he directs his quarrel. Mr. Krause is convinced that we would not believe measurements which went contrary to the axioms. On this point he may be right about a great many people, men who treasure propositions which are based upon ancient authority and closely integrated with all their other knowledge more dearly than their own thoughts. It should be different, however, with a philosopher. For a long time men vehemently denied the spherical shape of the earth, denied that it moved, and denied the existence of meteorites. It is right, of course, that the evidence used to establish points which are contrary to propositions maintained upon the basis of ancient authority should be all the stronger, the longer these propositions have been held factually true according to the experience of earlier generations. Ultimately, however, the facts, not some preconceived idea or the authority of Kant, should decide.

It is also true that if the axioms of geometry are natural laws, then like all other natural laws arrived at by induction, they should properly be thought of as only more or less probable. The desire to know exact laws is no proof that there are any. It is indeed strange that Mr. Krause, who rejects the results of scientific measurements because of their limited precision, has no qualms about accepting the estimations made by transcendental intuition—estimations which he needs in order to prove that we do not require any measurements to convince ourselves of the correctness of the axioms. This is, indeed, judging friend and foe by different standards! As if every circle made by even the poorest set of mathematical instruments were not more exact than the best estimations made with the eye alone, even ignoring the question (which my opponent does not even consider) whether the latter are innate and given a priori or are also acquired.

The expression *measure of curvature,* in its application to space of three dimensions, has caused great excitement among philosophical writers. This name designates a certain magnitude defined by Riemann, a magnitude which, calculated for surfaces, is the same as what Gauss called the measure of curvature of surfaces. Geometricians have retained this name as a short designation for the general case of more than two dimensions. The quarrel here is wholly and only about the name of a defined mathematical concept.

Appendix 3: The Applicability of the Axioms to the Physical World

Here I wish to show the consequences to which we are forced if Kant's hypothesis concerning the transcendental origin of the geometric axioms is correct. Further, I wish to assess in detail the value which this direct knowledge of the axioms would have for our judgments of relations in the objective world.

I

In this first section I shall retain the realistic hypothesis and use its language. I shall also assume that the things we perceive as objective really do exist and that they act on our senses. I do this only in order to be able to use the simple, understandable language of daily life and science and thus to express what I mean in a manner comprehensible to non-mathematicians. I shall drop the realistic hypothesis in later paragraphs and repeat the same discussion in abstract language, without making any special assumptions concerning the nature of reality.

We must first distinguish between *equality,* or the congruence of spatial magnitudes which is dependent upon the assumed transcendental intuition of spatial magnitudes, and *equivalence,* or the congruence of spatial magnitudes which is confirmed by measurements made with the help of physical instruments.

I call those spatial magnitudes which, under the same conditions and during equal periods of time, persist in and undergo the same physical processes *physically equivalent.* The procedure most frequently employed (with suitable precautions) in determining physically equivalent spatial magnitudes is the movement of rigid bodies, such as a pair of compasses or a ruler, from one place to another. It is a universal result of all our experience that if the equivalence of two spatial magnitudes can be determined by some method of physical measurement, the same method can be used to establish whether any other physical systems near it in space are equivalent.

Physical equivalence is thus a completely determined, single-valued, objective property of spatial magnitudes; and there is obviously nothing to prevent us from investigating, by observation and experimentation, how the physical equivalence of one such pair of spatial magnitudes is dependent upon the physical equivalence of some other pair. These procedures give us a kind of geometry which, for the purposes of our present investigation, I shall call *physical geometry,* in order to distinguish it from the geometry which is based upon the hypothesis of the transcendental intuition of space. A rigorously developed system of physical geometry is obviously possible and has fully the character of a natural science.

Already these first steps lead us to empirical propositions which correspond to the axioms of geometry–if, instead of the transcendental equality of geometric figures, we mean their physical equivalence.

As soon as we have found a method suitable for determining whether the distances between two pairs of points are equal (that is, physically equivalent) to each other, we can also investigate the special case in which three points, a, b, and c, lie in such a way that no point other than b can be found which is the same distance from a and c as b is. We then say that the three points lie in a straight line.

We can then consider three points, A, B, and C, which are all equidistant from one another and thus are the corners of an equilateral triangle. Then we can find two new points, b and c, both equidistant from A and such that b lies in the straight line AB and c in the straight line AC. Now the question arises: is the new triangle Abc, like the triangle ABC, equilateral; that is, does $bc = Ab = Ac$? In Euclidean geometry the answer is yes; in spherical geometry the answer is that $bc > Ab$ if $Ab < AB$; and in pseudospherical geometry the answer is that, under the same condition, $bc < Ab$. We have already reached a point where a factual decision among axioms must be made.

I have chosen this simple example because we are concerned here only with the measurement of the equality or inequality of the distance between points and with the definiteness or indefiniteness of the position of certain points, and because no complex geometrical figures, only straight lines and planes, must be constructed. The example shows that physical geometry indeed has empirical propositions functioning as the axioms.

As far as I can see, not even the followers of Kant's theory of knowledge can deny that it would be possible to establish an empirical geometry–if we did not already have one–in the way described. In such a geometry we would be concerned only with observable empirical facts and their laws. The science established in this way would be a geometry independent of the nature of physical bodies in space only insofar as the assumption that physical equivalence holds in general for all physical processes proves to be true.

Kant's followers, however, claim that in addition to such a physical geometry, there is a *pure geometry* which has its foundations solely in a transcendental intuition, and further, that this pure geometry is the very one that has been developed scientifically. In this geometry we have nothing to do with physical bodies and their movement. On the contrary, according to this theory, we can form through an inner intuition—without learning anything about them through experience—ideas of absolutely unchanging, immovable geometric figures (solids, surfaces, and lines) which, without the movement necessary to make physical bodies coincide with one another, can stand in the relation of equality and congruence with one another.

I must stress that this inner intuition of the straightness of lines and of the equality of distances and angles must be absolutely precise. Otherwise we would not be justified in deciding whether two straight lines, infinitely extended, cut one another only once or, like the great circles on the surface of a sphere, twice. Nor would we be justified in maintaining that any straight line which cuts one of two parallel lines lying in a common plane must necessarily cut the other one also. One may not substitute imperfect visual measurements for transcendental intuition, which demands absolute precision.

Let us suppose now that we do have such a transcendental intuition of geometric figures and of their equality and congruence, and that by referring to satisfactory evidence we are really able to justify the claim that we have such an intuition. Then indeed a system of geometry could be developed that would be independent of all properties of physical bodies—a pure, transcendental geometry. This geometry, of course, would also have its axioms. It is clear, however, even according to Kantian principles, that the propositions of this hypothetically pure geometry need not necessarily correspond to those of physical geometry. One system refers to the equality of spatial magnitudes in inner intuition; the other to physical equivalence. The latter obviously depends upon the empirical properties of natural bodies and not just upon the organization of our mind.

It is thus necessary to try to determine whether the two kinds of equality are necessarily always the same. Experience cannot decide this. Is there any meaning in asking whether two pairs of compass points span equal or unequal distances according to transcendental intuition? I cannot make any sense out of this and, as far as I can understand them, I believe we may conclude that the most recent followers of Kant cannot either. As was said earlier, visual measurement of spatial magnitudes must not be substituted for transcendental intuition.

Is it possible to deduce something from the propositions of pure geometry, showing that the distances between two pairs of compass points are equal? In order to do this, geometric relations between these distances and other geometric magnitudes, which we must know to be equal directly by means of transcendental intuition, must be known. Since we can never know these relations directlv, however, we can never make the required deduction.

If the proposition that both kinds of spatial equality are identical cannot be established by experience, it must be a metaphysical proposition expressing a necessity of thought. Such a proposition, however, would specify not only the form of empirical knowledge but also its content, as in the case of the two equilateral triangles discussed above. This would conflict with Kant's principles, however, for pure intuition and thought would then accomplish more than Kant was inclined to believe they could.

Let us suppose, finally, that in physical geometry a series of general empirical propositions have been found which are the same in asserted content as the axioms of pure geometry. At the most, it would follow from this that correspondence between the physical equivalence of geometric figures and their equality in pure spatial intuition is a permissible hypothesis which leads to no contradiction. This would not, however, be the only possible hypothesis. Physical space and the space of intuition can also be related in the way that real space is related to its image in a convex mirror.[11]

That the physical and transcendental geometries need not necessarily agree follows from the fact that we can actually represent them as not agreeing. In an earlier paper I have shown how such a lack of agreement can be demonstrated.[12] Let us assume that physical measurements are those of a pseudospherical space. Sense impressions in such a space, given a stationary observer and stationary objects of observation, would be the same as those in Beltrami's spherical model in Euclidean space, with the observer located at the center of the projection sphere. As soon as the observer changes his position, however, the center of the sphere necessarily changes with him, and the entire projection shifts. If the observer's spatial perceptions and judgments of magnitudes had been formed either by transcendental intuition or as a result of previous experience in Euclidean space, he would now have the impression that as he himself moved, all the objects around him also shifted in a definite way, and that objects in different directions expanded or contracted by different amounts.

In a similar way, though in accordance with different quantitative rela-

11. [See Chapter 8, pp. 259–260.]
12. [See *ibid.*, pp. 260–261.]

tions, as we move about in our objective world, we see the relative perspective positions and the apparent sizes of objects change in different ways depending upon their distance. We can tell from the way our optical images change, however, that the objects around us are not actually changing their relative positions and sizes, as long as their perspective displacements correspond exactly to certain laws established by our previous experience with stationary objects. On the other hand, we may infer the movement of objects, given any deviations from these laws.

As an exponent of the empirical theory of perception, I believe that anyone, in passing from Euclidean to pseudospherical space, would at first believe that he saw apparent movements of the objects around him but that very soon he would learn to adjust his judgments of spatial relations to the new conditions.

This is, to be sure, an assumption which is based only upon an analogy to what we already know by sense perception; it cannot be established by experiment. So let us assume instead that an individual's judgments of spatial relations cannot be changed because they are connected with innate forms of intuition. He would, nevertheless, quickly discover that the movements he believed he saw were only apparent movements, since they would always be reversed when he returned to his original position. A second individual, moreover, would be able to substantiate that everything remained at rest while the first one was moving. Thus, even if it is not clear to unreflective intuition, scientific investigations can establish what spatial relations are physically constant. This is analogous to our knowing by means of scientific investigations that the sun remains stationary and the earth rotates, in spite of the fact that the earth appears to remain stationary and the sun to revolve around it every twenty-four hours.

The assumed transcendental, a priori intuition has now been reduced, however, to the level of an illusion of the senses, an objectively false appearance from which we must free ourselves and which we must try to forget, just as we do in the case of the apparent motion of the sun. There would be a continual disagreement between what appears spatially equal according to innate intuition and what is shown to be so in the objective phenomena themselves. All our scientific and practical interests, of course, are bound up with the latter.

The transcendental form of intuition would represent physically equivalent spatial relations in the same manner that a flat map represents the surface of the earth; that is, it would represent very small pieces and strips correctly but larger ones perforce incorrectly. There would therefore be, not only a question of the *general way* in which things were represented or in which they appeared, necessitating some modifications in the contents represented, but also a question of the *specific relation* between appearance and content, since there would be a correspondence between the two in restricted regions but none in more extended ones.

The consequences which I draw from these considerations are as follows: If there really is an innate, fixed form of intuition of space which includes the axioms of geometry, the objective scientific application of these axioms to the world of experience is justified only if it can be shown by observation and experiment that equal distances in transcendental intuition are also physically equivalent. This condition is in agreement with Riemann's stipulation that the measure of curvature of the space in which we live be determined empirically by measurement.

Previous measurements of the value of this measure of curvature have resulted in no noticeable deviation from zero. Thus we may regard Euclidean geometry as factually correct within the present limits of precision of measurement.

II

The discussion in the preceding paragraphs has been conducted entirely from the objective, realistic point of view of the natural sciences, according to which the ultimate goal of science is conceptual understanding of natural laws, while knowledge gained through intuition is either of only heuristic value or to be discarded as a false appearance.

Professor Land believes that in my previous discussions I confused the concepts of *objective* and *real*—that in my statement that geometric propositions can originate from and be confirmed by experience, it was presupposed without foundation "that empirical knowledge is acquired by simple importation or by counterfeit, and not by peculiar operations of the mind, solicited by varied impulses from an unknown reality."[13] If Professor Land had studied my works on the sensations of the senses, he would have known that I have fought all my life against such an assumption as the one he has ascribed to me. I have not spoken in my essays about the distinction between objective and real because it seems to me that in this investigation such a distinction has no importance whatsoever. In order to justify this belief, I shall now drop what is hypothetical in the realistic interpretation and show that the propositions and arguments previously set forth still make perfectly good sense, and

---

13. *Mind, A Quarterly Review,* Vol. II, No. 5, p. 46.

that one is still justified in investigating the physical equivalence of geometric figures and in determining by experience whether they are equivalent or not.

The single assumption that I shall retain is the law of causality–the belief that the ideas which we think of as perceptions occur according to fixed laws, so that if different perceptions force themselves upon us, we are justified in inferring differences in the conditions under which they originated. We know nothing, of course, of these conditions themselves, the ultimate reality which lies behind all phenomena; any opinions we may entertain about them are to be considered only as more or less probable hypotheses. The principle of causality, on the other hand, is the fundamental principle of all of our thinking; if we give it up, we give up all claim to be able to think about these or any other matters.

I emphasize again that no assumptions will be made about the nature of the conditions which give rise to ideas. The hypothesis of subjective idealism is just as admissible as the realistic point of view, whose language we have used up to this point. We may even assume that all our perceptions are only a dream, although of course a highly consistent one, in which each idea is developed out of another according to fixed laws. Under this assumption, the reason why some new perception is experienced is to be sought only in the established relationships of perceptions and ideas in the dreamer's mind, or perhaps also in the relationships among ideas, perceptions, and acts of volition. What we call natural laws according to the realistic hypothesis are, according to the idealistic, laws which relate series of perceptual ideas.

We find, as a fact of consciousness, that we believe we perceive objects at definite places in space. The idea that an object appears at one specific place and not at another must depend upon the real conditions which produce the idea. We may infer that other conditions would have to exist in order to produce the perception of the same object at another place. Thus there must exist some sort of relation or complex of relations which determines at what place in space an object is going to appear. To be concise I shall call these *topogenous moments.* We know nothing of their nature. We know only that the occurrence of spatially different perceptions presupposes differences among topogenous moments.

In addition, there must be other causes in the sphere of the real which make us believe that we perceive, at the same place but at different times, different material things with different properties. I shall call these *hylogenous moments.* I choose these new names in order to avoid any confusion that might result from the connotations of words in ordinary use.

If we perceive some relation between spatial magnitudes and assert that it exists, the factual content of the assertion is only that between certain topogenous moments (the real nature of which remains unknown to us) a certain relation exists (the nature of which also remains unknown). Because of this, Schopenhauer and many followers of Kant have come to the erroneous conclusion that there is never a real content to our perception of spatial relations, that space and its relations are only transcendental appearances without anything real corresponding to them. We are certainly justified, however, in applying to our spatial perceptions the same analysis that we apply to such sensuous signs as, for example, colors. Blue is only a particular sensation; that we see blue at a certain time and in a certain direction, however, must have a basis in reality. If we see red in that same direction at another time, the real basis must have changed.

If we observe that the most diverse physical processes can occur in congruent parts of space during equal periods of time, this means that in the sphere of the real, equal aggregates and sequences of specific hylogenous moments are coming into being and existing in connection with specific groups of different topogenous moments, those equivalent parts of space. And if experience teaches us that any combination or any sequence of hylogenous moments which can exist in combination with a group of topogenous moments can also exist in combination with any physically equivalent group of other topogenous moments, that is a proposition which has real content, and topogenous moments undoubtedly influence the course of real processes.

In the example concerning the two equilateral triangles, we were concerned only with equality or inequality—that is, with the physical equivalence or nonequivalence of the distances between points—and with the definiteness or indefiniteness of the topogenous moments of certain points. The concepts of definiteness and equivalence, used normally in connection with familiar processes, can thus be applied to objects whose nature is completely unknown. From this I conclude that the science which I have called physical geometry contains propositions with real content and that its axioms are determined, not by the form of the understanding, but by relationships in the real world.

Still, this does not justify us in declaring that a geometry based on transcendental intuition is impossible. One might maintain, for example, that the intuition of the equality of two spatial magnitudes can be produced directly and without physical measurement by the influence of topogenous moments upon our consciousness. It would follow that in their directly perceptible psychic effects, certain aggregates of topogenous moments are equivalent.

All Euclidean geometry can be deduced from a formula which expresses the distance between two points as a function of their rectangular coordinates. If we assume that the intensity of psychic effects, whose equality appears to us as equality in the distance between two points, is dependent upon any three functions whatsoever of the topogenous moments of each point—just as the distance between two points in Euclidean space is a function of the three coordinates of each point—then the system of pure geometry constructed on this basis would satisfy the axioms of Euclid, whatever the topogenous moments of the real world and their physical equivalence might really be.

It is clear in this case again, however, that the correspondence between the psychic and physical equivalence of spatial magnitudes cannot be decided by the form of intuition alone. If a correspondence between the psychic and the physical should prove to be the case, this would either be a natural law—just as it is a natural law that the straight line described by a ray of light corresponds to one made by a stretched string—or, as I pointed out in my earlier popular lectures, it would have to be considered a pre-established harmony between the world of ideas and the real world.

I hope I have now shown that the demonstrations which were presented in Part I of this appendix, using the language of the realistic hypothesis, are valid also without the use of that hypothesis.

When we apply geometry to the facts of experience, where we are always concerned with physical equivalence, only the propositions of the science which I have called physical geometry can be used. To anyone who infers the axioms of geometry from experience, our traditional geometry is indeed physical geometry—a geometry based upon a large number of random experiences, rather than upon a systematically or methodically executed investigation. It should be mentioned that this was Newton's belief, which he expressed in the Introduction to the *Principia*: "Geometry has its foundation in mechanical practice, and is indeed nothing but that part of general mechanics which is based upon and precisely determined by the art of measurement."

The belief that our knowledge of the axioms of geometry is based upon transcendental intuition, on the other hand, is (a) an unproved hypothesis; (b) an unnecessary hypothesis, for it explains nothing in the empirical world which cannot be explained without its help; and (c) a hypothesis which is completely inapplicable in the explanation of the real world, for the propositions established by it with respect to relations in the real world can be applied only after their objective validity has been established by experience.

Kant's theory of a priori forms of intuition is a very fine and clear expression of an important point. These forms, however, must be empty and free from all restrictions in order to be able to encompass any content whatsoever which may enter a particular form of perception. The axioms of geometry limit the form of intuition of space so that not every conceivable content can be incorporated in it, particularly if geometry is to be applicable to the external world. If we omit the axioms, there can be no objections to the theory of the transcendental nature of the form of intuition of space. Kant in his *Critique* was not critical enough; he concerned himself there with a thesis in mathematics, the critical work concerning which lies within the province of the mathematicians.

15.

THE MODERN DEVELOPMENT OF FARADAY'S CONCEPTION OF ELECTRICITY [1881]

The Faraday Lecture, delivered before the Fellows of the Chemical Society in London on April 5, 1881

As I have the honor of speaking to you in memory of the great man who, from the very place where I stand, has so often revealed to his admiring auditors the most unexpected secrets of nature, I hope at the outset to gain your assent if I limit my exposition to that side of his activity which I know the best from my own experience and studies: I mean the theory of electricity. The majority, indeed, of Faraday's own researches were connected directly or indirectly with questions regarding the nature of electricity, and his most important and most renowned discoveries lay in this field. The facts which he discovered are universally known. Every physicist at present is acquainted with the rotation of the plane of polarization of light by magnetism, with dielectric tension and diamagnetism, and with the measurement of the intensity of galvanic currents by the voltameter, while induced currents act on the telephone, are applied to paralyzed muscles, and nourish the electric light.

Nevertheless, the fundamental conceptions by which Faraday was led to these much admired discoveries have not received an equal amount of consideration. They were very divergent from the trodden path of scientific theory and appeared rather startling to his contemporaries. His principal aim was to express in his new conceptions only facts, with the least possible use of hypothetical substances and forces. This was really an advance in general scientific method, destined to purify science from the last remnants of metaphysics. Faraday was not the first and not the only man who has worked in this direction, but perhaps nobody else at his time did it so radically. But every reform of fundamental and leading principles introduces new kinds of abstract notions, the sense of which the reader does not catch in the first instance.

Under such circumstances it is often less difficult for a man of original thought to discover new truth than to discover why other people do not understand and do not follow him. This difficulty must increase in Faraday's case because he had not gone through the same common course of scientific education as the majority of his readers. Now that the mathematical interpretation of Faraday's conceptions regarding the nature of electric and magnetic forces has been given by Clerk Maxwell, we see how great a degree of exactness and precision was really hidden behind the words which to Faraday's contemporaries appeared either vague or obscure; and it is in the highest degree astonishing to see what a large number of general theorems, the methodical deduction of which requires the highest powers of mathematical analysis, he found by a kind of intuition, with the security of instinct, without the help of a single mathematical formula. I have no intention of blaming his contemporaries, for I confess that many times I have myself sat hopelessly looking upon some paragraph of Faraday's descriptions of lines of force, or of the galvanic current being an axis of power, and so on.

A single remarkable discovery may, of course, be the result of a happy accident and may not indicate the possession of any special gift on the part of the discoverer; but it is against all rules of probability that the train of thought which has led to such a series of surprising and unexpected discoveries as were those of Faraday should be without a firm, although perhaps hidden, foundation of truth. We must also in his case acquiesce in the fact that the greatest benefactors of mankind usually do not obtain a full reward during their lifetime and that new ideas need the more time for gaining general assent, the more really original they are and the more power they have to change the broad path of human knowledge.

Faraday's electrical researches, although embracing a great number of apparently minute and disconnected questions, all of which he has treated with the same careful attention and conscientiousness, really always aim at two fundamental problems of natural philosophy—the one, more regarding the nature of the forces termed *physical*, or of forces working at a distance; the other, in the same way, regarding *chemical* forces, or those which act from molecule to molecule, and the relation between these and the first.

I shall give you only a short exposition on the degree of development which has been reached in the present state of science with regard to the first of these problems. The discussion of this question among scientific men is not yet finished, although, I think, it approaches its end. It is entangled with many geometric and mechanical difficulties. How these are to be solved, and what are the arguments pro and contra, I cannot undertake to explain in a short public lecture with any hope of gaining your scientific conviction. I can therefore give only a short statement of this side of the question, representing my own opinions; but I must not conceal the fact that several men of great scientific merit, principally among my own countrymen, do not yet agree with me.

The great fundamental problem which Faraday called up anew for discussion was the existence of forces working directly at a distance without any intervening medium. During the last and the beginning of the present century, the model after the likeness of which nearly all physical theories were formed was the force of gravitation acting between the sun, the planets, and their satellites. It is known with how much caution and even reluctance Sir Isaac Newton himself proposed his grand hypothesis, which was destined to become the first great and imposing example of the power of true scientific method. We need not wonder that Newton's successors attempted at first to gain the same success by introducing analogous assumptions into all the various branches of natural philosophy. Electrostatic and magnetic phenomena especially appeared as near relations to gravitation, because electric and magnetic attractions and repulsions, according to Coulomb's measurements, diminish in the same proportion as gravity with increasing distance.

But then came Oersted's discovery of the motions of magnets under the influence of electric currents. The force acting in these phenomena had a new and very singular character. It seemed as if this force would drive a single isolated pole of a magnet in a circle around the wire conducting the current, on and on, without end, never coming to rest. And although it is not possible really to separate one pole of a magnet from the other, Ampère succeeded in producing such continuous circular motions by making a part of the current itself movable with the magnet.

This was the starting point for Faraday's researches on electricity. He saw that a motion of this kind could not be produced by any force of attraction or repulsion, working from point to point. The first motive which guided him seems to have been an instinctive [conception] of the law of conservation of energy, which many attentive observers of nature had entertained before it was brought by Joule to a precise scientific definition. If the current is able to increase the velocity of the magnet, the magnet must react on the current. So he made the experiment and discovered induced currents. He traced them out through all the various conditions under which they ought to appear. He found that an electromotive force striving to produce these currents arises wherever and whenever magnetic force is generated or destroyed. He concluded that in a part of space traversed by magnetic force there ought to exist a peculiar state of tension and that every change of this tension produces electromotive force.

This unknown hypothetical state he called provisionally the electrotonic state, and he was occupied for years and years in finding out what this electrotonic state was. He first discovered in 1838 the dielectric polarization of electric insulators subject to electric forces. Such bodies, under the influence of electric forces, exhibit phenomena perfectly analogous to those observed in soft iron under the influence of the magnetic force. Eleven years later, in 1849, he was able to demonstrate that all ponderable matter is magnetized under the influence of sufficiently intense magnetic force, and at the same time he discovered the phenomena of diamagnetism, which indicated that even space, devoid of all ponderable matter, is magnetizable. The most simple explanation of these phenomena, indeed, is that diamagnetic bodies are less magnetizable than a vacuous space or than the luminiferous ether filling that space. In this way real changes corresponding to that hypothetical electrotonic state were demonstrated.

And now, with quite a wonderful sagacity and intellectual precision, Faraday performed in his brain the work of a great mathematician without using a single mathematical formula. He saw with his mind's eye that magnetized and dielectric bodies ought to have a tendency to contract in the direction of the lines of force and to dilate in all directions perpendicular to the former, and that by these systems of tensions and pressures–in the space which surrounds electrified bodies, magnets, or wires conducting electric currents–all the phenomena of electrostatic, magnetic, and electromagnetic attraction, repulsion, and induction could be explained, without recurring at all to forces acting directly at a distance. This was the part of his path where so few could follow him. Perhaps a Clerk Maxwell, a second man of the same power and independence of intellect, was needed to reconstruct in the normal methods of science the great building, the plan of which Faraday had conceived in his mind and attempted to make visible to his contemporaries.

Nobody can deny that this new theory of electricity and magnetism, originated by Faraday and developed by Maxwell, is in itself consistent, is in perfect and exact harmony with all the known facts of experience, and does not contradict any one of the general axioms of dynamics, which have been hitherto considered the fundamental truths of all natural science because they have been found valid, without any exception, in all known processes of nature. A confirmation of great importance was given to this theory by the circumstance, demonstrated by Clerk Maxwell, that the qualities which it must attribute to the imponderable medium filling space are able to produce and sustain magnetic and electric oscillations, propagating like waves and with a velocity exactly equal to that of light. Several parts even of the theory of light are deduced with less difficulty from this new theory than from the well-known undulatory theory of Huygens, which ascribes to the luminiferous ether the qualities of a rigid elastic body.

Nevertheless, the adherents of direct action at a distance have not yet ceased to search for solutions of the electromagnetic problem. The motive forces exerted upon each other by two wires conducting galvanic currents had long ago been reduced in a very ingenious way, by Ampère, to attracting or repelling forces belonging to the linear elements of every current. The intensity of these forces is considered to depend, not only on the distance of both parts of the current, but also in a rather complicated manner on the angles which the directions of the two currents make with each other and with the straight line joining them both. Ampère was not acquainted with induced currents, but the phenomena of these could be derived from the law of Ampère, connecting it with the general law, deduced by Faraday from his experiments, that the current induced by the motion of a magnet or of another current always resists this motion.

The general mathematical expression of this law was established by Professor Neumann, of Königsberg. It gave directly, not the value of the forces, but the value of their mechanical work, the value of what mathematicians call an electrodynamic potential; and it reduced electromagnetic phenomena to forces acting, not from point to point, but from one linear element of a current to another. Linear elements of a wire conducting a galvanic current are, of course, complicated structures compared with atoms. I have myself elaborated several mathematical papers to prove that this formula of Professor Neumann was in harmony with all the known phenomena exhibited by closed galvanic circuits and that it did not come into contradiction with the general axioms of mechanics in any case of electric motion. I succeeded in finding an experimental method of observing electrostatic effects of electromagnetic induction under conditions in which closed circuits could not be generated. This experiment decided against the supposition that Neumann's theory was complete so long as only the electric motions in metallic or fluid conductors were considered as active currents, but it was in accordance with the theory of Faraday and Maxwell, who supposed that from the extremities of conducting bodies, where an electric charge collects, electric motion is continued through the insulating media separating them.

Other eminent men have tried to reduce electromagnetic phenomena to forces acting directly between distant quantities of the hypothetical electric fluids, with an intensity which depends not only on the distance but also on the velocities and accelerations of those electric quantities. Such theories have been proposed by Professor W. Weber, of Göttingen, by Riemann, the too early deceased mathematician, and by Professor Clausius, of Bonn. All these theories explain very satisfactorily the phenomena of closed galvanic currents. But applied to other electric motions, they all come into contradiction with the general axioms of dynamics.

The hypothesis of Professor Weber makes the equilibrium of electricity unstable in any conductor of moderate dimensions and renders possible the development of infinite quantities of work from finite bodies. I do not find that the objections brought forward at first by Sir W. Thomson and Professor Tait in their *Treatise on Natural Philosophy* and discussed and specialized afterward by myself have been invalidated by the discussions going on about this question. The hypothesis of Riemann, which he did not himself publish during his lifetime, labors under the same objection and is at the same time in contradiction to Newton's axiom, which established the equality of action and reaction for all natural forces.

The hypothesis of Professor Clausius avoids the first objection but not the second, and the author himself has conceded that this objection could be removed only by the assumption of a medium filling all space, between which and the electric fluids the forces acted.

The present development of science shows then, I think, a state of things very favorable to the hope that Faraday's fundamental conceptions may in the immediate future receive general assent. His theory, indeed, is the only one existing which is at the same time in perfect harmony with the facts as far as they are observed and does not, beyond the reach of facts, lead into any contradiction to the general axioms of dynamics.

Clerk Maxwell himself has developed his theory only for closed conducting circuits. I have endeavored during the last few years to investigate the results of this theory also for conductors not forming closed circuits. I can already say that the theory is in harmony with all the observations we have on the phenomena of open circuits: I mean (1) the oscillatory discharge of a condenser through a coil of wire, (2) my own experiments on electromagnetically induced charges of a rotating condenser, and (3) Mr. Rowland's observations on the electromagnetic effect of a rotatory disc charged with one kind of electricity.

The deciding assumption which removes the theoretical difficulties is that introduced by Faraday, who assumed that any electric motion in a conducting body which charges its surface with electricity is continued in the surrounding insulating medium as beginning or ending dielectric polarization with an intensity equivalent to that of the current. A second inference from this supposition is that forces working at a distance do not exist—or are, at least, unimportant—when compared with the tensions and pressures of the dielectric medium.

It is not at all necessary to accept any definite opinion about the ultimate nature of the agent which we call electricity. Faraday himself avoided as much as possible giving any affirmative assertion regarding this problem, although he did not conceal his disinclination to believe in the existence of two opposite electric fluids. For our own discussion of the electrochemical phenomena, to which we shall now turn, I beg permission to use the language of the old dualistic theory, which considers positive and negative electricity as two imponderable substances, because we shall have to speak principally of relations of quantity.

We shall try to imitate Faraday as well as we can by keeping carefully within the domain of phenomena and, therefore, need not speculate about the real nature of that which we call a quantity of positive or negative electricity. Calling them substances of opposite sign, we imply with this name nothing else than the fact that a positive quantity never appears or vanishes without an equal negative quantity appearing or vanishing at the same time in the immediate neighborhood. In this respect they behave really as if they were two substances which cannot be either generated or destroyed but which can be neutralized and become imperceptible by their union.

I see very well that this assumption of two imponderable fluids of opposite qualities is a rather complicated and artificial machinery and that the mathematical language of Clerk Maxwell's theory expresses the laws of the phenomena very simply and very truly with a much smaller number of hypothetical implications. But I confess I should really be at a loss to explain, without the use of mathematical formulas, what he considers a quantity of electricity and why such a quantity is constant, like that of a substance. The original, old notion of substance is not at all identical with that of matter. It signifies, indeed, that which behind the changing phenomena lasts as invariable, which can be neither generated nor destroyed, and in this oldest sense of the word we may really call the two electricities substances.

I prefer the dualistic theory because it expresses clearly the perfect symmetry between the positive and negative side of electric phenomena, and I keep the well-known supposition that as much negative electricity enters where positive goes away, because we are not acquainted with any phenomena which could be interpreted as corresponding with an increase or a diminution of the total electricity contained in any body. The unitary theory, which assumes the existence of only one imponderable electric substance and ascribes the effects of opposite kind to ponderable matter itself, affords a far less convenient basis for an electrochemical theory.

I now turn to the second fundamental problem aimed at by Faraday, the connection between electric and chemical force.

Already, before Faraday went to work, an elaborate electrochemical theory had been established by the renowned Swedish chemist, Berzelius, which formed the connecting link of the great work of his life, the systematization of the chemical knowledge of his time. His starting point was the series in which Volta had arranged the metals according to the electric tension which they exhibit after contact with each other. Metals easily oxidized occupied the positive end of this series, those with small affinity for oxygen the negative end. Metals widely distant in the series develop stronger electric charges than those near each other. A strong positive charge of one metal and a strong negative of the other must cause them to attract each other and to cling to each other. The same faculty of exciting each other electrically was ascribed by Berzelius to all the other elements; he arranged them all into a series, at the positive end of which he placed potassium, sodium, barium, calcium, etc.; at the negative end, oxygen, chlorine, bromine, etc. Two atoms of different elements coming into contact are supposed to excite each other electrically, like the metals in Volta's experiment. Berzelius' conceptions about the distribution of opposite electricities in the molecules, and his deductions regarding the intensity of these forces, were not very clear and not in harmony with the laws of electric forces which had already been developed by Green and Gauss. A fundamental point, which Faraday's experiment contradicted, was the supposition that the quantity of electricity collected in each atom was dependent on their mutual electrochemical differences, which Berzelius considered the cause of their apparently greater chemical affinity.

His theory of the binary character of all chemical compounds was also connected with this electrochemical theory. Two elements, he supposed—one positive, the other negative—could unite into a compound of the first degree, a basic oxide or an acid; two such compounds, into a compound of the second degree, a salt. But there was nothing to prevent one atom of every positive element from uniting as directly with two, three, or even seven of another negative element as with one. The same was assumed by Berzelius for negative elements. The modern experience of chemistry directly contradicts these statements. But although the fundamental conceptions of Berzelius' theory have been forsaken, chemists have not ceased to speak of positive and negative constituents of a compound body. Nobody can overlook that such a contrast of qualities as was expressed in Berzelius' theory really exists, well developed at the extremities, less evident in the middle terms of the series, and playing an important part in all chemical actions, although often subordinated to other influences.

When Faraday began to study the phenomena of decomposition by the galvanic current, which of course were considered by Berzelius as among the firmest supports of his theory, he put a very simple question—the first question, indeed, which every chemist speculating about electrolysis ought to have thought of. He asked, what is the quantity of electrolytic decomposition if the same quantity of electricity is sent through several electrolytic cells? By this investigation he discovered that most important law, generally known under his name but called by him the law of definite electrolytic action.

When he began his experiments, neither Daniell's nor Grove's battery was known, and there were no means of producing currents of constant intensity; the methods of measuring this intensity were also in their infancy. This may excuse his predecessors. Faraday overcame this difficulty by sending the same current of electricity for the same time through a series of two or more electrolytic cells. He proved at first that the dimensions of the cell and the size of the metallic plates through which the current entered and left the cell had no visible influence upon the quantity of the products of decomposition. Cells containing the same electrolytic fluid between plates of the same metals gave always the same quantity after being traversed by the same current. Then he compared the amount of decomposition in cells containing different electrolytes, and he found it exactly proportional to the chemical equivalents of the elements, which were either separated or converted into new compounds.

Faraday concluded from his experiments that a definite quantity of electricity cannot pass a voltametric cell containing acidulated water between electrodes of platinum without setting free at the negative electrode a corresponding definite amount of hydrogen and at the positive electrode the equivalent quantity of oxygen, one atom of oxygen for every pair of atoms of hydrogen. If, instead of hydrogen, any other element capable of replacing hydrogen is separated from the electrolyte, this is done also in a quantity exactly equivalent to the quantity of hydrogen which would have been evolved by the same electric current. According to the modern chemical theory of quantivalence, therefore, the same quantity of electricity passing through an electrolyte either sets free or transfers to other combinations always the same number of units of affinity at both electrodes; for instance, instead

of $\left.\begin{matrix}H\\H\end{matrix}\right\}$, either $\left.\begin{matrix}K\\K\end{matrix}\right\}$, or $\left.\begin{matrix}Na\\Na\end{matrix}\right\}$, or $Ba\}$, or $Ca\}$, or $Zn\}$,

or $Cu\}$ from cupric salts,

or $\left.\begin{matrix}Cu\\Cu\end{matrix}\right\}$ from cuprous salts, and so on.

The simple or compound halogens separating at the other electrodes are equivalent, of course, to the quantity of the metallic element with which they were formerly combined.

According to Berzelius' theoretical views, the quantity of electricity collected at the point of union of two atoms ought to increase with the strength of their affinity. Faraday demonstrated by experiment that, insofar as this electricity came forth in electrolytic decomposition, its quantity did not at all depend on the degree of affinity. This was really a fatal blow to Berzelius' theory.

Since that time our experimental methods and our knowledge of the laws of electrical phenomena have made enormous progress, and a great many obstacles have now been removed which entangled every one of Faraday's steps and obliged him to fight with the confused ideas and ill-applied theoretical conceptions of some of his contemporaries. The original voltameter of Faraday, an instrument which measured the quantity of gases evolved by the decomposition of water in order to determine with it the intensity of the galvanic current, has been replaced by the silver voltameter of Poggendorff, which permits of much more exact determinations by the quantity of silver deposited from a solution of silver nitrate on a strip of platinum. We have galvanometers which not only indicate that there is a galvanic current but likewise measure its electromagnetic intensity very exactly and in a very short time, and do this as well for the highest as for the lowest degrees of intensity. We have electrometers, like the quadrant electrometer of Sir W. Thomson, able to measure differences of electric potential corresponding to less than one hundredth of a Daniell cell. As for the frequently used term *electric potential,* a term introduced by Green, you may translate it as signifying the electric pressure to which the positive unit of electricity is subject at a certain place. We need not hesitate to say that the more experimental methods were refined, the more completely were confirmed the exactness and generality of Faraday's law.

In the beginning Berzelius and the adherents of Volta's original theory of galvanism, based on the effects of metallic contact, raised many objections against Faraday's law.

By the combination of Nobili's astatic pairs of magnetic needles with Schweigger's multiplier, a coil of copper wire with numerous circumvolutions, galvanometers became so delicate that they were able to indicate the electrochemical equivalent of currents so feeble as to be quite imperceptible by all chemical methods. With the newest galvanometers you can very well observe currents which would have to last a century before decomposing one milligram of water, the smallest quantity that is usually weighed on chemical balances. You see that if such a current lasts only some seconds or some minutes, there is not the slightest hope of discovering its products of decomposition by chemical analysis. And even if it should last a long time, the minute quantities of hydrogen collected at the negative electrode may vanish because they combine with the traces of atmospheric oxygen absorbed by the liquid. Under such conditions a feeble current may continue as long as you like without producing any visible trace of electrolysis, not even of galvanic polarization, the appearance of which can be used as an indication of previous electrolysis.

Galvanic polarization, as you know, is an altered state of the metallic plates which have been used as electrodes during the decomposition of an electrolyte. Polarized electrodes, when connected by a galvanometer, give a current which they did not give before being polarized. By this current the plates are discharged again and returned to their original state of equality. The most probable explanation of this polarization is that molecules of the electrolyte, charged with electricity, are carried by the current to the surface of the metal, itself charged with opposite electricity, and are retained there by electric attraction. That really constituent atoms of the electrolyte partake in the production of galvanic polarization cannot well be doubted, because this state can be produced and also destroyed purely by chemical means. If hydrogen has been carried to an electrode by the current, contact with the atmospheric oxygen removes the state of polarization.

The depolarizing current is indeed a most delicate means of discovering previous decomposition. But even this may fail if the nascent polarization is destroyed by an intervening chemical action, like that of the oxygen of the air. To avoid this, delicate experiments on this subject cannot be performed except in vessels carefully purified of all gases.

I have lately succeeded in doing this in a far more perfect way than before by using a hermetically sealed cell (Fig. 1) which contains water acidulated with sulphuric acid. Two platinum wires, *b* and *c,* and a third platinum wire, *a,* which in the interior is connected with a spiral of palladium, can be used as electrodes. The tube, before it was closed, had been connected with an air-water pump, and at the same time oxygen was evolved from *b* and *c* by two Grove's elements; the hydrogen carried to the palladium wire, *a,* was occluded in the metal. In this way the liquid in the tube is washed out with oxygen under low pressure and freed from all other gases. After the closing of the tube, the remaining small traces of electrolytic oxygen combine slowly with the hydrogen of the palladium. Traces of hydrogen occluded in the platinum wires *b* and *c* can be transferred by a feeble electromotive force into the palladium; and even new quantities of electrolytic gases, evolved after closing the tube, can be removed again by a Daniell cell, which carries hydrogen to the palladium, where it is occluded, and oxygen to *b* and *c,* where it combines with hydrogen, as long as traces of this gas are dissolved in the

Fig. 1

liquid. The rest of the oxygen absorbed by the liquid combines with the occluded hydrogen.

I have ascertained with this apparatus that under favorable conditions one can observe the polarization produced during a few seconds by a current which would decompose only one milligram of water in a century.

But even if the appearance of galvanic polarization should not be acknowledged by opponents as a sufficient indication of previous decomposition, it is not difficult at present to reduce the indications of a good galvanometer to absolute measure, to calculate the amount of decomposition which ought to be expected according to Faraday's law, and to verify that in all the cases in which no products of electrolysis can be discovered, their amount is too small for chemical analysis.

Products of decomposition cannot appear at the electrodes without the occurrence of motions of the constituent molecules of the electrolyte throughout the whole length of the liquid. On this point the majority of Faraday's predecessors were already agreed, but they differed from one another as soon as they came to the question what those notions were. Faraday saw very clearly the importance of this problem and again appealed to experiment. He filled two cells with an electrolytic fluid, connecting them by a thread of asbestos wetted with the same fluid, in order to determine separately the quantity of all the chemical constituents transferred to the one and the other extremity of the electrolytic conductor. You know that he proposed for these atoms or groups of atoms transported by the current through the fluid the Greek word *ions* ("the travelers"); and comparing the current of positive electricity with a stream of water, he called *cations* those atoms which go down the stream in the same direction with the positive electricity to the cathode, the metallic plate through which this electricity left the fluid. The *anions,* on the contrary, go up the stream to the anode, the metal plate which is the source of the current of $+E$. Cations generally are atoms which are substitutes of hydrogen; anions are halogens.

This subject has been studied very carefully and for a great number of liquids by Professor Hittorff, of Münster, and Professor G. Wiedemann, of Leipzig. They found that generally the velocities of the cation and the anion are different. Professor F. Kohlrausch, of Würzburg, has brought to light the very important fact that in diluted solutions of salts, including hydrates of acids and hydrates of caustic alkalis, every ion under the influence of currents of the same density moves on with its own peculiar velocity, independently of other ions moving at the same time in the same or in opposite directions.

Among the cations, hydrogen has the greatest velocity; then follow potassium, ammonium, silver, sodium, and afterward the bivalent atoms of barium, copper, strontium, calcium, magnesium, zinc; near the latter appears univalent lithium. Among the anions, hydroxyl (OH) is the first; then follow the other univalent atoms, iodine, bromine, cyanogen, chlorine, the compounds NO_3, ClO_3, and the bivalent halogens of sulphuric and carbonic acid; after these, fluorine and the halogen of acetic acid ($C_2H_3O_2$). The only exception to this rule is the difference observed between the decomposition of univalent and bivalent compounds. Generally the velocity of any ion when separated from a bivalent mate is less than when separated from one or two univalent mates.

It seems possible that the majority of molecules SO_4H_2 may be divided electrolytically into SO_4 and H_2; some of them, on the other hand, into SO_4H and H. By the latter, some hydrogen would be carried backward, and therefore the velocity of the total amount might appear diminished.

If both ions are moving, we shall find liberated at each electrode (1) that part of the corresponding ion which has been newly carried to that side: (2) another part which has been left by the opposite ion, with which it had been formerly combined. The total amount of chemical motion in every section of the fluid is, therefore, represented by the sum of the equivalents of the cation gone forward and of the anion gone backward, in the same way as in the dualistic theory of electricity the total amount of electricity flowing through a section of the conductor corresponds to the sum of positive electricity going forward and of negative electricity going backward.

Thus established, Faraday's law tells us that through each section of an electrolytic conductor we have always equivalent electrical and chemical motion. The same definite quantity of either positive or negative electricity moves always with each univalent ion, or with each unit of affinity of a multivalent ion, and accompanies it during all its motions through the interior of the electrolytic fluid. This quantity we may call the electric charge of the atom.

I beg to remark that hitherto we have spoken only of phenomena. The motion of electricity can be observed and measured. Independently of this, the motion of the chemical constituents can also be measured. Equivalents of chemical elements and equivalent quantities of electricity are numbers which express real relations of natural objects and actions. That the equivalent relation of chemical elements depends on the pre-existence of atoms may be hypothetical; but we have not yet any other theory sufficiently developed which can explain all the facts of chemistry as simply and as consistently as the atomic theory developed in modern chemistry.

Now, the most startling result of Faraday's law is perhaps this. If we accept the hypothesis that the elementary substances are composed of atoms, we cannot avoid concluding that electricity also, positive as well as negative, is divided into definite elementary portions, which behave like atoms of electricity. As long as it moves about in the electrolytic liquid, each ion remains united with its electric equivalent or equivalents. At the surface of the electrodes, decomposition can take place if there is sufficient electromotive force, and then the ions give off their electric charges and become electrically neutral.

The same atom can be charged in different compounds with equivalents of positive or of negative electricity. Faraday pointed out sulphur as being an element which can act either as anion or as cation. It is an anion in sulphide of silver, a cation perhaps in strong sulphuric acid. Afterward he suspected that the deposition of sulphur from sulphuric acid might be a secondary result. The cation may be hydrogen, which combines with the oxygen of the acid and drives out the sulphur. But if this is the case, hydrogen recombined with oxygen to form water must retain its positive charge, and it is the sulphur which in our case must give off positive equivalents to the cathode. Therefore this sulphur of sulphuric acid must be charged with positive equivalents of electricity. The same may be applied to a great many other instances. Any atom or group of atoms which can be substituted by secondary decomposition for an ion must be capable of giving off the corresponding equivalent of electricity.

When the positively charged atoms of hydrogen or any other cation are liberated from their combination and evolved as gas, the gas becomes electrically neutral; that is, according to the language of the dualistic theory, it contains equal quantities of positive and negative electricity. Either every single atom is electrically neutralized, or one atom, remaining positive, combines with another charged negatively. This latter assumption agrees with the inference from Avogadro's law, that the molecule of free hydrogen is really composed of two atoms.

Now arises the question: are all these relations between electricity and chemical combination limited to that class of bodies which we know as electrolytes? In order to produce a current of sufficient strength to collect enough of the products of decomposition without producing too much heat in the electrolyte, the substance which we try to decompose ought not to offer too much resistance to the current. But this resistance may be very great, and the motion of the ions may be very slow—so slow, indeed, that we should need to allow it to go on for hundreds of years before we should be able to collect even traces of the products of decomposition. Nevertheless, all the essential attributes of the process of electrolysis could subsist. In fact we find the most various degrees of conducting power in various liquids. For a great number of them, down to distilled water and pure alcohol, we can observe the passage of the current with a sensitive galvanometer. But if we turn to oil of turpentine, benzene, and similar substances, the galvanometer becomes silent. Nevertheless, these fluids also are not without a certain degree of conducting power. If you connect an electrified conductor with one of the electrodes of a cell filled with oil of turpentine, the other with the earth, you will find that the electricity of the conductor is discharged unmistakably more rapidly through the oil of turpentine than if you take it away and fill the cell only with air.

We may in this case also observe polarization of the electrodes as a symptom of previous electrolysis. Connect the two pieces of platinum in oil of turpentine with a battery of eight Daniells, let it stand twenty-four hours, then take away the battery, and connect the electrodes with a quadrant electrometer. It will indicate that the two surfaces of platinum, which were homogenous before, produce an electromotive force which deflects the needle of the electrometer. The electromotive force of this polarization has been determined in some instances by Mr. Picker in the laboratory of the University of Berlin; he has found that the polarization of alcohol decreases with the proportion of water which it contains, and that that of the purest alcohol, ether, and oil of turpentine is about 0.3, that of benzene 0.8 of a Daniell element.

Another sign of electrolytic conduction is that liquids placed between two different metals produce an electromotive force. This is never done by metals of equal temperature or by other conductors which, like metals, let electricity pass without being decomposed. The production of an electromotive force is observed even with a great many rigid bodies, although very few of them allow us to observe electrolytic conduction with the galvanometer, and even these only at temperatures near their melting points. I remind you of the galvanic pile of Zamboni, in which pieces of dry paper are intercalated between thin leaves of metal. If the connection lasts long enough, even glass, resin, shellac, paraffin, sulphur—the best insulators we know—do the same. It is nearly impossible to prevent the quadrants of a delicate electrometer from being charged by the insulating bodies by which they are supported.

In all the cases which I have quoted, one might suspect that traces of humidity absorbed by the substance or adhering to their surface were the electrolytes. I show you, therefore, this little Daniell's cell (Fig. 2) constructed by my former assistant, Dr. Giese, in which a solution of sulphate of copper, with a platinum wire, *a,* as an electrode, is enclosed in a bulb of glass hermetically sealed. This is surrounded by a second cavity, sealed in the same way, which contains a solution of zinc sulphate and some amalgam of zinc, to which a second platinum wire, *b,* enters through the glass. The tubes *c* and *d* have served to introduce the liquids and have been sealed afterward. It is, therefore, like a Daniell cell in which the porous septum has been replaced by a thin stratum of glass. Externally all is symmetrical at the two poles; there is nothing in contact with the air but a closed surface of glass, through which two wires of platinum penetrate. The whole charges the electrometer exactly like a Daniell cell of very great resistance, and this it would not do if the septum of glass did not behave like an electrolyte, for a metallic conductor would completely destroy the action of the cell by its polarization.

All these facts show that electrolytic conduction is not at all limited to solutions of acids or salts. It will, however, be rather a difficult problem to find out how far the electrolytic conduction is extended, and I am not yet

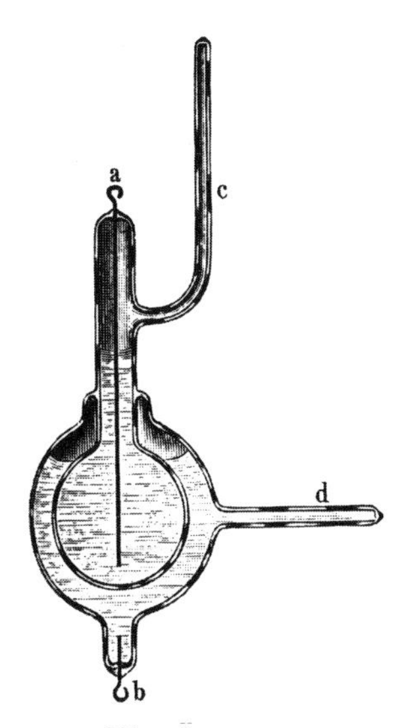

Fig. 2

prepared to give a positive answer. What I intended to remind you of was only that the faculty to be decomposed by electric motion is not necessarily connected with a small resistance to the current. It is easier for us to study the cases of small resistance, but the illustration which they give us about the connection of electric and chemical force is not at all limited to the acid and saline solutions usually employed.

Hitherto we have studied the motions of ponderable matter, as well as of electricity, going on in an electrolyte. Let us now study the forces which are able to produce these motions. It has always appeared somewhat startling to everybody who knows the mighty power of chemical forces and the enormous quantity of heat and mechanical work which they are able to produce, how exceedingly small is the electric attraction at the poles of a battery of two Daniell cells, which nevertheless is able to decompose water. One gram of water, produced by burning hydrogen with oxygen, develops so much heat that this heat, transformed by a steam engine into mechanical work, would raise the same weight to a height of 1,600,000 meters. And on the contrary we have to use the most delicate contrivances to show that a gold leaf or a little piece of aluminum hanging on a silk fiber can be at all moved by the electric attraction of the battery. The solution of this riddle is found if we look at the quantities of electricity with which the atoms appear to be charged.

The quantity of electricity which can be conveyed by a very small quantity of hydrogen, when measured by its electrostatic forces, is exceedingly great. Faraday saw this and endeavored in various ways to give at least an approximate determination. He ascertained that even the most powerful batteries of Leyden jars, discharged through a voltameter, give scarcely any visible traces of gases. At present we can give definite numbers. The electrochemical equivalent of the electromagnetic unit of the galvanic current has been determined by Bunsen and more recently by other physicists. This determination was followed by the very difficult comparison of the electromagnetic and electrostatic effects of electricity, accomplished at first by Professor W. Weber and afterward, under the auspices of the British Association, by Professor Clerk Maxwell. The result is that the electricity of one milligram of water, separated and communicated to two balls one kilometer distant, would produce an attraction between them equal to the weight of 26,800 kilograms. [The amount of electricity contained in one milligram of water would be twice as much, and the attraction of both quanta four times as much, that is, equal to the weight of 102,180 kilograms. (Added in 1884 to the German translation.)]

As I have already remarked, the law that the intensity of the force is inversely proportional to the square of the distance, and directly proportional to the quantities of the attracting and attracted masses, holds good as well in the case of gravitation as in that of electric attraction and repulsion. We may, therefore, compare the gravitation acting between two quantities of hydrogen and oxygen with the attraction of their electric charges. The result will be independent of the size and the distance of these quantities. We find that the electric force is as great as the gravitation of ponderable masses, being 71,000 billion times greater than that of the oxygen and hydrogen containing these electric charges.

The total force exerted by the attraction of an electrified body upon another charged with opposite electricity is always proportional to the quantity of electricity contained on the attracting as on the attracted body. Although, therefore, the attracting forces exerted by the poles of a little battery able to decompose water on such electric charges as we can produce with our electric machines are very moderate, the forces exerted by the same little apparatus on the enormous charges of the atoms in one milligram of water may very well compete with the mightiest chemical affinity.

If we now turn to investigate how the motions of the ponderable molecules are dependent upon the action of these forces, we must distinguish two different cases. At first we may ask, what forces are wanted to call forth motions of the ions with their charge through the interior of the fluid? Secondly, what are wanted to separate the ion from the fluid and its previous combinations?

Let us begin with the case in which the conducting liquid is surrounded everywhere by insulating bodies. Then no electricity can enter, and none can go out through its surface; but positive electricity can be driven to one side, negative to the other, by the attracting and repelling forces of external electrified bodies. This process, going on as well in every metallic conductor, is called electrostatic induction. Liquid conductors behave quite like metals under these conditions. Great quantities of electricities are collected, if large parts of the surfaces of the two bodies are very near each other. Such an arrangement is called an electric condenser. We can arrange electric condensers in which one of the surfaces is that of a liquid, as Messrs. Ayrton and Perry have done lately. The water-dropping collector of electricity, invented by Sir W. Thomson, is a peculiar form of such a condenser, which can be charged with perfect regularity by the slightest electromotive force perceptible only to the most sensitive electrometers. Professor Wüllner has proved that even our best insulators, exposed to electric forces for a long time, are ultimately charged quite in the same way as metals would be charged in an instant. There can be no doubt that even electromotive forces less than 1/100 Daniell produce perfect electric equilibrium in the interior of an electrolytic liquid.

Another somewhat modified instance of the same effect is afforded by a voltametric cell containing two electrodes of platinum, which are connected with a Daniell cell the electromotive force of which is insufficient to decompose the electrolyte. Under this condition the ions carried to the electrodes cannot give off their electric charges. The whole apparatus behaves, as was first emphasized by Sir W. Thomson, like a condenser of enormous capacity. The quantity of electricity, indeed, collected in a condenser under the same electromotive force is inversely proportional to the distance of the plates. If this is diminished to 1/100th, the condenser takes in one hundred times as much electricity as before. Now, bringing the two surfaces of platinum and of the liquid into immediate contact, we reduce their interval to molecular distances. The capacity of such a condenser has been measured by Messrs. Varley, Kohlrausch, and Colley. I have myself made some determinations which show that oxygen absorbed in the fluid is of great influence on the apparent value. By removing all traces of gas, I have got a value a little smaller than that of Kohlrausch, which shows that if we divide equally the total value of the polarization between two platinum plates of equal size, the distance between the two strata of positive and negative electricity—the one lying on the last molecules of the metal, the other on those of the fluid—ought to be 1/10,000,000th (Kohlrausch 1/15,000,000th) of a millimeter. We always come nearly to the same limit when we calculate the distances through which molecular forces are able to act, as already shown in several other instances by Sir W. Thomson.

Owing to the enormous capacity of such an electrolytic condenser, the quantity of electricity which enters into it, if it is charged even by a feeble electromotive force, is sufficiently great to be indicated easily by a galvanometer. What I now call charging the condenser, I have before called polarizing the metallic plate. Both, indeed, are the same process, because electric motion is always accompanied in the electrolytes by chemical decomposition.

Observing the polarizing and depolarizing currents in a cell like that represented in Fig. 1, we can observe these phenomena with the most feeble electromotive forces of 1/1000 Daniell, and I found that down to this limit the quantity of electricity entering into the condenser was proportional to the electromotive force by which it was collected. By taking larger surfaces of platinum, I suppose it will be possible to reach a limit much lower than that.

If any chemical force existed, besides that of the electric charges, which could bind all the pairs of opposite ions together and required any amount of work to be vanquished, an inferior limit ought to exist to such electromotive forces as are able to attract the ions to the electrodes and to charge these as condensers. No phenomenon indicating such a limit has as yet been discovered, and we must therefore conclude that no other force resists the motions of the ions through the interior of the liquid than the mutual attractions of their electric charges. These are able to prevent the atoms of the same kind which repel each other from collecting at one place, and atoms of the other kind attracted by the former from collecting at any other part of the fluid, as long as no external electric force favors such distribution. The electric attraction, therefore, is able to produce an equal distribution of the opposite constituent atoms throughout the liquid, so that all parts of it are neutralized electrically as well as chemically.

On the other hand, as soon as an ion is to be separated from its electric charge, we find that the electric forces of the battery meet with a powerful resistance, the overpowering of which requires a good deal of work to be done. Usually the ions, losing their electric charges, are at the same time separated from the liquid; some of them are evolved as gases, others are deposited as rigid strata on the surface of the electrodes, like galvanoplastic copper. But the union of two constituents having powerful affinity to form a chemical compound always produces, as you know very well, a great amount of heat, and heat is equivalent to work. Conversely, decomposition of the compound substance requires work, because it restores the energy of the chemical forces which has been spent by the act of combination. Oxygen and hydrogen separated from each other contain a store of energy, for when the hydrogen is burned in the oxygen, they unite, form water, and develop a great amount of heat. In the water the two elements are contained, and their chemical attraction continues to work as before to keep them firmly united, but it can no longer produce any change, any positive action. We must reduce the combined elements into their first state—we must separate them, applying a force which is capable of vanquishing their affinity—before they are ready to renew their first activity. The amount of heat produced by the chemical combination is the equivalent of the work done by the chemical forces brought into action. It requires the same amount of work to separate the compound and to restore hydrogen and oxygen uncombined. I have already given the value of this amount calculated as a weight raised against the force of gravity.

Metals uniting with oxygen or halogens produce heat in the same way—some of them, like potassium, sodium, and zinc, even more than an equivalent quantity of hydrogen; less oxidizable metals, like copper, silver, and platinum, less. We find, therefore, that heat is generated when zinc drives copper out of its combination with the compound halogen of sulphuric acid, as is the case in a Daniell cell.

If a galvanic current passes through any conductor, a metallic wire or an electrolytic fluid, it evolves heat. Dr. Joule was the first who proved experimentally that if no other work is done by the current, the total amount of heat evolved in a galvanic circuit during a certain time is exactly equal to that which ought to have been generated by the chemical actions which have been performed during that time. But this heat is not evolved at the surface of the electrodes where these chemical actions take place; it is evolved in all the parts of the circuit, proportionally to the galvanic resistance of each part. From this it is evident that the heat evolved is an immediate effect, not of the chemical action, but of the galvanic current and that the chemical work of the battery has been spent to produce only the electric action.

To keep up an electric current through an electric conductor, indeed, requires work to be done. New stores of positive electricity must be continually introduced at the positive end of the conductor, the repulsive force acting upon them having to be overcome; negative electricity, in the same way, into the negative end. This can be done by mere mechanical force, with an electric machine working by friction, or by electrostatic or by electromagnetic induction. In a galvanic current it is done by chemical force, but the work required remains the same.

If we apply Faraday's law, a definite amount of electricity passing through the circuit corresponds with a definite amount of chemical decomposition going on in every electrolytic cell of the same circuit. According to the theory of electricity, the work done by such a definite quantity of electricity which passes, producing a current, is proportionate to the electromotive force acting between both ends of the conductor. You see, therefore, that the electromotive force of a galvanic circuit must be, and is indeed, proportional to the heat generated by the sum of all the chemical actions going on in all the electrolytic cells during the passage of the same quantity of electricity. In the cells of the galvanic battery, chemical forces are brought into action able to produce work; in cells in which decomposition is occurring, work must be done against opposing chemical forces; the rest of the work done appears as heat evolved by the current, as far as it is not used up to produce motions of magnets or other equivalents of work.

You see, the law of the conservation of energy requires that the electromotive force of every cell must correspond exactly with the total amount of chemical forces brought into play, not only the mutual affinities of the ions, but also those minor molecular attractions produced by the water and other constituents of the fluid. These minor attractions have lately formed the subject of most valuable and extended calorimetric researches by Messrs. Andrews, Thomson, and Berthelot. But even influences too minute to be measured by calorimetric methods may be discovered by measuring the electromotive force. I have myself deduced from the mechanical theory of heat the influence which the quantity of water contained in a solution of metallic salts has on the electromotive force. The chemical attraction between salt and water can be measured in this instance by the diminution of the tension of the aqueous vapors over the liquid, and the results of the theoretical deduction have been confirmed in a very satisfactory manner by the observations of Dr. James Moser.

Hitherto we have supposed that the ion with its electric charge is separated from the fluid. But the ponderable atoms can give off their electricity to the electrode and remain in the liquid, being now electrically neutral. This makes scarcely any difference in the value of the electromotive force. For instance, if chlorine is separated at the anode, it will at first remain absorbed by the liquid; if the solution becomes saturated, or if we make a vacuum over the liquid, the gas will rise in bubbles. The electromotive force remains unaltered. The same may be observed with all the other gases. You see in this case that the change of electrically negative chlorine into neutral chlorine is the process which requires so great an amount of work, even if the ponderable matter of the atoms remains where it was.

On the other hand, if the electric attraction does not suffice to deprive the ions collecting at the surface of the electrodes of their electric charge, you will find the cation attracted and retained by the cathode, the anion by the anode, with a force far too great to be overpowered by the expansive force of gases. You may make a vacuum as perfect as you like over a cathode polarized with hydrogen, or an anode polarized with oxygen; you will not obtain the smallest bubble of gas. Increase the electric potential of the electrodes, so that the electric force becomes powerful enough to draw the electric charge of the ions over to the electrode, and the ions will be liberated and free to leave the electrode, passing into the gaseous state or spreading in the liquid by diffusion. One cannot assume, therefore, that their ponderable matter is attracted by the electrode; if this were the case, this attraction ought to last after discharge as before. We must conclude, therefore, that the ions are drawn to the electrode only because they are charged electrically.

The more the surface of the positive electrode is covered with negative atoms of the anion, and the negative with the positive ones of the cation, the more is the attracting force of the electrodes exerted upon the ions of the liquid diminished by this second stratum of opposite electricity covering them. Conversely, the force with which the positive electricity of an atom of hydrogen situated at the surface of the electrode itself is attracted toward the negatively charged metal increases in proportion as more negative electricity collects before it on the metal and more ions of hydrogen behind it in the fluid.

The electric force acting on equal quantities of electricity situated at the inside of one of the electric strata of a condenser is proportional to the electromotive force which has charged the condenser, and inversely proportional to the distance of the charged surfaces. If these are 1/100th of a millimeter apart, it is one hundred times as great as if they are one millimeter apart. If we come, therefore, to molecular distances, like those calculated from the measurement of the capacity of polarized electrodes, the force is ten million times as great and becomes able, even with a moderate electromotive force, to compete with the powerful chemical forces which combine every atom with its electric charge and hold the atoms bound to the liquid.

Such is the mechanism by which electric force is concentrated at the surface of the electrodes and increased in its intensity to such a degree that it becomes able to overpower the mightiest chemical affinities we know of. If this can be done by a polarized surface, acting like a condenser charged by a very moderate electromotive force, can the attractions between the enormous electric charges of anions and cations be an unimportant and indifferent part of chemical affinity?

In a decomposing cell the ions resist external forces striving to separate them from their electric charges. Let the current go in the opposite direction, and you will have an opposite effect. In a Daniell cell neutral zinc enters as cation into the electrolyte, taking with it only positive electricity and leaving its negative electricity to the metallic plate. At the copper electrode, positive copper separates from the electrolyte and is neutralized, giving off its charge to the electrode. But the Daniell cell in which this goes on does work, as we have seen. We must conclude, therefore, that an equivalent of positive electricity, on charging an atom of zinc, does more work than the same equivalent does on charging an atom of copper.

You see, therefore, if we use the language of the dualistic theory and treat positive and negative electricities as two substances, the phenomena are the same as if equivalents of positive and negative electricity were attracted by different atoms and perhaps also by the different values of affinity belonging to the same atom with different force. Potassium, sodium, and zinc must have strong attraction to a positive charge; oxygen, chlorine, and bromine, to a negative charge.

Do we perceive effects of such an attraction in other cases? Here we come to the much discussed question of Volta's assumption that electricity is produced by contact of two metals. About the fact there can be no doubt. If we produce metallic contact between a piece of copper and a piece of zinc, opposed to each other like the two plates of a condenser and carried by insulating rods of shellac, we find that after contact the zinc is charged positively, the copper negatively. This is just the effect we ought to expect if zinc has a higher attracting force to positive electricity, this force working only through molecular distances. I have proposed this explanation of Volta's experiments in my little pamphlet on the conservation of energy, published in 1847. All the facts observed with different combinations of metallic conductors are perfectly in harmony with it. Volta's law of the series of tension comprising all metallic conductors is easily deduced from it. If only metals come into play, their galvanic attractions produce instantaneously a state of electric equilibrium, so that no lasting current can occur. Electrolytic conductors, on the contrary, are decomposed chemically by every motion of electricity through their surface. Electric equilibrium, therefore, will not be possible before this decomposition has been finished, and till that stage is reached, the electric motion must continue. This point has been accentuated already by Faraday as the essential difference between the two classes of conductors.

The original theory of Volta was incomplete in an essential point because he was not acquainted with the fact of electrolytic decomposition. His original conception of the force of contact is, therefore, in contradiction to the law of conservation of energy; and even before this law was established and enunciated with scientific precision, there were many chemists and physicists, among them Faraday, who had the right instinct that this could not be the true explanation. The opponents of Volta's opinions tried to give chemical explanations also of those experiments of his which were carried out exclusively with metallic conductors. They might be oxidized by the oxygen of the air, and the amount of oxidation required for a very slight electric charge was so infinitesimal that no chemical analysis could ever discover it—so small that even in the highest vacuum, and in the purest specimens of hydrogen or nitrogen with which we might surround the plates, there was oxygen enough to continue the effect for years. From this point of view the chemical theory cannot be refuted.

On the other hand, the so-called chemical theory of Volta's fundamental experiments was rather indefinite. It scarcely did more than tell us: here is the possibility of a chemical process, here electricity can be produced. But which kind, how much, to which potential, remained indefinite. I have not found in all the papers which have been written for the defense of the chemical theory a clear explanation of why zinc opposed to copper in liquids, where zinc really is oxidized and dissolved, becomes negative and why in air and other gases it becomes positive, if the same cause—namely, oxidation—is at work. The hypothesis, on the contrary, of a different degree of affinity between the metals and the two electricities gives a perfectly definite answer. I do not see why an actual chemical process should be wanted to charge the zinc and copper on contact. But you see that the forces, which according to their hypothesis produce the electric effect, are the same as those which must be considered the cause of a main part of all chemical actions.

Again, the electric charges produced by contact of zinc and copper are very feeble. They have become measurable only with the help of the latest improvements introduced into the construction of electrometers by Sir W. Thomson; but the cause of their feeble intensity is evident. If you bring into narrow contact two plain and well-polished plates of zinc and copper, the quantity of electricity collected at both sides of their common surface is probably very great; but you cannot observe it before having separated the plates. Now, it is impossible to separate them at the same instant over the whole extent of their surface. The charge which they retain will correspond with the inclined position which they have at the moment when the last point of contact is broken; then all the other parts of the surfaces are already at a distance from one another infinitely greater than molecular distance, and conduction in metals always establishes nearly instantaneously the electric equilibrium corresponding to the actual situation. If you wish to avoid this discharge during the separation of the plates, one of them must be insulated; then indeed we get a far more striking series of phenomena, those belonging to electricity of friction.

Friction, probably, is only the means of producing a very close contact between the two bodies. If the surfaces are very clean and free from air, as for instance in a Geissler tube, the slightest rolling contact is sufficient to develop the electric charge. I can show you two such tubes exhausted so far that very high electric tension is necessary to make the vacuum luminous, one containing a small quantity of mercury, the other the fluid compound of potassium and sodium. In the first the negative metal is intensely negative relatively to glass, in the second the metal is on the positive extremity of Volta's series; the glass proves to be more positive also in this case, but the difference is much smaller than with mercury, and the charge is feeble.

Faraday very often recurs to this to express his conviction that the forces termed chemical affinity and electricity are one and the same. I have endeavored to give you a survey of the facts connected with the question and to avoid as far as possible the introduction of hypotheses, except the atomic theory of modern chemistry. I think the facts leave no doubt that the very mightiest among the chemical forces are of electric origin. The atoms cling to their electric charges, and opposite electric charges cling to each other; but I do not suppose that other molecular forces are excluded, working directly from atom to atom. Several of our leading chemists have lately begun to distinguish two classes of compounds, viz., molecular aggregates and typical compounds, the latter being united by atomic affinities, the former not.

Electrolytes belong to the latter class. If we conclude from the facts that every unit of affinity is charged with one equivalent either of positive or of negative electricity, they can form compounds, being electrically neutral only if every unit charged positively unites under the influence of a mighty electric attraction with another unit charged negatively. You see that this ought to produce compounds in which every unit of affinity of every atom is connected with one and only one other unit of another atom. This, as you will see immediately, is the modern chemical theory of quantivalence, comprising all the saturated compounds. The fact that even elementary substances, with few exceptions, have molecules composed of two atoms, makes it probable that even in these cases electric neutralization is produced by the combination of two atoms, each charged with its full electric equivalent, not by neutralization of every single unit of affinity.

Unsaturated compounds with an even number of unconnected units of affinity offer no objection to such a hypothesis; they may be charged with equal equivalents of opposite electricity. Unsaturated compounds with one unconnected unit, existing only at high temperatures, may be explained as dissociated by intense molecular motion of heat in spite of their electric attractions. But there remains one single instance of a compound which, according to the law of Avogadro, must be considered unsaturated even at the lowest temperature, namely, nitric oxide (NO), a substance offering several very uncommon peculiarities, the behavior of which will be perhaps explained by future researches.

But I abstain from entering into further particulars; perhaps I have already gone too far. I would not have dared to do it, had I not felt myself sheltered by the authority of that great man who was guided by a never-erring instinct of truth. I thought that the best I could do for his memory was to recall to the minds of the men by whose energy and intelligence chemistry has undergone its modern astonishing development, what important treasures of knowledge lie still hidden in the works of that wonderful genius. I am not sufficiently acquainted with chemistry to be confident that I have given the right interpretation, the interpretation which Faraday himself would have given, if he had been acquainted with the law of chemical quantivalence. Without the knowledge of this law I do not see how a consistent and comprehensive electrochemical theory could be established. Faraday did not try to develop a complete theory of this kind. It is as characteristic of a man of high intellect to see where to avoid going further in his theoretical speculations for want of facts, as to see how to proceed when he finds the way open. We ought therefore to admire Faraday also in his cautious reticence, although now, standing on his shoulders, and assisted by the wonderful development of organic chemistry, we are able, perhaps, to see further than he did. I shall consider my work of today well rewarded if I have succeeded in kindling anew the interest of chemists in the electrochemical part of their science.

16.

AN EPISTEMOLOGICAL ANALYSIS OF COUNTING AND MEASUREMENT [1887]

An essay contributed to the Festschrift *honoring Eduard Zeller in 1887**

ALTHOUGH counting and measurement are the foundations of the most fruitful, the most certain, and the most precise scientific methods we know, comparatively little work has been done on their epistemological foundations. From the philosophical side, the strict followers of Kant must adhere to his system as it was developed within the perspectives and knowledge of his period in history. Above all, they must hold that the axioms of arithmetic are propositions which are given a priori, propositions which more specifically fix or define the transcendental intuition of time in the same way that the axioms of geometry define more specifically the transcendental intuition of space. In both cases, this interpretation cuts short any further investigation or analysis of these propositions.

In earlier essays I have tried to establish that the axioms of geometry are not propositions given a priori, but propositions that must be confirmed or refuted by experience. I emphasize here once more that I do not reject Kant's conception of space as a transcendental form of intuition. In my opinion, only a single, unjustified point of secondary importance must be discarded from his system, though it is a point, to be sure, whose rejection has disastrous consequences for the metaphysical efforts of his successors.

Since the empirical theory, which I have advocated, interprets the axioms of geometry not only as not undemonstrable but indeed as requiring justification, it must clearly take the same position concerning the origin of the axioms of arithmetic, which are related to the form of intuition of time in the same way that the axioms of geometry are related to the form of intuition of space.

---

*A few bibliographical footnotes are omitted.

In the past, mathematicians have taken the following propositions as axioms and made them the center of their analyses:

Axiom I. If two quantities are equal to a third, they are equal to each other.

Axiom II. The associative law of addition (following H. Grassmann's notation):

$$(a + b) + c = a + (b + c).$$

Axiom III. The commutative law of addition:

$$a + b = b + a.$$

Axiom IV. Equals added to equals give equals.

Axiom V. Equals added to unequals give unequals.

In addition to the other mathematicians whose work I am familiar with and who hold a similar philosophical position, Hermann and Robert Grassmann have entered into an investigation of this subject. (I shall follow their procedures throughout this discussion in carrying out arithmetic proofs.) They combine Axioms II and III to form a single one, which I shall call *Grassmann's axiom of addition:*

$$(a + b) + 1 = a + (b + 1).$$

From this, by mathematical induction, they derive both of the general propositions given above. In this way, as I hope to show in the following pages, the proper foundation has been laid for the theory of addition of pure numbers.

With respect to the objective application of arithmetic to physical quantities, in addition to the concepts of *quantity* and *equality* (whose meaning in the factual realm must remain unexplained for the moment) there is a third important concept, that of *unit;* in addition, it would seem to me an unnecessary restriction on the scope of validity of the propositions under consideration if physical quantities were treated, as a matter of course, only as if they were aggregates of units.

Among the newer mathematicians, E. Schroeder has followed the Grassmann brothers on essential points but has gone more deeply into several important questions. While earlier mathematicians usually interpreted the fundamental concept of *number (Zahl)* to mean a collection or *total number (Anzahl)* of objects, they could not detach themselves completely from the laws of the relations of these objects, and they took it simply as a fact that the total number *(Anzahl)* of a group of objects was independent of the sequence or order in which the objects were counted. Schroeder is, as far as I have found, the first to recognize that there is a problem hidden at this point. In addition, he recognized–in my opinion, correctly–that there is also a problem for psychology here, and further, that the empirical properties which objects must possess if they are to be counted must be defined or specified.

In addition to the works of these men, discussions of the present subject–the concept of quantity–are also to be found in Paul du Bois-Reymond's *Allgemeiner Functionentheorie* (Tübingen, 1882) and A. Elsas' *Über die Psychophysik* (Marburg, 1886). Both books, however, are occupied with special issues and do not contain full discussions of the foundations of arithmetic. Both authors believe that it is possible to derive the concept of quantity from that of line, the former in an empirical manner, the latter in a strict Kantian fashion. What I take exception to in the latter position has already been mentioned, and I have explained it in previous essays. Du Bois-Reymond ends his discussion with a paradox in which two opposed points of view, both entangled in contradictions, are equally possible. Since du Bois-Reymond is a highly ingenious mathematician, one who has investigated with special interest the deepest foundations of his science, I have taken courage from his final results in setting forth my own thoughts on this subject.

In order to characterize briefly at the beginning the point of view which I believe leads to simple, consistent derivations and to a resolution of the paradox just mentioned, let me say the following:

I consider arithmetic, or the theory of pure numbers, a method built upon purely psychological processes–a method by means of which one learns the consistent application of a system of signs (that is, numbers) unlimited in extension and in the possibility of refinement. Arithmetic investigates which of the various kinds of combinations of, or operations with, these signs lead to the same final result. Among other things, it is concerned with how to replace extremely complicated calculations, some of which cannot even be terminated in any finite period of time, with others which are quite simple.

Aside from their use as tests for the inner consistency of our thinking, such calculations would really be only a kind of ingenious play with imaginary objects–something which du Bois-Reymond derisively compares with problems concerning the knight's move on a chessboard–if they did not admit of such extraordinarily useful applications. By means of the system of numbers we can give descriptions of the relations of real objects which, where it is practicable, may reach any required degree of precision. Moreover, in a great many cases where natural bodies operate or interact under the control of known natural laws, the number system permits us to determine by calculation the magnitude of all measurable values in advance of their actual occurrence.

One must now ask, however: What is the objective meaning of the fact that relations between and among real objects can be expressed as quantities by the use of concrete numbers, and under what conditions can this be done? This question, as we shall find, resolves itself into two simpler ones: (1) What is the objective meaning of the fact that we consider two objects equal in some respect? and (2) What property must the physical combination of two objects have in order that we may combine some attribute of the objects *additively* and thus consider this attribute a quantity which can be expressed by a concrete number? (Concrete numbers are numbers which may be treated as being constituted out of their respective parts or units by addition.)

The Regular Number Series

Counting is a procedure which is based upon the fact that temporal sequences of conscious acts are retained in the memory. Basically, we may consider the numbers a series of arbitrarily chosen signs, only one definite sequence of which is accepted as regular or, to use the common mode of expression, as natural. The term *natural* originally was applied to the number series only in connection with counting, that is, only in connection with the determination of the total number *(Anzahl)* of actual objects in given groups. As we throw down one after another of a group of objects being counted, the numbers follow one after another in their regular sequence by a natural process. This has nothing to do, however, with the actual sequence of the number symbols, for just as the symbols are different in different languages, so their sequence is only conventionally specified.

The important point is that some definite sequence be inalterably settled upon as the normal or regular one. This then serves as a norm or law—one laid down by the men who developed the language, that is, by our ancestors. I emphasize this point since the alleged naturalness of the sequence of numbers is part of the ordinary incomplete analysis of the concept of number. Mathematicians call this regular sequence of numbers the *positive whole numbers.*

The number series is much more firmly fixed in our memory than any other series, undoubtedly because of the extremely frequent repetition of the series. Because it is so firmly fixed, whenever we wish to establish other series in our memory, we do so by relating them to it; that is, we use the numbers as *ordinal numbers.*

Uniqueness of the Series

Moving forward and moving backward in the number series are not equivalent to each other; on the contrary, like the sequence of perceptions in time, they are actually quite different procedures. This is in contrast to motion along lines, which exist in space and remain unchanged in time, so that neither direction of movement has any priority over the other.

In actuality, every present mental event–whether it be a perception, a feeling, or an act of volition, along with the related memory images of past events–has the same effect upon our consciousness. This is not the case with future events, however, for these are not yet present in consciousness. We also know, or are aware of, the difference between a present mental event and the memory images which exist beside it. The present content of consciousness is distinguished from the memory images, in the intuition of time, as the successor to them. This relation, of course, is not reversible.

Since all the ideas which enter into consciousness are of necessity interwoven with this sequence, ordering in the time sequence is the inescapable form of our inner intuition.

Meaning of the Notation

As suggested in the preceding discussion, each number is determined only by its place in the regular series.

We use the symbol *one* for that member of the series with which we start. *Two* is the number which follows immediately–that is, without the interposition of another number–after *one* in the regular series. *Three* is the number which follows immediately after *two.* And so on.

There is no reason for breaking this series in any way or for returning in it to a symbol used earlier. Indeed, the decimal system makes it possible, by combining only ten different number symbols, to continue the series in a simple, easily understood manner without repeating any combination of number symbols.

The numbers which follow a given member in the regular series we call *higher;* those which precede it, *lower.*[1] In this way a complete disjunction results, which is grounded in the nature of the temporal sequence. We can express it as

Axiom VI. If two numbers are different, one of them must be higher than the other.

Addition of Pure Numbers

In order to formulate general propositions concerning numbers, I shall use the familiar notation of algebra. Letters of the lower-case Latin alphabet will be used to designate numbers, each letter always designating the same number in a given theorem or in some particular computation.

SYMBOL EXPLANATION: If any number is designated by a letter, for example, a, the following number in the regular series will be designated by $(a + 1)$.

For the present, the symbol $(a + 1)$ will have no other meaning. In general, however, and as is customary, the parentheses indicate that the numbers enclosed between them should be reduced to a single number before any other operations involving them are carried out.

The sign of equality, $a = b$, when we are dealing with pure numbers, means that a is the same number as b. Thus if $a = b$ and $c = b$, it follows immediately that $a = c$, for the two equations express the fact that both a and c are the same number as b. In the theory of pure numbers, this confirms the validity of Axiom I for the series of whole numbers.

Counting of Numbers

We shall now assume that the regular series of numbers has been established and is given. We may consider its members to be a series of ideas given in our consciousness—a series whose order, starting from an arbitrarily chosen member, can be represented by the normal series of numbers beginning with *one.*

DEFINITION: I designate by $(a + b)$ that number of the main series to which I go if I count $(a + 1)$ as *one,* $[(a + 1) + 1]$ as *two,* and so forth, until I have counted to b.

The description of this procedure is expressed in the following equation (Grassmann's axiom of addition):

1. I avoid here the terms *greater* and *smaller,* which are more suitable for the concept of total number *(Anzahl).* More on this later.

$$(a + b) + 1 = a + (b + 1). \tag{1}$$

EXPLANATION: This equation states that if I had counted from $(a + 1)$ as *one* to b and thus had found the number represented by $(a + b)$, and if I counted further, I would move in the second sequence to $(b + 1)$, and in the first to the number following $(a + b)$, namely to $[(a + b) + 1]$. Thus I can represent that which was expressed in the definition as $[(a + 1) + 1]$ by $[a + (1 + 1)]$, or $(a + 2)$; and $[(a + 2) + 1]$ by $(a + 3)$, and so on without limit.

In the language of arithmetic this procedure is called addition; the number $(a + b)$ is called the sum of a and b, and a and b are called the summands. It should be noted that in the procedure just outlined, the quantities a and b do not play equal roles; thus it must be proved that they can be interchanged without altering the sum. This will be done later. Meanwhile, if we keep this point in mind, we can make use of this notation and say that the expression $(a + b)$ indicates that b is to be added to a, and that $(a + b)$ is the sum of a and b, the order of b after a being retained for the moment. Thus, a may be called the first, and b the second, summand. Corresponding to this, in a consistent application of this notation, any number $(a + 1)$ may be considered to be the sum of a and the number *one.*

The process of addition just outlined will always result in a single member of the regular sequence of numbers. Further, given the same numbers a and b, this resultant number will always be the same, for each step by which we obtained the sum $(a + b)$ is an advance in the series of positive whole numbers to the next stage, from $(a + b)$ to $[(a + b) + 1]$ and from b to $(b + 1)$. Each step can be made, and each, according to our assumptions concerning the inalterable preservation of the sequence of numbers in our consciousness, must always give the same result.

Thus there will always be one number and one only which corresponds to the number $(a + b)$. This proposition corresponds to the content of Axiom IV, if it is applied to the pure numbers and to the process of addition just described.

Further, it follows from the description given of the process of addition that $(a + b)$ is necessarily different from a and that it is higher than a if b is a positive whole number.

If c is a number higher than a, counting forward stepwise from a I shall necessarily reach c, and I can count what number c is, starting from a. If it is the bth, then

$$c = (a + b).$$

For later reference we shall call this proposition:

Axiom VII. If a number c *is higher than another number* a, c *can be represented as the sum of* a *and an unknown positive whole number* b.

THEOREM I—concerning the sequence of the execution of several processes of addition (Grassmann's associative law). *If a number* c *is added to the sum* (a + b), *the same result is obtained as when the sum* (b + c) *is added to the number* a.

Expressed as an equation:

$$(a + b) + c = a + (b + c). \tag{2}$$

PROOF: As Equation (1) states, this proposition is true for $c = 1$. It must be shown that if it is true for any arbitrary value of c, it is also true for the following value, that is, for $(c + 1)$.

According to Equation (1),

$$\begin{aligned}[(a + b) + c] + 1 &= (a + b) + (c + 1)\\ [a + (b + c)] + 1 &= a + [(b + c) + 1]\\ &= a + [b + (c + 1)].\end{aligned}$$

The last line follows according to Equation (1).

Thus, if Equation (2) is true for some value of c, the left-hand expressions of the first two equations above are the same number, and thus

$$(a + b) + (c + 1) = a + [b + (c + 1)];$$

that is, the proposition is true for $(c + 1)$.

Since, as we noted a moment ago, it is true for $c = 1$, it is also true for $c = 2$. If it is true for $c = 2$, it is true for $c = 3$, and so on without limit.

COROLLARY: Since both sides of Equation (2) have the same meaning, we can omit the parentheses and use the following notation:

$$a + b + c = (a + b) + c = a + (b + c). \tag{2a}$$

We still cannot change the order of *a, b,* and *c* in these expressions, however, for we have not established the admissibility of such a change.

Generalization of the Associative Law

First let us generalize the expression given in Equation (2a). Let

$$R = a + b + c + d + \ldots + k + 1$$

designate a sum in which the individual additions in the series are carried out in the order in which they are written. Using this same notation,

$$m + R = m + a + b + c + d + \ldots + k + 1,$$

while

$$m + (R) = m + (a + b + c + d + \ldots + k + 1).$$

In addition, according to the meaning of this notation,

$$(R) + m = R + m.$$

(Other capital letters of the Latin alphabet will be used in the same way as R.)

Thus

$$R + b + c + S = [(R) + b + c] + S,$$

since these expressions mean the same thing. Further, according to Equation (2a),

$$(R) + b + c = (R) + (b + c).$$

Thus

$$\begin{aligned} R + b + c + S &= [R + (b + c)] + S \\ &= R + (b + c) + S; \end{aligned}$$

that is, instead of adding all members of the sequence in order, we can first combine any two arbitrarily selected consecutive members within the sequence.

After this has been done, the sum of, say, $(b + c)$ may be expressed by using only a single number, and we can continue in this same manner by combining any other pair of consecutive numbers.

Thus, the order in which the additions just characterized are executed can be altered any number of times without changing the final result.

THEOREM II (Grassmann's commutative law). *If, in a sum of two summands, one of the summands is the number* one, *the order of the two summands may be reversed.*

Expressed as an equation:

$$1 + a = a + 1. \tag{3}$$

PROOF: The equation is true for $a = 1$. It must now be shown that, if it is true for any value of a, it is also true for $(a + 1)$.

According to Equation (1),

$$(1 + a) + 1 = 1 + (a + 1).$$

According to the assumption, Equation (3) holds for a, and from this it follows that

$$(1 + a) + 1 = (a + 1) + 1.$$

From these two equations we have

$$1 + (a + 1) = (a + 1) + 1, \tag{3a}$$

which was to be proved.

Since the law is true for $a = 1$, it is also true for $a = 2$, and since it is true for $a = 2$, it is also true for $a = 3$, and so on without limit.

THEOREM III. *In any sum of two summands, the order of the summands can be changed without changing the sum of the numbers.*

Expressed as an equation:

$$a + b = b + a. \tag{4}$$

According to Theorem II, this proposition is true for $b = 1$. If it is true for a specific value of b, it is also true for $(b + 1)$. For, according to Theorem I,

$$(a + b) + 1 = a + (b + 1),$$

and according to our assumption,

$$\begin{aligned}(a + b) + 1 &= (b + a) + 1 = 1 + (b + a) \\ &= (1 + b) + a = (b + 1) + a.\end{aligned}$$

Of the last three members of these equations, the first and the last follow from Theorem II, Equation (3), and the second from Theorem I, Equation (2). It follows that

$$a + (b + 1) = (b + 1) + a,$$

which was to be proved.

From the proposition

$$a + 1 = 1 + a,$$

it thus follows that

$$a + 2 = 2 + a,$$

and from this that

$$a + 3 = 3 + a,$$

and so on without limit.

PROOF OF AXIOM V: As was indicated in Axiom VII, if a and f are different numbers, we can always find a positive whole number b such that

$$(a + b) = f.$$

Then

$$c + f = c + (a + b) = (c + a) + b.$$

According to this equation, $(c + a)$ is necessarily different from $(c + f)$; that is: *different numbers added to the same number give different results.*

Since, according to Theorem III,

$$c + f = f + c,$$
$$a + c = c + a,$$

the last conclusion may also be written:

$$(f + c) = (a + c) + b;$$

that is, *the same number added to different numbers gives different results.*

From this an important proposition for the theory of subtraction and for the theory of equations follows, namely, the proposition that two numbers must be identical if, when they are added to the same number, the same result is obtained.

Exchange in the Order of an Arbitrary Number of Elements

With the kind of counting for the purpose of addition just outlined, two series of numbers in their normal sequence were joined to one another in pairs, so that $(n + 1)$ was coordinated with 1, $(n + 2)$ with 2, and so on. Only the starting points of the two sequences in the number series were different.

We shall now consider the general case of the coordination of the elements of two sequences, one of which has a fixed order and thus can be represented by the number symbols, while the other has an order that may be changed. We shall use the letters of the Greek alphabet as symbols for the latter. These letters, to be sure, also have an order, the one which is impressed upon our memory as the letters are given in the traditional arrangement of the alphabet. We shall use them, however, only as one of several readily available symbol systems, which is distinguished from the others by characteristics that are purely conventional. Our clear recollection of this series will make it easy for us to vary it arbitrarily. We require, however, that none of the elements be omitted and none be repeated as we make alterations in their order. We can ensure this most easily in our memory if we lay down the rule that the group shall contain as elements all letters which precede some particular letter, for example, χ, in the usual order of the alphabet.

Transposition of Two Consecutive Elements of a Series

Suppose that two consecutive numbers, n and $(n + 1)$, are coordinated with two elements, for example, ϵ and ζ, in a paired series. Since the elements are arbitrary, n could as readily have been coordinated with ζ as with ϵ. There are thus two possibilities, the original one:

$$\begin{matrix} n & (n+1) \\ \epsilon & \zeta \end{matrix}$$

and the transposition:

$$\begin{matrix} n & (n+1) \\ \zeta & \epsilon \end{matrix}$$

If we replace the first of these correlations with the second and leave all the other coordinated pairs of one number and one letter unchanged, no number in the series will lack a coordinated letter and no letter a coordinated number, and no letters will be repeated or omitted as a result of the transposition. If the series which contained the first of the two coordinated pairs just mentioned was unbroken and without repetitions prior to the transposition, it will also be unbroken and without repetitions afterward.

By a suitable execution of such transpositions of consecutive elements, without repetitions or a break in the series, any arbitrarily chosen element can be made the first in the series. Thus, if the element chosen, ζ, is the nth, it can be exchanged with the $(n - 1)$st, then with the $(n - 2)$nd, and so on, in such a way that its place number always gets lower, until finally it reaches the lowest place number in the series, that is, one.

In this way, any element in the series whose place number is higher than m can be made the mth member of the group, without any breaks or repetitions. In this case, those numbers of the series whose place numbers are less than m will remain in the same position.

Hence, *by a continuous exchange of consecutive members of a group, without omitting elements and without repeating them, any possible sequence of its members can be achieved.* I can specify arbitrarily which shall be the first member of the series and make it the first according to the procedure just described. Then I can specify which shall be the second and again carry out the procedure; as I do this, the element placed first will remain in position. Then I can specify the third element, and so on, up to the last.

THEOREM IV. *Attributes of a series of elements that remain unaltered when any two arbitrarily chosen consecutive elements are interchanged with each other, are not altered by any possible interchange in the order of the elements.*

This leads to the generalization of the commutative law of addition. (As with the generalization of the associative law, capital letters will be used to represent sums of any arbitrary number of numbers.) According to the generalized law of association,

$$R + a + b + S = R + (a + b) + S.$$

According to the commutative law for two summands,

$$a + b = b + a.$$

Since $(a + b)$ is the same number as $(b + a)$,

$$\begin{aligned} R + a + b + S &= R + (b + a) + S \\ &= R + b + a + S. \end{aligned}$$

This last line follows according to the associative law.

Since we can thus arbitrarily interchange any two consecutive elements in the sum without changing the amount of the sum, we can, according to Theorem IV, interchange all the elements with one another and place them in any arbitrarily chosen order without changing the amount.

The five fundamental axioms of addition have now been established and derived from the concept of addition which we have presented. It must still be established that this concept coincides with that which results from an inquiry concerning the total number *(Anzahl)* of a group of enumerable objects.

This inquiry leads us first to the concept of the number of elements in a group. If I need the complete number series from *one* to n in order to coordinate one number with each element of a group, I call n the total number of elements in the group. The previous discussion in connection with Theorem IV showed that regardless of changes in the sequence of these elements, the total number of elements remains the same, provided that omission and repetition of members is avoided.

This proposition is applicable to real objects, which for our purposes we shall name α, β, γ, and so on. If these objects are to be enumerable and if the result of the counting of them is to be correct, they must satisfy certain factual conditions: they must not disappear or blend with one another, no one may divide into two, and no new ones may appear. Thus each Greek letter must always name and correspond continuously to one and only one object. Whether these conditions are realized for a particular class of objects naturally can only be determined by experience.

That *the total number of members of two groups which have no members in common is equal to the sum of the total numbers of the individual groups,* according to the concept of addition previously given, follows immediately from the method of counting just described. In order to find the total number, one of the two groups must first be counted. If it has p members, the first member of the other group will have the number $(p + 1)$, the second $(p + 2)$, and so on. Hence the total number of members in the two groups combined is found by exactly the same process of counting used to determine the sum of the two numbers which indicate the total number of elements in each group.

Thus the concept of addition discussed earlier does indeed coincide with the concept which results from determining the total number of elements in several groups of enumerable objects. It has the advantage, however, in that it can be established without reference to external experience.

We have now established the set of axioms of addition for the concepts of *number* and *sum,* insofar as these concepts arise in inner intuition. These axioms are necessary as a foundation for arithmetic. In addition, the agreement between the results of this kind of addition and those which are found in counting external, enumerable objects has also been established.

The theory of subtraction and multiplication can be developed from here without further difficulty, since the difference $(a - b)$ can be defined as that number which we must add to b in order to get a as the sum and multiplication can be defined as the addition of a number of equal numbers. For the latter I need only refer to the Grassmann brothers, who defined the multiplication of pure numbers by the two equations,

$$1a = a,$$
$$(b + 1)a = ba + a.$$

In reference to subtraction, it should also be mentioned that we can continue the numbers as a series of signs continuously in the descending direction, since we can go from 1 backward to 0 and from there to −1, −2, and so on. These new signs are treated just as the positive whole numbers were earlier. Thus the difference between two numbers always has one and only one value; it is uniquely determined.

Because of what is to follow, the agreement as well as the difference between the laws of addition and the laws of multiplication must still be discussed. The associative and commutative laws are true for both operations. As we have seen,

$$(a + b) + c = a + (b + c)$$
$$a + b = b + a.$$

In a similar fashion,

$$(ab)c = a(bc)$$
$$ab = ba.$$

A difference between the fundamental characteristics of the two operations makes its appearance, however, when we combine by each of them a number n of equal numbers a. In an additive combination, this results in the product, na, for which the commutative law holds:

$$na = an.$$

In the multiplication of n equal factors, on the other hand, we get the power a^n, in which a and n take on the special characteristic of not being interchangeable with each other without changing the value of the expression.

Finally, with certain combinations there is an analogy between these two operations. Thus

$$(a + b)c = (ac) + (bc)$$
$$c^{a+b} = c^a c^b.$$

The analogy does not hold, however, if the numbers in this last expression are reversed, for $(a + b)^c$ and $a^c b^c$ do not necessarily have the same value.

Concrete Numbers

As was mentioned earlier, in order to be able to count objects in general, that is, objects which may be quite different from one another, we need only some means by which they can be identified.

Of much more extensive applicability and much greater importance is the counting of objects that are similar to one another in some specific way. We call such objects the *units* of the enumeration. The number of these units we call a *concrete number (benannte Zahl);* the specific kind of unit in question we call the *designation (Benennung)* of the number.

As we have already seen, a total number *(Anzahl)* is divisible into elements, which are additively collected into the whole. The sum of two concrete numbers with the same designation is the total number of all of their elements and thus necessarily a concrete number with the same designation. If we compare two different groups with different total numbers, we call that to which the higher number belongs the greater and that to which the lower belongs the smaller. If they both have the same number, we say they are equal.

Objects or attributes of objects which can be compared in order to determine which is greater and which is smaller, or whether they are equal, we call *quantities (Grössen).* If we can express a quantity by a concrete number, we call this the *value* of the quantity, and the procedure by which the concrete number is determined we call the *measurement* of the quantity. In many empirical investigations we make measurements using arbitrary units or units which are determined by the instruments used. In such cases the values obtained are only *relation numbers,* and they remain such until the units in question are reduced to the familiar *absolute units* of physics. These well-known units are not defined in a purely conceptual fashion; on the contrary, they have been established in connection with specific natural bodies (weights, scales) and specific natural processes (the day, pendulum oscillations). That they are quite generally familiar has nothing to do with the concept and procedure of measurement; on the contrary, that they are well known is a purely accidental characteristic.

In the following paragraphs we shall investigate the conditions under which we can express quantities by concrete numbers; that is, we shall investigate the conditions under which we are able to find their values and what sort of factual knowledge we gain in doing this.

We must first, however, consider the concepts of equality and quantity, according to their objective meanings.

Physical Equality

The special relation which may exist between the attributes of two objects for which we use the term *equality* is characterized by Axiom I: *If two quantities are equal to a third, they are equal to each other.* It is apparent that the relation of equality is reciprocal, for from

$$a = c,$$
$$b = c,$$

it follows that $a = b$ as well as that $b = a$.

The equality of two objects with respect to some attribute is the exceptional case. Such cases are observed empirically only when the two objects, coinciding or interacting under appropriate conditions, have some particular effect which does not occur as a rule with other pairs of objects possessing the attribute in question.

We call the procedure by which we place the two objects in the appropriate conditions for observing the effect specified, that is, for determining whether it occurs or not, the *method of comparison.* If the method of comparison is to yield reliable information concerning the equality or inequality of the two objects with respect to an attribute, the result of the comparison must be dependent exclusively upon the fact that the objects possess that attribute to some specified degree (presupposing, of course, that the method of comparison is properly carried out).

From the axioms stated above it follows, first of all, that *the result of the comparison must remain unchanged if the two objects are interchanged with each other.* Moreover, it follows that if two objects, a and b, prove to be equal and if upon further investigation using the same method of comparison a also proves equal to a third object, c, then in a corresponding comparison of b and c they too must be equal.

These are conditions which all methods of comparison must satisfy. Only such methods of comparison as satisfy these conditions can be used to establish equality. Given these conditions, the proposition that *equal quantities may be substituted for one another* is true for the effects which we observe in order to establish equality.

The equality of other effects or relations may be connected uniformly—that is, by a scientific law—to some initial equality, so that in these respects too the objects may be substituted for one another. We usually express this in speech by objectifying the ability to produce the effect observed during the comparison as an attribute of the objects. We ascribe to those objects that have been found equal the same amounts of the attribute, and we consider the other ways in which equality is maintained to be the effects of it or empirically dependent upon it alone. Such objectification means only that the attributes of objects which have proved equal using some method of comparison may be substituted for one another in subsequent cases without a change in the result.

Quantities whose equality or inequality are determined by the same method of comparison are called similar *(gleichartig).* Since by abstraction we separate the attribute whose equality or inequality is to be determined from all the other attributes or properties of objects, there are in different objects only differences in the quantity of this abstracted attribute.

I shall illustrate the meaning of these abstract propositions by some familiar examples.

Weight: If I put two arbitrarily selected bodies on the two pans of an accurate balance, the weights will generally not be the same, and one of the pans will sink. I can, however, find certain pairs of bodies *a* and *b* which, when placed on the balance, do not disturb its equilibrium.

If I then interchange *a* with *b,* the balance must remain in equilibrium. This is the well-known test of whether a balance is accurate or not, that is, whether the equilibrium of the balance indicates that the weights are equal.

Further, it holds that if the weight of *a* is equal, not only to the weight of *b,* but also to the weight of *c,* then *b* is also equal to *c.* The equilibrium of weights on an accurate balance is thus the basis of a method for determining a kind of equality.

The bodies whose weights we compare, moreover, may consist of quite different materials and be of very different form and volume. Their weight is only an attribute separated from them by abstraction; if we can ignore all the other properties of the bodies, we can think of them as weights and the weights as quantities. In practice, this has meaning only if we are able to observe or induce processes in which only this one attribute is significant, as is the case, for example, with bodies which can be substituted for one another while a balance remains in equilibrium. This is also the case when we measure the inertia of bodies. However, the inference that bodies of equal weight also have the same inertia and thus that they can be substituted for one another insofar as inertia is concerned, may be drawn, not from the concept of equality, but only from our knowledge of the natural laws involving this particular concept.

The distance between two points: The simplest geometric relationship that can be treated quantitatively is the distance between a pair of points. To insure that the distance has a stable value, at least for the period of the measuring operation, the points must have a fixed connection, as, for example, two points of a pair of compasses. The well-known procedure for comparing the distances between two pairs of points consists in investigating whether or not they can be brought to congruence. That this method is capable of determining equality, that congruence always reoccurs given any position of the points in space and given the interchange of the pairs of points in any manner whatsoever, and that two pairs of points which are congruent to a third are congruent to each other—all of these propositions have been established by experience. It is in this way that we form the concept of equal distances.

From here one can proceed to the concepts of straight line and the length of a line. Think of two fixed points through which a line passes. A straight line is one such that it is impossible to change the locus of any point in it without changing the distance between that point and at least one of the fixed points. A curved line, on the other hand, can be rotated axially about the two points; in this case there is a change in the position of every given point in the line but not in the distance between it and the fixed points.

We consider the lengths of two finite straight lines to be equal if their respective end points have equal distances and thus can be made congruent. The lines then also fall congruent. The concept of length contains something more than does the concept of distance. Let two pairs of points of different distances, *ab* and *ac,* fall on *a;* think of them as being laid out in a straight line such that a segment of this line includes both; then either *b* falls in the line *ac* or *c* in the line *ab.* In this way a difference results to which the concepts of greater or smaller are related; the concept of distance, however, contains only that of equality or inequality.

Time measurement presupposes that physical processes can be found which, under the same given conditions, repeat themselves with perfect regularity. These processes—such as the day, the oscillation of a pendulum, and sand and water clocks—if they begin simultaneously, also end at the same moment. The only justification for believing that with the repetition of these processes, the periods are equal, is the fact that all the various methods of time measurement, when carefully carried out, always lead to the same results. If two such processes, *a* and *b,* begin and end together, they take place not only in equal but in the same time. With respect to time there is no difference between them and no possibility of substituting one for the other. Further, if a third process, *c,* begins at the same moment as *a* and ends at the same moment, it will also do so with the simultaneous process *b.*

We compare the *brightness* of two visual fields when we relate them to each other in such a way that every difference between them except a difference in brightness may be ignored and then see whether there is still a perceptible boundary between them. If we restrict ourselves to small differences, we compare *pitch* by the phenomenon of beats, which are absent if the tones are the same. We compare the *intensity of electric currents* by the galvanometer, which remains at rest if they are equal. And so on.

Thus, in establishing various kinds of equality, very different physical processes are utilized. All the conditions mentioned above must be satisfied, however, if any process is to establish a kind of equality. Axiom I ("If two quantities are equal to a third, they are equal to each other") is thus not a law of objective significance but a criterion of what physical relations we may consider equal.

An example of this axiom being based directly upon a mechanical operation is the grinding of plane glass surfaces. If two surfaces are worn away by continuous rotation one against the other, both may become spherical–one concave, the other convex. However, if three are ground together, with interchanging, they will all finally become plane. This is how the edges of high-quality metal rulers are made straight, that is, by wearing three of them against one another.

Additive Physical Combination of Quantities of the Same Kind

As it has been treated so far, comparison of two quantities provides information as to whether or not they are equal, but it does not give a measure of the magnitude of the difference between them in case they are different. If quantities are completely specified by concrete numbers, however, the greater of the two numbers will be the sum of the smaller plus the difference between them. We must now investigate the conditions which must be satisfied in order that we may rightly consider a physical combination of the same kind of quantity additive.

As we are here concerned with it, the method of combination will depend upon the kind of quantity to be combined. For example, we add weights by placing them in the same pan of a balance. We add periods of time by allowing the second period to begin at the exact moment that the first ends. We add lengths by joining them in a specific way, that is, in a straight line. And so on.

The three conditions are:

1. *Similarity of the sum and the summands.* Since the sort of combination we are considering has to do with quantities of a specific kind, the result will not be changed if one or more of the quantities is replaced by equal quantities of the same kind, for the result will have the same numerical value and the same designation as before. Further, the result of the combination, if it is to represent the sum of the combined quantities, must be a number with the same designation. Thus, *the invariance of the result of a combination involving a substitution of parts must be determined by using the same method of comparison used to establish the equality of the parts exchanged.* This is the factual meaning of the requirement that the sum of similar quantities must be similar to the summands.

Thus, for example, in the addition of weights I can replace individual pieces by others of equal weight but made of different materials. The sum of the weights will still be the same; other physical attributes, however, will be changed.

2. *Commutative law.* The result of the addition is independent of the order in which the summands are combined. This must hold true for all physical combinations that are to be considered additive.

3. *Associative law.* The combination of two quantities of the same kind occurs physically when, in place of the two, one undivided quantity of the same kind, equal to their sum, is substituted. In this way, any two or more quantities can be combined additively before all the others. In weighing, for example, a five-gram piece can be substituted for five individual gram pieces. The result of the larger combination will not be changed just because some of the quantities to be combined have been replaced by another which is equal to their sum.

Finally, it can be shown that if the first two conditions are satisfied, the third will also be satisfied. Assume that the elements in a sequence are arranged one after another in the order in which they are to be combined, so that each element may be added to the result of the combination of those preceding it, as was indicated earlier for the addition of $a + b + c + \ldots$. If any of these quantities must be combined prior to the others, we can set these in the first and second positions according to the commutative law (which is here assumed to be valid), and they can then be combined before the others without altering the result. Then, according to the first condition mentioned above, we can replace the result of this combination by a single object which, considered as a quantity of the same kind, is of equal magnitude. After this we can bring the next two quantities or sums of quantities to be combined to the first two positions, and we can continue this process until all the elements in the sequence have been combined. In none of these operations do we alter the magnitude of the final result of the combinations.

A physical combination of quantities of the same kind may be considered additive if the result of the combination, which is also a quantity of the same kind, is not altered either by the interchange of the elements or by the substitution of equal quantities of the same kind.

If we have found a method of combining some given quantities in the way just described, we can also determine which of the quantities are larger and which are smaller. The result of the additive combination, the whole, is greater than any of the parts out of which it is formed. With the simplest quantities, such as time, length, and weight, with which we have been familiar since our earliest youth, we have never doubted which are greater and which are smaller because even as children we knew additive methods for combining them. It should be noticed here, however, that the methods of comparison described above indicate only whether the quantities are equal or unequal. If two quantities are equal, all functions of them formed in the same way by calculation will also be equal. Some of these functions will increase as the quantities increase; others will decrease. Which of them will permit of an additive physical combination, however, must be established by experience.

Cases where two kinds of additive combination are possible are therefore very instructive. One example is the combining of electric circuits. We determine by exactly the same method of comparison whether two wires have the same resistance, w, relative to equal conductivities, λ, for $w = 1/\lambda$. We add resistances if we combine the wires one after another in series, so that the electricity conducted through them must pass through one after the other. We add conductivities if we place the wires in parallel and join all their beginnings and all their ends. By doing these things, we objectify as physical quantities two different functions of the same variables. (If a wire has a greater resistance, it has a smaller conductivity, and vice versa.) Thus, in the two different combinations, the question as to which function is greater and which is smaller will be answered in opposite ways.

As another example, let two electric condensers (Leyden jars) be combined first in parallel and then in series. In the first case we add the capacities and in the second the voltage (potential) for the same charge. The first increases when the latter decreases.

Again, we must not be surprised if the axioms of arithmetic turn out to be true of natural processes, for we recognize as additive only such physical combinations as conform to the axioms of arithmetic.

Divisibility of Quantities and Units

Up to this point we have not had to reduce quantities to units. The concept of *quantity,* as well as the concepts of *equality* and *addition,* can be established without such an analysis. We get the greatest simplification in the expression of quantities, however, if we reduce them to units and express them by concrete numbers.

Quantities which can be added can also be divided. If any of the quantities hitherto considered can be thought of as having been combined additively from a number of equal parts, according to the method of addition used for quantities of that kind, then—since, according to the associative law of addition, only the final value is important—it can be represented as the sum of these parts. It can then be given a concrete number, and other quantities of the same kind can be expressed as sums of other numbers of these same parts. The value of actual quantities of the same kind can then be communicated to anyone who knows the part chosen as the unit merely by stating the number.

If a quantity cannot be expressed without a remainder by using a given unit, the unit can be divided further in a familiar manner, and in this way a unit can be obtained for any quantity up to any desired degree of precision. Complete precision, to be sure, can only be achieved for expressions containing rational numbers.

Irrational ratios may appear in connection with real objects and processes. Their values can never be given with complete accuracy using numbers; they can be fixed only within arbitrarily narrow limits. This restriction to a certain range of values is sufficient for all calculations involving functions whose values vary less and less with smaller and smaller changes in the quantities upon which they depend, until finally the variations fall below any given finite amount. It is sufficient, in particular, for all calculations involving differentiable functions of irrational numbers. There are also discontinuous functions, however, and a knowledge of the narrowly drawn limits between which irrational values lie is not sufficient for calculations involving these functions. For this reason, the representation of irrational numbers by our number system is always incomplete. In geometry and physics, however, we are not yet faced with such forms of discontinuity.

Determination of the Values of Quantities (Physical Constants, Coefficients)

In addition to the quantities previously discussed, which may be recognized as such directly because they may be combined additively, there are a number of others, expressed by either concrete or abstract numbers, for which no procedures are as yet known for combining them additively. These quantities make their appearance each time a regularity between additive quantities concerns a process which is affected by the peculiarities of a specific substance or body or by the way the process is initiated.

The law of refraction of light, for example, shows that a definite relation exists between the sines of the angles of incidence and refraction of a light ray of specific wave length which enters a given transparent medium from another medium. The number which expresses this relation, however, is different for different transparent media and thus represents a property of a medium, namely, its index of refraction. Other such quantities include specific weight, heat conductivity, electric conductivity, and heat capacity. With each of these, there are values (the integration constants of dynamics) which remain constant during the uninterrupted course of processes involving finite systems of bodies.

Little by little physics has succeeded—by multiplication, raising to a power, and their inverse operations—in reducing all these values to units which are in turn related to the three fundamental measuring units of time, length, and mass. The difference between these values and additive quantities is not strictly maintained in the language of physicists and mathematicians. Frequently, because they are expressed by concrete numbers, the values are called quantities too. The term *coefficient,* however, indicates their physical nature much better. The difference, however, is not intrinsic, for occasionally new discoveries can lead to the additive combination of such coefficients, whereupon they become one of the quantities which can be established directly. (In some measure this distinction corresponds well with what earlier metaphysicians meant by the distinction between *extensive* and *intensive quantities.* Du Bois-Reymond called the first *linear quantities* and the second *nonlinear.*)

It follows from the analysis given above that additive quantities must be available before coefficients can be determined. An equation which expresses a natural law can provide the actual value of a coefficient only if all the other quantities appearing in the equation have some specific value. Thus we must know the value of additive quantities before the value of nonadditive quantities can be established.

Addition of Heterogeneous Quantities

In physics a major role is played by objects which, as a result of several different methods of comparison, possess two or three or even several quite different quantities, each of which can be added in the same kind of physical combination of the objects. To this class, for example, belong the very large number of quantities which also have a spatial orientation, that is, quantities which have a specific magnitude and at the same time a specific direction. These quantities can be represented as composed of components of fixed direction (two in a plane, three in space). In general, the relations among these quantities are simplest if the components are arranged along three rectilinear coordinate axes and are combined to produce a resultant in the way prescribed by the law of the parallelogram of forces.

To this class also belong the quantities describing the movement of a point in space, its velocity, its acceleration, and the moving force corresponding to its acceleration. In addition, rotational velocities, vanishingly small rotations, magnetic moments, the velocities of flow of viscous fluids, electricity, and heat, and other quantities belong to this class.

Components having the same direction can be combined additively, with the resultant expressing the sum. All physical combinations of quantities whose effect depends solely upon the magnitude and direction of the final resultant can be treated as the result of the kind of additive combination that was described by Gauss for the two-dimensional representation of imaginary numbers, by H. Grassmann for the addition of geometric segments in more than two dimensions, and by Sir William Hamilton in the theory of quaternions. With each of these, the commutative law must be satisfied; thus we can combine vanishingly small rotations of a fixed body about two different axes into a resultant rotation. This can also be done for rotational velocities but not for finite rotations, since in the latter case it is no longer the same whether the rotation takes place first about axis a and then about b, or vice versa.

Something similar to the addition of components occurs with the mixture of colored lights. As far as our sense impressions are concerned, a quantity of colored light can be represented as the resultant of the combination of quantities of lights of three suitably chosen fundamental colors. Thus the result of the mixture of certain amounts of each of several colors affects our eyes the same as would certain amounts of light of the three fundamental colors acting in combination. The possibility of geometric representation of the law of color mixture through the construction of so-called centers of gravity, a method first presented by Newton, rests upon this principle.

Multiplication of Concrete Numbers

A concrete number, ax, where x designates the kind of unit and a the total number, can be multiplied by a pure number n. This falls quite simply under the definition given above of a product as the sum of n equal summands a. Since the sum of similar summands is a quantity of the same kind, the product, na, is a quantity of the same designation as a. The commutative law holds for this product insofar as

$$n(ax) = a(nx),$$

that is, a can be considered a pure number and the new concrete number, nx, formed.

Equally, the law of multiplication of a sum can be given immediately:

$$(m + n)(ax) = m(ax) + n(ax)$$
$$n(ax + bx) = n(ax) + n(bx).$$

Thus the multiplication of concrete numbers by pure numbers remains completely within the scope of the definitions and propositions presented above concerning the multiplication of pure numbers.

It is different, however, with the multiplication of two or more different concrete numbers. These products have a meaning only in certain cases, namely, where physical combinations of the units under consideration are possible. Such combinations must satisfy the three laws of multiplication:

$$ab = ba,$$
$$a(bc) = (ab)c,$$
$$a(b + c) = ab + ac.$$

The best known examples of such multiplicative combinations from geometry are the values for the areas and volumes of parallelograms and parallelepipeds, expressed as the product of two or three lengths, that is, one side and one or two altitudes. Physics, however, has a number of products involving different kinds of units and also provides examples involving quotients, powers, and roots. If we use l to designate a length, t a time, and m a mass, we have the following designations:

Surface	l^2
Volume	l^3
Velocity	l/t
Force	$m \cdot l/t^2$
Work	$m \cdot l^2/t^2$
Pressure on a surface	$m/l \cdot t^2$
Tension in a surface	m/t^2
Density	m/l^3
Magnetic moment	$\frac{1}{t}\sqrt{ml}$
Magnetic force	$\frac{1}{t}\sqrt{m/l}$
Etc.	

Most of these combinations depend upon the determination of coefficients. Many of them, however, can also be added physically, as, for example, velocities, currents, forces, pressures, and resistances. All these units defined by multiplication are not the same in kind as those out of which they are produced, and they are meaningful only in relation to specific geometric or physical laws.

Mention should be made here of a special kind of multiplication of directed quantities, which H. Grassmann discussed in his theory of dimensions *(Ausdehnungslehre).* It is also basic to the theory of quaternions. This kind of multiplication obeys a different commutative law, namely,

$$ab = -ba,$$

which provides great simplification in notation, if not in the calculation of the value of the resultant of quantities having different directions.

In this kind of calculation, the product of two segments is the area of a parallelogram of which they are the sides; the surface of the parallelogram, however, is treated as positive on one side and negative on the other. Looking at the surface from one side, in going from side *a* to side *b,* I must pass through the angle turning to the right; looking at it from the other side, I must go in the opposite direction, turning to the left from *b* to *a.* Upon this rests the difference between the product *ab* and the product *ba.*

It is sufficient here to have mentioned this form of calculation and to have indicated its place within the theory of pure numbers, for my aim in this essay has been only to show the meaning of and the justification for calculations with pure numbers and to indicate the possibility of their application to physical quantities.

In conclusion, the fact that we are able to represent a physical relation as a quantity rests always upon empirical knowledge of the concurrence or interaction of various physical processes. It is for this reason that I conceive of the congruence of two spatial quantities, insofar as this relates to physical bodies, as I did in my earlier work on the axioms of geometry—that is, as a physical relation which must be established empirically. We must *first* know the method of comparison of the quantities in question, by which their nature is characterized, and *secondly* know either the methods of additive combination of these quantities or the natural law in which they appear as coefficients, in order to be able to express them by concrete numbers.

The great simplification and clarity which we achieve by reducing the multiplicity of things and processes around us to quantitative relations is rooted deeply in the nature of our way of comprehending things. When we form the concept of a class, we bring together that which is the same in each of the objects belonging to the class. When we express a physical relation by using concrete numbers, we also ignore everything about objects different from (but connected in reality with) the properties or relations covered by a concept. The objects are thought of only as instances of a class, and their usefulness in investigations is dependent solely on the fact that they are such instances. Thus, in the quantities formed in this way, there remains only the most accidental of differences, which is represented by a concrete number.

17.

AN AUTOBIOGRAPHICAL SKETCH [1891]

An address delivered on the occasion of his Jubilee in Berlin, 1891

IN the course of the past year, most recently on the occasion of the celebration of my seventieth birthday and during the ensuing festivities, I have been overwhelmed with honors and with demonstrations of respect and goodwill to a degree that could never have been foreseen. His Majesty the Emperor has raised me to the highest rank in the civil service; the Kings of Sweden and Italy, my former sovereign the Grand Duke of Baden, and the President of the French Republic have conferred Grand Crosses upon me. Many academies, not only of the sciences but also of the arts, academic faculties, and learned societies, spread over the entire world from Tomsk to Melbourne, have sent me diplomas and beautifully decorated ceremonial addresses, expressing, in terms which I cannot read without a feeling of embarrassment, their recognition of my scientific endeavors and their appreciation of them. My native town, Potsdam, has conferred its freedom upon me. To all these must be added the countless individuals–scientific and personal friends, students, and others personally unknown to me–who have sent their congratulations in telegrams and in letters.

Nor is this all. You wish to make my name a banner, as it were, of a magnificent institution which, founded by lovers of science of all nations, is to encourage and promote scientific inquiry in all countries. Science and art are, indeed, at the present time the only remaining bonds of peace among civilized nations. Their ever-greater development is a common aim of all, effected by the common work of all and for the common good of all. A great and sacred work! The founders of the institution even wish to devote their gift to the promotion of those branches of science which I have pursued in my own career and thus to present my work, despite its shortcomings, to future generations almost as a model of scientific investigation. This is the proudest honor which you could confer upon me, showing as it does that you have accorded me your unqualified favorable judgment. It would border on insolence on my part, however, to accept this honor without the implicit expectation that the judges of future generations will not be influenced in their judgment by considerations of a personal nature.

Even this transient body, within which I have passed my life, you have had represented in marble by a master of the first rank, so that I shall appear to future generations more ideal in form than I do to the present one. A master of the etching needle, however, has ensured that faithful portraits of me will be distributed among my contemporaries.

I shall not forget that everything you have done is an expression of the sincerest and warmest goodwill. I am most deeply indebted to you for it. I must, however, be excused if the ultimate effect of these many honors is rather more surprising and confusing to me than intelligible. I find, in my own evaluation of the work I have tried to do, no estimate which would leave a balance in my favor such as the one you have drawn. I know how simply everything I have done has been brought about, how scientific methods worked out by my predecessors have naturally led to certain results, and how frequently a fortunate circumstance or a lucky accident has helped me.

But the chief difference between us appears to be this: some things which I have seen growing slowly from small beginnings, through months and years of tedious and frequently groping work, seem to you to have sprung suddenly like a fully equipped Pallas from the head of Jupiter. Your judgment has usually been influenced by a feeling of surprise; mine has not. Indeed, mine has at times—perhaps frequently—been lowered by the fatigue of the work and by vexation over all of the futile steps I have taken.

My colleagues, as well as the public at large, evaluate a scientific or artistic work on the basis of its utility, its instructiveness, or the pleasure which it affords. An author is more inclined to base his evaluation on the labor a work has cost him, and it is but rarely that both kinds of judgment agree. Indeed, we can see from occasional statements of some of the most celebrated men, especially artists, that they assign small value to achievements which seem to us inimitable, compared with others which were difficult for them and yet which appear much less successful to readers and observers. I need only mention Goethe, who once stated to Eckermann that he did not value his poetic works as highly as the work he had done in the theory of color.

The same may have happened to me, although to a more modest degree, if I may trust your assurances and those of the authors of the addresses which have reached me. Permit me, therefore, to give you a short account of the manner in which my work came to take the special direction it has taken.

During my first seven years I was a delicate boy, confined for long periods to my room and often to bed; nevertheless, I had a strong inclination toward several occupations and activities. My parents busied themselves a good deal with me, while picture books and games, especially games with wooden blocks, filled the rest of my time. In addition, reading came fairly early, and this, of course, greatly increased the range of my occupations. A defect among my mental powers showed itself, however, almost as early: I had a poor memory for unrelated facts. The first indication of this was, I believe, the difficulty I had in distinguishing between left and right. Later, when I began the study of languages at school, I had greater difficulty than others in learning vocabularies, irregular grammatical forms, and peculiar forms of expression. I could barely master history as it was then taught to us. To learn prose by heart was martyrdom. This defect has, of course, grown and has been a vexation to me in my later years.

When there were small mnemotechnical methods, even such as are afforded by the meter and rhyme of poetry, learning by rote and retaining what I had learned were simpler. I easily remembered the poems of great masters, though I had more trouble with the somewhat artificial verses of authors of the second rank. I think this is probably due to the natural flow of thought in great poems, and I am inclined to think that an essential element of aesthetic beauty is to be found in this flow. In the higher classes of the Gymnasium I could recite several passages from the *Odyssey*, a rather large number of Horace's odes, and large stores of German poetry. In this I was like our early ancestors, who were not able to write and hence expressed their laws and history in verse so as to be able to learn them by heart.

What a man does easily he usually does willingly, so from the beginning I was a great admirer and lover of poetry. This inclination was encouraged by my father, who, while he had a strict sense of duty, was also of an enthusiastic disposition and impassioned for poetry, especially for the classic period of German literature. He taught German in the upper classes of the Gymnasium and read Homer with us. Under his guidance we had alternately to do themes in German prose and metrical exercises–"poems," as we called them. Even if most of us remained indifferent poets, we learned better in this way than in any other I know of, how to express what we had to say in the most varied manner.

The most perfect mnemotechnical method, however, is the knowledge of the laws of phenomena. I first began to learn such laws in geometry. From the time of my childhood playing with wooden blocks, the relations among spatial dimensions were well known to me by actual perception. I knew well, and without much reflection, what sorts of figures would be produced when bodies of regular shape were placed next to one another. When I began the scientific study of geometry, all the facts which I was supposed to learn were perfectly familiar to me, much to my teacher's astonishment. As nearly as I can remember, this occurred at the elementary school attached to the Potsdam Training College, the school I attended up to my eighth year. Rigorous scientific methods, on the other hand, were new to me, and with their help I saw disappear the difficulties which had hindered me in other subjects.

One thing was lacking in geometry: it dealt exclusively with abstract spatial forms, while I delighted in complete reality. As I grew bigger and stronger, I traveled about a good deal in the neighborhood of my native town of Potsdam with my father or my schoolfellows, and I developed a great love of nature. This is perhaps the reason why the first fragments of physics which I learned in the Gymnasium engrossed me much more completely than pure geometric and algebraic studies. Here was a rich and multifarious region, with the mighty fullness of nature, to be brought under the dominion of abstract laws. And, in fact, that which initially excited my interest was the intellectual mastery over nature, at first so unfamiliar to us, which could be achieved through the logical form of laws. But this, of course, soon led to the recognition that knowledge of the laws of natural processes is the magic key which places ascendency over nature in the hands of its possessor. In this order of ideas I felt myself at home.

I plunged with pleasure and great zeal into the study of all the books on physics I found in my father's library. They were very old-fashioned: phlogiston still held sway, and galvanism had not grown beyond the voltaic pile. A young friend and I tried, with our limited means, all sorts of experiments about which we had read. The action of acids on our mothers' stores of linens was investigated thoroughly; otherwise we had but little success. Most successful, perhaps, was our construction of optical instruments, using the spectacle glasses that were to be had in Potsdam and a small botanical lens belonging to my father. The limitation of our means during these early studies was valuable in that I was compelled always to vary my plans for experiments in all possible ways, until I got them in a form in which I could carry them out. I must confess that many times while the class was reading Cicero or Virgil, both of whom I found very tedious, I was calculating under the desk the path of light rays in a telescope. Even at that time I discovered some optical laws, not ordinarily found in textbooks, but which I afterward found useful in constructing the ophthalmoscope.

That is how it happened that I entered upon that special line of study to which I have subsequently adhered—a line which, under the conditions I have mentioned, grew into an absorbing desire, amounting even to a passion. This desire to govern reality by acquiring an understanding of it—or, what I think is only another expression for the same thing, to discover the causal connection of phenomena—has directed me throughout my life. The strength of this desire may be the reason why I have found no satisfaction in apparent solutions to problems as long as I felt there were still obscure points in them.

And now I was to go to the university. At that time physics was not considered a profession at which one could make a living. My parents were compelled to be very economical, and my father explained to me that he knew of no way I could study physics other than by taking up the study of medicine in the bargain. I was by no means averse to the study of living nature and assented to this without much difficulty. Moreover, the only influential person in our family, the late Surgeon General Mursinna, had been a physician, and this relationship was a point in my favor when I applied for admission to our army medical school, the Friedrich-Wilhelm Institute, which very materially helped the poorer students to get through their medical course.

In my studies I came immediately under the influence of a profound teacher, the physiologist Johannes Müller, who also introduced du Bois-Reymond, Brücke, Ludwig, and Virchow to the study of anatomy and physiology at that time. With respect to the crucial question of the nature of life, Müller still struggled between the older, essentially metaphysical view and the new scientific view which was then being developed. The conviction, however, that nothing could replace the knowledge of facts forced itself upon him with increasing certainty, and it may be that his influence over his students was the greater because he still so struggled.

Young people are eager to attack at once the most profound problems, and I took up the perplexing question of the nature of vital force. Most physiologists at that time had adopted G. E. Stahl's solution to the difficulty, namely, that while it is indeed the physical and chemical forces of the organs and substances of the living body which act in it, there is also an indwelling life soul, or vital force, which controls the activities of these forces. After death the free action of these physical and chemical forces produces decomposition, but during life their action is continually being regulated by the life soul. I had the feeling that there was something contrary to nature in this explanation; it took a good deal of effort, however, to state my misgivings in the form of a definite question.

Finally, in the last year of my career as a student, I realized that Stahl's theory treated every living body as a *perpetuum mobile*. I was fairly well acquainted with the controversies over the subject of perpetual motion, as I had heard it discussed by my father and by our mathematics teachers during my school days. In addition, while a student at the Friedrich-Wilhelm Institute I was helping in the library, and in my spare moments I looked through the works of Daniel Bernoulli, d'Alembert, and other mathematicians of the last century. I thus came to the questions: What relations must exist among the various natural forces for perpetual motion to be possible, and do these relations actually exist? In my memoir "The Conservation of Force" my aim was merely to provide a critical examination of these questions and to present the facts for the benefit of physiologists.

I was quite prepared for the experts to say simply, "We know all that. What is this young doctor thinking about who considers himself called upon to explain it all to us so fully?" To my astonishment, however, the authorities on physics with whom I came in contact received it quite differently. They were inclined to deny the correctness of the law and, because of the heated fight in which they were engaged against Hegel's philosophy of nature, to treat my essay as a fantastic piece of speculation. Only the mathematician Jacobi recognized the connection of my line of thought with that of the mathematicians of the preceding century, defended my investigations, and protected me from misconception. I also met with enthusiastic applause and practical help from my younger friends, especially Emil du Bois-Reymond. They soon brought the members of the newest physical association of Berlin over to my side. I knew little at that time about Joule's researches on the subject and nothing at all about those of Robert Mayer.

Connected with this study were some smaller experimental researches on putrefaction and fermentation, in which I was able to furnish evidence that, contrary to Liebig's hypothesis, they were not merely chemical decompositions, occurring spontaneously or brought about with the aid of atmospheric oxygen. Specifically, I showed that alcoholic fermentation was bound up with the presence of yeast spores, which are produced only by reproduction. There was, further, my work on metabolism during muscular activity, which afterward was connected with work on the development of heat during muscular activity, a phenomenon which was to be expected according to the law of the conservation of force.

These researches were sufficient to attract the attention of Johannes Müller, as well as that of the Prussian Ministry of Instruction, and this led to my being called to Berlin as Brücke's successor and immediately afterward to the

University of Königsberg. The army medical authorities, with gracious liberality, very readily agreed to release me from the obligation of further military service and thus made it possible for me to take up a scientific position.

At Königsberg I had to lecture on general physiology and pathology. A university professor undergoes a very valuable training in being compelled, not only to lecture each year over the whole range of his science, but to do so in such a manner as to satisfy and convince the intelligent students in his audience, the ones who will be the leading men of the next generation. This requirement yielded two valuable results. In preparing my course of lectures, I hit first of all upon the possibility of the ophthalmoscope and then upon a plan for measuring the rate of propagation of excitations in nerves.

The ophthalmoscope is probably the most popular of my scientific accomplishments, and I have already told the ophthalmologists how luck really played a more important role than my skill in its discovery. I had to explain to my students Brücke's theory of ocular illumination. In this theory, Brücke himself was actually within a hair's breadth of inventing the ophthalmoscope. He merely neglected to ask the question: To what optical image do the rays which come from the illuminated eye belong? For the purposes he had in view at that time, it was not necessary to ask this question. If he had asked it, he would have been just the man to answer it as quickly as I could, and the plan of the ophthalmoscope would have been given. I turned Brücke's theory over in my mind in various ways to see how I could best explain it to my audience, and in doing so I hit upon the question.

I knew well, from my medical studies, the difficulties which ophthalmologists had with the conditions then included under the name of amaurosis, and I at once set about constructing the instrument by means of spectacle glasses and cover glasses used in microscopic work. At first it was difficult to use, and without a firm theoretical conviction that it had to work, I might not have persevered. In about a week, however, I had the great pleasure of being the first man ever to see a living human retina clearly before him.

The construction of the ophthalmoscope had a most decisive effect on my position in the eyes of the world. From that time on I met with immediate recognition from the authorities and my colleagues, and with an eagerness to satisfy my wishes. Thus I was able to follow far more freely the impulses of my desire for knowledge. I must, however, say that I attribute my success in great measure to the fact that, possessing some geometric understanding and equipped with a knowledge of physics, I had the good fortune to be thrown into medicine, where I found in physiology a virgin territory of great fertility. Furthermore, I was led by my knowledge of vital processes to questions and points of view which are usually foreign to pure mathematicians and physicists.

Up to that time I had only been able to compare my mathematical ability with that of my fellow students and medical colleagues; that I was for the most part superior to them in this respect did not, perhaps, say very much. Moreover, mathematics was always regarded at school as a subject of secondary importance. In Latin composition, on the other hand, which then determined the palm of victory, half of my fellow students were ahead of me.

In my own mind my researches were simple logical applications of the experimental and mathematical methods which had been developed in the sciences and which, by slight modifications, could easily be adapted to the particular problems at hand. Colleagues and friends who, like myself, devoted themselves to the physical aspect of physiology made discoveries no less surprising.

In the course of time, however, matters could not remain at that stage. Problems which could be solved by established methods had gradually to be handed over to the students in my laboratory, and I had to turn to more difficult researches, where success was uncertain, where standard methods left the investigator stranded, or where the method itself had to be worked out.

In these regions closer to the boundaries of our knowledge I have also succeeded in many things experimental and mathematical—I do not know if I may add philosophical. With respect to the first, like anyone who has attacked many experimental problems, I had come to be a person of experience, acquainted with many plans and devices, and my youthful habit of considering things geometrically had developed into a kind of mechanical intuition. I felt, intuitively as it were, how stresses and strains were distributed in any mechanical arrangement. This is a faculty also met with in experienced mechanics and machinists; I had the advantage over them, however, in that I was able to make especially important and complicated relations clear by means of theoretical analysis.

I have also been in a position to solve several problems in mathematical physics, some of which the great mathematicians since the time of Euler had worked on in vain—for example, problems concerning vortex motion and the discontinuity of motion in fluids, the problem of the motion of sound waves at the open ends of organ pipes, and so on. But the pride which I might have felt about the final result of these investigations was considerably lessened by my knowledge that I had only succeeded in solving such problems, after many erroneous attempts, by the gradual generalization of favorable examples and by a series of fortunate guesses. I would compare myself to a mountain climber who, not knowing the way, ascends slowly and toilsomely and is often compelled to retrace his steps because his progress is blocked; who, sometimes by reasoning and sometimes by accident, hits upon signs of a fresh path, which leads him a little farther; and who finally, when he has reached his goal, discovers to his annoyance a royal road on which he might have ridden up if he had been clever enough to find the right starting point at the beginning. In my papers and memoirs I have not, of course, given the reader an account of my wanderings, but have only described the beaten path along which one may reach the summit without trouble.

There are many people of narrow vision who admire themselves greatly if once they have had a good idea—or even think they have had one. An investigator or an artist who is continually having a great number of them is undoubtedly a privileged being and is recognized as a benefactor of humanity. But who can count or measure such mental flashes? Who can follow the hidden paths by which ideas are connected?

What man does not know
Or has not thought of
Wanders in the night
Through the labyrinth of mind.

Was von Menschen nicht gewusst
Oder nicht bedacht,
Durch das Labyrinth der Brust
Wandelt in der Nacht.

I must say that those regions in which one does not have to rely on lucky ideas and accidental discoveries have always been the most agreeable to me as fields of work.

As I have often found myself in the unpleasant position of having to wait for useful ideas, I have had some experience as to when and where they come to me which may perhaps be useful to others. They often steal into one's train of thought without their significance being at first understood; afterward some accidental circumstance shows how and under what conditions they originated. Sometimes they are present without our knowing whence they came. In other cases they occur suddenly, without effort, like an inspiration. As far as my experience goes, they never come to a tired brain or at the desk.

I have always had to turn my problems about in my mind in all directions, so that I could see their turns and complications and think them through freely without writing them down. To reach that stage, however, was usually not possible without long preliminary work. Then, after the fatigue of the work had passed away, an hour of perfect bodily repose and quiet comfort was necessary before the fruitful ideas came. Often they came in the morning upon waking, as is suggested in the words of Goethe I have just quoted, and as Gauss also noted.[1] But, as I once stated at Heidelberg, they were most apt to come when I was leisurely climbing about on wooded hills in sunny weather. The slightest quantity of alcohol seemed to frighten them away.

Such moments of fruitful thought were indeed very delightful, but not the times when the redeeming ideas did not come. For weeks or months I was

> Like a beast upon a barren heath
> Dragged in a circle by an evil spirit,
> While all around are pleasant pastures green.

> *dem Tier auf dürrer Haide*
> *Vor einem bösen Geist in Kreis herumgeführt*
> *Und rings uhmer ist schöne grüne Weide.*

Often only a sharp attack of migraine released me from this strain and set me free to follow other interests.

I have also entered another region to which I was led by investigations of sensation and sense perception, namely, the theory of knowledge. Just as the physicist must examine the telescope and galvanometer with which he is working in order to get a clear conception of what he can attain with them and of how they may deceive him, so it seemed to me necessary to investigate our powers of thought. Here also we are concerned only with a series of factual questions to which definite answers can and must be given. We have specific sense impressions, as a consequence of which we know how to act. The observable results of actions usually agree with what was anticipated; sometimes, however, as in cases of subjective impressions, they do not. These are all objective facts, and it is possible to find the lawful relations among them. My principal conclusions were that sensory impressions are only signs of the constitution of the external world and that the interpretation of these signs must be learned by experience.

My interest in questions raised in the theory of knowledge was implanted in me in my youth, when I often heard my father, who was strongly influ-

1. "The law of induction discovered January 1835, at 7 A.M., before rising" (*Works*, V, p. 609).

enced by Fichte's idealism, argue with those of his colleagues who followed Kant or Hegel. Hitherto I have had but little reason to be proud of my investigations. For everyone on my side I have found about ten opponents; in particular I have aroused all the metaphysicians—even the materialists—and all people of hidden metaphysical tendencies. The addresses of the last few days, however, have revealed a host of friends of whom I was not aware, so that in this respect too I am indebted to this ceremony for pleasure and fresh hope. Philosophy has, to be sure, for nearly three thousand years been the battleground for the most violent differences of opinion, and it is not to be expected that these can be settled during a single lifetime.

It has been my aim to explain to you how the history of my scientific endeavors and achievements, as far as they go, appears when seen from my own point of view. Perhaps you will now understand why I am surprised at the unusual amount of praise you have heaped upon me. With respect to my own estimate of myself, my achievements have had primarily the following value: they have provided a measure of what I must still attempt. They have not, I hope, led me to self-admiration. I have often enough seen how injurious for a scholar an exaggerated sense of self-importance may be, and I have always known that a rigorous self-criticism of my own work and of my own capabilities is the protection and palladium against this fate. It is really only necessary, however, to keep one's eyes open for what others can do and for what one cannot do, to avert that great danger. Moreover, as regards my own work, I do not believe that I have ever finished correcting the last proof of a memoir without finding, in the course of the next twenty-four hours, a few points which I could have done better or more carefully.

Finally, with respect to the thanks that you believe you owe me, I should be unjust if I said that the good of humanity stood before me from the beginning as the conscious object of my labors. In reality it was my desire for knowledge which drove me and which caused me to employ in scientific research all the time that was not required by my official duties and by my family. These two responsibilities did not, however, necessitate any real deviation from the ends toward which I was striving. My office required that I prepare myself to deliver lectures in the university; my family, that I establish and maintain my reputation as a scientist. The state—which provided my maintenance, my scientific appliances, and a large part of my free time—had in my opinion acquired thereby the right to expect that I communicate all that I had discovered with its help faithfully, completely, and in a suitable fashion to my fellow citizens.

The writing out of scientific investigations is, for the most part, a troublesome undertaking; at least it has been so for me—to the highest degree. I have rewritten many parts of my memoirs four to six times and have continued to revise their order until I was at least fairly satisfied. An author gains greatly by such careful recording of his work. It compels him, more thoroughly even than the lecturing at the university which I have already mentioned, to make the severest criticism of every sentence and every inference. I have never considered an investigation finished until it was completely written without any logical deficiencies.

At the same time, the most expert of my friends have stood before my mind as my conscience. I have asked myself whether they would approve of my work. They hovered before me as the embodiment of the scientific spirit of an ideal humanity, and they furnished me with a standard.

During the first half of my life, when I still had to work to establish myself, I would not say that higher ethical motives were not present, along with a desire for knowledge and a feeling of duty as a servant of the state. It was difficult, however, to be certain of their presence as long as egoistic motives to work were still there. This is perhaps the case with most investigators. Afterward, however, when an assured position has been attained and those who have no inner impulse toward science may cease their labors, a higher conception of one's relation to humanity does influence those who continue to work. They gradually learn from their own experience how the thoughts which they have expressed, either through their writings or through oral instruction, continue to act on other men and possess, as it were, an independent life.

They learn, moreover, how their thoughts, further developed by their students, acquire a deeper significance and a more definite form and how, reacting on their originator, they furnish him with fresh instruction. An individual's ideas, those which he himself has conceived, are of course more closely connected than strange ones with his mental perspective and he feels more encouragement and satisfaction when he sees them more fully developed. A kind of parental affection for such mental children finally springs up, and this leads him to care and struggle for the furtherance of his mental offspring as he does for his real children.

But, at the same time, the whole intellectual world of civilized humanity presents itself to him as a continuous, developing whole, whose duration seems eternal when compared with the life of a single individual. He sees that through his small contributions to the building up of science, he is in the service of, and is connected by close bonds of affection to, something everlasting and sacred. His work thereby seems to him more sanctified. Anyone can, perhaps, apprehend this theoretically, but actual personal experience is undoubtedly necessary to develop this idea into a strong feeling.

The world, which is not apt to believe in ideal motives, calls this feeling love of fame. But there is a decisive criterion by which the two motives can be distinguished. Just ask the question: Ignoring all external advantages and disadvantages, is it all the same to you whether the results which you have obtained in your investigations are recognized as belonging to you or not? The reply to this question is easiest in the case of those who are directors of laboratories. They usually must furnish to each student the fundamental idea for a piece of research, along with a number of more or less ingenious proposals for overcoming experimental difficulties. All this passes as the student's work, and ultimately it appears in his name when the research is finished. Who can afterward decide what one or the other has accomplished? And how many teachers are there who in this respect are devoid of jealousy?

Thus, gentlemen, I have been in the happy position that, in freely following my own inclinations, I have been led to useful, instructive investigations for which you now praise me. I am extremely fortunate in that I am praised in so high a degree and honored by my contemporaries for a course of work which is to me the most interesting I could pursue. But my contemporaries have afforded me great and essential help, quite apart from relieving me of the care for my own existence and that of my family and providing me with external means. I have found in them a standard of the intellectual capacity of man, and by their sympathy for my work they have evoked in me a vivid image of the universal intellectual life of humanity, an image which enables me to see the value of my researches in a higher light. Under these circumstances, I can only regard the thanks which you wish to accord me as a gift, given unconditionally and without thought of any return.

18.

GOETHE'S ANTICIPATION OF SUBSEQUENT SCIENTIFIC IDEAS [1892]

An address before the Goethe Society in Weimar in 1892

It is the gracious custom of the Goethe Society, in its endeavor to discover and evaluate correctly Goethe's many influences, to give representatives of the most diverse kinds of scientific and literary activity the opportunity to discuss the relationships between their own areas of thought and those of the great poet. Men who, like Goethe, have absorbed the total contents of the culture of their period without becoming limited in the strength and natural independence of their feelings–who, in finding the right path among the obstacles of life, need only follow their own warm, inborn sympathy for all the movements of the human spirit–have in our time become rare indeed. They will probably become still more rare in the future. His candor and the health of his spirit stand out all the more admirably due to the fact that he lived in a very artificial age, when even the longing to return to nature took an unnatural form. His example, therefore, is for us an immensely valuable measure of what is original and genuine in the spiritual nature of man, a measure by which we should never fail to judge ourselves and our own narrowly circumscribed endeavors.

Once before, at the beginning of my scientific career, I sought to give a report on Goethe's work in the natural sciences. At that time I was mainly interested in defending the scientific point of view of the physicist against the accusations which had been made by the poet. Goethe received far more ready acceptance then among the educated and cultured classes than did the young natural sciences, whose title to entrance into the same circle with the other disciplines, which had become venerable through long tradition, was looked upon with no little distrust.

Since then, forty years of productive development on the part of the sciences has passed in Europe. During this period the natural sciences have proved, through the transformation they have wrought in all the practical affairs of life, the reliability and fruitfulness of their fundamental principles. They have also gained comprehensive vantage points from which a profoundly different overview of nature, organic as well as inorganic, can be seen. One need only think of Darwin's theory of the origin of species and of the law of the conservation of force, which in themselves are surely sufficient to cause one to reflect upon previous principles and submit them to a new examination.

There is something more in these developments of special interest to me personally. My course of studies turned early toward physiological problems, specifically toward the laws of nerve action or activity. In this area it was impossible for me to ignore questions concerning the origin of sense perceptions. Just as the chemist, from the very beginning of his professional work, must investigate the accuracy and reliability of his balance and the astronomer those of his telescope, so the natural sciences as a whole must examine the mode of operation of those instruments which are the source of all our knowledge, that is, the human sense organs. Everyone knows that what are called illusions of the senses occur, and one must try to learn as much about their origin as is necessary to be able to avoid them.

Previous philosophy provided no help whatsoever on these issues. Even Kant–who, for those who followed him, saved the faculty of the understanding from earlier difficulties within the theory of knowledge–included everything which lies between pure sensations, on the one hand, and the formation of an idea of a spatially extended object perceived in time, on the other, in an act which he called an intuition (*Anschauung*). For him and for his followers, an intuition was solely and totally the effect of a natural mechanism which, apart from its end result, could not be subjected to any psychological investigation. (The end result, being simply an idea, could be treated like all ideas in accordance with certain formal principles.)

As soon as it came to be accepted that correct perceptions could be obtained through the senses, the subsequent path of investigation, that is, the inductive method of the natural sciences, was prescribed. The chief emphasis fell upon finding the natural laws of phenomena and upon the necessity of expressing them in precisely defined terms. The initial, only partially adequate attempt to formulate a natural law is rightly called a hypothesis. The observable consequences of such a hypothesis must be sought out and compared with the facts under the most varied conditions. Expressing an apparent or presumptive law in words makes it possible for it to be communicated to many people; thus many people can contribute to the testing of it. It also makes it possible, over an extended period of time, for a very large number of cases to be considered, so that along with the confirmations attention is drawn to real or apparent exceptions. Eventually such an imposing amount of observational data will be assembled that, at least in the areas where it has been exhaustively checked, no doubt can seriously be entertained any longer concerning the correctness of the law.

This is a long, arduous undertaking; and its success, I must emphasize once more, depends essentially upon the possibility of expressing the law in words which have precisely defined meanings. We are already able at the present time, however, to reduce large areas of natural phenomena, especially those involving the simpler relations of inorganic nature, to precisely defined, widely known laws.

Whoever comes to know the law of some phenomenon gains, not only knowledge, but also the power to intervene on appropriate occasions in the the course of nature and to compel nature to work according to his will and for his benefit. Moreover, he acquires an insight into the future course of the phenomenon. In truth, he acquires the power or ability which in superstitious ages was sought by prophets and magicians.

There is still another way, besides that of science, to acquire insight into the complicated workings of nature and the human spirit and to communicate this insight to others in such a way that they, too, will be fully convinced of the truth of what has been communicated. Such a way is provided by artistic representations. It should not be difficult to convince you that, at least in some forms of art, this can indeed be successfully accomplished. We shall have to discuss more fully later whether such success is restricted to particular branches of art or whether it can be achieved in all.

Think of some masterwork of tragedy. You see human emotions and passions develop and intensify, and finally they give rise to noble or horrible deeds. Throughout you know that, under the given circumstances and conditions, the events must occur just as they are presented by the poet. You believe that you, too, in a similar situation, would feel the same impulses and behave in the same way. You come to know the depth and power of feelings which are never aroused in the quiet of everyday life, and you leave the theater with a deep conviction of the truth and validity of the thoughts and emotions which have been represented, although at the same time you have never doubted that what you have seen was only a theatrical production.

The truth which you recognize is only the *inner truth* of the mental and emotional processes represented and the truth of what follows from them. It is their agreement with what you yourself have come to know about the development of such moods and emotions; it is the correctness of the representation of the invariant or lawful course of these states of feeling and mind. The artist must have had this knowledge—and the member of the audience, too, at least insofar as he recognizes it when it is presented to him.

Where does such knowledge come from, knowledge of regions in which the efforts of science have so far been least fruitful—that is, knowledge of the movements of the spirit, of personal traits and characteristics, and of the decisions and resolutions of individual men? It certainly has not been gained along the arduous paths of science nor by means of reflective thought. On the contrary, whenever an author begins to reflect and tries to assist himself with philosophical insights, the audience almost immediately become disenchanted and critical. They feel that some surrogate has appeared in place of the artist's living imagination.

The artist himself can tell us little about where or how various images come to him. Even the most able artists become aware only gradually, through the effects of their work, that they can do something which the majority of other men cannot do. Frequently they assign less importance to the kind of activity which led them to do the inexplicable than to subsidiary matters which gave them a good deal of trouble. In this connection, Goethe once told Eckermann that he thought he had accomplished something more important in the theory of color than he had in his poetry, and I myself once heard Richard Wagner say that he valued his verse much more highly than he did his music.

We do not know how to refer to this kind of mental activity, which occurs so effortlessly, so quickly, and with so little reflection, except by using the term *intuition* and speaking of a special *artistic intuition.* The concept of intuition, however, is almost always characterized in a negative way. According to the terminology of philosophy, it is the opposite of thought, that is, the opposite of the conscious relating of ideas which have already been conceptualized through the abstraction of similar properties. In this view, sensuous intuitions come to exist instantly, without reflection, and without mental effort, as long as the corresponding sense impressions affect us. No act of the will can oppose them; it seems to us that the perceptions of objects are determined completely by the sense impressions which produce them, so that the same impressions always excite the same ideas.

The artistic power of imagination, of course, does not always operate with immediate or existing sense impressions. It works much more, especially in the art of poetry, with memory images. In this respect, however, these are in no way different from immediate sensuous images. As I have already noted, earlier theories did not allow of analysis of sensuous intuition; they are therefore of no help to us in understanding artistic intuition.

We are certainly justified, however, in taking exception to the belief that both ordinary sensuous intuitions and artistic intuitions are totally independent of the influence of experience. And experience is the result of processes which fall within the domain of conceptual thought.

It should be noticed, first of all, that quite often extremely fast decisions or resolutions, which are not determined solely by the sense impressions being experienced at the moment, flash through our minds. These occur especially upon the sudden appearance of danger; they also occur, however, with the appearance of favorable opportunities that must be seized quickly. In general, all cases in which we praise someone's "presence of mind" are to be included here. As a rule, awareness of some danger is not the result of an especially compelling sense impression; rather, it is due to judgment based upon earlier experience. Thus there can be no doubt that the speed with which an idea comes to mind does not weigh in favor of its origin in a physiological mechanism, nor for its independence of the results of previous experience.

The other supposedly distinguishing characteristic of a sensuous intuition—that is, that the ideas of objects given by intuition are dependent only upon the sense impressions being experienced at the time—excludes, of course, the influence of immediate experience of variable or changing relations in the external world. It does not exclude, however, experience of invariant relations, those which are repeated again and again in the same way and which, if they should be part of a new sense experience, must take place there in the same way as all their predecessors. Obviously, all relations governed by strict natural laws fall within this category.

A shadow, for example, will fall upon a lighted surface only if the body casting the shadow lies on the same side of the surface as that from which the light comes. In any painted representation, therefore, one of the most important techniques for making the relative positions of objects intelligible is to plan carefully the shadows that are cast. It is easily possible with stereoscopic pictures to produce a situation in which the ideas (those based upon immediate sense impressions) of the positions of bodies in the background of the pictures and at various distances from the eyes are repressed because of an improperly placed shadow, so that the correct spatial intuition is not experienced.

In general, the influence of the laws of perspective, shading, masking of the outlines of distant objects, aerial perspective, and so on, on the spatial significance of our visual images is extremely great. This influence can be understood, however, only by reference to the effects of previous experience, even though it makes itself felt just as surely and just as quickly in an image as do the colors and the outlines of objects.

Thus, in my opinion, there can be no doubt that elements which are the result of experience are important in the formation of our ideas of objects in immediate sense perception. Physiological investigations of the relation of our perceptions to the sensations from which they originate provide hundreds of examples of this. To be sure, in individual cases it is often difficult to separate clearly what must be attributed to the physiological processes of the nerves and what has been contributed by experience of the invariant laws of space and of nature. I am inclined myself to assign the largest possible role to the latter.

Even the little that we now know about the laws of our memory allows us to imagine how such effects might come about. It is well known to all of us how the repetition of similar sequences of similar impressions strengthens their traces in our memory; this was the method we used during our school days in the rote learning of aphorisms, poems, and grammatical rules. Deliberate repetition is more reliable, but memory images are also strengthened without our intentionally doing anything. We have already mentioned that something which is repeated in the same manner without exception—and thus something which becomes fixed through repetition—is linked by a law of nature, that is, by a necessary chain of cause and effect, to other effects and events.

On the other hand, we may rightly conclude that the effects of sequences of events which result from the accidental combination of changing circumstances will fade from our memory and, in general, will finally disappear. It is just these accidental circumstances, however, which distinguish the specific instances of a lawful process from one another. When our memory of these circumstances fades, we lose the means of separating one case from another in our memory and of recalling them individually. We retain the knowledge of what is regular or invariant, but we lose sight of the particular characteristics of the cases from which our knowledge was derived. Finally we are unable to give an account, either to ourselves or to others, of how we arrived at such knowledge. We know only that things have always been so, that we have never seen them otherwise.

We are able to acquire this kind of knowledge of the most diverse objects and relations, beginning in childhood with the simplest spatial relations and with the effects of gravity, and obviously continuing with maturity. For an attentive observer with acute senses, this knowledge can be extended without limit, as long as law and order prevail in nature and among mental operations.

These reflections, which I have applied here first to the case of sensuous intuitions, can also be extended to artistic intuitions. True, they come effortlessly, they burst forth suddenly, and the possessor does not know whence they come; but it in no way follows from these facts that they do not contain results which are drawn from experience and memory and which thus can be embraced under a law. In experience and memory we can point to a positive source of the artistic power of imagination—one which, in contrast to the "free play of fancy" celebrated by the poets of the Romantic school, is also completely adequate to account for the rigorous consistency of even the greatest work of art.

Since artistic intuitions are not established in the same way as conceptual thought, it is not possible to define them in words. When one wants to stress this difference, one can only refer to the kind of knowledge of regular relations which originates in these artistic intuitions as *an acquaintance with the type of phenomenon* in question.

Just as the diversity of sense impressions is greater than the verbal descriptions which can be given of objects, so naturally artistic representations turn out to be much richer, finer, and more full of life than scientific descriptions. To this asset may be added the swift evocation of memory images, given a satisfactory starting point. It is therefore possible for an artist to present a listener or spectator with a great deal of material in a short period of time or in a picture of only moderate size.

As I wished to begin by emphasizing the fact that art as well as science can represent and convey the truth, I have restricted myself to the most prominent example, that is, the art of tragedy. It may perhaps be asked whether what has been said is valid for the other branches of art as well. It seems to me undeniable that an artist can be successful in his work only if he possesses a good knowledge of the lawful behavior of the phenomena he wishes to represent, as well as of their effects on a listener or spectator. Whoever has not yet come to know the finer effects of art, especially those of the pictorial arts, allows himself to be misled into thinking of absolute fidelity to nature as the most essential criterion of a good picture or statue. In this respect a well-made photograph would obviously be superior to all the drawings, etchings, and engravings of the greatest masters, yet we soon learn to recognize how much more expressive the latter are.

This fact is also a clear indication that an artistic representation is, not a copy of a particular case, but a representation of the *type* of phenomenon in question. Here we draw close to the very controversial question of the nature—indeed, the mystery—of beauty in art. We shall not, however, try to consider this fully today; we shall touch upon it only insofar as it is connected with our theme, which is the artistic representation of something true.

Through a consideration of the need for beauty and depth of expression, it becomes clear that there are other demands on an artist besides what can be accomplished by copying individual cases. These demands can be satisfied only by transforming the individual cases, while at the same time not deviating from the lawfulness of the type. The closer an artist's intuitive image is to the type, the freer he will be to respond to the requirements of beauty and expressiveness.

Often in a work of art this transformation goes so far that in accessory matters, fidelity to nature is intentionally ignored, provided that in this way a heightening of beauty or expressiveness can be achieved. As examples I need mention only meter and rhyme in poetry and the addition of music to the texts of dramas or songs. In contrast to the contents of poetry, the ordinary forms of language are external, unimportant, or even ugly accessories. They are entirely conventional and change with translation into another language. Rhythm and rhyme give some external order to these forms, along with a kind of musical movement in which delay, acceleration, or interruption can be quite striking. If, on the stage, we give emphasis to language by means of song, we destroy all fidelity to nature, but by the richer, finer, and more expressive movements of the sound we gain certain advantages in the representation of the movements of the spirit.

Since the question of how, in the widest circles of art, facility of expression coincides with the demands both of beauty and of the purest representation of types, has already been discussed so often and so exhaustively, I believe it is quite sufficient to mention it only briefly here.

In my book on the sensations of tone I have tried to indicate that in music, too, the more or less harmonic effect of intervals in melody and harmony is related to specific perceptible phenomena, the overtones, which limit the harmonic intervals all the more clearly and precisely in proportion to their simplicity and purity.

Investigations concerning the sensations of the sense of vision indicate that certain moderate degrees of brightness, those which are most comfortable to us in vision, favor both the perception of the finest distinctions in the modeling of spatial forms and the perception of the smallest objects. These investigations also indicate that a certain balance of colors is necessary if the eyes are not to be disturbed by colored afterimages.

Above all, we must not disdain the sensations of our senses as constituents of beauty. In the long course of generations, nature has so formed our bodies that we find pleasure in an environment in which our perceptual activity can develop in the freest and surest way.

I also consider the powerful influence of beauty on our memory a factor contributing to ready understanding and comprehension. Poetry is more easily remembered than prose. Nations which were not yet literate, or within which only a few people were able to write, naturally tried to preserve their history, sayings, laws, and moral rules in verse. One cannot forget a beautiful building, picture, or song. A melody can fix itself so strongly in our minds that we have trouble breaking free from it.

I believe that an essential part of the power of beauty is due to its effect on the memory. Even as we first begin to contemplate something beautiful, we come quickly to a clear idea of the whole. We then continue to look at and consider the individual parts in a quiet, peaceful manner, all the while feeling completely oriented concerning the relations of the parts to the whole.

We have now arrived at the point where the path of the scientist and that of the artist begin to separate. The fact that the artist's memory is more delicate and accurate for those phenomena which interest him, that is, for the particulars of experience, than is the memory of the majority of other men, can be shown by many examples. A landscape painter must be able to fix an image of rapidly fading light and shifting weather conditions firmly in memory. He must also retain in memory the most fleeting of moments, such as illumination due to the moon (by which it is impossible to paint) or the rolling of the waves of the sea, in order to bring us under his spell by a painting. What he can retain by means of quick preliminary sketches is really very slight; for the majority of things he must rely on his memory images of what happened.

The musician's memory appears even more astonishing to us. He knows how to reproduce countless compositions on his instrument without having a score in front of him. Most remarkable of all is the orchestra conductor who, without a score, can conduct countless symphonies, the individual notes of which must number in the millions. I believe I am not mistaken when I assume that what he has in his memory is, not the individual notes and rests, but rather the phrases of a composition, along with their order, their relationships, and the various changes of timbre. I believe, too, that conductors are able to elicit what they wish to hear with sureness and swiftness only insofar as they are able to bring to full consciousness their image of a complete score, since this is clearly necessary in order to give their musicians the correct directions.

An inclusive, accurate memory does not have the same importance in scientific as in artistic work, for what we can express in words we can also express in writing. Only the first creative thoughts, those which of necessity occur before any verbal expression, are formed and emerge in the same way in both kinds of activity. Indeed, such thoughts can occur only in the way an artistic intuition does, that is, as an insight into a new lawful regularity. Such a creative thought consists in the discovery of a previously unknown uniformity in the way some phenomenon recurs in a series of similar cases.

We use the word *wit* to refer to the power of discovering similarities which were previously unsuspected. Our ancestors used this word in a serious sense; it meant a sudden insight, one which could not be reached methodically through reflection, but which appeared rather as a sudden stroke of good fortune. For the same reason, the word for poet in ancient Latin was the same as that for seer. A sudden insight was considered to be like a divination, a kind of godlike gift.

At times a happy accident can also be an aid and can reveal some unknown relation. Such an occurrence can hardly be useful, however, if the person to whom it occurs has not already had experience sufficient to give him some degree of assurance concerning the correctness of an apparent uniformity. Goethe's story of his discovery of the vertebral structure of the head after coming upon the sheep's skull in the sands of the Lido at Venice appears to me typical of this kind of discovery. In one of his own versions of the story he mentions this incident as the initial discovery; in another it was only the confirmation of a truth already known.

I have now presented to you my basic thoughts concerning the relationship between science and art. In what is to follow we shall turn specifically to the work of Goethe. (Goethe was not the only artist who carried out scientific investigations. To mention only one other, there is Leonardo da Vinci; he was more interested, however, in practical problems in engineering and optics, areas in which he developed comprehensive insights.)

The region in which Goethe has gained the greatest fame and in which his merits are most easily and clearly evident is that of animal and plant morphology. Here he reached the conclusion that a common architectonic plan, one which is consistently carried out even in apparently insignificant details, underlies the bodily structure of various animal and plant forms. This was a subject which was especially open to the artistic mode of comprehension, and it was clearly fortunate that this conclusion was first reached and firmly held by artistic intuition. At Goethe's time scientific anatomy and zoology were prevented by a prejudice—a belief in the inalterability of organic forms—from moving in the direction in which he wanted to go and even from understanding his intuitions once they were explained.

Goethe himself, to be sure, did not know quite what to say concerning the origin and significance of these similarities in form. Characteristically, he wrote in "The Metamorphosis of Plants":

> All forms are alike and none is like another,
> And so the choir hints of a secret law,
> Of a sacred riddle. Oh, my lovely friend, would that I,
> With equal fortune, could give you the solution!
>
> *Alle Gestalten sind ähnlich und keine gleicht der andern*
> *Und so deutet das Chor auf ein geheimes Gesetz,*
> *Auf ein heiliges Rätsel. O, könnt ich Dir, liebliche Freundin,*
> *Überliefern zugleich glücklich das lösende Wort!*

It was Darwin who first found the solution, after he had freed himself from the prejudices of his predecessors. On the basis of the many, many well-known examples of transformations of forms under the hands of men who breed animals, Darwin showed that conditions similar to those deliberately produced by breeders act also on wild animals and that in the course of generations these conditions can effect important transformations in animal forms. I trust, however, that I need not pursue this subject in this assembly. It concerns one of the greatest revolutions in biology, one which has excited the most general attention and has been thoroughly discussed among educated and cultured people. I shall also forego discussing this subject further on the present occasion because one of the most active and adroit exponents of the theory of evolution [Ernst Heinrich Haeckel, 1834-1919] is at the university here. In any case, this part of Goethe's work has been discussed frequently and in great detail. In the recently published volume of the *Goethe-Jahrbuch* Professor K. Bardeleben has described all the poet's work in this direction.

In another region of natural science, the theory of color, Goethe's endeavors were less fortunate. I have already given a detailed report on the reasons for his failure in my essay on his researches in the natural sciences. Essentially, he failed because, with the comparatively inadequate apparatus he had at his disposal, he simply could not possibly have observed the most crucial facts. He never had completely pure, monochromatic light and thus would not believe in its existence.

Men like Sir David Brewster, who were much more experienced and skilled than Goethe in optical investigations and who were equipped with the best instruments, also foundered because of the difficulties in obtaining completely pure, simple, spectral colors. Brewster, too, formulated an incorrect theory of colors. Like Goethe, he maintained that it was not differences in the refraction of light rays which determine the colors of the spectrum; he maintained, rather, that there were three different kinds of light, red, yellow, and blue, each of which appears with all degrees of refraction. Brewster was deceived because he did not know that transparent bodies are always somewhat cloudy—a fact upon which Goethe built his entire theory—and that because of this cloudiness, a diffused light is cast over the observer's field of view.

In order to investigate certain phenomena described by Brewster which seemed to contradict Newton's theory, I was forced to employ more carefully purified colored light than any that Newton, Goethe, or Brewster had known. I finally achieved my goal, although not without difficulty. I believe, however, that it would not be profitable here for me to discuss more fully the defectiveness of Goethe's experiments, the sources of error which he overlooked, his misunderstanding of Newton's statements, and so on. This would provide only the smallest amount of significant new insight into this misguided research by the poet.

Goethe declared his firm conviction that one must search for an *archetypal phenomenon* (*Urphänomen*) in each of the branches of physics and then relate all other phenomena, in all their diversity, to it. Within physics the objections he raised were directed against the abstraction of cognitively empty concepts, upon which theoretical physics at his time tended to rely. Matter, for example, was held to be in essence without forces and thus without properties, yet in special cases it was considered the bearer of indwelling forces. These forces, moreover, if they were imagined to be separated from matter, had the supposed capacity to effect changes, yet they were without any points of contact for any possible kind of action. Goethe would have nothing to do with such inconceivable metaphysical abstractions, and one must admit that his opposition to them was not unjustified. These abstractions, although they had been used meaningfully and without contradiction by the great theoretical physicists of the seventeenth and eighteenth centuries, contained a seed of the most confused misunderstanding, which grew in muddled and superstitious heads. I need only mention the belief in animal magnetism; and in connection with the theory of vital forces, forces separated from matter also played a disastrous role.

At the present time, it should be noted, physics has taken the road upon which Goethe wished to travel. Any possible immediate consequence of his influence, however, was painfully vitiated by his incorrect interpretation of the examples he chose and by his bitter polemic against the physicists which ensued. It is very much to be regretted that he did not know the wave theory of light, which had already been formulated by Huygens. This would have given him a more legitimate and intuitively a more satisfactory archetypal phenomenon than the unsuitable process which he chose for this purpose, the colors of turbid media. (In external nature, to be sure, the phenomenon of turbidity is widely manifested. Both the blue of the heavens and the red of the evening sky are instances of it.) At that time Newton's corpuscular theory of light had, of course, to make many clumsy and artificial assumptions, especially in order to explain the polarization and interference of light which had just been discovered. Because of this, it has now largely been discarded by physicists, who have turned much more toward Huygens' wave theory.

Mathematical physics received the impetus for its change in viewpoint—from a stress on abstractions to a stress on phenomena—without any apparent influence from Goethe. The change was due primarily to Faraday, who was self-taught and, like Goethe, an enemy of abstract concepts, with which he was unable to work. His entire interpretation of physics rested upon his intuition of phenomena; he too tried to eliminate from his explanations everything which was not an immediate expression of observable facts. Perhaps Faraday's wonderful facility in discovering new phenomena is related to his freedom from bias and from the theoretical presuppositions of previous science. In any case, the number and importance of his discoveries was sufficient to guide others, at first the most able of his fellow countrymen, into the same path.

German scientists also soon followed. Gustav Kirchhoff, for example, began his textbook of mechanics with the statement that the task of mechanics is to "describe completely and in the simplest possible way the movements occurring in nature." In my opinion, what Kirchhoff meant by a description "in the simplest manner" is not far from Goethe's archetypal phenomenon.

The most prominent among earlier mathematical physicists, of course, were not far from the same approach. Newton and his contemporaries found great difficulty in representing forces which act at a distance through empty space, just as more recently Faraday and his followers have raised objections to this idea and have actually dismissed electromagnetic action at a distance from physics.

It is not difficult to express even the fundamental law of motion of the heavenly bodies in the way required by Goethe for an archetypal phenomenon, that is, so that it refers only to observable facts: *If ponderable bodies are present together in space, each one undergoes a continuous acceleration in its movement toward the others, the magnitude of the acceleration depending upon the masses of the bodies and their distances apart in accordance with Newton's law.* It is presupposed that the concept of acceleration has already been explained, as well as the meaning ascribed to the simultaneous existence and combination of several different velocities and accelerations. The masses of bodies, their velocities, and their accelerations are all observable, measurable phenomena. The law just expressed refers only to these, yet it contains the kernel out of which all that part of astronomy which deals with the movements of the stars can be developed. You can also see, however, that on the whole such a form of expression seems awkward and artificial.

According to Newton himself, in expressing his fundamental conception of the law of gravitation, anything that goes beyond a phenomenon is introduced only as an *image.* For him, the heavenly bodies move *as if* they were being drawn toward one another by a force of attraction of a certain magnitude. Goethe used the word *image* in a similar way but in a favorable sense when, in his history of the theory of colors, he discussed the opinions of Roger Bacon, the English monk. Any emphasis on the old scholastic assumption of some kind of similarity between cause and effect, which more recent science no longer recognizes, was avoided, of course, in Goethe's presentation.

It is clear that Schiller, too, had the insight necessary to express the law of gravitation. He wrote in the poem "The Walk" ("Der Spaziergang"):

> The wise man
> Seeks for the familiar law amidst the awesome multiplicity
> of accidental occurrences,

Seeks for the eternal Pole Star amidst the constant flight
of appearances.

Der Weise
Sieht das vertraute Gesetz in des Zufalls grausenden Wundern,
Sucht den ruhenden Pol in der Erscheinungen Flucht.

A law of nature actually has still another significance. It is not only a guide for our empirical understanding; it also rules the course of some natural process, whether we are aware of it or not and whether we wish it to or not—indeed, quite often against our wishes and to our sorrow. We must recognize it as the expression of a power which is ready to act at any moment when the conditions for its appearance are realized. In this sense, we consider it a *force,* and since it is ready and able at any moment to produce some effect, we ascribe *existence* to it. On this fact, in my opinion, also rests the conception of a force as the *cause* (*Ursache*) of the changes which occur under its influence; it is *the existent which lies hidden behind the changes in phenomena.*[1]

All these conceptual or linguistic expressions have their full justification as long as they refer to matters of fact open to observation. Moreover, if they are used rightly, they are to be preferred, for these abstract forms of expression permit much more concise statements than does a description of an archetypal phenomenon formulated hypothetically. The fact that the use of abstract concepts can result in the most fantastic nonsense issuing from the mouths of ignorant people, people who do not know the basic meaning of an expression, is not a peculiarity of theoretical physics.

Naturally, it would be a mistake to believe that a deeper insight into the nature of things can be gained through these abstract expressions. Goethe observed in his *Poems in Prose*:

> If at last I let my mind come to rest with some archetypal phenomenon, it is nevertheless only resignation. There is a great difference, however, in whether I resign myself at the limits of human endeavor or within some hypothetical limitation of my own narrow individualism.

Again:

> The immediate awareness of some archetypal phenomenon places us in a kind of anxiety. We feel our inadequacy. Only through the eternal play of the empirical are we enlivened, do we feel joy.

While we must acknowledge to the fullest the poet's healthy sensibility

1. The term *Sache* corresponds in meaning to the Latin *res*, from which the terms *real* and *reality* are derived. Here it denotes *the enduring, the effective.*

and deep insight on this point, we must not overlook the fact that in what he sought to achieve in the theory of color there were certain gaps which would not have been tolerated in scientific treatment of the subject. In his *Theory of Color* he discusses frequently and in detail how, according to his opinion, blue and yellow light originate. He always worked with the images of bright and dark surfaces. In his judgment, if these images shifted relative to each other, the light of one would pass through the other, and the latter would affect the transmitted light in the same way that a turbid medium does (a conception, incidentally, which places a serious demand on the reader's imagination). Nowhere, however, does he discuss how, according to his ideas, blue and yellow light are to be distinguished from each other. It suffices for him to state that both come to be somewhat shadowy due to their passage through some substance. He evidently does not feel obligated to state how to distinguish the shading in the blue from that in the yellow, and that of both from their mixture, which he considered green to be.

It is just at this point that Newton's theory, and still more Huygens' wave theory, gives the most detailed explanations, which can be established by means of the most precise measurements. These theories, moreover, applied in astronomy, have led to the determinations of the orbits of the most distant double stars, which was once believed impossible. (It is, of course, the number of light vibrations in a given period of time which determines color, just as it is the number of sound vibrations in a given period of time which determines pitch.)

Obviously an optical image, which gives rise to a clearly formed intuition of a material object or visual field, was in the last analysis what could be represented intuitively; it was therefore the limit of Goethe's interest. The means by which such sensuous intuitions were obtained, on the other hand, was of little interest to him. He was very vague about how the sensations produced in the eye are related to the objective agent, that is, the light, which produces them and whose presence and nature is in turn indicated by the sensations.

These issues were strongly emphasized by his friends. He reports that at their urging he studied Kant and found much stimulation in the *Critique of Judgment,* on which he found himself in close agreement with Schiller. He obviously could not be completely sympathetic to the *Critique of Pure Reason:*

> I gladly applaud all my friends who agree with Kant that, while all our knowledge originates in experience, it does not all arise out of experience.

> The entrance pleased me, but I could not enter the labyrinth; sometimes I was hindered by the gift of poetry and sometimes by common sense; nowhere did I feel myself enlightened.

The aesthetic impression which Kant's world of the *ding an sich* made on him he depicted with unmistakable irony upon the occasion of Faust's journey to the Mothers:

> No Space around them, Place and Time still less;
> Only to speak of them embarrasses.
>
> Naught shalt thou see in endless Void afar,—
> Not hear thy footstep fall, nor meet
> A stable spot to rest thy feet.[2]
>
> *Um sie kein Ort, noch weniger eine Zeit,*
> *Von ihnen sprechen ist Verlegenheit.*
>
> *Nichts wirst Du seh'n in ewig leerer Ferne,*
> *Den Schritt nicht hören, den Du tust,*
> *Nichts Festes finden, wo Du ruhst.*

Physiological investigations of the sense organs and of their activities have at last produced results which agree in essential points (at least, those which I consider essential) with Kant. Indeed, in the region of physiology they provide the most obvious analogies to Kant's transcendental aesthetic. From the standpoint of the natural sciences, however, objections must be raised to the line which Kant drew to separate the facts of experience from the a priori forms of intuition. We may assume that Goethe would not have felt himself hindered by what he called "common sense" from joining us in insisting on the necessity of shifting this line, so that the fundamental propositions of the theory of space will also be included under the facts of experience.

Forms of intuition, such as those which Kant sought to establish for the total sphere of our ideas, are to be found also for the perceptions of the individual senses. The optic nerves, for example, experience everything to which they are sensitive in the form of sensations of light in the field of vision. It need not be external light, however, which excites them. A blow, pressure on the eyes, electric currents passing through the head, changes in blood pressure—all can produce sensations in these nerves. In all these cases, however, the sensations aroused are sensations of light, and they produce the same impressions as those originating in external light.

---

2. *Faust, Part II.* Translated by Bayard Taylor. All subsequent translations from *Faust* in this chapter are by Taylor.

Similarly, a blow, pressure, a strain, or an electric current can excite the skin, in which case we experience them as tactile sensations. Indeed, the rays of the sun, which appear to the eyes as light, arouse the sensation of warmth in the skin. Depending upon whether the tongue or the ear is excited, we can arouse taste or auditory sensations by means of electric currents.

The proposition, much discussed in recent years, that the most fundamental distinctions among our sensations are dependent, not upon the means of excitation, but only upon the sense organs which are excited, also follows from these facts. We give recognition to the basic nature of these distinctions when we speak of man's five different senses. No comparison whatsoever is possible among the sensations of the different senses; there are no relations of similarity or dissimilarity among them. The fact that we see an object as a colored visual image depends only upon our eyes, although the specific colors we see depend upon the kind of light which it transmits to us. This general principle was established by Johannes Müller, the physiologist, and is called the law of the specific energies of the senses. Careful, specific comparisons of the qualities of sensations of just one sense with the qualities of the means of excitation, moreover, show that similar color impressions can result from different mixtures of light and that these impressions do not necessarily correspond with any physical property of light.

I have therefore come to believe that the relation between sensations and their objects must be so formulated that sensations are thought of only as *signs* of the actions of objects. In order to have a sign relation it is only necessary that the same sign always be given for the same object. No kind of similarity is necessary between a sign and its object, just as no similarity is necessary between spoken words and the objects we designate by them.

We cannot call sense impressions "images," for an image represents like by means of like. In a statue, bodily form is represented by means of bodily form; in a drawing, the perspective appearance of objects is represented by means of a similar total visual image; in a painting color is represented by means of color. Sensations can be images only of temporal relations in the course of events. Numbers, too, should be included under the heading of temporal relations; more can be accomplished with them, however, than can be accomplished with mere signs.

Goethe knew a good bit about various subjective visual phenomena; some of them he discovered himself. At best, however, he came to know the law of the specific energies of the senses only in an incomplete form through the study of Schopenhauer. He rejected what Kant and the elder Fichte presented on this subject because it was related to other matters which were unacceptable to him. How astonished we are, then, when, at the conclusion of *Faust,* we find the thought of the Holy Spirit, which beholds the eternal truth directly, sketched in the words of the Chorus Mysticus as follows:

All things transitory
But as symbols are sent.

Alles Vergängliche
Ist nur ein Gleichnis.

That is, what occurs in time and what we perceive through the senses, we know only in symbols. I hardly know how more pregnantly to express the final result of our physiological theory of knowledge.

Earth's insufficiency
Here grows to Event.

Das Unzulängliche
Hier wird's Ereignis.

All knowledge of the laws of nature is inductive, and no induction is ever totally complete. We feel (sharing the poet's confession quoted above) our inability to penetrate further as a kind of anxiety. The event which is about to occur finally justifies the results of earthly thought.

The Indescribable,
Here it is done.

Das Unbeschreibliche,
Hier ist's getan.

The indescribable, that is, what cannot be grasped in words, we know only in the form of artistic representations, only in images. For the blessed, it is reality.

With this, the poet's epistemological point of view reaches its culmination, and the concluding stanza turns toward a higher region. The final lines point to the elevation of all spiritual activity in the service both of mankind and of the moral ideals, which are symbolized by the Eternal Feminine.

The deeper we seek to penetrate into the internal workings of the poet's thought, the weaker become the clues which he provides for us to follow. If we ourselves have been led to the same goal as he, however, we will surely be aware of where the connecting links fail and where the whole seems doubtful.

Faust escaped from the unhappy position of having thought and knowledge turned in upon itself, from the position where he could not hope to gain the full possession of the truth and did not know how to grasp reality, to the *deed.* Before he made the pact with Mephistopheles, Goethe placed him—obviously with the intention of preparing for certain developments in Part Two—in the scene in which Faust undertook to translate the Gospel According to St. John. Faust stumbled over the frequently discussed concept of the Logos: "In the beginning was the Word." The word is only a symbol for its meaning. The meaning of a word is a concept or, if it refers to occurrences, a law of nature. A law of nature, as we have seen, if it is to be thought of as existing and causally effective, must be considered a force. In this transition from *word* to *meaning* to *force,* which Faust made in his translation, lies an interrelated reformulation of a concept. But *force,* too, did not satisfy Faust, and so his thought made a huge leap:

> The Spirit aids me: now I see the light!
> "In the Beginning was the *Act,*" I write.
>
> *Mir hilft der Geist, auf einmal seh ich Rat*
> *Und schreibe getrost: Im Anfang war die Tat.*

The reference in the Gospel is, above all, to the fundamental nature of the Creative Spirit. Faust, however, sought his own peace of mind and found hope for it in these thoughts, which filled the diabolical poodle with the deepest displeasure. Thus I do not believe that Goethe wished to present Faust as stirred only by a theoretical interest in the act of world creation, but wished more to show his subjective drive for the truth.

The epistemological or theoretical counterpart of this scene lies in the fact that the difficulties of the philosophical schools, as they try to justify our belief in the existence of an external reality, will remain unresolved as long as they start with only passive observation of the external world. These schools cannot extricate themselves from the world of appearances as long as they fail to recognize that the actions which result from our wills form an absolutely necessary part of our basic experience.

We have seen that our sense impressions are only a language of signs which inform us about the external world. We must learn to understand this system of signs, and this understanding comes only as we observe the consequences of our actions and learn to distinguish between those changes in our sense impressions which follow our acts of will and those which occur independently of our wills. I have discussed this elsewhere ("The Facts of Perception") and have shown how in this way we come to a knowledge of reality. It would lead us too far into the most abstract circles of thought to pursue this subject here. It must suffice to note that a theory of human knowledge based solely upon the physiology of the senses is inadequate; men must act in order to be sure of reality.

I must mention still another allegorical figure of Goethe's, the Earth Spirit in *Faust.* (I have already referred to him on an earlier occasion.) His words, in which he sketches his own being, correspond so completely to another of the most recent theories within the natural sciences that it is difficult to rid oneself of the thought that it is really meant. The Spirit says:

In the tides of Life, in Action's storm,
A fluctuant wave,
A shuttle free,
Birth and the Grave,
An eternal sea,
A weaving, flowing
Life, all-glowing,
Thus at Time's humming loom 't is my hand prepares
The garment of Life which the Deity wears!

In Lebensfluten, in Tatensturm
Wall ich auf und ab,
Webe hin und her.
Geburt und Grab,
Ein ewiges Meer,
Ein wechselnd Weben,
Ein glühend Leben:
So schaff ich am sausenden Webstuhl der Zeit
Und wirke der Gottheit lebendiges Kleid.

We now know that the universe has an indwelling store of energy, or motive force capable of doing work, which can neither be increased nor decreased. It can appear in the most manifold, ever-changing forms—now as a raised weight, now in the energy of moving masses, now as heat or chemical affinity, and so on. We know too that these changes in form occur in the realm of living beings, as well as in that of lifeless matter.

The germ of this insight into the constancy of the total amount of energy was already present in the eighteenth century, and Goethe could well have been quite familiar with it. Comparison with other contemporary (1780) statements of his, however, makes it more probable that the Earth Spirit should be thought of as the precursor of organic life upon the earth, as the words "*Ein glühend Leben*" clearly suggest. These two interpretations do not necessarily conflict with each other; both Robert Mayer and I were led to the generalization of the law of the conservation of energy through considerations of the general character of vital processes.

To be sure, we can no longer restrict the constant store of energy to the earth but must include at least the sun. A presentiment of a poet, however, does not have to be exact in all particulars.

In conclusion, we may perhaps gather the results of our reflections together in the following way: When he was dealing with problems which could be resolved through the intuition-images of poetic divinations, Goethe showed himself capable of the highest achievements. When only well-established inductive methods could help, he foundered. When he was concerned, however, with the highest questions of the relations of knowledge to reality, his healthy hold on reality guarded him from going astray and led him confidently to insights which extend to the limits of human knowledge.

19.

THE ORIGIN AND CORRECT INTERPRETATION OF OUR SENSE IMPRESSIONS [1894]

An article published in the Zeitschrift für Psychologie und Physiologie der Sinnesorgane, *VII (1894), pp. 81-96*

EARLY philosophers and psychologists usually subsumed all perceptual images (*Wahrnehmungsbildern*) which occur instantaneously, without reflection, and in the same way with every individual, under the concept of *perception* and interpreted them as immediate products of the structure and functioning of the nervous system. The possible assistance of the so-called lower psychical processes, such as the various memory processes, was completely neglected.

Our understanding of our native language is, however, an instructive example of how the normal meanings of frequently repeated perceptions can come to mind with lightning quickness, without the least reflection, and yet with irrevocable certainty. This understanding is *not* innate in us. Each of us has unquestionably *learned* his native language–learned it by using it, that is, by frequently repeated experiences. Children of our nationality, if they are born outside the frontiers of this country and raised among people speaking a foreign language, will learn that language and will become just as confident in its use as we are in using ours. The fully developed language of a civilized nation is, of course, such a richly developed, flexible means of expressing the most varied and finest shades of thought that, in this respect, it may be compared favorably with the great diversity of the forms of objects which we see around us.

The example of language is also instructive in another respect, for it indicates how an individual can acquire a consistent, reliable understanding of a system of signs which must seem to him completely arbitrary (even if comparative philologists do know how to recognize patterns of interrelations among its individual roots). A native language is learned only through the use of words. A child hears the common names of objects mentioned whenever the objects are shown or presented to him, and he hears the same perceptible changes in the external world referred to always with the same names. In this way, the oftener and the more consistently each such experience is repeated, the more firmly the word is fixed to the object in his memory.

The repetition, moreover, need not always be the same in all respects. A single name may also be attached to a class of objects which are only similar, or to a class of similar processes. In this way, the names of the ideas of classes of intuition-images are learned, as well as the areas in which names for various modifications of the ideas are customarily applied. All this is done solely through the use of language; only in exceptional cases is it supported by conceptual definitions.

Because of these processes, which we are familiar with from daily experience and which are repeated as we master the vocabulary of any foreign language we may later learn, we know that the more frequently we repeat or apply words or hear them applied, the more fully their meanings are imprinted. In the beginning the individual cases in which we heard them applied are retained in our memory. Later, however, when the number of these cases gets larger, so that we can separate them in our memory from the accidental circumstances and the chronological order in which they occurred, there remains only the final result of our experience—the fact that specific words are used to signify specific sets of similar objects or processes. We can no longer, however, give an account of the individual cases by which we have come to this knowledge. Moreover, we do not know why we sometimes use one modification of a concept but hesitate to use another.

I conclude from these reflections that by the frequent repetition of similar experiences we can indeed succeed in establishing and fixing more and more firmly an invariant connection (that is, one which always recurs) between two different perceptions or ideas such as the sound of a word and a visible or sensible intuition. These need have no natural connection with each other, nor need we remember the specific instances by which we came to this knowledge, that is, the individual observations upon which it is based.

Finally, we find that, not only with our own native language, but also with a fully mastered foreign language we can attain a degree of comprehension such that we are able to understand in a moment, without reflection or deliberation, the meaning of what is being said to us, and we are able to follow the subtlest, most varied shadings of thought and feeling. If, however, we tried to say how we acquired this knowledge, we could explain it only in the form of a general proposition: we always found that certain words were used in certain ways.

We know, however, as a general principle of the operation of our memory, that impressions which are repeated very frequently in the same way and which under similar conditions are always linked in the same combination leave permanent traces of themselves and their combinations behind. We know, too, that these combinations recur much more surely and quickly to consciousness than do those which result from changing or accidental combinations.

The same principle is evident in a very large number of other cases. Among the most universal and exceptionless of these are combinations of observed phenomena resulting from laws of nature, which require either their simultaneous occurrence or their uniform sequence in a specific period of time. While simultaneous occurrences or particular sequences which are the result of lawless accident will, to be sure, be repeated occasionally, they will not be repeated without exception. Between these extremes fall cases involving changing or even contradictory consequences, which work against and ultimately rule out the exclusive predominance of any one combination, so much so that the variability—or that which, in different occurrences of the processes, is not the expression of a conformity to law—is fixed even more surely and inalterably than that which is the result of a law.

When we learn a language, what appears to us lawful or uniform is only a rule chosen and followed by men, a rule to which we cannot attribute the constancy and inalterability of a law of nature. Moreover, words for very similar objects are not necessarily similar to one another; on the contrary, they may show the most complete lack of order and the most astonishing differences. Our observations of the behavior of objects, with respect both to one another and to our organs of sense and movement, are incomparably more numerous and self-consistent. Thus it is no cause for wonder if we come to a much more complete knowledge of the ways in which objects are normally related to one another, and of how they appear in different positions and during different movements, than can possibly be reproduced in language.

Language is much too poor for the exact description of the many sense impressions which even a single object, especially one of somewhat irregular or complicated form, affords the eye and the hand. To describe such impressions in words, moreover, would be an enormously lengthy, time-consuming occupation—one which we obviously need not carry out if we have had the intuition-image of the object impressed upon us. Where this has been done, as well as in cases where no verbal description is possible, the sense impressions without any verbal expression are quite sufficient. With their help we can recognize even the most subtly varied phenomena, such as the features of a human face, sometimes after a very short initial observation and a very long interval of time.

There can be no doubt that in such cases we retain in our memory the sense impressions which an object has made upon us—in this case, with sufficient individuality to be able many years later to distinguish with certainty a particular human face from those of all other men.

If we retain such an intuition-image of an object, an image given only through sense impressions, we usually say that we are acquainted with (*kennen*) the object, in contrast to having knowledge (*wissen*) expressed in language. Such an acquaintance need not be restricted to single perspective images; it can encompass and unite the totality of perspective images which can be experienced one after another by observations from different points of view. We do, in fact, retain ideas of the material forms of well-known objects which represent such a totality of perspective images. Being acquainted with the material form of an object, we are able to represent clearly in our minds all the perspective images we expect to see when we look at it from different sides; and we are startled if an image we actually see does not correspond to our expectations, as can happen, for example, when a change in the form of an object accompanies changes in its position. Consider, too, how extraordinarily sensitive an attentive observer is to mistakes in a drawing of a man or a horse, or to small errors in perspective drawings which are meant to represent regular architectonic structures. Indeed, it frequently happens that one notices a rather small error in perspective before noticing an error of the same size in a right angle forming part of the same drawing.

The material form of a solid object provides a greater variety of lawful, or invariant, relations among its various parts and dimensions than do any individual perspective images of the object. From these relations, changes in its perspective images with given changes of position can be more surely deduced than from just the impressions of a spatially extended image encompassing the totality of all individual external views. A single perspective view does not provide the necessary data for a completely definite, unambiguous idea of the total form of an object, along with its different appearances from different sides. Thus in this region, too, ideas based upon simple, invariant conformity to law prove to be those which provide the most certain intuitions.

These facts and relations stand out very clearly when we consider stereoscopic pictures. Specifically, if one looks at a pair of stereoscopic pictures of figures which are somewhat complicated in form, such as a regular polygon or the model of a crystal, the attempt to bring the three-dimensional form to unity out of the two pictures often fails at first because the two eyes do not glide easily along on either of the pairs of corresponding lines. The eyes shift before the correct idea of the object's form has been obtained. When the form has finally been perceived, however, the two lines of vision move over all parts of the figures with the greatest speed and assurance. Thus the law that one must follow both lines of vision continuously in order to grasp the complete form of an object and that the eyes must remain on corresponding points of the two drawings, holds good here.

How young children first acquire an acquaintance with or knowledge of the meaning of their visual images is easily understood if we observe them while they busy themselves with playthings. Notice how they handle them, consider them by the hour from all sides, turn them around, put them into their mouths, and so on, and finally throw them down or try to break them. This is repeated every day. There can be no doubt that this is the school in which the natural relations among the objects around us are learned, along with the understanding of perspective images and the use of the hands.

Observation of young children also indicates that they do not have this knowledge during the first weeks of their lives. If they had been born with any sort of instinctive knowledge, one would expect that it would be knowledge, first of all, of the form of the mother's breast and of the movements by which they can turn toward that visual image. There is obviously no such knowledge. One sees an infant become animated when it is brought into a position to nurse and, seeking restlessly, turn its head this way and that in order to find the breast. During the first few days, however, the child turns away just as often as toward it, even though the breast can easily be recognized. At this early age the child knows neither how to interpret its visual images nor how to direct its movements.

Similarly, if we hold the flame of a candle before a child one or two weeks old, we see the child become restless and turn its eyes here and there, clearly with the intention of staring at the bright flame. As soon as it has found the correct position of its eyes, it follows the slower movements of the flame with its gaze. In the beginning a child does not know how to turn its gaze confidently toward a flame somewhat to the side of its visual field. After two or three weeks, however, it succeeds in doing this fairly quickly. Only much later is it successful in grasping an object with its hands.

I conclude from these facts that the meaning of some of the simplest, most important visual images for a human infant must be learned. They are not given to the child, a priori and independently of experience by some innate mechanism. We need not go here into the extent to which a similar conclusion may be drawn concerning newborn animals. The mental activities of animals are probably limited by instincts to narrow paths. Young animals can conduct themselves more surely in restricted ways, ways that would not serve for the later development of human beings who have the power of choice.

I would not have brought up these matters, which have been discussed so fully before, if there were not a very widespread and stubborn prejudice to be met with here. It is a prejudice which, it seems to me, has its origin in an incorrect interpretation of the concepts of intuition (*Anschauen*) and thought (*Denken*).

The term *thought* should rightly be applied only to those combinations of ideas for which a person is able to formulate explicitly the individual propositions from which inferences are drawn, to verify their reliability, and finally to connect them consciously in making the inferences. The term *intuition,* on the other hand, should be used to designate the occurrence of ideas wherein only a sense impression is experienced, the idea of an object coming subsequently to consciousness without the mediation of any further conceptualization. According to this distinction, we must be clearly conscious of the intermediate steps in an inference if they are to come under the scrutiny of logic (taken in the narrow sense), which tests to be certain that the premises have been used correctly and that the reasoning results in valid inferences. It is necessary, too, for all the premises of an inference to be brought to consciousness and formulated with complete clarity. Any member of a chain of ideas for which a conscious formulation of this kind is not available is not a part of a logical proof; at most it is only an axiomatic premise selected without question from memory.

It would obviously be false, however, to try to maintain that we have no knowledge other than that which is developed from sense perception by logical or conceptual thought. Indeed, the example mentioned above of learning a language, as well as learning skills and learning to understand changing visual images, shows that knowledge can be gained without deliberate reflection. It can, in fact, reach the highest degree of certainty and refinement, without any possibility of testing the validity of the inductions later by recalling the time and place at which particular observations were made. These observations, moreover, may have been such that, on the whole, they cannot be adequately described in words. They can perhaps be reproduced with complete accuracy only by evoking the memory of some previous sense impressions.

Thus the memory images of pure sense impressions can also be used as elements in combinations of ideas, where it is not necessary or even possible to describe those impressions in words and thus to grasp them conceptually. A large part of our empirical knowledge of the natural relations among the objects around us obviously originates in this way. The blending of the many perspective images of an object into the idea of a three-dimensional form seems to me an especially clear example of the kind of combination of sensual intuitions which corresponds to an inference. The idea of the material form represents or stands for all the perspective images, which in turn can be derived from it by a sufficiently powerful geometric imagination. Even views not previously perceived, such as those which result when cross-sectional cuts are made in any one of a number of directions, are derivable from such an idea. Indeed, the idea of a three-dimensional figure has no content other than the ideas of the series of visual images which can be obtained from it, including those which can be produced by cross-sectional cuts.

In this sense, we may rightly claim that the idea of the stereometric form of a material object plays the role of a concept, formed on the basis of the combination of an extended series of sensuous intuition-images. It is a concept, however, which, unlike a geometric construct, is not necessarily expressible in a verbal definition. It is held together or unified only by the clear idea of the laws in accordance with which its perspective images follow one another.

I have sought to prove that an effortless intuition of the normal sequence of lawfully related perceptions can be attained through experience, provided, of course, that the experience is rich enough. This process—which, insofar as we can tell, is the result of the involuntary, unconscious action of our memory—is nevertheless able to produce combinations of ideas whose results are in all essentials in agreement with the results of conscious thought. As has already been mentioned, the more frequently the impressions which we receive through our senses are repeated in a similar fashion and in the same sequence, the more they are strengthened. Those impressions, on the other hand, which occur accidentally will fade and, as a rule, will finally be extinguished, unless they are emphasized and deepened through some special effort.

As has also been noted, the aspects of the pérceived phenomena which will be strengthened in this way with time are those which correspond to the effects of a natural law of the observed processes. The idea that some pattern of events will continue to occur in the same way in which it was initially perceived is all the more firmly fixed, the more frequently and without exception we have experienced it.

The expectation we come to feel is similar to the conclusion of an inductive inference, and such an inference may mislead us if it is not based upon an adequate number of observed cases. Animals make the same kind of inductive inferences, and they make more incorrect ones than we do. This can often be seen in their behavior, as when they draw back from some object which is similar in appearance to one by which they have earlier been burned.

In the first edition of my *Handbook of Physiological Optics* (1866) I referred to instances of this kind of inductive inference, which are based upon acquaintance with the uniform, or lawful, behavior of the objects around us, as *unconscious inferences.* At least to a certain extent, I still find this term admissible and appropriate, for on the whole these associations of perceptions in memory do occur in such a way that they are not noticed when they take place, or are noticed only the way any familiar process is noticed as one which has been seen before. At the most, when something happens which has been only infrequently observed, our memory of earlier cases, including the conditions under which they occurred, may be partially conscious. The mental processes then have a stronger analogy to conscious thought.

Inductive inferences are never as reliable as valid deductive inferences executed consciously. There is a clear distinction between conscious scientific knowledge and the kind of knowledge gained unconsciously through the accumulation of experience of objects and processes. With scientific knowledge it is possible to make a complete survey of all the instances relevant to some inference, either by examining written reports, or by direct observations, or if necessary by experiments (that is, observations of instances produced deliberately). With experiments it is advisable to try to produce instances which, insofar as their circumstances are concerned, are different from all those previously observed. Knowledge gained through daily experience, with all its accidents, does not usually have the range and completeness which it is possible to obtain with experiments. Such knowledge may be equally reliable only when it concerns processes of which there are an enormous number of instances with comparatively little variation or change.

We usually refer to incorrect inductive inferences concerning the meaning of our perceptions as *illusions of the senses.* For the most part they are the result of incomplete inductive inferences. Their occurrence is largely related to the fact that we tend to favor certain ways of using our sense organs—those ways which provide us with the most reliable and most consistent judgments about the forms, spatial relations, and properties of the objects we observe. In vision, for example, we usually bring objects which excite our interest into focus on the points of the two retinas which permit the most precise vision; then we run our gaze over the outstanding points and lines of the objects. In this way we learn their characteristics, while at the same time protecting our eyes against the formation of disturbing afterimages.

There are a large number of normal regularities in the movements of the eyes—regularities which are not based upon any automatic mechanisms in the muscles or nerves but which can be varied at will by any observer, provided he has learned to give the appropriate innervations. This proves that the normal, regular movements of the eyes are the results of habit, not automatic responses built into the organization of our bodies. Such habits are, of course, often very deeply rooted and not at all easy to overcome. Movements which vary from the normal require decidedly more effort and tire us more quickly. (This is a general characteristic of unusual movements of any of our muscles—that is, movements produced by inappropriate innervations, or movements which work against others and thus cause more strain than do those which are common and the result of extensive practice.)

Unusual perceptions, concerning whose meaning we have no trained knowledge, occur with unusual positions and movements of our sense organs, and incorrect interpretations of these perceptions may result. We can, in fact, lay down the general rule that with abnormal positions and movements of the eyes, the intuitions which occur are those of the objects which would actually have to exist in order to produce the same perceptions under the conditions of normal vision. The intuitions which result when light rays deviate from a straight line before they strike the eyes, as in reflection and refraction, fall under the same rule. In these cases, however, the illusions are recognized as such. To repeat: the image we experience is always that of an object, or an apparent distribution of light in the field of vision, such as would have to exist in order for objective light to produce the same visual image under normal conditions.

The degree of deception in such illusions can vary greatly. Think, for example, of the images reflected by a good plane mirror which hangs on a wall in such a way that it is impossible to see behind it. Such a mirror gives one of the most perfect optical illusions that can be imagined, yet even animals are seldom led into making mistakes by such mirror images. Children, to be sure, will glance for a moment, if they can, behind a mirror, and they will amuse themselves with the movements of the images. They realize rather quickly, however, that they are perceiving an image which does not correspond with reality. They also soon learn to accept a mirror image as their own portrait. To preserve the illusion for even a short period of time, it is necessary to mask carefully or hide the edges of a mirror in which an individual sees himself.

Most other illusions of the senses are also quickly recognized as such. Most observers know how to change an unusual kind of observation into one that is common and normal, thus causing the illusion both to be recognized and to disappear. Only when there is no time to make this change may an actual mistake be made for a moment. This may occur, for example, in connection with the flash of light caused by a blow on the eye. The majority of illusions are recognized by noticing that an image does not correspond completely with reality and by comparing it with images which different objects produce under conditions of normal vision.

One can fix or retain the particular form of an illusion in memory only by describing (to himself or some other person) the object which would really have to be there to produce a similar image with normal vision. The expression, "I see what looks like . . .," is the correct way of describing this perception. Only with the resumption of normal perception will one be completely clear in his own mind about the illusion which has been experienced.

In the case of a subjective phenomenon, whose cause is located at a fixed place on the eyeball, its movement with the eye when the gaze is shifted is a sign which is very quickly understood; the phenomenon is thereby recognized as subjective. However, since our interest is primarily in understanding the external world, we ordinarily turn our attention from subjective appearances as soon as they are recognized as subjective. Indeed, there is usually some difficulty even in observing them and directing our attention toward them. This difficulty, moreover, increases to a high degree with the increased sensitivity of those places on the retina which are not overexposed to light, compared with those places where sensitivity is decreased by continuous exposure. It is chiefly to this process that we can trace the gradual disappearance of strong stationary images in the eyes.

Difficulty in concentrating our attention on a specific part of a perception plays a role here, although deliberate effort seems to have a definite effect. I refer to the experiment which I have described earlier (section 28 of the *Handbook of Physiological Optics*) involving the momentary electric illumination of a completely darkened field upon which a sheet of paper with large printed letters has been placed. Before the electric discharge, an observer is aware of nothing but a moderately lighted pinhole in the paper. This point is firmly fixed in vision and serves for general orientation in the direction of the darkened field. The electric discharge illuminates the printed sheet for only a moment, during which time the image of the paper is visible. The image then remains visible as an afterimage for a very short period of time. The duration of the perceptibility of the image is limited, in effect, to the duration of the afterimage.

Eye movements of any measurable magnitude cannot occur during the flash; and during the short period of the afterimage, its position on the retina cannot change. I found it possible, however, to plan in advance the part of the darkened field to one side of the bright pinhole (upon which attention was continuously fixed) which I would perceive in indirect vision; and with successive electric illuminations I was able to recognize groups of letters in every part of the field, though with some gaps among them. As a rule, I saw more letters with stronger illumination than with weaker. I could not, however, perceive the letters on the periphery of the field, nor could I always perceive those near the point of fixation. With each new electric spark I could, while always focusing on the pinhole, direct my perception to a different region of the field and read a group of letters there.

These observations seem to prove that without moving our eyes and without changing accommodation, we can, by a deliberate effort, concentrate our attention on the sensations of a specific part of the peripheral system of nerves and, at the same time, ignore all other parts.

In ordinary observation, of course, we intentionally direct our attention to particular parts of the field of vision or to a general region of perception. The direction of our gaze and the accommodation of the eyes follow our interest, and this can be so ordered and controlled that our attention is always focused precisely on the central pit of the retina, regulating its direction simply by regulating the movements of our eyes. Indeed, it is very difficult—and requires a great deal of practice, if we wish to learn—to turn our attention to the images on the lateral or peripheral parts of the retina, as is indicated to some extent in almost all the phenomena of this kind which have been previously described.

Among effects or conditions to which interest and attention are easily drawn, the following may be mentioned: (*a*) increased intensity of phenomena, particularly when it interferes with the visibility of objects, and (*b*) rapid changes in differences of brightness among closely adjacent parts of the field of vision, including the movements of surfaces within the field, the movements of shadows due to changes in the direction of lighting, and the movements of entoptic objects. Because of the decreasing effect of the negative afterimage, changes of brightness always produce more intense impressions than does light of constant intensity. This can explain in part the increase in attention which results. The immediate impression in consciousness, however, is more that every sudden change in the lateral part of the field of vision raises a question concerning the reason for the change, and for that reason the gaze is ordinarily directed toward the place where it has been noticed.

In general, objective interest has a powerful influence on the direction of attention and can govern it almost completely. Think of your behavior in reading: your gaze, accommodation, and attention all follow the words of the line and continue to move back and forth without interruption or disturbance, at least if the reading material is interesting. For the most part, the influence of objective interest coincides with the influence of the will, since our intentions are usually linked most easily and most frequently to desire, that is, to interests. Furthermore, daily experience teaches that deliberate direction of attention has a tiring effect on the brain, even when there is no muscular activity connected with it.

The final results of the experience and reflections just presented may, I believe, be summarized as follows:

1. In human beings we find reflex movements and instincts as effects of innate organizations. Instincts act in the interest of the pleasure of some impressions and in the avoidance of the discomfort of others.

2. Inductive inferences, executed by the unconscious activity of the memory, play a commanding part in the formation of intuitions.

3. It may be doubted that there is any indication whatsoever of any other source or origin for the ideas possessed by a mature individual.

20.

INTRODUCTION TO THE LECTURES ON THEORETICAL PHYSICS (INTRODUCTION AND PART I) [1894]

The volume entitled Introduction to the Lectures on Theoretical Physics *(1894) contains an Introduction and two Parts. ("Part II, The Mathematical Principles," is very similar to the essay "An Epistemological Analysis of Counting and Measurement")*

INTRODUCTION

1. Philosophy and the Natural Sciences

DUE to the influence of the philosophy of identity of Schelling and Hegel, the relationship between philosophy, on the one hand, and the natural sciences, on the other, was not a happy one during the first half of the present century. Basically the cause of the trouble was the profound antithesis between the methods each used to justify its position vis-à-vis the other. The discord, however, did not continue long with its initial bitterness. Through a rapid series of brilliant discoveries, the natural sciences proved to everyone that they contained a healthy seed of exceptional fertility. Attention and recognition, even by their principal opponents, could no longer be denied them.

It is natural that men at all times should strive to gain a knowledge of the order of the total universe, even if only in a schematic way. Since, during the first half of the century, the investigation of natural phenomena was proceeding slowly, and since it was apparent that by the powers of thought alone, very general laws could be found in mathematics on the basis of the most general daily experience, men tried to find such general laws by so-called pure thought, or speculation, even in the regions occupied by the empirical sciences. The illusions and errors which are unavoidable along this path to necessary truth—for here abstractions and grammatical expressions are treated as realities, and the results of untested experience are looked upon as necessities of thought—have long since brought philosophy into disrepute.

In condemning philosophy, however, we have tended to go too far, for even in the natural sciences philosophy is surely justified as the critique of methods. We must always investigate the instruments with which we work. The critique of methods can be neglected only as long as we are able to confine ourselves to the application of methods which have already been proven valid by their results. If scientific investigations enter regions where it is uncertain whether the difficulties encountered are to be ascribed to the subject matter or to the inadequacy of the method being used, critical analysis must make its appearance. For this reason, scientists have recently begun to discuss a number of problems of a philosophical nature. It has quickly become evident that discussions of individual, unrelated points are of little value; philosophical discussions must be carried out systematically and built upon really adequate foundations.

Since in the systematic presentation of physics—the science of the general properties of physical objects, which is not improperly called natural philosophy by the English-speaking peoples—it is more appropriate than in any other branch of the natural sciences to start with the most general point of view, I shall develop first the general logical and epistemological principles of the methodology of the empirical sciences. In my presentation I shall deviate somewhat from the general practice, however, for ordinarily in discussing the individual branches of science in the universities, not much time is devoted to the logical principles which lie at the foundation of the investigations which are undertaken.

For my personal justification I must point out that through long practice I have gained extensive knowledge of the problems to be considered. I also have had some experience regarding the easiest ways to reach solutions of the difficulties found in scientific work. Since the philosophers, to whom I would gladly go for counsel, begin their investigations with knowledge which can already be expressed in words, and since they know nothing at all (or only by hearsay) about the prior process of collecting factual data of experience, I have come to rely upon myself, and I shall frequently set things down in my own personal way.

2. *The Physical Sciences*

Before we proceed to a more detailed presentation of the methodological principles of the physical sciences, however, we must first consider the boundaries and contents of the sciences for which the following considerations are meant to serve as an introduction.

We may define physics as the science of the general properties of objects or bodies (*Naturkörper*). Among these general properties are included, not only those which are common to all objects without exception, but also those which belong only to very extensive classes of objects. We also use the term *physical* to refer to those sciences which are similar to physics conceptually, even though they are concerned with only a narrowly limited class of objects and sometimes with only one specific object or system of objects. Chemistry, for example, has the task of investigating the special properties by which the individual elements and their combinations are distinguished from one another. In other branches of the physical sciences there is little discussion of differentiating properties. Investigations are directed, rather, toward the changes observed in individual bodies or systems of bodies. Astronomy, for example, belongs to this class; it studies the forces and movements which we perceive in the heavenly bodies. Physical geography is concerned with the phenomena of the earth–phenomena both of the whole and of vastly extended parts of it. Meteorology treats of the phenomena of, and the transformations in, our atmosphere. These are all special but definite members of the class of physical sciences. They correspond to physics in method and so, in the broadest sense of the word, may be characterized as physical.

Physical science in the narrow sense, or what we earlier called *physics,* is the science which is concerned with and investigates the general properties of objects–more accurately, the properties common to *all* objects. It is divided into two main parts, which ordinarily are treated separately: *theoretical physics* and *experimental physics.* Essentially these are separated because of the kind of intellectual labor required to carry out investigations in each. As long as attention was paid primarily to the external form of the intellectual work in the individual sciences, without close attention to the essential nature of this work, there was an inclination to separate theoretical physics (which may also be called *mathematical physics*) sharply and strictly from experimental. Practice, too, separates these branches of physics–or, more precisely, these ways of treating physics–inasmuch as the students of one are more inclined toward experimental investigations, for which they have special ability, and tend to have the kind of imagination necessary to discover new, instructive experiments, while students of the other satisfy themselves more with the theoretical or mathematical side of the subject. Both groups are fully justified in what they do. It must not be forgotten, however, that mathematical knowledge and exercise in the mathematical treatment of physical problems are important for anyone who devotes himself to physics, whether he directs his energies mainly toward experimentation or toward calculation.

As one penetrates more deeply into physics, one will benefit, at least from a purely practical point of view, by deciding which direction to follow and thus where to devote most of his energy. It must be emphasized here at the beginning, however, that experimental physics, when completely divorced from mathematical physics, is a very narrowly limited science and offers very little insight into physical phenomena. Similarly, mathematical physics without experimental physics is for the most part a crippled, unfruitful science. One cannot really produce theories about natural processes before one has come to know these processes through one's own experience.

Ordinarily, in the systematic development of the various sciences, not much time is devoted to a consideration of the logical principles which lie at the foundation of the investigations which are undertaken. In physics, however, at least to a certain extent this seems to be necessary.

PART I: THE METHODOLOGICAL PRINCIPLES

3. *Critique of the Old Logic*

Logic, or the study of scientific thought, after it had been developed by Aristotle, was handed down through the scholastic philosophers of the Middle Ages. For the most part it has remained unchanged down to our own time. It is concerned, as was mentioned a moment ago, only with knowledge insofar as it is expressed in words and appears in the form of judgments. Grammatically, a judgment is a sentence which connects two concepts, the subject and the predicate; new judgments are obtained by syllogisms.

In order to make inferences we must have a universal proposition (which is called the major premise of the syllogism) and another proposition (called the minor premise) which need refer only to particular objects or, under certain conditions, only to a single object. The major, however, can be a premise in a general, completely valid syllogism only if it expresses a universal proposition which is true of all objects of a certain class.

In logic, as it is usually developed as a part of philosophy, there is ordinarily no information about the source of the major and minor premises. It is *presupposed* that the truth of these propositions can be ascertained. Discussion is devoted strictly to the various kinds of relations between them and to the validity or invalidity of the inferences which may be drawn from given major and minor premises. Thus ordinary logic is restricted to an analysis of the procedures by which one can obtain new propositions from known and given propositions—that is, analysis of the inference of conclusions in syllogisms. It provides no information as to how we arrive at the original propositions, the major and minor premises, which as a rule are just given by an outside authority.

The deduction of such a conclusion is never, of course, the discovery of new knowledge. This becomes obvious when we realize that, as a rule, the major premise cannot be stated with certainty unless we already know that the object named in the minor premise (that is, the subject of the minor premise) belongs under the major and thus that the major is true of that object. If one has not established the major premise himself and does not possess independent evidence for its truth, nothing can be inferred in a syllogism except the fact that the subject of the minor premise belongs under the major, which is precisely what one must know before one can set down the universal major premise in the first place.

In this sense, the name *logic,* which originally meant simply the art of speech, is completely appropriate. In the vast majority of cases, the original propositions from which one begins are given orally or in writing. Traditional logic is in reality nothing but a set of directions for expressing these propositions correctly, so that they possess the meaning desired; auditors who should, or who would like to, understand the propositions can thus connect the proper meaning to them. Logic, in short, is nothing more than directions for one person to speak correctly and for another to attribute the correct meaning to the propositions stated. In the entire series of logical operations, no new knowledge is generated.

In fundamental opposition to this, we in the natural sciences must obtain knowledge of what was previously unknown, knowledge which no one can give us on his own authority. At least, it is precisely this knowledge which forms the chief part of the natural sciences and their most important element. Hence the mental operations which we must carry out in the deliberations of the natural sciences vary from—and, indeed, display a fundamental difference from—the mental operations which were discussed in traditional logic. It is therefore necessary here to offer a short exposition of the logical operations which we actually do carry out in the investigations of the physical sciences.

4. *Concepts and Their Connotations*

The goal of the physical sciences is to comprehend (*begreifen*) natural phenomena. *To comprehend,* however, means to form concepts. If we wish to learn how to form concepts, arranging them in the order of superior and subordinate, traditional logic offers us the following procedure: First we combine objects which are the same in some respect to form a class. Then we specify the set of characteristics that will be used to distinguish the objects in the class; this is usually called the *definition* of the class. Giving a definition, therefore, consists in specifying the complex of properties which are necessarily present in all members of a class. When we have found such a complex of properties belonging to all members of some class but absent from all objects assigned to other classes, the boundaries of this class of objects, separating the members of it from all other objects, are determined. The objects have been collected under a concept.

This is the usual description of the formation of concepts, according to the teachings of traditional logic. John Stuart Mill, however, has called attention to a fundamental difference among the various kinds of properties which are used to define concepts. In natural history we very frequently find, besides those characteristics which are necessary and sufficient for the definition of a class or kind, and thus for a concept, other properties which also appear in all individuals of the specific class or kind. We can make this clear by an example.

Let us choose the concept *mammal.* We can delimit the class of mammals by the definition: "Mammals are animals which are born alive and suckled by their mothers." This definition excludes all other animals–all birds, amphibia, and so on. We find, however, that besides these characteristics of birth and early form of nourishment, there exists another set of properties, involving anatomical construction, which are the same among all mammals. We find, namely, that they are all warm-blooded; that they all have a double circulatory system, in which the blood, before it has traversed its entire path, has passed twice through the heart; and further, that there exist certain common characteristics in the formation of the bones in the ear, in the lower jaw, and so forth. All these general characteristics can also be used as a means for marking out mammals, since they do not appear among birds, amphibia, fish, and so on.

Thus, if a definition of a class which we are to designate with a common name, is to be formulated, we must distinguish between two different kinds of characteristics. One kind is necessary and sufficient to form the definition, to delimit the class, and to fix the name. At the same time, however, there may be other properties which are also present in all the individuals of the class but which are not necessary to define that class. If, for example, we discover an animal that was born alive and suckled by its mother, we declare it a mammal. We do not demand–(at least, not for the purpose of determining its name and class) that such anatomical characteristics as the formation of the bones of the ear and the formation of the heart also be investigated. John Stuart Mill was the first to make this important distinction–to separate those characteristics which belong to the definition of the concept and are necessary to establish the definition from those which are also always present in the individuals grouped under the concept. The latter he called the *connotations* of the concept.

This is a very basic and important distinction, which plays an essential role wherever conceptual definitions are employed, for if connotations can be found for the objects which are comprehended under a certain concept, we can make general statements. When we say, for example, "All mammals possess at least three ear bones, have a double blood circulation, and are warm-blooded," these three assertions about mammals are connotations.

It is a fact–and the reason for it can easily be explained–that in the classification which lies at the basis of human speech, the classification which determines naming, concepts which have connotations are preferable. It would not help us much to throw together arbitrary definitions based upon arbitrarily selected characteristics. We could, for example, place all plants with blue flowers in a class and give this class a special name. The definition of such a class, however, would not admit of any connotations, for we could not assert any general statements about these objects other than those already found in the definition, namely, that they are plants and that they have blue flowers. All universal propositions which we could formulate with respect to such a class would be tautologically necessary propositions. Thus, if we are going to formulate new concepts, we must try to obtain some which possess at least one connotation; to formulate any other kind would be fruitless.

Our language has already been formed in this way, of course, for even without strict scientific study we come to know through daily observation a great many relations of the kind just described. We cannot always give an account of the origin of this knowledge; indeed, we are rarely conscious of how we acquired it. With careful and intelligently observant peoples, however, the influence of ordinary observation on the formation of language must of necessity have been very great. We may generalize in part by saying that civilized peoples have finely constructed languages, that is, languages in which there are many concepts which can be used in universal statements and which thus have connotations.

5. Generic Concepts and Natural Laws

The kind of concept formation just discussed is used rather extensively in the so-called natural history sciences. For the most part, these sciences are concerned only with describing the objects which exist around us in our natural environment, especially those which, through the generative processes of animals, are always produced in the same natural form. There are always a great many instances of each of these classes. Thus in these sciences our main task is to describe static conditions. Only in recent times have we begun to study the variations which occur among such natural forms in the course of several generations.

The physical sciences, on the other hand, are concerned primarily with the changes which natural bodies or objects undergo. Our earlier statement that physical science seeks to determine the general properties of objects is not a contradiction, for we become aware of the properties of the bodies considered in physics only when we perceive changes. In the physical sciences we seek in general to determine what changes occur, what external influences and causes must exist in order for these changes to take place, and what must exist in order to prevent these changes. To comprehend these processes of *change* which we observe among the objects in the external world, we must carry out in the physical sciences a procedure completely analogous to the procedure of concept formation with respect to natural *forms.* We must combine cases to form classes in such a way that, besides the changes which serve as a definition (that is, which fix or characterize a specific class of processes), there are other changes which accompany these processes.

This is really the same conceptual task; only the form of the linguistic product is different. We can give linguistic expression to a class encompassing the conditions and course of some process of change only in the form of a natural law. For example, "two ponderable bodies which are at a finite distance from each other in space undergo an acceleration, each in the direction of the other"–that is, they move with increasing velocities toward each other. This is a phenomenon which always occurs if nothing hinders the movement (such as a third body causing the same acceleration but in a different direction). It is thus a phenomenon which can always be expressed in the same way, that is, in the form of a natural law.

In the natural history parts of zoology, botany, etc., we seek to include classes of bodies under concepts and to look for their connotations. In the physical sciences we have exactly the same logical task, except that here the concepts refer to processes. Our task, in other words, is to form classes of changes or processes such that, in addition to the observed invariant relation which corresponds to the definition of a concept, there are other regular processes analogous to the connotations of a concept.

Another group of natural processes can be included under and expressed by the following proposition: "A ray of light which passes through the boundary of two heterogeneous transparent media undergoes a refraction, and the deviation from the original path is given by a specific trigonometric function." Just as the first natural law mentioned above refers to changes in the positions of material bodies, the second refers to changes in the direction of a ray of light. We can multiply such examples very considerably; indeed, we can express all regularities in the processes of nature as laws, each with its specific factual content. To do this, we have only to state precisely the conditions under which a specific phenomenon takes place and then specify exactly how the process will continue. This is nothing but a description of the processes we actually perceive.

6. *Natural Law, Force, and Cause*

As a rule, in the linguistic formulation of a law we deviate from the formulation of natural laws just indicated in that we form abstractions and, indeed, introduce *verba substantia.* For example, we express the first of the laws mentioned above by stating that between any two material bodies at a finite distance from each other in space there exists a continuous force of attraction of a certain magnitude. Instead of the simple description of the phenomenon of movement, we introduce an abstraction–the force of attraction. In a factual sense, to be sure, we express nothing more and nothing less by this abstraction than what is contained in the mere description of the phenomenon. By stating the law in a form which makes use of the concept of force, we assert only that the phenomenon of two bodies approaching each other will occur whenever the required conditions are satisfied.

The concept of a certain force expresses nothing factual, of course, except that as long as the force is operative, or the conditions for its operation are realized, a certain phenomenon will be observed. In a sense it is thus an empty abstraction; however, it is one which, if rightly understood, actually describes the phenomenon taking place. On the other hand, we must stress that with the introduction of the concept of force into the formulation of a law, something enters which is rather obviously hypothetical, something which is not really given by the facts. As long as its meaning was not completely clear, this mode of expression caused many errors. More specifically, the abstract substantive *force* was thought to denote something actually existent, and some men believed that they were entitled to make statements about the real properties of forces. These statements, however, if true, either were tautologies or had only an apparently real content.

As a consequence, many physicists in recent times have tried not to introduce into physics anything that is not a pure expression of the facts; they have also tried to arrange the expression of the facts in such a way that nothing hypothetical enters. Faraday was the first to turn in this direction. He did not go through the normal course of instruction of a trained physicist, but acquired the greatest part of his knowledge by genuine intelligence and insight. Originally a poor printer's helper, he began to read the books he was assisting with and, in this way, rapidly increased his knowledge. Later he came into contact with the famous chemist Humphry Davy, whom the intelligent boy interested so much that he at once took him on as a helper, or better, as a laboratory assistant. From this position Faraday rose (Davy soon having transferred some of his lectures to him) to a position of great authority in the world.

The secret of his very significant scientific discoveries—and especially of the originality of his way of looking at things—is to be found in the fact that he did not know the usual paths of science but from the beginning taught himself to seek an understanding of phenomena in his own way. Faraday found the hypothetical element in the interpolation of forces very objectionable. It happened that he could not represent certain properties which physicists ascribed to forces, and from the beginning he carried on a fight in particular against the idea of action at a distance through empty space. For those phenomena which interested him most (magnetic and electric forces), he succeeded in discovering and establishing an interpretation to replace the concept of action at a distance, which had been previously accepted but which could not explain these phenomena.

Because of the unique way in which he visualized things, Faraday was forced to set up his own terminology. He was not fortunate in this, however, for he was unable to make clear the methodological nature of what he was trying to do. His contemporaries did not know how to follow him in his representations, and a long time passed—almost until the following generation—before mathematicians appeared who could figure out the meaning of what he said. Among the most prominent of these was Maxwell. After Faraday and Maxwell it was Sir William Thomson (now Lord Kelvin) who sought to avoid all figurative and abstract expressions in mathematical physics. More recently Gustav Kirchhoff has followed these men. In the preface to his textbook of mechanics Kirchhoff states flatly that the task of mechanics is to "describe completely and in the simplest possible way the movements occurring in nature." What is essential for him is only an exact description of phenomena.

I would add to the position expressed by Kirchhoff that the most complete and simplest description can be given only when we formulate the laws which lie at the foundation of phenomena. The abstract mode of expression has its advantages in formulating these laws, and justification for using abstract substantives can be found in the very nature of things. When we formulate a law, we give expression to an experience which, we may rightly assume, will always be repeated as often as the conditions under which it occurs are present. As we all know and as we can confirm by experience at any moment, what a law expresses takes place independent of our intellect and independent of our wishes. We know that we cannot, just by the ideas we have at the same time, bring any influence to bear on the course of external processes. Using a terminology still frequently found in philosophical discussions, we may characterize the phenomenon which occurs in accordance with a law as a Non-Ego, something which is not dependent upon our intellect, our consciousness, our will, or our wishes. When it occurs, we can only consider it a fact.

The more exact and searching our investigations are, the more we become convinced that the occurrence of such a phenomenon is dependent only upon certain objective conditions and is completely independent of occurrences in our mind. Thus we must recognize a law as expressing something which occurs completely independent of our mind and our wishes, and we must recognize further that the phenomenon specified by a law may occur at any moment whatsoever, provided only that the necessary conditions are satisfied. This power (*Macht*), which without our intervention and our intellect can produce such definite effects, must be recognized as something that exists continuously and is present and ready to work at any moment—indeed, as something powerful (*Mächtiges*), since its effects can occur contrary to our will and our desire. In this there is more than the mere comprehension of phenomena as phenomena.

We usually designate things—those which continue in existence, which prove to be powerful, and which determine the processes of the external world without our needing to trouble with them—by names which indicate that they are actual, existing things. If I have understood the situation correctly, referring to a force as an existing agent is therefore completely justified. Only by using this mode of expression can we state that a law which we have discovered is always ready to operate and may show its power at any moment. This obviously is the real reason why men have adopted the substantive mode of expression and, instead of speaking of *laws of forces*—although (this cannot be repeated often enough) that term is closer to the factual meaning—have preferred to speak substantially of *a force acting* in a specific way. This mode of expression factually contains nothing more and has no real content other than when we say, "The law will manifest itself in every case where the conditions necessary for its appearance are realized." Of this hypothetical substantive, for that is what we must consider a force to be, we know nothing at all except that it lies in its nature to produce a specific effect.

It should next be noted that because of the very great multiplicity of natural phenomena, we admit the most diverse kinds of forces. We consider the acceleration of material bodies toward each other the effect of a force of attraction which the bodies exert reciprocally. In the refraction of light we speak of a force of refraction, which we ascribe to transparent media. We find that certain sets of metals and conducting fluids produce electric currents, and when we measure the strength of these currents and determine more precisely the conditions under which they appear and upon which their strength depends, we speak of the electromotive force of galvanic batteries. And so on. In short, we admit many different kinds of forces—and in doing so, we at first set no limits. Later, when we come to investigate these forces more carefully, we discover that there are certain connections and relationships among them.

It is easy to see that these abstractions, *forces* and the *material objects* to which we ascribe them, cannot be separate from each other. If we are considering forces of movement, we characterize that which is moveable as a mass or as matter. A force without matter has no meaning; it would correspond to a law describing changes where there is nothing to be changed. Such a law would be contradictory and self-refuting. There would be just as little sense in speaking of material objects without forces; such objects could not undergo any changes, since change always presupposes the existence of a force. These simple observations show immediately that the abstractions *material object* and *force* cannot be separated. They have definite meanings only in relation to and in combination with each other.

Many errors have occurred in science because the real meaning of the word *force* was forgotten. Simply because the word is a substantive, that which it expresses was taken to be a real thing having *independent* existence. For a long time, especially in the physiological sciences, there was talk of the possibility of a vital force, detached from and not dependent upon matter. The human soul was even considered a force which did not adhere in matter, hence separate from objects in which changes could occur. This notion, however, lost its meaning as soon as the origin of the concepts *force* and *matter* became clear. Such erroneous notions—and the fact that in the development of science they were connected with all sorts of false and inadmissible ideas—gave rise to distrust of this entire mode of expression. Many men, wishing to avoid these errors completely in their own works, began to avoid all abstract concepts and to reduce their treatment of natural processes to pure descriptions.

Although there is some variation in meaning, another term is important in connection with the present discussion. It is the term *cause* (*Ursache*). Insofar as it is that which brings about a change, we also refer to a force as the cause of the change. If we consider the etymology of the word *Ursache,* we find that the prefix *ur* means "hidden behind phenomena," while *Sache* is the designation for "something that exists continuously." Thus, according to its etymology, the word *Ursache* (which I use here precisely and literally) means *that existent something (Bestehende) which lies hidden behind the changes we perceive.* It is the hidden but continuously existent basis of phenomena. This corresponds completely with what we have just said concerning the concept of force, for we saw that a force can be understood as something continuously existent only if it is ready at any instant to effect a change.

The concept of a law of phenomena implies everything that can be expressed by using this additional terminology. We have made use of it only to stress the concepts of *continuous existence* and *continuous capacity to produce effects.* If we accept that all changes in nature follow necessarily from causal conditions—that is, that every change in nature has a sufficient cause—then on the basis of the analysis given above we can express this proposition in a somewhat different way: "All changes in nature occur according to law." Since we can comprehend natural phenomena only if they are uniform (*gesetzlich*) in all their parts and only if they always occur uniformly, we can also formulate the law of causality in the following way: "All natural phenomena are comprehensible." With the statement of the law of the uniformity of natural phenomena, we also assert implicitly that laws exist continuously and are always ready to produce their effects. Given this, we are justified in considering such laws the causes (in the sense of the word given above) of phenomena.

The law of causality is a proposition which we cannot establish by experience. If we do not already have it, we cannot conclude that any phenomenon is definitely the effect of the cause. We must make use of the law in order to arrive at the ideas of force and cause in the first place. The law of causality, therefore, is an a priori proposition dependent upon the formal conditions of our understanding, for (as was just said) we cannot arrive at the idea of a particular cause, nor can we recognize one, if we do not approach nature with the idea that it is always possible to discover them.

When we begin to search for causes, we find that there are indeed large areas of phenomena within which we can point them out and within which we can demonstrate complete, strict conformity to law. It must be stressed, however, that our knowledge of natural phenomena is not sufficient to conclude that the proposition "All natural phenomena are comprehensible" is an empirical one. We must instead recall that for large classes of natural phenomena, above all for organic processes in both animals and plants, it would be very risky, before the most complete knowledge of their causes is obtained, to infer a general conceivability a posteriori from the great fields of natural phenomena already conceptualized. In other areas, such as meteorological phenomena, while they have not been reduced entirely to causes, it is quite clear that they eventually will be completely understood.

7. Hypotheses: The First Elements of Laws

Our task is to search for the laws of natural phenomena, and for this there is no method other than the observation of such phenomena. Ordinarily, however, laws do not appear without exceptions. On the contrary, the conditions under which natural changes take place are in general so extraordinarily complicated that even when we use experiments, which are set up for the very purpose of discovering laws, we are seldom in a position to recognize a law in its simplest mode of operation and to see it clearly before us in immediate experience.

Usually, therefore, it is advisable to formulate a law on a trial basis and then to investigate to see whether it proves correct in all the cases in which we can test it by experiment. On the whole, this is a slow, laborious process. Thus the initial formulation of a law must always have the character of a hypothesis. It cannot be taken as completely certain; rather, it requires verification, which in most cases is a very time-consuming task. Even though the method of investigation of natural phenomena leads of necessity through hypotheses, however, it must never be forgotten that these hypotheses are meant to serve only as a basis for the subsequent formulation of laws. It is incorrect to believe that we can hold on to an initial hypothesis indefinitely.

A certain power of invention is always required in the formulation of a hypothesis. It is almost a guess, and in general only those who are qualified to guess formulate hypotheses which amount to anything. In stating a hypothesis, it is advisable to do so in such a way that in any specific case to which it is applied, there will always be a definite, completely unequivocal result.

An investigation is generally carried out in the following manner: A scientist makes a series of particular observations in some area of physics and then attempts on a trial basis to collect these observations under a specific law. He then considers whether there are still further instances–instances which he has already seen in the course of his own experience or which are known to him–which fall under his definition, that is, within the class of cases to which the trial law he has formulated refers. Afterward, still other instances will come to mind which have not been verified but which ought to fall under the law. Wherever possible, these new instances, whose results are unknown, are brought about by experiment in order to determine whether they actually do occur according to the law.

In scientific investigations carried out according to methodological principles, the more an investigation can be directed to specific, individual cases, the more a hypothesis will be employed. The scientific element in this procedure is to be found in the fact that we try to obtain the most complete knowledge possible of a given class of natural phenomena, in part by observing directly what our experience offers and in part by deliberately seeking to produce new instances, especially such as undoubtedly belong within the range of a law but which deviate in some essential respect from those known to us from previous observation.

8. The Completeness of Scientific Experience and Its Practical Significance

The procedure just described for establishing a generally valid law, after starting from a tentatively suggested hypothesis, culminates in a search for the most complete possible knowledge of all individual instances. When a law has been established for all cases, we have an additional advantage in that we gain a certain kind of control, in part through the completeness of our understanding of the instances and in part because we have allowed none which contradict our law to escape us. The more careful we have been with respect to a law and the more we are accustomed to insist that it be confirmed in all cases, the more striking will be any instance which contradicts it. If we were testing it only in random cases, we might easily overlook one that was anomalous. The very fact that we compare all those we experience with the law, however, means that in time we shall reach the highest degree of assurance that there are none which contradict it.

Genuine science, carried on in accordance with the proper methods, provides far better knowledge of individual occurrences and their consequences than any sort of dilettantish knowledge–knowledge which has accumulated by chance in the experience of individual men–could possibly do. Quite often one hears talk about the completely abstruse statements which the natural sciences supposedly make about a given phenomenon. More specifically, in sociopolitical and moral science one very frequently finds false, nonsensical denunciations of what is allegedly asserted and taught in the natural sciences. As a rule, the hidden and only rarely recognized reason for such errors lies in the confusion of experimental hypotheses with firmly established laws. Often people have the idea that the natural sciences set up unfounded hypotheses and are then led into error by them. As we have just observed regarding genuine scientific methods, however, science is nothing but methodologically and deliberately completed and purified experience–indeed, experience which is much more extensive and much more certain than the fortuitously accumulated experience of any single individual. Because of this, we may trust genuine science more and ascribe greater value to its results.

After we solve our problem of finding the specific laws of natural phenomena, we must seek and, if possible, find still more inclusive laws, under which many more individual instances are comprehended. The more we succeed in this, the richer our knowledge of the individual cases will be. If we know the laws for an area of phenomena, we know in advance how to anticipate the consequences if certain conditions are realized. In many cases we can even choose the conditions so that desirable consequences will result. Thus a knowledge of laws has, not only a theoretical value for human insight and for our understanding of the interconnections of things, but also an enormous practical value. Our entire control over nature and natural forces, especially as this has developed during the last century, arises from this knowledge of laws.

As long as we possess knowledge of laws, we are able to do two important things. First, we can know in advance what will happen under given conditions. This, as is well known, is the only true gift of prophecy given to man.

Any prophecy which is based upon the complete knowledge of laws and the conditions under which they will act in given cases is a reliable one; it really foretells what will occur. Secondly, and more important, by means of the knowledge of laws we are able to make natural forces work for us as we wish. The entire recent development of industry, and with it the complete change in the form of human life and human activity, is dependent essentially upon our control over the forces of nature.

BIBLIOGRAPHY

The following bibliography of von Helmholtz' works is arranged chronologically. The most readily available source of each item in the original is given, and the English title is added whenever a translation of the item exists. The following abbreviations are used:

VR, I. *Vorträge und Reden,* Volume I. 5th ed.; Braunschweig; 1903.
VR, II. *Vorträge und Reden,* Volume II. 5th ed.; Braunschweig, 1903.
WA, I. *Wissenschaftliche Abhandlungen,* Volume I. Leipzig, 1882.
WA, II. *Wissenschaftliche Abhandlungen,* Volume II. Leipzig, 1883.
WA, III. *Wissenschaftliche Abhandlungen,* Volume III. Leipzig, 1895.

1842

"De Fabrica Systematis nervosi Evertebratorum." Inaugural-Dissertation. Berlin, November 2, 1842. (*WA*, II, 663-679.)

1843

"Ueber das Wesen der Fäulniss und Gährung." (*WA*, II, 726-734.)

1845

"Ueber den Stoffverbrauch bei der Muskelaction." (*WA*, II, 735-744.)

1846

"Wärme, physiologisch." (*WA*, II, 680-725.)

1847

"Bericht über die Theorie der physiologischen Wärmeerscheinungen für 1845." (*WA*, I, 3-11.)

"Ueber die Erhaltung der Kraft." (*WA*, I, 12-75.) ("The Conservation of Force: A Physical Memoir.")

1848

"Ueber die Wärmeentwicklung bei der Muskelaction." (*WA*, II, 745-763.)

"Bericht über ‚die Theorie der physiologischen Wärmeerscheinungen' betreffende Arbeiten aus dem Jahre 1846," *Fortschritte der Physik im Jahre 1846* (Berlin), II, 259-260.

1850

"Ueber die Fortpflanzungsgeschwindigkeit der Nervenreizung." (*WA*, III, 1-3.)

"Notes sur la vitesse de la propagation de l'agent nerveux dans les nerfs rachidiens," *Comptes Rendus,* XXX, 204-206.

"Messungen über den zeitlichen Verlauf der Zuckung animalischer Muskeln und die Fortpflanzungsgeschwindigkeit der Reizung in den Nerven." (*WA*, II, 764-843.)

"Ueber die Methoden, kleinste Zeittheile zu messen, und ihre Anwendung für physiologische Zwecke." (*WA*, II, 862-880.) ("On the Methods of Measuring Small Intervals of Time, and Their Application to Physiological Purposes.")

"Bericht über ‚die Theorie der physiologischen Wärmeerscheinungen' betreffende Arbeiten aus dem Jahre 1847," *Fortschritte der Physik im Jahre 1847* (Berlin), III, 232-245.

1851

"Deuxième note sur la vitesse de la propagation de l'agent nerveux," *Comptes Rendus,* XXXIII, 262-265.

"Beschreibung eines Augenspiegels zur Untersuchung der Netzhaut im lebenden Auge." (*WA*, II, 229-260.) ("Description of an Ophthalmoscope for the Investigation of the Retina of the Living Eye.")

"Ueber den Verlauf und die Dauer der durch Stromesschwankungen inducirten elecktrischen Ströme." (*WA*, III, 554-557.)

"Ueber die Dauer und den Verlauf der durch Stromesschwankungen inducirten elektrischen Ströme." (*WA*, I, 429-462.)

1852

"Messungen über Fortpflanzungsgeschwindigkeit der Reizung in den Nerven, Zweite Reihe." (*WA*, II, 844-861.)

"Die Resultate der neueren Forschungen über thierische Elektricität." (*WA*, II, 886-923.)

"Ueber die Natur der menschlichen Sinnesempfindungen." (*WA*, II, 591-609.)

"Ueber Herrn D. Brewster's neue Analyse des Sonnenlichts." (*WA*, III, 558-561.)

"Ueber Herrn D. Brewster's neue Analyse des Sonnenlichts." (*WA*, II, 24-44.) ("On Brewster's New Analysis of Solar Light.")

"Ueber die Theorie der zusammengesetzten Farben." (*WA*, II, 3-23.) ("Theory of Compound Colors.")

"Ein Theorem über die Vertheilung elektrischer Ströme in körperlichen Leitern." (*WA*, III, 562-564.)
"Bericht über die theoretische Akustik betreffende Arbeiten vom Jahre 1848." (*WA*, I, 233-250.)
"Bericht über ‚die Theorie der physiologischen Wärmeerscheinungen' betreffende Arbeiten aus dem Jahre 1848," *Fortschritte der Physik im Jahre 1848* (Berlin), IV, 222-223.
"Ueber eine neue einfachste Form des Augenspiegels." (*WA*, II, 261-279.)

1853

"Ueber eine bisher unbekannte Veränderung am menschlichen Auge bei veränderter Accommodation." (*WA*, II, 280-282.)
"Ueber einige Gesetze der Vertheilung elektrischer Ströme in körperlichen Leitern mit Anwendung auf die thierisch-elektrischen Versuche." (*WA*, I, 475-519.)
"Ueber Goethes naturwissenschaftliche Arbeiten." (*VR*, I, 23-47.) ("The Scientific Researches of Goethe.")

1854

"Bericht über die theoretische Akustik betreffende Arbeiten vom Jahre 1849." (*WA*, I, 251-255.)
"Erwiderung auf die Bemerkungen von Hrn. Clausius." (*WA*, I, 76-93.)
"Ueber die Wechselwirkung der Naturkräfte und die darauf bezüglichen neuesten Ermittelungen der Physik." (*VR*, I, 49-83, 401-417.) ("The Interaction of Natural Forces.")
"Ueber die Geschwindigkeit einiger Vorgänge in Muskeln und Nerven." (*WA*, II, 881-885.)

1855

"Ueber die Zusammensetzung von Spectralfarben." (*WA*, II, 45-70.)
"Ueber das Sehen des Menschen." (*VR*, I, 85-117.)
"Ueber die Empfindlichkeit der menschlichen Netzhaut für die brechbarsten Strahlen des Sonnenlichts." (*WA*, II, 71-77.)
"Zusatz zu einer Abhandlung von E. Esselbach über die Messung der Wellenlänge des ultravioletten Lichtes." (*WA*, II, 78-82.)
"Bericht über ‚die Theorie der Wärme' betreffende Arbeiten aus dem Jahre 1852," *Fortschritte der Physik im Jahre 1852* (Berlin), VIII, 369-387.
"Ueber die Accommodation des Auges." (*WA*, II, 283-345.)

1856

"Ueber die Erklärung des Glanzes." (*WA*, III, 4-5.)
"Zuckungscurven von Froschmuskeln." (*WA*, III, 6.)
"Ueber die Combinationstöne odor Tartinischen Töne." (*WA*, III, 7-9.)
"Ueber die Bewegungen des Brustkastens." (*WA*, II, 953-954.)

"Ueber Combinationstöne." (*WA*, I, 256-262.)
"Ueber Combinationstöne." (*WA*, I, 263-302.)
Handbuch der physiologischen Optik, Park I. Leipzig. (*Handbook of Physiological Optics*, Part I.)
"Bericht über ‚die Theorie der Wärme' betreffende Arbeiten aus dem Jahre 1853," *Fortschritte der Physik* (Berlin), IX, 404-432.

1857

"Ein Telestereoskop." (*WA*, III, 10-12.)
"Das Telestereoskop." (*WA*, II, 484-491.)
"Ueber die Vokale." (*WA*, I, 395-396.)
"Die Wirkungen der Muskeln des Armes." (*WA*, II, 955-957.)
"Bericht über ‚die Theorie der Wärme' betreffende Arbeiten aus dem Jahre 1854," *Fortschritte der Physik* (Berlin), X, 361-398.
"Ueber die physiologischen Ursachen der musikalischen Harmonie." (*VR*, I, 119-155.) ("The Physiological Causes of Harmony in Music.")

1858

"Ueber die subjectiven Nachbilder im Auge." (*WA*, III, 13-15.)
"Ueber Integrale der hydrodynamischen Gleichungen, welche den Wirbelbewegungen entsprechen." (*WA*, I, 101-134.) ("On the Integrals of the Hydrodynamic Equations Which Express Vortex Motions.")
"Bericht über ‚die Theorie der Wärme' betreffende Arbeiten aus dem Jahre 1855," *Fortschritte der Physik* (Berlin), XI, 361-373.
"Ueber die physikalische Ursache der Harmonie und Disharmonie," *Amtlicher Bericht über die 34. Versammlung deutscher Naturforscher und Aerzte zu Carlsruhe im September 1858*, pp. 157-158.
"Ueber Nachbilder," *ibid.*, pp. 225-226.

1859

"Ueber die Klangfarbe der Vocale." (*WA*, I, 397-407.) ("On the Quality of Vocal Tones.")
"Ueber Luftschwingungen in Röhren mit offenen Enden." (*WA*, III, 16-20.)
"Ueber Farbenblindheit." (*WA*, II, 346-349.)
"Theorie der Luftschwingungen in Röhren mit offenen Enden." (*WA*, I, 303-382.)
"Bericht über ‚die Theorie der Wärme' betreffende Arbeiten aus dem Jahre 1856," *Fortschritte der Physik* (Berlin), XII, 343-359.

1860

"Ueber die Contrasterscheinungen im Auge." (*WA*, II, 350-351.)
"Ueber musikalische Temperatur." (*WA*, I, 420-423.)
"Ueber die Bewegung der Violinsaiten." (*WA*, I, 410-419.) ("On the Motion of the Strings of a Violin.")

Handbuch der physiologischen Optik, Part II. Leipzig. (*Handbook of Physiological Optics,* Part II.)

"Ueber Klangfarben." (*WA*, I, 408–409.)

"Ueber Reibung tropfbarer Flüssigkeiten" (with G. von Piotrowski). (*WA*, I, 172-222.)

1861

"Zur Theorie der Zungenpfeifen." (*WA*, I, 388-394.)

"Ueber eine allgemeine Transformationsmethode der Probleme über elektrische Vertheilung." (*WA*, I, 520-525.)

"The Application of the Law of the Conservation of Force to Organic Nature." *Proceedings of the Royal Institute.* (III, 347-357.) Also in *WA*, III, 565-580.

1862

"Ueber das Verhältniss der Naturwissenschaften zur Gesammtheit der Wissenschaften." (*VR*, I, 157-185.) ("The Relation of the Natural Sciences to Science in General.")

"Ueber die Erhaltung der Kraft." (*VR*, I, 187-229.) ("The Conservation of Force.)

"Ueber die arabisch-persische Tonleiter." (*WA*, I, 424–426.)

"Ueber die Form des Horopters, mathematisch bestimmt." (*WA*, II, 420-426.)

1863

"Ueber den Einfluss der Reibung in der Luft auf die Schallbewegung." (*WA*, I, 383-387.)

"Ueber die Bewegungen des menschlichen Auges." (*WA*, II, 352-359.)

"Ueber die normalen Bewegungen des menschlichen Auges." (*WA*, II, 360-419.)

Die Lehre von den Tonempfindungen als physiologische Grundlage für die Theorie der Musik. 1st ed., Braunschweig.

1864

"On the Normal Motions of the Human Eye in Relation to Binocular Vision." (*WA*, III, 25–43.)

"Lectures on the Conservation of Energy, Delivered at the Royal Institution." *Medical Times and Gazette* (London), I, 385-388, 415–418, 443-446, 471–474, 499-501, 527-530.

"Bemerkungen über die Form des Horopters." (*WA*, II, 478–481.)

"Ueber den Horopter." (*WA*, III, 21-24.)

"Versuche über das Muskelgeräusch." (*WA*, II, 924-927.)

"Ueber den Horopter." (*WA*, II, 427–477.)

1865

"Ueber den Einfluss der Raddrebung der Augen auf die Projection der Retinalbilder nach Aussen." (*WA*, II, 482-483.)

"Ueber Eigenschaften des Eisen." (*WA*, I, 94-98.)
"Ueber stereoskopisches Sehen." (*WA*, II, 492-496.)
"Ueber die Augenbewegungen." (*WA*, III, 44-48.)
Populäre wissenschaftliche Vorträge, Number I. Braunschweig.
Die Lehre von den Tonempfindungen als physiologische Grundlage für die Theorie der Musik. 2nd ed., Braunschweig.
"Eis und Gletscher." (*VR*, I, 231-263, 418-422.) ("Ice and Glaciers.")

1866

"Ueber den Muskelton." (*WA*, II, 928-931.)
"Ueber die thatsächlichen Grundlagen der Geometrie." (*WA*, II, 610-617.)
"On the Regelation of Ice," *Philosophical Magazine, Fourth Series*, XXXII, 22-23.

1867

Handbuch der physiologischen Optik, Part III. Leipzig. (*Handbook of Physiological Optics*, Part III.)
"Mittheilung, betreffend Versuche über die Fortpflanzungsgeschwindigkeit der Reizung in den motorischen Nerven des Menschen, welche Herr N. Baxt aus Petersburg im physiologischen Laboratorium zu Heidelberg ausgeführt hat." (*WA*, II, 932-938.)
"Ueber die Mechanik der Gehörknöchelchen." (*WA*, II, 503-514.)
"Sur la production de la sensation du relief dans l'acte de la vision binoculaire." (*WA*, III, 581-586.)

1868

"Die neueren Fortschritte in der Theorie des Sehens." (*VR*, II, 265-365.) ("Recent Progress in the Theory of Vision.")
"Ueber discontinuirliche Flüssigkeitsbewegungen." (*WA*, I, 146-157.) ("On Discontinuous Movements of Fluids.")
"Sur le mouvement le plus générale d'un fluide." (*WA*, I, 135-139.)
"Sur le mouvement des fluides." (*WA*, I, 140-144.)
"Résponse à la note de M. J. Bertrand du 19. octobre." (*WA*, I, 145.)
"Ueber die thatsächlichen Grundlagen der Geometrie." (*WA*, II, 610-617.)
"Ueber die Thatsachen, die der Geometrie zum Grunde liegen." (*WA*, II, 618-639.)

1869

"Zur Theorie der stationären Ströme in reibenden Flüssigkeiten." (*WA*, I, 223-230.)
"Ueber die physiologische Wirkung kurz dauernder elektrischer Schläge im Innern von ausgedehnten leitenden Massen." (*WA*, I, 526-530.)
"Ueber elektrische Oscillationen." (*WA*, I, 531-536.)
"Ueber die Schallschwingungen in der Schnecke des Ohres." (*WA*, II, 582-588.)

"Ueber das Ziel und die Fortschritte der Naturwissenschaft." (*VR,* II, 367-398.) ("The Aim and Progress of Physical Science.")

"Ueber das Heufieber. (Als briefliche Mittheilung enthalten in einer Abhandlung von C. Binz: pharmakologische Studien über das Chinin)," *Virchow's Archiv für pathologische Anatomie,* XLVI, 100-102.

"Die Mechanik der Gehörknöchelchen und des Trommelfelles." (*WA,* II, 515-581.) ("The Mechanism of the Ossicles of the Ear and Membrana Tympani.")

1870

"Ueber die Gesetze der inconstanten elektrischen Ströme in körperlich ausgedehnten Leitern." (*WA,* I, 537-544.)

"The Axioms of Geometry," *The Academy, I,* 128-131.

"Neue Versuche über die Fortpflanzungsgeschwindigkeit der Reizung in den motorischen Nerven der Menschen, ausgeführt von N. Baxt aus Petersburg." (*WA,* II, 939-946.)

"Ueber die Theorie der Elektrodynamik," Essay I: "Ueber die Bewegungsgleichungen der Elektricität für ruhende leitende Körper." (*WA,* I, 545-628.)

Die Lehre von den Tonempfindungen als physiologische Grundlage für die Theorie der Musik. 3rd ed., Braunschweig. (*The Theory of the Sensations of Tone as a Physiological Basis of the Theory of Music.*)

"Ueber den Ursprung und die Bedeutung der geometrischen Axiome." (*VR,* II, 1-31, 381-383.) ("The Origin and Meaning of Geometric Axioms (I).")

"Vorrede" to the German translation of J. Tyndall's *Faraday as a Discoverer* (Braunschweig), pp. v-xi.

1871

Populäre wissenschaftliche Vorträge, Number II. Braunschweig.

"Vorrede" to Part I of the German translation of W. Thomson and P. G. Tait's *Treatise on Natural Philosophy,* Volume I (Braunschweig), pp. x-xii.

"Ueber die Fortpflanzungsgeschwindigkeit der elektrodynamischen Wirkungen." (*WA,* I, 629-635.)

"Ueber die Zeit, welche nöthig ist, damit ein Geschichtseindruck zum Bewusstsein kommt. Resultate einer von Herrn N. Baxt im Heidelberger Laboratorium ausgeführten Untersuchung." (*WA,* II, 947-952.)

"Zum Gedächtnis an Gustav Magnus." (*VR,* II, 33-51.) ("In Memory of Gustav Magnus.")

"Ueber die Entstehung des Planetensystems." (*VR,* II, 53-91.) ("The Origin of the Planetary System.")

"Optisches über Malerei." (*VR,* II, 93-135.) ("The Relation of Optics to Painting.")

1872

"Ueber die Theorie der Elektrodynamik," [Essay II.](*WA,* I, 636-646.)

"Ueber die galvanische Polarisation des Platins," *Tageblatt der 45. Versammlung deutscher Naturforscher und Aerzte zu Leipzig im August 1872,* pp. 110-111.

1873

"Vergleich des Ampère'schen und Neumann'schen Gesetzer für die elektrodynamischen Kräfte." (*WA,* I, 688-701.)

"Ueber ein Theorem, geometrisch ähnliche Bewegungen flüssiger Körper betreffend, nebst Anwendung auf das Problem, Luftballons zu lenken." (*WA,* I, 158-171.)

"Ueber galvanische Polarisation in gasfreien Flüssigkeiten." (*WA,* I, 823-834.) ("On Galvanic Polarization in Fluids Free from Gases.")

"Ueber die Grenzen der Leistungsfähigkeit der Mikroskope." (*WA,* II, 183-184.)

"Ueber die Theorie der Elektrodynamik," Essay II: "Kritische." (*WA,* I, 647-687.)

1874

"Ueber die Theorie der Elektrodynamik," Essay III: "Die elektrodynamischen Kräfte in bewegten Leitern." (*WA,* I, 702-762.)

"Die theoretische Grenze für die Leistungsfähigkeit der Mikroskope." (*WA,* II, 185-212.)

"Kritische zur Elektrodynamik." (*WA,* I, 763-773.)

"Zur Theorie der anomalen Dispersion." (*WA,* II, 213-226.)

"On the Later Views of the Connection of Electricity and Magnetism," *Annual Report of the Smithsonian Institution for 1873,* pp. 247-253.

"Induction und Deduction." "Vorrede" to Part II of W. Thomson and P. G. Tait's *Treatise on Natural Philosophy,* Volume I (Braunschweig). (*VR,* II, 413-421.)

"Vorrede" and "Kritische Beilage" to the German translation of J. Tyndall's *Fragments of Science* (Braunschweig). (*VR,* II, 422-434.) ("The Endeavor to Popularize Science.")

1875

"Versuche über die im ungeschlossenen Kreise durch Bewegung inducirten elektromotorischen Kräfte." (*WA,* I, 774-790.)

1876

"Wirbelstürme und Gewitter." (*VR,* II, 137-163.)

"Bericht betreffend Versuche über die elektromagnetische Wirkung elektrischer Convection, ausgeführt von Hrn. Henry A. Rowland," (*WA,* I, 791-797.) ("Report on Experiments on the Electromagnetic Effect of Electrical Convection, Carried out by Mr. Henry A. Rowland.")

"Bericht über Versuche des Hrn. Dr. E. Root aus Boston, die Durchdringung des Platins mit elektrolytischen Gasen betreffend." (*WA,* I, 835-839.)

("Report on the Experiment of Dr. E. Root of Boston Concerning the Penetration of Platinum by Electrolytic Gases.")

Populäre wissenschaftliche Vorträge, 2nd edition. Braunschweig.

1877

"Das Denken in der Medicin." (*VR,* II, 165-190, 384-386.) ("Thought in Medicine.")

"Ueber die akademische Freiheit der deutschen Universitäten. Rektoratsrede vom 15 Oktober 1877." (*VR,* II, 191-212.) ("On Academic Freedom in German Universities.")

"Ueber galvanische Ströme, verursacht durch Concentrationsunterschiede; Folgerungen aus der mechanischen Wärmetheorie." (*WA,* I, 840-854.) ("On Galvanic Currents Caused by Differences in Concentration: Deductions from the Mechanical Theory of Heat.")

Die Lehre von den Tonempfindungen als physiologische Grundlage für die Theorie der Musik. 4th ed., Braunschweig.

1878

"Telephon und Klangfarbe." (*WA,* I, 463-474.)

"Ueber die Bedeutung der Convergenzstellung der Augen für die Beurtheilung des Abstandes binocular gesehener Objecte." (*WA,* II, 497-500.)

"Ueber den Ursprung und Sinn der geometrischen Sätze; Antwort gegen Herrn Professor Land." (*WA,* II, 640-660.) ("The Origin and Meaning of Geometric Axions [II].")

"Die Thatsachen in der Wahrnehmung." (*VR,* II, 213-247, 387-406.) ("The Facts of Perception.")

Review of Lord Rayleigh's *Theory of Sound, Nature, XVII,* 237-239; *XIX,* 117-118.

1879

"Ueber elektrische Grenzschichten." (*WA,* III, 49-51.)

"Studien über elektrische Grenzschichten." (*WA,* I, 855-898.) ("Studies of Electric Boundary Layers.")

1880

"Ueber Bewegungsströme am polarisirten Platina." (*WA,* I, 899-921.)

1881

"Vorbemerkung zu einer nachgelassenen Abhandlung von Franz Boll: Thesen und Hypothesen zur Licht- und Farbenempfindung," *Du Bois-Reymond's Archiv, 1881,* pp. 1-3.

"Ueber die auf das Innere magnetisch oder diëlektrisch polarisirter Körper wirkenden Kräfte." (*WA,* I, 798-820.)

"The Modern Development of Faraday's Conception of Electricity." (*WA,* III, 52-87. *VR,* II, 249-291, 407-410.)

Popular Lectures on Scientific Subjects, First Series, edited by E. Atkinson. London.
"Note on Stereoscopic Vision," *Philosophical Magazine, Fifth Series, XI,* 507-508.
"Eine elektrodynamische Waage." (*WA*, I, 922-924.)
"Ueber die elektrischen Maasseinheiten nach den Berathungen der elektrischen Congresses, versammelt zu Paris, 1881." (*VR*, II, 293-309, 411-412.)
"Ueber galvanische Polarisation des Quecksilbers und darauf bezügliche neue Versuche des Hrn. Arthur König." (*WA*, I, 925-938.)

1882

"Die Thermodynamik chemischer Vorgänge." (*WA*, II, 958-978.)
"Zur Thermodynamik chemischer Vorgänge." (*WA*, II, 979-992.)
"Ueber absolute Maassysteme für elektrische und magnetische Grössen." (*WA*, II, 993-1005.)
"Bericht über die Thätigkeit der internationalen elektrischen Commission," *Verhandlungen der physikalischer Gesellschaft zu Berlin, Sitzung von 17. November 1882,* I.
Wissenschaftliche Abhandlungen, Volume I (Leipzig).

1883

"Bestimmung magnetischer Momente durch der Waage." (*WA*, III, 115-118.)
"Zur Thermodynamik chemischer Vorgänge. Folgerungen, die galvanische Polarisation betreffend." (*WA*, III, 92-114.)
Wissenschaftliche Abhandlungen, Volume II (Leipzig).

1884

"On Galvanic Currents Passing Through a Very Thin Stratum of an Electrolyte." (*WA*, III, 88-91.)
"Studien zur Statik monocyklischer Systeme." (*WA*, III, 117-141, 163-172, 173-178.)
"Ueber die Beschlüsse der internationalen Conferenz für elektrische Maasseinheiten," *Verhandlungen der physikalischer Gesellschaft in Berlin vom 9. Mai 1884,* III, 26-28.
"Verallgemeinerung der Sätze über die Statik monocyklischer Systeme," *Berlin Sitzungsberichte,* December 18, 1884, pp. 1197-1201.
"Principien der Statik monocyklischer Systeme." (*WA*, III, 142-162, 179-202.)
Vorträge und Reden. 2 vols., Braunschweig.

1885

Review of Sir William Thomson's *Mathematical and Physical Papers,* Volumes I, II. (*WA*, III, 587-596.)

Handbuch der physiologischen Optik. 2nd ed., Section 1; Hamburg.

1886

Handbuch der physiologischen Optik. 2nd ed., Sections 2, 3; Hamburg.
"Ueber die physikalische Bedeutung des Princips der kleinsten Wirkung." (*WA*, III, 203-248.)
"Rede beim Empfang der Gräfe-Medaille." (*VR*, II, 311-320.)
"Wolken- und Gewitterbildung." (*WA*, III, 287-288.)

1887

"Zur Geschichte des Princips der kleinsten Action." (*WA*, III, 249-263.)
"Versuch um die Cohäsion von Flüssigkeiten zu zeigen." (*WA*, III, 264-266.)
"Joseph Fraunhofer. Rede bei der Gedankfeier, zur hundertjährigen Wiederkehr seines Geburtstages (6. März 1887)." (*VR*, II, 321-333.)
"Weitere Untersuchungen, die Elektrolyse des Wassers betreffend." (*WA*, III, 267-281.)
"Zählen und Messen, erkenntnisstheoretisch betrachtet." (*WA*, III, 356-391.) ("An Epistemological Analysis of Counting and Measurement.")
Handbuch der physiologischen Optik. 2nd ed., Section 4; Hamburg.
"Mittheilung zu dem Bericht über die Untersuchung einer mit der Flüssigkeit Pictet arbeitenden. Eismaschine erstattet von Hrn. Dr. Max Corsepius." (*WA*, III, 282-286.)

1888

"Ueber atmosphärische Bewegungen." (*WA*, III, 289-308.)
"Ueber das Eigenlicht der Netzhaut," *Verhandlungen der physikalischer Gesellschaft zu Berlin vom 2. November 1888,* VII, 85-86.
"Zur Erinnerung an R. Clausius," *Verhandlungen der physikalischer Gesellschaft zu Berlin vom 11. Januar 1889,* VIII, 1-6.
"Ueber atmosphärische Bewegungen," Second Report: "Zur Theorie von Wind und Wellen." (*WA*, III, 309-332.)
Handbuch der physiologischen Optik. 2nd ed., Section 5; Hamburg.

1890

"Die Störung der Wahrnehmung kleinster Helligkeitsunterschiede durch das Eigenlicht der Netzhaut." (*WA*, III, 392-406.)
"Die Energie der Wogen und des Windes." (*WA*, III, 333-355.)
"Suggestion und Dichtung," *Deutsche Dichtung,* IX, 125. (Also in K. E. Franzos, *Die Suggestion und die Dichtung* [Berlin, 1892], p. 69-71.)

1891

"Bemerkunger über die Vorbildung zum akademischen Studium," *Verhandlungen über Fragen des Höhern Unterrichts* (Berlin), pp. 202-209, 763-764.

"Versuch einer erweiterten Anwendung des Fechner'schen Gesetzes im Farbensystem." (*WA*, III, 407–437.)
"Versuch, das psychophysische Gesetz auf die Farbenunterschiede trichromatischer Augen anzuwenden." (*WA*, III, 438–459.)
"Kürzeste Linien im Farbensystem." (*WA*, III, 460–475.)
Popular Lectures on Scientific Subjects, Second Series, edited by E. Atkinson. London and New York.
"Autobiographisches. Tischrede bei der Feier des 70. Geburtstages." (*VR*, I, 1-21.) ("An Autobiographical Sketch.")

1892

"Das Princip der kleinsten Wirkung in der Elektrodynamik." (*WA*, III, 476-504.)
Handbuch der physiologischen Optik. 2nd ed., Sections 6, 7; Hamburg.
"Goethe's Vorahnungen kommender naturwissenschaftlicher Ideen." (*VR*, II, 335-361.) ("Goethe's Anticipation of Subsequent Scientific Ideas.")
"Elektromagnetische Theorie der Farbenzersteuung." (*WA*, III, 505-525.)

1893

"Zusätze und Berichtigungen zu dem Aufsatze: ‚Elektromagnetische Theorie der Farbenzerstreuung.'" (*WA*, III, 523-525.)
"Adresse an Hrn. E. du Bois-Reymond bei Gelegenheit seines 50 jährigen Doctorjubiläums verfasst im Auftrage der Königl. Akademie der Wissenschaften zu Berlin," *Berliner Sitzungsberichte,* February 16, 1893, pp. 93-97.
"Folgerungen aus Maxwell's Theorie über die Bewegungen des reinen Aethers." (*WA*, III, 526-535.)
"Gustav Wiedemann beim Beginn des 50. Bandes seiner *Annalen der Physik und Chemie* gewidnet." *Wiedem. Ann.*, L, iii-xl.

1894

"Ueber den Ursprung der richtigen Deutung unserer Sinneseindrücke." (*WA*, III, 536-553.) ("The Origin and Correct Interpretation of Our Sense Impressions.")
"Vorwort" to Heinrich Hertz's *The Principles of Mechanics* (Leipzig). (*VR*, II, 363-378.)
Handbuch der physiologischen Optik. 2nd ed., Section 8; Hamburg.
"Nachtrag zu dem Aufsatze: Ueber das Princip der kleinsten Wirkung in der Elektrodynamik." (*WA*, III, 597-603.)

Posthumous

Wissenschaftliche Abhandlungen, Volume III. Leipzig, 1895.
Vorlesungen über die Dynamik discreter Massenpunkte, edited by Otto Krigar-Menzel. Leipzig, 1897.

Vorlesungen über die elektromagnetische Theorie des Lichts, edited by Arthur König and Carl Runge. Leipzig, 1898.

Vorlesungen über die mathematischen Principien der Akustik, edited by Arthus König and Carl Runge. Leipzig, 1898.

Einleitung zu den Vorlesungen über theoretische Physik, edited by Arthur König and Carl Runge. Leipzig, 1903.

Vorlesungen über Theorie der Wärme, edited by Franz Richarz. Leipzig, 1903.

Vorlesungen über Elektrodynamik und Theorie des Magnetismus, edited by Otto Krigar-Menzel and Max Laue. Leipzig, 1907.